SOMATIC EMBRYOGENESIS IN WOODY PLANTS

FORESTRY SCIENCES

Volume 46

*The titles published in this series are listed at the end of this volume.*

# Somatic Embryogenesis in Woody Plants

## *Volume 2 – Angiosperms*

*Edited by*:

S. MOHAN JAIN
*Department of Plant Production, University of Helsinki, Helsinki, Finland*

PRAMOD K. GUPTA
*Weyerhaeuser Inc., Tacoma, Washington, U.S.A.*

and

RONALD J. NEWTON
*Department of Forest Science, Texas A & M University, College Station, Texas, U.S.A.*

SPRINGER-SCIENCE+BUSINESS MEDIA, B.V.

**Library of Congress Cataloging-in-Publication Data**

Somatic embryogenesis in woody plants / edited by S. Mohan Jain, Pramod K. Gupta, Ronald J. Newton.
p. cm. -- (Forestry sciences ; v. 46)
Includes index.
ISBN 978-94-010-4220-8 ISBN 978-94-011-0491-3 (eBook)
DOI 10.1007/978-94-011-0491-3
1. Trees--Micropropagation. 2. Woody plants--Micropropagation. 3. Plant tissue culture. I. Jain, S. Mohan. II. Gupta, Pramod K. III. Newton, Ronald J. IV. Series: Forestry sciences (Dordrecht, Netherlands) ; v. 46.
SD403.5.S66 1995
635.9'77--dc20 94-18512

ISBN 978-94-010-4220-8

Printed on acid-free paper

# Table of Contents

# General Preface

The quality of human life has been maintained and enhanced for generations by the use of trees and their products. In recent years, ever rising human population growth has put tremendous pressure on trees and tree products; growing awareness of the potential of previously unexploited tree resources and environmental pollution have both accelerated development of new technologies for tree propagation, breeding and improvement. Biotechnology of trees may be the answer to solve the problems which cannot be solved by conventional breeding methods. The combination of biotechnology and conventional methods such as plant propagation and breeding may be a novel approach to improving and multiplying in large number the trees and woody plants.

So far, plant tissue culture technology has largely been exploited in the propagation of ornamental plants, especially foliage house plants, by commercial companies. Generally, tissue culture of woody plants has been recalcitrant. However, limited success has been achieved in tissue culture of angiosperm and gymnosperm woody plants. A number of recent reports on somatic embryogenesis in woody plants such as Norway spruce (*Picea abies*), Loblolly pine (*Pinus taeda*), Sandalwood (*Santalum album*), *Citrus*, Mango (*Mangifera indica*), etc., offer a ray of hope of: a) inexpensive clonal propagation for large-scale production of plants or "emblings" or "somatic embryo plants", b) protoplast work, c) cryopreservation, d) genetic transformation, and e) artificial or manufactured seed production. In the future, with the basic biology for better understanding the genetic control of somatic embryogenesis and in embryo development and maturation with the help of molecular biology, it may be possible for us to have better control over the induction of somatic embryogenesis. For cost effective large-scale production of elite trees, robotic and automation technology will interface with somatic embryogenesis in the 21st century.

The rapid progress of somatic embryogenesis and its prospects for potential applications to improving woody plants prompted us to edit this book in three volumes. Moreover, most of the research information in this field on woody plants is scattered in national and international meeting proceedings, refereed journals, biotechnology books, etc. There is a lack of availability of a comprehensive work on somatic embryogenesis in woody plants including both angiosperms and gymnosperms. We were all convinced that such a treatise was needed and would be extremely useful to researchers and students. Dr. Jain contacted Kluwer Academic Publishers, The Netherlands to work on this book project. The positive response of the publisher encouraged us to proceed further.

In our present endeavour, we have attempted to bring all the research information on somatic embryogenesis in woody plants into three volumes.

In Volume 1, we included review articles on different aspects and applications of somatic embryogenesis such as historical, biochemical, molecular, anatomical and physiological, protoplasts, cryopreservation, manufactured seed (artificial seed), genetic transformation, somaclonal variation, bioreactors, mutation, and future uses in woody plantations. In the introductory chapter, we have taken the views of a forest geneticist on somatic embryogenesis in relation to woody plant improvement. What lies ahead in the application of this technology for commercial purposes and the establishment of germplasm banks, has been discussed. Both Volumes 2 and 3 cover selected important angiosperm and gymnosperm woody plants, respectively. Each selected woody plant has been described in detail covering botany and genetics, importance and geographical distribution, breeding problems, *in vitro* propagation and problems, initiation of embryogenic cultures, culture maintenance, embryo development, embryo germination, and field trials (if any). We have also encouraged the authors to incorporate their recent data including genetic transformation, cryopreservation etc. in their manuscripts. These volumes are designed as the key reference works, providing detailed information on all aspects of somatic embryogenesis for beginners as well as experienced research workers.

We have invited an international and diverse group of contributors from several countries such as Australia, Austria, Canada, Czech Rep., Denmark, England, Finland, France, India, Ireland, Italy, Japan, Slovakia, Spain, Sweden, Taiwan, USA, and Yugoslavia. Our invited authors belong to academic institutes, universities and industries, and they are leading research scientists in somatic embryogenesis research work.

The selection of chapters and their authors was initially done by Dr. S.M. Jain and the final list was prepared by including the suggestions of Dr. Gupta and Prof. Newton. Our final list of chapters with their authors, was reviewed by anonymous reviewers of Kluwer Academic Publishers and the project was finally approved. All the correspondence with the invited authors and the publisher, any further additions or deletions of the chapters, initial and final review of the manuscripts including editing, the selection of reviewers, and finally the mailing of the reviewed manuscripts were done by Dr. Jain. All the manuscripts have been critically reviewed by one–two persons and revised according to the referees' comments.

S. Mohan Jain
P.K. Gupta
R.J. Newton

# Preface to Volume 2

This volume contains 24 chapters on somatic embryogenesis in selected important angiosperm woody plants including bamboo (general review), another chapter covers *Bambusa oldhamii*, *B. beecheyana* and *Sinocalamus latiflora*, birch (*Betula pendulum*), *Citrus*, cacao, coconut, *Eucalyptus*, rubber (*Hevea brasiliensis*), walnut (*Juglans regia*), *Populus* spp., *Quercus* spp., olive (*Olea europaea*), grape, *Salix*, *Camellia* spp., *Rosa* spp., white ash (*Fraxinus americana*), mango (*Mangifera indica*), neem (*Azadirachta indica*), horse chestnut, apple (*Malus*), date palm, eastern redbud (*Cercis canadensis*). In general, all the chapters initially cover the introduction of the plant describing botany and genetics, importance and geographical distribution, breeding problems and *in vitro* propagation and problems. This is followed by initiation of embryogenic cultures, culture maintenance, embryo development, embryo germination, field trials (if any), and concluding remarks. There was no limit set for the number of photos or figures or the length of the manuscript. All the contributory authors were encouraged to include their recent data.

All the submitted manuscripts to this volume were reviewed by one–two persons and revised accordingly. We are thankful to all the reviewers for their help in reviewing the manuscripts well in time.

We express our gratitude to all contributory authors to this volume for sending their manscripts in time.

S. Mohan Jain
P.K. Gupta
R.J. Newton

# Acknowledgements

I am very grateful to the Finnish Academy of Science, for providing me with a grant for editing this book.

My thanks are due to my friends Prof. R.J. Newton and Dr. Pramod K. Gupta, co-editors of this book, for their promptness in responding to me whenever I needed their help. They were extremely helpful in deciding the final list of chapters and their authors, and in reviewing the manuscripts. It has been my greatest pleasure to work with Ron and Pramod on this project, and certainly we have formed an excellent and highly efficient team.

I would like to acknowledge with great appreciation Prof. M.R. Ahuja, Dr. J. Blake, Dr. J.M. Bonga, Dr. J. Finer, Prof. Larry Fowke, Dr. P.K. Gupta, Prof. J. Janick, Prof. S.C. Minocha, Prof. R.J. Newton, Prof. J. Preece, Dr. R.C. Saxena, Prof. R.H. Smith, Dr. David Thompson, and Dr. R.N. Trigiano for critically reviewing the manuscripts promptly, and to all the contributory authors for sending their manuscripts well in time. However, we had some anxious moments due to delay in some manuscripts.

I wish to express my thanks to my colleagues Prof. Eija Pehu, Mr. Tapio Poutala, Mr. Matti Teittinen, and Mr. Tapani Pulkki in our Plant Production Department for their timely help and assistance.

Also, with great love and affection, I want to thank my daughters Sarita and Sonia, and my wife Marja-Liisa for their unceasing patience and understanding while I was working on these volumes.

Finally, I express my deepest sense of appreciation to Adrian C. Plaizier of Kluwer Academic Publishers, The Netherlands, for giving us the opportunity to work on this exciting book project. Adrian has always been cooperative and helpful, and gave me useful advice.

Book Project Leader
S. Mohan Jain

# Abbreviations

| | |
|---|---|
| 2,4-D | 2,4-dichlorophenoxyacetic acid |
| 2,4,5-T | 2,4,5-trichlorophenoxyacetic acid |
| 2iP | $N^6$ (2-isopentyl)adenine |
| ABA | Abscisic acid |
| BA | N6-benzyladenine |
| BAP | benzylaminopurine |
| BLG | Verhagen & Wann (1989) |
| $BM_3$, $BM_4$ and $BM_5$ | Gupta & Pullman (1991) |
| CBB | Coomassie Brillant Blue |
| CD | Campbell & Durzan (1975) |
| CP | cotyledons protoplasts |
| DCR | Gupta & Durzan (1986) |
| DMSO | dimethyl sulfoxide |
| DW | dry weight |
| EP | embryogenic cell suspension |
| ESM | Embryonal suspensor masses |
| FDA | Fluorescein diacetate |
| FW | fresh weight |
| G-medium, DCR | Gupta & Durzan medium (1986) |
| GD | Gresshoff & Doy (1972) |
| H medium | Halperin medium (1965) |
| HBLG | half strength Verhagen & Wann (1989) |
| HLP | half strength von Arnold & Eriksson (1981) |
| HLM | half strength Litvay *et al.* (1981) |
| HRH | High relative humidity |
| IAA | indoleacetic acid |
| IBA | indolebutyric acid |
| KIN, K | kinetin |
| LM | Litvay *et al.* (1981) |
| $LN_2$ | liquid nitrogen |
| LP,AE,VE | von Arnold & Eriksson (1981) |
| MCM | Borman (1981) |
| MES | 2-(N-morpholino)ethanesulfonic acid |
| MI | mitotic index |
| MS | Murashige & Skoog |
| NAA | naphthaleneacetic acid |
| NEC | non-embryogenic callus |
| NEPHGE-PAGE | non-equilibrium pH gel electrophoresis |
| OP | osmotic pressure |
| $P_6$ | Teasdale *et al.* (1986) |

| | |
|---|---|
| PAR | photosynthetically active radiation |
| PCV | packed cell volume |
| PEG | polyethylene glycol |
| PG | phloroglucinol |
| 6 PGD 1 | 6-phosphogluconate dehydrogenase |
| PVP | polyvinylpyrrolidone |
| RAPD | Random amplified polymorphic DNA |
| RFLP | Restriction fragment length polymorphism |
| RPM | revolution per minute |
| SDS-PAGE | sodium dodecyl sulphate polyacrylamide gel electrophoresis |
| SE | somatic embryos |
| SH | Schenck & Hildebrandt (1972) |
| TAG | triacylglycerols |
| TCA | trichloroacetic acid |
| TDZ | thidiazuron |
| WPMG | Jain *et al.* (1989) |
| X-GLUC | 5-bromo-4-chloro-3-indolyl glucuronide |
| Zea | zeatin |

# 1. Somatic embryogenesis in *Citrus* species

S.S. Gosal, M.I.S. Gill and H.S. Grewal

**Contents**



## 1. Introduction

The citrus group of plants belongs to family Rutaceae and is comprised of six genera including *Citrus*, *Clymenia*, *Eremocitrus*, *Fortunella*, *Microcitrus* and *Poncirus*. Among these, the genus *Citrus* is of much economic importance for bearing edible fruits and serving as rootstocks. Swingle and Reece (1967) recognized mandarin (*Citrus reticulata* Blanco), pummelo (*C. grandis* Osbeck), lime (*C. aurantifolia* Christm.), citron (*C. medica* L.), lemon (*C. limon* L. Burn. f.) and grapefruit (*C. paradisi* Osbeck) as important *Citrus* species. The true citrus fruits are native to Asiatic area extending from the Himalayan foothills of India to North Central China and Philippines in the East, Burma, Thailand, New Guinea, Northern Australia and New Caledonia in the South-East. Now *Citrus* is cultivated in the tropical and sub-tropical regions of the world extending from equator to latitudes 40° North and South. World production of citrus fruits continues on an upward trend. Total world *Citrus* production in 1991 amounted to 8551 thousand metric tons (FAO, 1992). Highest production has been in Asia (4036 thousand metric tons) followed by Europe (2052 TMT), South America (1087 TMT), Africa (750 TMT) and N. America (575 TMT). Citrus fruits are consumed in different forms such as fresh fruits, juice squashes and jams.

In general, the *Citrus* species are diploid with a possessing chromosome number of 2n = 18. Using various conventional breeding approaches, *Citrus* improvement has been rather difficult because of high heterozygosity, self-incompatibility, sterility and occurrence of nucellar embryony in many species. However, innovative approaches of plant biotechnology, offer a rich scope for creation, conservation, utilization and characterization of genetic variability for *Citrus* improvement and rapid cloning of elite germplasm. Somatic embryogenesis, i.e., development of embryoids from cultured somatic cells/tissues, is an important aspect first discovered by Steward *et al.* (1958) in carrot. Since then, various *Citrus* species have exhibited frequent

*S. Jain, P. Gupta & R. Newton (eds.), Somatic Embryogenesis in Woody Plants, Vol. 2*, 1–21.

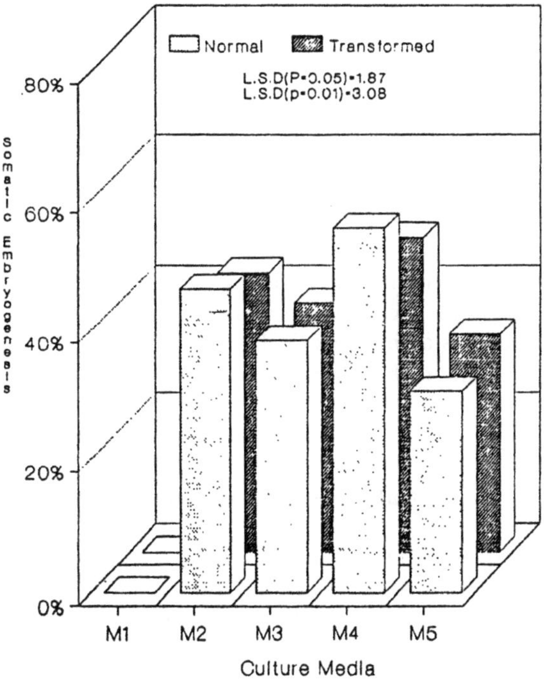


*Figure 1.* Per cent callus cultures derived from immature embryos of Local Sangtra, showing somatic embryogenesis on different media. M1 – MS (1962) basal medium; M2 – MS + NAA (10 mg/l) + Kin (1 mg/l) + vitamins (10×); M3 – MS + NAA (10 mg/l) + Kin (0.5 mg/l); M4 – MS + NAA (10 mg/l) + Kin (0.5 mg/l) + Sucrose (5%); M5 – MS + NAA (15 mg/l) + Kin (5 mg/l).

somatic embryogenesis in culture. Excellent reviews have been published by Button and Kochba (1977), Spiegel-Roy and Vardi (1984), Tisserat *et al.* (1979) and Wann (1988). Presently, somatic embryogenesis is being viewed as one of the important techniques of plant tissue culture for mass cloning of elite plants through production of synthetic seeds. Besides, embryogenic cell cultures constitute an important material for protoplast isolation and genetic transformation using biolistic methods. The various aspects of somatic embryogenesis in different *Citrus* species are presented in Table 1.

Tisserat *et al.* (1979) interpreted that the internal state of explant cells hold a prime importance for somatic embryogenesis, whereas, the exogenous application of growth regulators direct the already determined pattern of their development. Sharp *et al.* (1980, 1982) and Evans *et al.* (1981) hypothesised that direct embryogenesis proceeds from pre-embryogenic determined cells (PEDCs) in which embryogenic development begins even prior to explanting, thus requiring only little exogenous initiative of growth regulators for cell division and expression of embryogenesis. However, in the direct embryogenesis, redetermination of different cells leading to callus induction-proliferation and the development of an embryogenically determined state, require growth regulators for re-entry of cells into mitosis and determination of the embryogenic state. Thus, the induced embryogenically determined cells (IEDCs) under the influence of growth regulators develop into the embryos from which the plantlets can be obtained. In *Citrus*, both direct as well as indirect somatic embryogenesis has been observed from various explants cultured *in vitro*. In general, embryogenesis takes place from rapidly dividing meristematic cells which are small in size, possessing dense cytoplasm, large nuclei with prominent enlarged nucleoli, small vacuoles and profusion of starch grains (Williams and Maheswaran, 1986). Different aspects of somatic embryogenesis in some citrus species have been presented in Table 1.

## 2. Culture initiation development

Besides genotype, chemical composition of the culture media and the nature of explant are important factors affecting the establishment of embryogenic cultures. In *Citrus*, somatic cell cultures were initiated using Murashige and Skoog (1962) as well as Murashige and Tucker (1969) media supplemented with varying concentrations of growth regulators and growth adjuvants (Table 1). The generally used growth regulators include auxins, cytokinins, gibberellic acid and abscisic acid. In addition to this, malt extract and casein hydrolysate have been frequently used as growth adjuvants for inducing somatic embryos in *Citrus* cultures. Besides, coconut milk and orange juice have also been used to induce somatic embryogenesis in *Citrus*. Among auxins, 2,4-D has been frequently used to induce embryogenic calli from nucellar and ovular explants of *C. limon* and *C. volckameriana* (Saad, 1975) and immature ovules (Moore, 1985a,b). Likewise, the basal medium supplemented with 2,4-D (0.5 mg/l) and kinetin (0.25 mg/l) has been reported to induce embryoids in *C. sinensis* (Pasqual *et al.*, 1988). Apart from 2,4-D, the naphthalene acetic acid (Chaturvedi and Mitra, 1975; Nito and Iwamasa, 1990) and indoleacetic acid (Kochba *et al.*, 1972) have also been used for the initiation of cultures in *Citrus*. In our laboratory, NAA (10 mg/l) proved better as compared to 2,4-D for establishing embryogenic callus cultures (Table 3) of *Citrus reticulata* (Gill *et al.*, 1991). The auxins on the other hand, have been reported to limit the frequency of somatic embryogenesis (Spiegel-Roy and Vardi, 1984) due to high endogenous levels of auxin and other growth substances. Embryogenesis in *Citrus* was thus induced, even on the medium lacking growth regulators (Kobayashi *et al.*, 1984; Sim *et al.*, 1988; Vardi *et al.*, 1986; Wu, 1990; Yang, 1983). Embryogenesis in some *Citrus* species has been inhibited by the exogenous application of auxins like IAA, NAA or cytokinins such as BA, kinetin and zeatin in the culture medium. The basal medium supplemented with cytokinins such as BA (Bhansali and Arya, 1978a,b), kinetin (Hidaka, 1984b; Hidaka and Omura, 1989; Kochba *et al.*, 1972; Pasqual *et al.*, 1988), zeatin (Chaturvedi and Mitra, 1975; Hidaka and Kajiura, 1988) have resulted in embryoid formation. On the contrary, the growth inhibitors such as ABA, has stimulated the embryoid formation in *Citrus* (Moore, 1985a). Likewise, the use of low ABA (0.04–4.0 $\mu$M) concentrations in the basal medium induced somatic embryogenesis in *Citrus* (Spiegel-Roy and Vardi, 1984). Besides ABA, the inhibitors of auxin synthesis such as HNB (7-aza-indole) as well as inhibitors of $GA_3$ synthesis such as Daminozide, CCC (2-chloro-ethyltrimethyl ammonium chloride) and ALAR (Succinic acid 2,2-methyl-hydrazide) have been reported to stimulate somatic embryogenesis in *Citrus* (Kochba and Spiegel-Roy, 1977; Kochba *et al.*, 1978a; Moore, 1985a).

The growth adjuvants have also been repeatedly used by various research workers to induce embryogenic cultures in *Citrus*. Among various growth adjuvants, malt extract (ME) has been widely used (Button and Bornman, 1971; Gmitter and Moore, 1986; Kochba *et al.*, 1972; Kochba *et al.*, 1974; Starrantino and Russo, 1980; Tusa *et al.*, 1983).

The orange juice has been used in combination with NAA and adenine sulphate for culturing the nucelli (Rangan *et al.*, 1968, 1969). The embryo formation has also been achieved from nucellar tissues of *C. sinensis* using orange juice as a supplement in the basal medium (Button and Bornman, 1971). The beneficial effect of orange juice in inducing embryogenesis may be attributed to the presence of ascorbic acid.

The use of carbohydrates such as sucrose, glucose, fructose, glactose, and lactose has been found to influence the production of embryogenic cultures in *Citrus*. Sucrose has extensively been used in *Citrus* tissue culture media as a source of carbon. Sucrose supported embryogenesis at lower levels (8–32 mM), however, galactose and lactose enhanced embryogenesis over a broad range (8–256 mM). Glucose and fructose were ineffective in stimulating embryogenesis in nucellar calli of *Citrus* cultivars viz. Villafranca lemon and Shamouti oranges. Galactose, lactose and raffinose stimulated embryogenesis considerably in *Citrus* (Kochba *et al.*, 1978b, 1982). The addition of lactose (5%) in the culture medium helped embryoid formation in long term callus cultures of *C. sinensis* cv. Liu Cheng (Zhang *et al.*, 1987). Likewise, embryoid formation from immature ovules has been obtained by supplementing the basal medium with lactose (5%) (Starrantino and Caponnetto, 1990). It has also been observed that the addition of glycerol (280 mM) in the culture medium resulted in transformation of nucellar calli into small green embryos within 3–4 weeks in *C. sinensis*, *C. paradisii* and *C. limon*. The use of glycerol in the culture medium has also been found beneficial in the ovule culture for inducing somatic embryogenesis (Nel, 1987). The supplementation of a sucrose-containing-medium with mannitol enhanced embryoid formation in the protoplast culture of *C. sinensis* (Kobayashi *et al.*, 1985).

The effect of radiation on initiation of embryogenic cultures has also been studied in *Citrus*. Pasqual and Ando (1990) reported that gamma-irradiation (0.5–12 kR) of the medium alone or the nucellus plus the medium resulted in similar effects in *C. sinensis*. Irradiation of the medium only increased callus proliferation at low doses (up to 2 kR), whereas irradiation of the nucellus at 2- and 4-kR increased the frequency of highly embryogenic calli.

In *Citrus* various tissues such as the nucellus, ovules (mature and immature), stem explants, juice vesicles, protoplasts and testa have been commonly used for initiation of tissue cultures. Maheshwari and Rangaswamy (1958), for the first time, cultured developing ovules and seeds before and after fertilization to obtain complete plants. Since then, ovules have been frequently used for the culture initiation in several citrus species such as *C. deliciosa* (Tusa *et al.*, 1983; Kobayashi *et al.*, 1984), *C. jambhiri* (Ben-Hayyim and Neumann, 1983), *C. limon* (Starrantino and Russo, 1980), *C. paradisi* (Kochba *et al.*, 1972; Ben-Hayyim and Neumann, 1983), *C. reticulata* (Wu *et al.*, 1990) and *C. sinensis* (Tusa *et al.*, 1983; Ben-Hayyim and Neumann, 1983; Kobayashi *et al.*, 1984; Moore, 1985a; Kochba *et al.*, 1972, 1974; Starrantino and Russo, 1980; Button and Bornman, 1971; Starrantino and Caponnetto, 1990; Wu *et al.*, 1990). Although, previously, it was thought that only the fertilized ovules responded to culture conditions (Maheshwari and Rangaswamy, 1958). However, it was observed that nucellar embryos also developed from nucellar explants of unfertilized ovules (Button and Bornman, 1971; Kochba *et al.*, 1972; Kochba and Spiegel-Roy, 1973). Thus, even the undeveloped ovules have been commonly used for the initiation of cultures (Starrantino and Russo, 1980; Starrantino and Caponnetto, 1990; Moore, 1985a; Gmitter and Moore, 1986).

Mitra and Chaturvedi (1972) reported that unpollinated ovaries of *C. aurantifolia* and *C. sinensis* produced embryoids and complete plants where ovary wall tissue was responsible for this response. We have obtained somatic embryogenesis in *Citrus reticulata* (cv. Kinnow and Local Sangtra) from calli initiated from immature embryos (Figs. 1 and 2).

Besides ovules, nucelli have been widely used for the culture initiation and induction of somatic embryogenesis in *Citrus*. The mature embryos of nucellar origin have been obtained in *Citrus* (Ohta and Furusato, 1957). Nucellar embryos were obtained from nucelli of monoembryonic *Citrus* cultivars so as to produce virus free lines (Rangan *et al.*, 1968; Sabharwal, 1963). The explants containing fully developed nucellar embryos developed multiple embryos and callus upon culture (Rangaswamy, 1961; Sabharwal, 1963). It has been reported that pseudobulbils and embryoids were obtained by culturing the nucelli on a MS medium supplemented with malt extract, adenine or with methylaminopurine (Starrantino *et al.*, 1977). Embryogenic calli have also been initiated from the nucelli and unfertilized ovules (Kochba and Spiegel-Roy, 1973; Mitra and Chaturvedi, 1972, Kochba *et al.*, 1982; Ben-Hayyim and Neumann, 1983; Pasqual *et al.*, 1988; Pasqual and Ando, 1990).

Cytologically diploid somatic embryoids have been obtained in *C. auran-*

*Table 1.* Somatic embryogenesis in *Citrus* species.

| Genotype | Explant | Culture medium | Growth response | Mode of embryogenesis | Reference |
|---|---|---|---|---|---|
| *Citrus aurantifolia* | Stem segment | MS + BA | Embryoid formation | ID | Bhansali and Arya, 1978a |
| *C. aurantifolia* | Anther | i) MS + BA(0.5 mg/l) + IAA (1 mg/l) | Embryoid formation | D | Chaturvedi and Sharma, 1985 |
| | | ii) MS medium without auxins | Plantlet formation | | |
| *C. aurantifolia × Feroniella lucida* | Protoplasts | – | Embryoid and plantlet formation | ID | Takayanagi *et al.*, 1992 |
| *C. aurantifolia × Swinglea glutinosa* | Protoplasts | – | Embryoid and plantlet formation | ID | Takayanagi *et al.*, 1992 |
| *C. aurantium* | Ovules | MT+Glycerol (280 mM) | Embryoid formation | ID | Ben-Hayyim and Neumann, 1983 |
| *C. aurantium* | Anther | MS + Sucrose (70 g/l) | Embryoid formation | D | Hidaka, 1984a |
| *C. aurantium* | Embryos | – | Embryoid and plantlet formation | ID | Beloualy, 1991 |
| *C. aurantium* | Nucellii | – | Embryoid formation | ID | Gavish *et al.*, 1992 |
| *C. deliciosa* | Ovules | MT + ME (1 g/l) | Embryoid formation | ID | Tusa *et al.*, 1983 |
| *C. deliciosa* | Ovules | MT medium | Embryoid formation | ID | Kobayashi *et al.*, 1984 |
| *C. grandis* | Stem callus | MS + Zeatin + NAA | Embryoid formation | ID | Chaturvedi and Mitra, 1975 |
| *C. grandis* | Endosperm (12–14 weeks post anthesis) | MT + BA + 2,4-D + CH + Kin + ME | Embryoid and plantlets formation | ID | Gmitter *et al.*, 1990 |
| *C. grandis* | Immature ovules | MT + ME (500 mg/l) + Sucrose (7.0 g/l) | Embryoid formation | D | Oh *et al.*, 1991 |
| *C. jambhiri* | Ovules | MT+Glycerol | Embryoid formation | – | Ben-Hayyim and Neumann, 1983 |
| *C. junos* | Immature ovules | – | Embryoid formation | D | Oh *et al.*, 1992 |
| *C. limon* | Undeveloped ovules | MS + ME (500 mg/l) | Embryoid and pseudobulbils | – | Starrantino and Russo, 1980 |

| | | | | | |
|---|---|---|---|---|---|
| *C. limon* | Nucellii | MS + Galactose + Lactose | Embryoid formation | D | Kochba *et al.*, 1982 |
| *C. limon* | Ovules | MT + Glycerol (280 mM) | Embryoid formation | ID | Ben-Hayyim and Neumann, 1983 |
| *C. madurensis* | Anthers | i) $N_6$ medium + BA (1 mg/l) + 2,4-D (0.1 mg/l) | Embryoid formation | ID | Chen, 1985 |
| | | ii) MS + BA (0.25 mg/l) + IAA (0.1 mg/l) | Plantlets | | |
| | | iii) MS + IBA (0.1–0.2 mg/l) | Root formation | | |
| *C. mitis* | Protoplasts | MS basal medium | Embryoid formation | ID | Sim *et al.*, 1988 |
| *C. paradisi* | Ovules and nucellii | MT + ME (500 mg/l) | Embryoid formation | ID | Kochba *et al.*, 1972 |
| | | MT + Kin + IAA | Plantlets formation | | |
| *C. paradisi* | Ovules | MT+Glycerol (280 mM) | Embryoid formation | ID | Ben-Hayyim and Neumann, 1983 |
| *C. reticulata* | Nucellus | MS + ADS (25 mg/l) + ME (500 mg/l) | Embryoid formation | – | Kochba *et al.*, 1974 |
| *C. reticulata* | Testas | Modified Murashige and Skoog medium | Embryoid formation | – | Yang, 1983 |
| *C. reticulata* | Protoplasts | i) MS + Zeatin (1 uM) | Embryoid formation | ID | Hidaka and Kajiura, 1988 |
| | | ii) MS + $GA_3$ (1 uM) + Sucrose (0.06 M) | Plantlets | | |
| *C. reticulata* | Ovules | MT medium without growth regulators | Embryoid formation | – | Wu *et al.*, 1990 |
| *C. reticulata* | Immature embryo and seedling explants viz. leaf, epicotyl, cotyledon and root | MS + NAA (10 mg/l) + Kin (0.5–1 mg/l) | Embryoid formation and plantlets | ID | Gill *et al.*, 1991, 1994 |

*Table 1.* Continued.

| Genotype | Explant | Culture medium | Growth response | Mode of embryogenesis | Reference |
|---|---|---|---|---|---|
| *C. sinensis* | Ovules | MS + Adenine (40 mg/l) + ME (400 mg/l) | Embryoid formation | – | Button and Bornman, 1971, 1974 |
| *C. sinensis* | Ovules and nucelli | MT + ME (500 mg/l) | Embryoid formation | – | Kochba *et al.*, 1972 |
| *C. sinensis* | Ovular callus | MT + $GA_3$ + ADS | Embryoid formation | – | Kochba *et al.*, 1974 |
| *C. sinensis* | Stem segment callus | MS + BA | Embryoid formation | ID | Bhansali and Arya, 1978a,b |
| *C. sinensis* | Undeveloped ovules | MS + ME (500 mg/l) + Agar (10 g/l) | Embryoid formation | ID | Starrantino and Russo, 1980 |
| *C. sinensis* | Nucellar callus | MS + Galactose + Lactose | Embryoid formation | D | Kochba *et al.*, 1982 |
| *C. sinensis* | Ovules | MT + ME (1 g/l) | Embryoid formation | ID | Tusa *et al.*, 1983 |
| *C. sinensis* | Ovules and nucellii | MT + Glycerol (280 mM) | Embryoid formation | ID | Ben-Hayyim and Neumann, 1983 |
| *C. sinensis* | Anthers | MS+Sucrose (10 g/l) | Embryoid formation | D | Hidaka, 1984a |
| *C. sinensis* | Ovules | MT medium | Embryoid formation | ID | Kobayashi *et al.*, 1984 |
| *C. sinensis* | Anthers | MS + Kin + IAA | Embryoid formation | D | Hidaka, 1984b |
| *C. sinensis* | Immature ovules | MS + 2,4-D | Embryoid formation | D | Moore, 1985a |
| *C. sinensis* | Protoplasts | MT + Suc (0.15 M) + Mannitol (0.25 M) | Embryoid formation | D | Kobayashi *et al.*, 1985 |
| *C. sinensis* | Embryo-derived callus | MT + CH (500 mg/l) | Embryoid formation | ID | Zhang *et al.*, 1987 |
| *C. sinensis* | Nucelli | MS + Kin (0.25 mg/l) + 2,4-D (0.5 mg/l) | Embryoid formation | – | Pasqual *et al.*, 1988 |

| | | | | | |
|---|---|---|---|---|---|
| *C. sinensis* | Protoplasts | MS + Zeatin (1 mM) | Embryoid formation | – | Hidaka and Kajiura, 1988 |
| | | MS + $GA_3$ (1 uM) + Suc. (0.06 M) | Plantlets | | |
| *C. sinensis* | Embryogenic irradiated callus | – | Embryoid formation | ID | Nito *et al.*, 1989 |
| *C. sinensis* | Unfertilized ovules | MT + BA (1.0 mg/l) + IAA (0.5 mg/l) | Embryoid formation | – | Gao *et al.*, 1990 |
| | | MT + IBA (1.0 mg/l) | Plantlets | | |
| *C. sinensis* | Endosperm (12–14 weeks post-anthesis) | MT + BA + 2,4-D + CH + Kin + ME | Embryoid and plantlets | – | Gmitter *et al.*, 1990 |
| *C. sinensis* | Immature ovules | MT + Lactose (50 g/l) | Embryoid formation | ID | Starrantino and Caponnetto, 1990 |
| *C. sinensis* | Nucellii | MS + Supplements + Irradiation (0.5–12 kR) | Embryoid formation | ID | Pasqual and Ando, 1990 |
| *C. sinensis* | Ovules | MT without growth regulators | Embryoid formation | ID | Wu *et al.*, 1990 |
| *C. unshiu* | Juice vesicles | MS + NAA (1 mg/l) | Embryoid formation | ID | Nito and Iwamasa, 1990 |
| | | MS + GA (1 mg/l) | Plantlets | | |
| *C. unshiu* | Protoplasts | MT + Lactose (5%) | Embryoid formation | ID | Kunitake *et al.*, 1991 |
| | | MT + GA (1 mg/l) + Suc (3%) + Gellan gum (0.2%) | Plantlets | | |
| *Citrus* spp. (5 monoembryonic and 17 polyembryonic cultivars) | Undeveloped ovules | MS + Daminozide (0.01 mg/l) | Embryoid formation | – | Moore, 1985a,b |

*Table 1.* Continued.

| Genotype | Explant | Culture medium | Growth response | Mode of embryogenesis | Reference |
|---|---|---|---|---|---|
| *Citrus* spp. (11 cultivars belonging to 7 species) | Protoplasts | MS + Kin (50 μM) + Sucrose (0.16 M) + ME (0.5 g/l) | Embryoid formation | ID | Hidaka and Omura, 1989 |
| | | MS + $GA_3$ (1 μM) + Sucrose (0.06 M) | Plantlet regeneration | | |
| *Citrus* spp. (8 cultivars) | Immature ovules | MS + ME + 2,4-D | Embryoid formation | ID | Gmitter and Moore, 1986 |
| Microcitrus | Protoplasts | Solid medium lacking growth regulators | Embryoid formation | ID | Vardi *et al.*, 1986 |

D – Direct somatic embryogenesis; ID – Indirect somatic embryogenesis; MS – Murashige and Skoog (1962); MT – Murashige and Tucker (1969); $N_6$ – Chu *et al.* (1975); 2,4-D – 2,4-dichlorophenoxyacetic acid; NAA – naphthaleneacetic acid; IBA – 3-indolebutyric acid; IAA – indole-3-acetic acid; $GA_3$ – gibberellic acid; BA – benzylaminopurine; KIN – kinetin; ADS – adenine sulphate; Suc – sucrose; CH – casein hydrolysate.

*Table 2.* Per cent callus cultures of *C. reticulata* (Kinnow) showing somatic embryogenesis after induction on different media.

| Culture media | Embryogenic calli (%) | | | | Mean |
|---|---|---|---|---|---|
| | Leaf-derived callus | Epicotyl-derived callus | Cotyledon-derived callus | Root-derived callus | |
| MS (1962) basal medium | 0.0 (0.0) | 0.0 (0.0) | 0.0 (0.0) | 0.0 (0.0) | 0.0 (0.0) |
| MS + NAA (2 mg/l) + Kin (0.5 mg/l) | 5.7 (13.4) | 10.0 (18.4) | 12.8 (20.6) | 0.0 (0.0) | 7.1 (13.1) |
| MS + NAA (5 mg/l) + Kin (0.5 mg/l) | 15.2 (23.0) | 17.9 (24.8) | 26.1 (31.0) | 12.9 (20.9) | 18.0 (24.9) |
| MS + NAA (10 mg/l) + Kin (0.5 mg/l) | 33.3 (35.2) | 31.1 (33.6) | 38.1 (37.8) | 17.7 (24.8) | 30.0 (32.8) |
| MS + 2,4-D (2 mg/l) + Kin (0.5 mg/l) | 23.6 (28.7) | 34.6 (35.8) | 25.0 (30.2) | 16.6 (23.9) | 24.9 (29.6) |
| MS + 2,4-D (5 mg/l) + Kin (0.5 mg/l) | 8.0 (16.5) | 15.3 (23.0) | 10.8 (18.7) | 0.0 (0.0) | 8.5 (14.5) |
| MS + 2,4-D (2 mg/l) + NAA (2 mg/l) + Kin (0.5 mg/l) | 20.0 (26.8) | 29.1 (32.7) | 34.2 (35.8) | 19.2 (26.1) | 25.6 (30.3) |
| Modified MT (1969) medium | 28.5 (32.4) | 40.0 (39.4) | 32.2 (34.5) | 23.0 (28.8) | 30.9 (33.7) |
| Mean | 16.7 (22.0) | 22.2 (25.9) | 22.4 (26.0) | 11.1 (15.5) | |

Figures in parenthesis are Arc Sine transformed means.
L.S.D. ($p = 0.05$) Explant = 2.48, Media = 3.50, Interaction (Explant × Media) = 7.01.
L.S.D. ($p = 0.01$) Explant = 4.04, Media = 4.70, Interaction (Explant × Media) = 9.43.

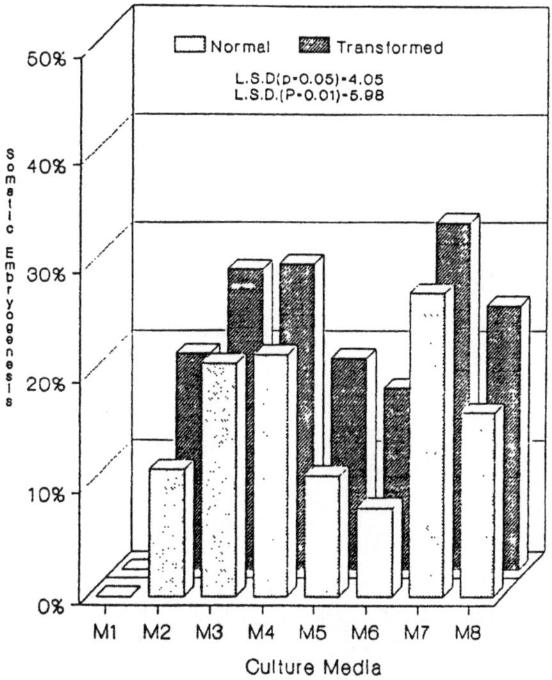


*Figure 2.* Per cent callus cultures derived from immature embryos of Kinnow, showing somatic embryogenesis on different media. M1 – MS (1962) basal medium; M2 – MS + NAA (2 mg/l) + Kin (0.5 mg/l); M3 – MS + NAA (5 mg/l) + Kin (0.5 mg/l); M4 – MS + NAA (10 mg/l) + Kin (0.5 mg/l); M5 – MS + 2,4-D (2 mg/l) + Kin (0.5 mg/l); M6 – MS + 2,4-D (5 mg/l) + Kin (0.5 mg/l); M7 – MS + NAA (2 mg/l) + 2,4-D (2 mg/l) + Kin (0.5 mg/l); M8 – Modified MT (1969) medium.

*tium* using anthers as a source material (Hidaka *et al.*, 1981). Furthermore, it has been observed that anthers of even closely related species viz., *C. aurantium*, *C. sinensis*, *P. trifoliata* possessed different cultural requirements regarding optimal sucrose concentrations, pH and temperature (Hidaka, 1984a) in order to produce embryoids. Likewise, in *C. madurensis*, *C. reticulata* × *C. grandis* (Chen, 1985), *C. sinensis* (Hidaka, 1984b) and *C. aurantifolia* (Chaturvedi and Sharma, 1985), embryoids have been obtained via anther culture.

The stem segment has also been used by several research workers to initiate callus (Bhansali and Arya, 1978a,b; Chaturvedi and Mitra, 1975; Gill *et al.*, 1991). We established embryogenic callus cultures from seedling segments viz., roots, epicotyls and cotyledons by culturing on MS medium supplemented with NAA (10 mg/l) and Kinetin (0.5 mg/l) in *Citrus reticulata* (Tables 2 and 3; Fig. 3). Apart from the stem segments, the testas (Yang,

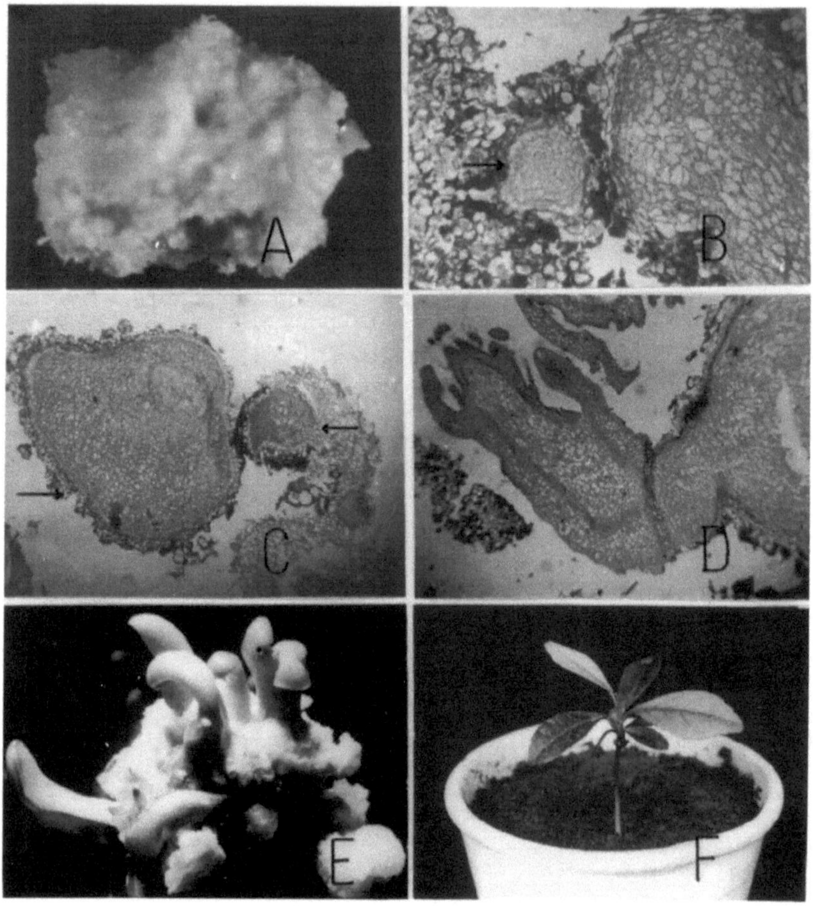


*Figure 3.* Induction of somatic embryogenesis and plant regeneration in *Citrus reticulata*. (A) Embryogenic callus obtained from seedling epicotyls cultured on MS + NAA (10 mg/l) + Kin (0.5 mg/l). (B-D) Showing different developmental stages of somatic embryos and their germination. (E) Regeneration of shoots from embryogenic calli transferred to half strength MS medium. (F) *In vitro* obtained complete plantlet, transferred to soil.

1983) and juice vesicles (Nito and Iwamasa, 1990) have also been used as a source tissue in *C. reticulata* and *C. unshiu* to obtain embryogenic cultures.

The embryoid formation in *Citrus* has also been widely reported from protoplast cultures. The embryoid formation from protoplasts is via callus formation. It has been reported in several citrus species such as *C. sinensis* (Hidaka and Kajiura, 1988; Kobayashi *et al.*, 1985), *C. mitis* (Sim *et al.*, 1988), *C. unshiu* (Kunitake *et al.*, 1991; Nito and Iwamasa, 1990), Microcitrus (Vardi *et al.*, 1986) and 7 other citrus species (Hidaka and Omura, 1989).

*Table 3.* Per cent callus cultures of *Citrus reticulata* (Local Sangtra) showing somatic embryogenesis after induction on media of different composition.

| Culture media | Embryogenic calli (%) | | | | Mean |
|---|---|---|---|---|---|
| | Leaf-derived callus | Epicotyl-derived callus | Cotyledon-derived callus | Root-derived callus | |
| MS + NAA (10 mg/l) + Kin (1 mg/l) + Vitamins (10 ×) | 63.6 (52.7) | 86.6 (70.1) | 60.0 (50.7) | 35.7 (36.7) | 61.4 (52.5) |
| MS + NAA (10 mg/l) + Kin (0.5 mg/l) | 53.5 (46.8) | 79.1 (63.0) | 50.0 (44.9) | 26.5 (31.0) | 52.2 (46.4) |
| MS + NAA (10 mg/l) + Kin (0.5 mg/l) + Sucrose (5%) | 61.1 (51.7) | 83.3 (68.0) | 57.6 (49.2) | 31.8 (34.1) | 58.4 (50.7) |
| MS + NAA (15 mg/l) + Kin (5 mg/l) | 66.6 (54.7) | 83.3 (66.5) | 58.3 (49.7) | 28.3 (32.0) | 59.1 (50.7) |
| Mean | 61.2 (51.4) | 83.0 (66.9) | 56.4 (48.6) | 30.5 (33.4) | |

Figures in parenthesis are Arc Sine transformed means.
L.S.D. ($p = 0.05$) Explant = 6.82, Media = NS, Interaction (Explant×Media) = NS.
L.S.D. ($p = 0.01$) Explant = 9.44, Media = NS, Interaction (Explant×Media) = NS.

### 3. Culture maintenance and embryo development

Maintenance of cultures in embryogenic form, over a longer period, has been an important aspect for utilizing the phenomenon of somatic embryogenesis in different studies. In *Citrus*, the cultures have been maintained in embryogenic form for years through periodic subculturing either on the fresh induction medium itself or on the media with reduced levels of growth regulators and other growth adjuvants. Genotype has been found to influence the maintenance of embryogenic cultures. Under similar cultural conditions the multiplication rate of embryoids was greater in case of *C. sinensis* than *C. aurantifolia* when cultures were maintained for one and a half year (Mitra and Chaturvedi, 1972). In the former case subculturing resulted into a "perennial bed" of continuously proliferating embryos, whereas in the later case the proliferation altogether ceased after a few transfers. In *C. grandis*, the stem-derived calli have exhibited the formation of numerous greenish globular bodies on Murashige and Skoog (1962) medium supplemented with NAA (0.1 mg/l) and zeatin (0.2 mg/l), which mature into embryoids during maintenance on the same medium (Chaturvedi and Mitra, 1975). The use of malt extract (500 mg/l) in the maintenance medium has proved beneficial for somatic embryogenesis (Gmitter *et al.*, 1990; Hidaka and Omura, 1989; Kochba *et al.*, 1972, 1974; Starrantino and Russo, 1980; Tusa *et al.*, 1983). However, the habituated callus of Shamouti orange (*C. sinensis*) has been found to lose its dependence on the malt extract for embryogenesis (Kochba and Spiegel-Roy, 1973). The addition of casein hydrolysate (CH) at concentrations ranging from 400–500 mg/l has also been reported to improve callus multiplication and embryoid formation in long term subcultured callus of *Citrus* (Rangaswamy 1958a,b; Zhang *et al.*, 1987). Likewise, the use of coconut milk (CM) in the maintenance medium has stimulated somatic embryogenesis (Button and Botha, 1975). The low concentration (10% v/v) of coconut milk has promoted the normal development of globular embryos (Sabharwal, 1963); however, the higher concentration (40% v/v) was harmful (Rangaswamy, 1961).

During maintenance, the irradiation of culture medium/calli has proven stimulatory for somatic embryogenesis at optimum doses. In Shamouti orange, the maximum stimulation of embryogenesis in ovular callus was obtained with gamma-irradiation doses of 12–16 kR with a distinct peak at 16 kR, whereas doses of 18–32 kR proved lethal (Spiegel-Roy and Kochba, 1973). The higher exposure (24–28 kR) decreased the callus growth and embryoid formation with lethal effect at 32 kR. Likewise, gamma irradiation has retarded the growth of calli and stimulated the embryoid formation; maximum embryoid formation occurred with 50-, 5-, 10-, 20- and 10 kR in Valencia, Yoshida oranges, Calamondin, *C. junos* and Satsuma, respectively (Nito *et al.*, 1989).

Duration in culture, has been found to effect the embryogenic competence of callus cultures. In *C. microcarpa* (Rangaswamy, 1961), the pseudobulbils produced on calli have been maintained for four years through continuous subculturing. Likewise, the friable embryogenic calli derived from ovules of sweet orange (*C. sinensis*) have been maintained for two years with good viability and embryogenic potential (Starrantino and Caponnetto, 1990). The embryogenic cultures of *C. sinensis* and *C. limon* have been maintained for more than three years through periodic transferring onto fresh medium (Starrantino and Russo, 1980). Somatic embryogenesis in embryo- and ovary-derived calli was decreased and subsequently ceased with prolonged subculture (Mitra and Chaturvedi, 1972). However, Spiegel-Roy and Vardi (1984) have mentioned that even 12 year-old *Citrus* calli have been found to possess the embryogenic competence.

### 4. Embryo germination and embling formation

Germination, the process in which the embryo is awakened, is necessary for getting complete plants and it requires specific physical conditions. Similarly, the somatic embryos, although developed artificially need precise cultural conditions for initiation and development of shoots and roots leading to complete plantlet (embling) formation.

In *C. microcarpa* (Rangaswamy, 1958, 1961) the nucellar-derived embryoids in culture have exhibited better maturation and subsequent germination upon culturing on White (1963) medium supplemented with casein hydrolysate (400 mg/l) and kinetin (2 mg/l). It has been reported that embryoid development occurred in *Citrus* when 6-methyl amino purines were present in the medium, and during embryoid germination roots appeared first followed by shoots (Starrantino *et al.*, 1978; Starrantino and Russo, 1980). Although the type and concentration of growth regulators is varied, in most instances the basal medium was supplemented with gibberellic acid in the somatic embryo germination medium. Somatic embryos of *C. sinensis* derived from ovules were germinated by supplementing $GA_3$ (1 mg/l) in the basal medium (Button and Bornman, 1971). However, it was further noted that the germination of somatic embryos derived from ovules and nucelli occurred upon subculturing them on a basal MT medium containing kinetin (0.1 mg/l), IAA (0.1 mg/l), $GA_3$ (1 mg/l) or coconut milk (Kochba *et al.*, 1972), while the higher kinetin/IAA ratio or coconut milk favoured stem elongation more than root formation. The lower kinetin/IAA ratio, on the other hand, favoured root formation and inhibited stem elongation. The presence of $GA_3$ in the germination medium stimulated rooting as well as stem elongation. Likewise, it is further reported (Kochba *et al.*, 1974) that $GA_3$ and adenine sulphate (ADS) significantly stimulated the rooting of small embryoids developed from ovular callus. Large embryoids with fully developed root zones, although, rooted well even on the basal medium. Furthermore, $GA_3$ and adenine sulphate (ADS) also enhanced shoot formation of embryoids which was initiated prior to rooting. The $GA_3$ (1 mg/l) alone used as supplement in the basal medium converted embryoids of *C. unshiu* into complete plantlets (Nito and Iwamasa, 1990).

In *C. sinensis* (Kobayashi *et al.*, 1985), embryoids developed from protoplasts have also exhibited better germination when Murashige and Tucker (1969) medium was supplemented with sucrose (5%) and $GA_3$ (1 mg/l). The embryoids of *C. sinensis* and *C. reticulata* developed from protoplasts via callus formation (Hidaka and Kajiura, 1988) formed emblings upon transfer to the medium containing $GA_3$ (1 uM) and sucrose (0.06 M). The replacement of sucrose by galactose (0.1 M) and sorbitol (0.1 M) resulted in better plantlet growth (Hidaka and Omura, 1989; Hidaka and Kajiura, 1988). However, the protoplast derived somatic embryos of *C. reticulata* germinated well upon supplementing the MT medium with $GA_3$ (1 mg/l), sucrose (3%) and gallan gum (0.2%).

In a few instances, the malt extract (ME) was also a supplement in the somatic embryoid germination medium. It is reported that although ME was beneficial for embryoid formation, its presence in the medium reduced their subsequent germination (Moore, 1985a). The use of lactose (50 g/l) and glycerol (20 mg/l) has also been made for embryoid germination and complete plantlet formation (Starrantino and Caponnetto, 1990). The calli derived from immature ovules developed into somatic embryos and complete plantlets upon transferring onto MT medium supplemented with lactose. Likewise, the development of complete plantlets from the ovular callus-derived somatic embryos has been reported (Nel, 1987) upon sub-culturing such embryos on MT medium fortified with glycerol (2.6%) and agar (1%). In our laboratory, we have achieved complete plants in *Citrus reticulata* cvs. Kinnow and local sangtra (Fig. 3) through embryoid germination on half-strength MS (1962) medium supplemented with BA (5 mg/l). In Kinnow, only 23.5% of the somatic embryos developed into emblings, whereas in local sangtra 89.2% of somatic embryos resulted into emblings.

## 5. Conclusions

Various factors affecting induction, maintenance and germination of somatic embryos in commercially important *Citrus* rootstocks and scion cultivars have been established. Different combinations and concentrations of auxins, cytokinins, sugars and other growth adjuvants enhancing somatic embryogenesis from wide variety of explants cultured *in vitro* have been determined. During the recent years, thidiazuron (TDZ) has been investigated as a most active cytokinin-like substance for woody plant species (Huetteman and Preece, 1993). Thidiazuron, in the range of 0.5–10 $\mu$M, has proven to be stimulatory for induction of somatic embryogenesis in other woody plant species. Likewise, reduction in the level of inorganic nitrogen in the culture medium and supplementation with amino acids/amides such as asparagine, glutamine and L-proline has induced somatic embryogenesis in a number of plant species. Coupling such factors with those already established in *Citrus* would certainly improve the frequency and quality of somatic embryos for scaling up their production in bioreactors and finally delivering them in large numbers to greenhouses/fields. Furthermore, the phenomenon of somatic embryogenesis will be highly rewarding in efficient recovery of genetically altered *Citrus* plants through protoplast fusion and genetic transformation using various vector and vectorless systems.

## 6. References

Beloualy, N., 1991. Plant regeneration from callus cultures of three *Citrus* rootstocks. Plant Cell Tissue and Organ Culture 24: 29–34.

Ben-Hayyim, G. and H. Neumann, 1983. Stimulatory effect of glycerol on growth and somatic embryogenesis in *Citrus* callus cultures. Z. Pflanzenphysiol. 110: 331–337.

Bhansali, R.R. and H.C. Arya, 1978a. Shoot formation in stem and root callus of *Citrus aurantifolia* (Christm) Swingle grown in cultures. Curr. Sci. 47: 775–776.

Bhansali, R.R. and H.C. Arya, 1978b. Tissue culture propagation of *Citrus* trees. In: Proc. Int. Soc. Citriculture 1: 135–140.

Button, J. and C.E.J. Botha, 1975. Enzymic maceration of *Citrus* callus and the regeneration of plants from single cell. J. Exp. Bot. 26: 723–729.

Button, J. and C.H. Bornman, 1971. Development of nucellar plants from unfertilized ovules of Washington Navel orange through tissue culture. Citrus Grower and Sub-tropical Fruit J. 453: 11–14.

Button, J. and C.H. Bornman, 1974. Fine structure of and embryoid development from embryogenic ovular callus of "Shamouti" orange (*Citrus sinensis* Osb.). J. Exp. Bot. 25: 446–457.

Button, J. and J. Kochba, 1977. Tissue culture in the *Citrus* industry. In: J. Reinert and Y.P.S. Bajaj (Eds.), Applied and Fundamental Aspects of Plant Cell, Tissue and Organ Culture, pp. 70–92. Springer-Verlag, Berlin.

Chaturvedi, H.C. and G.C. Mitra, 1975. A shift in morphogenetic pattern in *Citrus* callus tissue during prolonged culture. Ann. Bot. 39: 683–687.

Chaturvedi, H.C. and A.K. Sharma, 1985. Production of androgenic plants of *Citrus aurantifolia*. J. Plant Physiol. 119: 473–477.

Chen, Z.G., 1985. A study of induction of plants from *Citrus* pollen. Fruit Vars. J. 39: 44–50.

Chu, C.C., C.D. Wang, S.S. Sun, K.C. Yin and C.Y. Chu, 1975. Establishment of an efficient medium for anther culture of rice through comparative experiments on the nitrogen sources. Sci. Sin. 18: 659–668.

Evans, D.A., W.R. Sharp and C.E. Flick, 1981. Growth and behaviour of cell cultures – Embryogenesis and organogenesis. In: T.A. Thorpe (Ed.), Plant Tissue Culture: Methods and Applications in Agriculture, pp. 45–113. Academic Press, New York.

FAO, 1992. Quarterly Bulletin of Statistics, Vol. 5(3).

Gao, F., J.Z. Chen, S.C. Chen, J.R. Zhang and A.R. Wu, 1990. Production of virus-free nucellar plantlets from unfertilized ovules of *Citrus in vitro*. Acta Botanica Sinica 32: 505–509.

Gavish, H., A. Vardi and R. Fluhr, 1992. Suppression of somatic embryogenesis in *Citrus* cell cultures by extracellular proteins. Planta 186: 511–517.

Gill, M.I.S., B.S. Dhillon, Z. Singh and S.S. Gosal, 1991. Induction of high frequency somatic embryogenesis and plant regeneration in mandarins. In: J. Prakash and R.L.M. Pierik (Eds.), New Technologies and Applications, pp. 231–235. Kluwer Academic Publishers, Dordrecht, The Netherlands.

Gill, M.I.S., Z. Singh, B.S. Dhillon and S.S. Gosal, 1994. Somatic embryogenesis and plantlet regeneration on caluses derived from seedling explants of "Kinnow" mandarin (*Citrus nobilis* Lour×*Citrus deliciosa*). J. Hort. Sci. 69(2): 231–236.

Gmitter, F.G. Jr. and G.A. Moore, 1986. Plant regeneration from undeveloped ovules and embryogenic calli of *Citrus*: embryo production, germination and plant survival. Plant Cell Tissue and Organ Culture 6: 139–147.

Gmitter, F.G. Jr., X.B. Ling and X.X. Deng, 1990. Induction of triploid *Citrus* plants from endosperm calli *in vitro*. Theor. Appl. Genet. 80: 785–790.

Hidaka, T., 1984a. Effects of sucrose concentration, pH of media and culture temperature on anther culture of *Citrus*. Jap. J. Breeding 34: 416–422.

Hidaka, T., 1984b. Induction of plantlets from anthers of "Trovita" orange (*Citrus sinensis* Osbeck). J. Jap. Soc. Hort. Sci. 53: 1–5.

Hidaka, T. and I. Kajiura, 1988. Plantlet differentiation from callus protoplasts induced from *Citrus* embryos. Scientia Hort. 34: 85–92.

Hidaka, T. and M. Omura, 1989. Control of embryogenesis in *Citrus* cell culture : regeneration from protoplasts and attempts to callus bank. Kaju Shikenjo Hokoku, B. Okitsu-Bull. Fruit Tree Res. Stn., Series B, Okitsu 16: 1–17.

Hidaka, T., Y. Yamada and T. Shichijo, 1981. Plantlet formation from anthers of *Citrus aurantium* L. In: Proc. of the Int. Soc. Citriculture, Vol. I, Tokyo, Japan, pp. 153–155.

Huetteman, C.A. and J.E. Preece, 1993. Thidiazuron: a potent cytokinin for woody plant tissue culture. Plant Cell Tiss. Org. Cult. 33: 105–119.

Kobayashi, S., I. Ideka and M. Nakatani, 1984. Induction of nucellar callus from orange (*Citrus sinensis* Osb.) ovules and uniformity of regenerated plants. Bull. Fruit Tree Res. Stn., Japan (Ekitsu) 5: 43–54.

Kobayashi, S., I. Ideka and H. Uchimiya, 1985. Conditions for high frequency embryogenesis from orange (*Citrus sinensis* Osb.) protoplasts. Plant Cell Tiss. Org. Cult. 4: 249–259.

Kochba, J., J. Button, P. Spiegel-Roy, C.H. Bornman and M. Kochba, 1974. Stimulation of rooting of *Citrus* embryoids by gibberellic acid and adenine sulphate. Ann. Bot. 38: 795–802.

Kochba, J. and P. Spiegel-Roy, 1973. Effect of culture media on embryoid formation from ovular callus of "Shamouti" orange (*Citrus sinensis*). Z. Pflanzenzüchtg. 69: 156–162.

Kochba, J. and P. Spiegel-Roy, 1977. The effects of auxins, cytokinins and inhibitors on embryogenesis in habituated ovular callus of the "Shamouti" orange. Z. Pflanzenphysiol. 81: 283–288.

Kochba, J., P. Spiegel-Roy, H. Neumann and S. Saad, 1982. Effect of carbohydrates on somatic embryogenesis in subcultured nucellar callus of *Citrus* cultivars. Z. Pflanzenphysiol. 105: 359–368.

Kochba, J., P. Spiegel-Roy and H. Safron, 1972. Adventive plants from ovules and nucelli in *Citrus*. Planta 106: 237–245.

Kochba, J., P. Spiegel-Roy, H. Neumann and S. Saad, 1978a. Stimulation of embryogenesis in *Citrus* ovular callus by ABA, ethephon, CCC and Alar and its suppression by $GA_3$. Z. Pflanzenphysiol. 89: 427–432.

Kochba, J., P. Spiegel-Roy, S. Saad and H. Neumann, 1978b. Stimulation of embryogenesis in *Citrus* tissue culture by galactose. Naturwissensehaften 65: 261.

Kunitake, H., H. Kagami and M. Mii, 1991. Somatic embryogenesis and plant regeneration from protoplasts of Satsuma mandarin (*Citrus unshiu* Marc.). Scientia Hort. 47: 27–33.

Maheshwari, P. and N.S. Rangaswamy, 1958. Polyembryony and *in vitro* culture of embryos of *Citrus* and *Mangifera*. Indian J. Hort. 15: 275–282.

Mitra, G.C. and H.C. Chaturvedi, 1972. Embryoid and complete plants from unpollinated ovaries and ovules of *in vitro* emasculated flower buds of *Citrus* species. Bull. Torrey Botanical Club 99: 184–189.

Moore, G.A., 1985a. Factors affecting *in vitro* embryogenesis from undeveloped ovules of mature *Citrus* fruit. J. Amer. Soc. Hort. Sci. 110: 66–70.

Moore, G.A., 1985b. Somatic embryogenesis in *Citrus*. In: Biotechnologia en las Americas. II.

Aplicaciones en agricultura tropical, San Jose, Costa Rica, CONICIT, Session I, Univ. FL, Gainesville, USA. Paper No. 2, p. 3 (Abstr.).

Murashige, T. and D.P.H. Tucker, 1969. Growth factor requirements of *Citrus* tissue culture. In: Proc. Ist Int. Citrus Symp. 3, Riverside, 1968–69, pp. 1155–1161.

Murashige, T. and F. Skoog, 1962. A revised medium for rapid growth and bioassays with tobacco tissue cultures. Physiol. Plant. 15: 473–497.

Nel, M., 1987. *In vitro* culture of *Citrus* meristems. Information Bull., Citrus and Subtropical Fruit Res. Inst., South Africa 175: 9.

Nito, N., J.T. Ling, M. Iwamasa and Y. Katayama, 1989. Effects of gamma-irradiation on growth and embryogenesis of *Citrus* callus. J. Jap. Soc. Hort. Sci. 58: 283–287.

Nito, N. and M. Iwamasa, 1990. *In vitro* plantlet formation from juice vesicle callus of satsuma (*Citrus unshiu* Marc.). Plant Cell Tiss. Org. Cult. 20: 137–140.

Oh, S.D., W.S. Song, J.S. Kim and E.H. Park, 1991. *In vitro* micropropagation of yooza (*Citrus junos* Tanaka). I. Plant regeneration from callus induced from shoot tips. J. Korean Soc. Hort. Sci. 32: 87–96.

Oh, S.D., W.S. Song and M.S. Lee, 1992. Direct somatic embryogenesis and plant regeneration from immature ovules in Yooza (*Citrus junos* Sieb. Et. Tanaka). HortSci. 27: 694.

Ohta, Y. and K. Furusato, 1957. Embryo cultures in *Citrus*. Seiken Ziho 8: 49–54.

Pasqual, M. and A. Ando, 1990. Effect of gamma-irradiation on the production of nucellar callus of *Citrus sinensis* Osb. cv. Valencia *in vitro*. Hort. Abstr. 61: 1221, No. 10443.

Pasqual, M., A. Ando and J. Crocomo, 1988. Effects of growth regulators on *in vitro* embryogenesis of nucelli in *Citrus sinensis* cv. Valencia. Pesquisa Agropecuaria Brasileira. 23: 255–259.

Rangan, T.S., T. Murashige and W.P. Bitters, 1968. *In vitro* initiation of nucellar embryos in monoembryonic *Citrus*. HortSci. 3: 226–227.

Rangan, T.S., T. Murashige and W.P. Bitters, 1969. *In vitro* studies on zygotic and nucellar embryogenesis in *Citrus*. In: H.B. Chapman (Ed.), Proc. Int. Citrus Symp., Univ. California, Riverside, pp. 225–229.

Rangaswamy, N.S., 1958a. Culture of nucellar tissue of *Citrus in vitro*. Experientia 14: 111–112.

Rangaswamy, N.S., 1958b. *In vitro* culture of nucellus and embryos of *Citrus*. In: P. Maheshwari (Ed.), Proc. Modern Dev. Plant Physiol., New Delhi, pp. 104–106.

Rangaswamy, N.S., 1961. Experimental studies on female reproductive structures of *Citrus microcarpa* Bunge. Phytomorphology 11: 109–127.

Saad, S., 1975. Factors affecting the growth of lemon (*Citrus limon*) callus *in vitro* culture. M.Sc. Thesis. Hebrew Univ. Jerusalem.

Sabharwal, P.S., 1963. *In vitro* cultures of ovules, nucelli and embryos of *Citrus reticulata* Blanco var. Nagpuri. In: P. Maheshwari and N.S. Rangaswamy (Eds.), Proc. Plant Tissue and Organ Culture – A Symposium, Delhi, pp. 332–344.

Sharp, W.R., M.R. Sandahl, L.S. Caldas and S.B. Maraffa, 1980. The physiology of *in vitro* asexual embryogenesis. Hort. Rev. 2: 268–310.

Sharp, W.R., D.A. Evans and M.R. Soudahl, 1982. Applications of somatic embryogenesis to crop improvement. In: A. Fujiwara (Ed.), Proc. 5th Intl. Cong. Plant Tissue and Cell Culture. Japan Assoc. for Plant Tissue Culture, Tokyo, pp. 759–762.

Sim, G., C. Loh and G. Goh, 1988. Direct somatic embryogenesis from protoplasts of *Citrus mitis* Blanco. Plant Cell Rep. 7: 5–8.

Spiegel-Roy, P. and A. Vardi, 1984. *Citrus*. In: P.V. Ammirato, D.A. Evans, W.R. Sharp and Y. Yamada (Eds.), Handbook of Plant Cell Culture, pp. 355–372. MacMillan Publ., New York.

Spiegel-Roy, P. and J. Kochba, 1973. Stimulation of differentiation in orange (*Citrus sinensis*) ovular callus in relation to irradiation of the media. Rad. Bot. 13: 97–103.

Starrantino, A. and F. Russo, 1980. Seedling from undeveloped ovules of ripe fruits of polyembryonic *Citrus* cultivars. HortSci. 15: 296–297.

Starrantino, A. and P. Caponnetto, 1990. Effect of cytokinins on embryogenic callus formation from undeveloped ovules of orange. Acta Horticulturae 280: 191–194.

Starrantino, A., P. Spino and F. Russo, 1977. Nucellar embryogenesis and plantlet development *in vitro* from nucelli of several *Citrus* species. Giornale Botanico Italiano 111: 358.

Starrantino, A., P. Spino and F. Russo, 1978. Nucellar embryogenesis and the development of plantlets *in vitro* from nucelli of certain *Citrus* species. Giornale Botanico Italiano 112: 41–52.

Steward, F.C., M.O. Mapes and K. Mears, 1958. Growth and organized development of cultured cells. Amer. J. Bot. 45: 705–708.

Swingle, W.T. and P.C. Reece, 1967. The botany of *Citrus*. In: W. Reuther, H.J. Webber and L.D. Batchelor (Eds.), The Citrus Industry, pp. 190–480. Univ. Calif. Press, Berkeley.

Takayanagi, R., T. Hidaka and M. Omura, 1992. Regeneration of intergeneric somatic hybrids by electric fusion between *Citrus* and its wild relatives: Mexican lime (*C. aurantifolia*) and Java feroniella (*Feroniella lucida*) or tabog (*Swinglea glutinosa*). J. Jap. Soc. Hort. Sci. 60: 799–804.

Tisserat, B., E.B. Esan and T. Murashige, 1979. Somatic embryogenesis in angiosperms. Hort. Rev. 1: 1–78.

Tusa, N., G. Ceraci and G. Occorso, 1983. *In vitro* culture of ovules in two *Citrus* species. Hort. Abstr. 54 (1398): 135–136.

Vardi, A., D.J. Hutchison and E. Galun, 1986. A protoplast to tree system in Microcitrus based on protoplasts derived from a sustained embryogenic callus. Plant Cell Rep. 5: 412–414.

Wann, S.R., 1988. Somatic embryogenesis in woody species. In: J. Janick (Ed.), Hort. Rev. 10: 153–177.

White, P.R., 1963. The Cultivation of Animal and Plant Cells. Ronald Press, New York.

Williams, E.G. and G. Maheswaran, 1986. Somatic embryogenesis: factors influencing coordinated behaviour of cells as an embryogenic group. Ann. Bot. 57: 443–462.

Wu, J.H., J.S. Cheng, W.C. Zhang and S.Y. Wan, 1990. Studies on the ovule culture and somatic embryogenesis of *Citrus*. J. Fruit Sci. 7: 19–24.

Yang, N.B., 1983. Studies on organogenesis in *Citrus* and *Fortunella*. Zhiwu Shenglixue Tongxum. Plant Physiol. Comm. 6: 33–37.

Zhang, J.R., A.R. Wu and F. Gao, 1987. Studies on the capability of redifferentiation and the inheritance of the long-term subcultural callus from the embryos of Liu Cheng (*Citrus sinensis* Osbeck). Acta Horticulturae Sinica 14: 213–216.

# 2. Somatic embryogenesis in *Eucalyptus*

E.M. Muralidharan and A.F. Mascarenhas

**Contents**



## 1. Introduction

*Eucalyptus* species are amongst the most widely planted trees in the world (Zacharin, 1978). They have very rapid growth rates and they can adapt to a wide range of climates and soil conditions. These wide ranging characteristics of *Eucalyptus* have contributed to their popularity as an exotic plantation species throughout the world.

### 1.1. *Botany*

The estimated number of *Eucalyptus* spp. varies from 450 to more than 700 depending on the source of information. This genus is divided into three subgenera viz. Symphomyrtus, Monocalyptus and Corymbia.

The genus derives its name (from Greek eu – well and kalypto – I cover) from the characteristic lid or operculum which covers the unopened flowers. Flowers are bisexual and pollination is, generally, dependent on an insect or animal vector. *Eucalyptus* is a predominantly out breeding though most

*S. Jain, P. Gupta & R. Newton (eds.), Somatic Embryogenesis in Woody Plants, Vol. 2*, 23–40.

species are to some degree self-compatible. Fruits are woody capsules at maturity and contain several seeds, although some are not viable.

### 1.2. *Regeneration*

The regeneration of *Eucalyptus* occurs in nature by seed although vegetative propagation is frequently practiced, especially with damaged individuals. Seed propagation is inexpensive and very successful, but does not meet the goals of improving most of the trees.

### 1.3. *Distribution*

The native habitat of the genus is Australia (where it constitutes 95% of the forest trees) and the neighboring islands of New Guinea, Java, Timor and the Philippines (Zacharin, 1978). As an exotic plant, it has been introduced as early as the mid-nineteenth century to Europe, Africa and Asia, primarily as a source of firewood, pulp and timber.

### 1.4. *Uses*

*Eucalyptus* trees are a significant source of fuelwood, timber, and raw material for the paper and pulp industry and to a lesser extent for honey, tannins and essential oils.

*Eucalyptus citriodora*, the "lemon scented gum" (also known as "spotted gum"), is grown extensively in Brazil, Congo, Java and China both for its timber and the essential oil. The major oil components are citronellal (65–85%) and citronellol (15–20%), which are found in the leaves. When cultivated for oil, the trees are pollarded to induce production of more leaves rather than timber.

### 1.5. *Clonal propagation of* Eucalyptus

The increasing demand for *Eucalyptus* biomass has necessitated research with unconventional methods for the improvement of productivity. Clonal propagation of superior genotypes has been utilized as a means of rapid genetic improvement. The largest operational clonal forestry programs of any tree species are associated with *Eucalyptus* species and significant genetic gains have been reported in plantations derived from rooted cuttings (Zobel, 1993).

Clonal propagation enables retention of most of the genetic potential of elite selected plantations, including the non-additive components of genetic variance in new generations. With sexual propagation, non-additive combinations are mostly lost due to genetic segregation. Thereby, clonal propagation is advantageous for the propagation of outstanding individuals with favorable combinations of additive genetic variation. Clonally propagated elite trees have an advantage to devoid segregation which normally occurs during sexual propagation.

Large-scale deployment of improved planting stocks of selected clones is possible in a much shorter time with the use of vegetative methods rather than the conventional methods via seed in tree-improvement programmes operated through seed orchards. An advantage that clonal forestry offers is the chance to test the genotype × environment interaction of clones.

The risk of pests and diseases is much greater in monocultural plantations. This has raised the question as to what the number of clones and the mixtures is to be used in *Eucalyptus* plantations. Zobel (1993) concluded that unless a proper assessment of gain vs. risk is made, clonal forestry will fall short of the potential benefits. Furthermore, only 15 unrelated and well adapted clones of *E. grandis* and *E. urophylla* are sufficient with planting in intermixed blocks.

Commercially, cuttings are the preferred method for clonal propagation but most eucalyptus do not root when mature cuttings are used (Hartney, 1980). Coppicing and hedging improves rooting, but clones show great variation in rooting ability.

### 1.6. *Micropropagation*

Vegetative propagation of mature superior trees of many important *Eucalyptus* species has been unsuccessful, and this makes it commercially nonviable. Clonal propagation through *in vitro* methods has been undertaken to overcome this problem. The highest multiplication rate with *in vitro* techniques has been estimated to be 100,000 plants per year from a single bud (Gupta *et al.*, 1981). This figure is an obvious indicator of an advantage of using *in vitro* methods for the large-scale production of superior planting stocks. It can hold true especially for clones that perform very well. Normally, clones with poor rooting ability are excluded from clonal forestry programmes.

Micropropagation of several *Eucalyptus* species has been very well covered in several review articles (McComb and Bennett, 1986; Le Roux and Van Staden, 1991). Explants from both juvenile and mature trees have been used for micropropagation. Plantlets have been regenerated, mostly through an *in vitro* rooting stage followed by transfer to soil for hardening. Rajbhandary (1990) reported direct rooting of micropropagated shoots in soil. These results demonstrated a substantial amount of savings in overall expenses in plantlet production including labour costs. An increase in biomass production of plantlets (up to 49.5%) over seedling controls was recorded in the tissue culture-derived 34– month-old *Eucalyptus* plantations (Khuspe *et al.*, 1987).

### 1.7. *Somatic embryogenesis*

Several reports on *Eucalyptus* somatic embryogenesis have appeared in literature. Lakshmi Sita (1981) described the formation of embryo like structures in suspension cultures of *E. citriodora* and *E. grandis* derived from stem explants of mature trees. No further development of the embryos or plantlet regeneration was reported. Some other brief reports on *Eucalyptus* somatic embryogenesis have appeared (Ouyang *et al.*, 1981; Boulay, 1987; Qin Chang-Le and Kirby, 1990). Ouyang *et al.* (1981) obtained initially an "embryonal cell mass" from cultured seedling explants of "*E.* × *leichow*" and from which they differentiated somatic embryos and regenerated plants. Boulay (1987) also induced somatic embryogenesis in calli derived from hypocotyl and internodes explants of *E. gunii*. Furthermore, direct induction of somatic embryogenesis from cultured mature seed explants of *E. citriodora* and plantlet regeneration system have been developed (Muralidharan and Mascarenhas, 1987; Muralidharan *et al.*, 1989). Long-term embryogenesis could be maintained at a high frequency through secondary (repetitive) somatic embryogenesis without involvement of a callus phase (Muralidharan *et al.*, 1989).

In this chapter, we have described the induction, maintenance and regeneration of plantlets from a repetitive embryogenesis system in *E. citriodora*.

## 2. Culture initiation

### 2.1. *Explant*

Seeds of *Eucalyptus citriodora*, stored at 4°C for two years, were used to induce somatic embryogenesis. Our preliminary experiments showed that culture of seeds with the seed coat resulted in heavy leaching of brown pigments into the medium which prevented observation of the explants. In the subsequent experiments, the seed coat was removed.

### 2.2. *Sterilization*

Seeds were soaked in water for 1 h with frequent changes in order to remove most of the pigments. They were then washed with water containing a few drops of the detergent Teepol followed by three washes with distilled water. Washed seeds were treated with 0.05% $HgCl_2$ for 5 min and then rinsed with sterile distilled water 3 times.

The sterilized seeds were left soaking in the last change of distilled water for 24 h. From soaked seeds, the seed coat was removed with a fine tipped pair of foreceps without damaging the cotyledons. Decoated seeds were used as explants, which consisted of the embryo with intact cotyledons.

### 2.3. *Induction of somatic embryogenesis*

Decoated seeds were cultured for the induction of somatic embryogenesis on a series of media designed to test the effect of various auxins. Initially two basal media, MS (Murashige and Skoog, 1962) and B5 (Gamborg *et al.*, 1968) supplemented with different concentrations (0.5–10 mg/l) of auxins (2,4-D and NAA) and sucrose (30 g/l) were tested.

After 2 weeks of culture, induction of somatic embryogenesis was observed on explants cultured on the B5 medium supplemented with 3 mg/l NAA and on MS medium having 5 mg/l NAA. The best response in producing embryogenic cultures was 3.3% (1 òut of 30 explants) on both the media. However, these experiments could not be repeated. On other media, explants either germinated or merely expanded in size or gave rise to profuse rooting. The formation of somatic embryos could be observed when cultures were examined under a stereomicroscope. Embryos developed in clusters of 2–3 directly on the surface of the cotyledons without callus formation. A dark red colour was characteristic of most of the embryos, although translucent white embryos were also observed. When thin sections of somatic embryos were examined, the pigmentation was found mainly in the epidermal cells.

The induction of somatic embryogenesis with the above mentioned methods was highly inconsistent, although several combinations of basal media with various auxins have been tested. Factors responsible for induction of somatic embryogenesis thus remain to be elucidated. Seed source, genetic factors or storage conditions may be involved which could not be confirmed by experimentation.

It was observed that somatic embryos induced on B5 + NAA (3 mg/l) undergo prolific secondary embryogenesis and indicated the feasibility of maintenance of embryogenic cultures.

### 2.4. *Repetitive embryogenesis*

It was evident from early subcultures that the entire mass of somatic embryos formed in the embryogenic cultures was not produced directly from the explants, but rather from embryo budding or secondary embryogenesis. This proliferation of secondary embryos (Fig. 1a) did not show synchrony in development. This process continued as long as the cultures were subcultured regularly onto the fresh media containing NAA. Secondary embryos developed only at the radicle end of the primary embryos (Fig. 1b).

Cultures of pure embryos (developed via embryo-to-embryo proliferation) are distinct from embryogenic callus cultures. Since these types of cultures are maintained over several subcultures, they merit a separate nomenclature. It is to be assumed that in many plants, where repetitive embryogenesis has been described (Raj Bhansali *et al.*, 1990; Parrott and Bailey, 1993; Weissinger II and Parrott, 1993), cultures similar to *Eucalyptus* could be developed with the isolation and maintanence of repetitive embryogenic cultures. Reports on secondary cycles of embryo formation are rather frequent in the literature. Williams and Maheshwaran (1986) observed an apparent substantial similarity between secondary embryogenesis and continuous proliferation of pro-embryo like nodules in embryogenic callus.

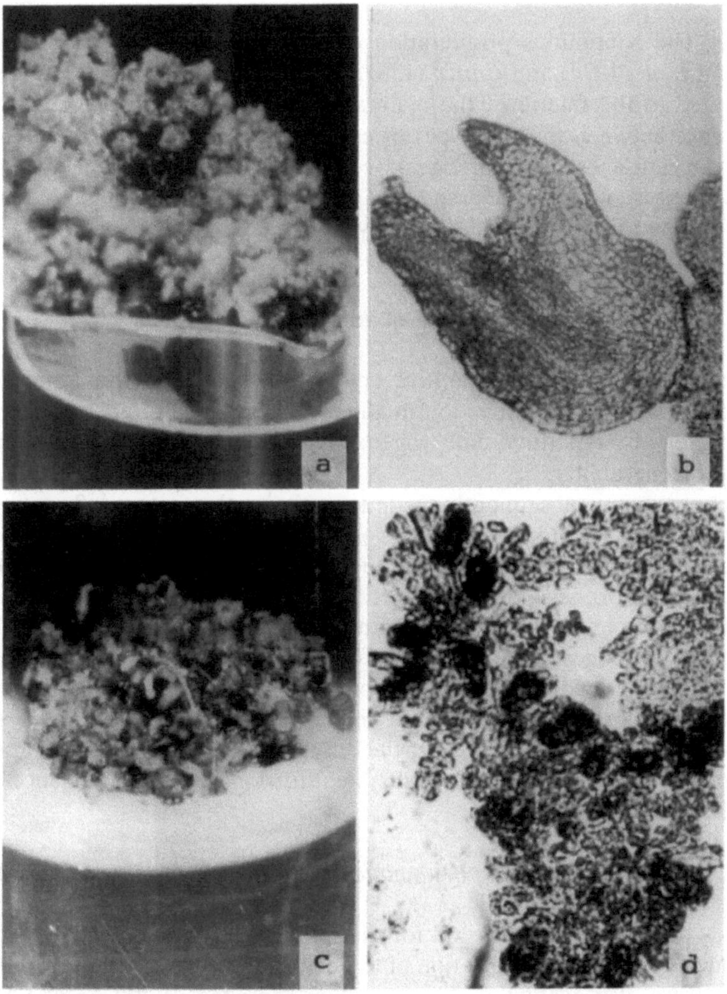


*Figure 1.* (a) The Embryogenic Mass (EM) culture of *E. citridora*. (b) A section through a secondary somatic embryo attached to the primary embryo tissues but with a closed radicular end (× 300). (c) Embryogenic callus (EC) of *E. citridora*. (d) Darkly staining proembryos in the EC (× 50).

Both repetitive embryogenesis and the formation of an embryogenic callus from primary embryos appear to be the manifestation of the same phenomenon of cells that escape from the integrated control that is required to maintain co-ordinated embryogenic behavior (Williams and Maheshwaran, 1986). The nature of the stimulus responsible for the control is still undefined although inter-cell communication through plasmodesmata appears to play a role. The continuous proliferation of pro-embryo like nodules in *Citrus* (Button *et al.*, 1974) and *Carica* (Litz and Conover, 1983) callus can also be attributed to the failure of the maintanence of the integrated control. The difference between the two types of cultures is probably only in the degree of organization of proembryogenic cells. The term "Embryogenic Mass" (EM) was used (Muralidharan *et al.*, 1987, 1989) to distinguish cultures proliferating through secondary embryogenesis.

## 3. Maintenance of embryogenic cultures

As described above, the repetitive embryogenesis occurs spontaneously in cultures grown on the induction medium. When subcultured regularly at 4 week intervals, repetitive embryogenesis could be maintained without any loss of regenerability.

Attempts were subsequently made to optimize the factors influencing the maintenance of secondary embryogenesis over long periods. This has resulted in the development of a novel method of maintaining a high frequency of somatic embryogenesis over long periods. Embryogenic cultures have been maintained for more than 9 years without any loss of regenerability. This is probably the longest period of time for which an embryogenic culture has been maintained without loss of embryogenic competence. Juretic and Jelaska (1991) reported the retention of embryogenic potential for 15 years in a callus line of pumpkin. However, the absence of a callus which is prone to variability, makes the EM an ideal system for long term maintenance in *E. citriodora*.

### 3.1. *Parameters effecting the maintenance of embryogenic mass (EM)*

Various parameters were tested for their effect on the maintenance of EM. The development and maturation of the embryos, as influenced by the treatments, were also evaluated. The inoculum consisted of 200 mg of EM per tube. Each treatment had undergone three 4-week-interval three subculture cycles and then were evaluated for: EM growth (increase in fresh weight), total number of embryos and number of mature embryos per 200 mg weight of EM.

### 3.2. *Effect of basal media*

Three different basal media viz. *MS* (Murashige and Skoog, 1962), *B*5 (Gamborg *et al.*, 1968), *W* (White, 1954) and *WPM* (Smith and McCown, 1983) were tested (Table 1). The growth (measured by fresh weight) of EM was higher in MS medium, whereas the total number of embryos formed and matured was higher in B5 medium. White medium did not support repetitive embryogenesis.

*Table 1.* Effect of basal medium on growth of EM.

| Basal medium | Final fresh weight (mg) | Increase (x times) | Total embryos/ 200 mg of EM | Mature embryos/200 mg of EM |
|---|---|---|---|---|
| B5 | 1064.6 ± 4.19 | 5.30 | 40.0 ± 0.63 | 14.8 ± 0.33 |
| MS | 1078.8 ± 2.88 | 5.39 | 35.6 ± 0.33 | 12.8 ± 0.33 |
| W | 281.4 ± 3.00 | 1.40 | 7.4 ± 0.87 | 6.2 ± 0.65 |
| WPM | 969.6 ± 3.71 | 4.80 | 30.6 ± 0.46 | 8.8 ± 0.45 |

Supplements: NAA – 3 mg/l; Sucrose – 3 g/l.

*Table 2.* Effect of auxins on maintenance of EM.

| Growth regulator | Conc. mg/l | Total embryos/ 200 mg of EM | Mature embryos/200 mg of EM |
|---|---|---|---|
| NAA | 0.5 | 34.8 ± 0.52 | 14.6 ± 0.54 |
| NAA | 1.0 | 36.8 ± 0.70 | 14.4 ± 0.21 |
| NAA | 5.0 | 42.0 ± 1.16 | 15.0 ± 0.28 |
| cpa | 0.5 | 21.8 ± 0.95 | 4.2 ± 0.59 |
| cpa | 1.0 | 18.8 ± 1.42 | 2.6 ± 0.67 |
| cpa | 5.0 | 18.2 ± 1.11 | 2.4 ± 0.21 |
| IAA | 0.5 | 32.6 ± 1.06 | 13.0 ± 0.40 |
| IAA | 1.0 | 34.6 ± 0.60 | 12.4 ± 0.21 |
| IAA | 5.0 | 35.8 ± 0.59 | 12.4 ± 0.67 |
| IBA | 0.5 | 34.6 ± 0.72 | 14.6 ± 0.36 |
| IBA | 1.0 | 36.8 ± 0.52 | 14.2 ± 0.59 |
| IBA | 5.0 | 39.6 ± 0.67 | 14.0 ± 0.86 |
| IPA | 0.5 | 31.8 ± 1.42 | 11.2 ± 0.33 |
| IPA | 1.0 | 33.2 ± 1.18 | 11.0 ± 0.28 |
| IPA | 5.0 | 34.0 ± 1.03 | 10.4 ± 0.22 |

Basal medium: B5 minerals and vitamins.
Supplements: Sucrose – 3 g/l.

### 3.3. *Effect of auxins*

The effect of auxins on repetitive embryogenesis was studied. The three concentrations (0.5, 1.0 and 5.0 mg/l) of auxins including: α-naphthalene-acetic acid (NAA); 2,4,5-chlorophenoxyacetic acid (cpa); indoleacetic acid (IAA); indolebutyric acid (IBA); indolepropionic acid (IPA): 2,4-dichloro-phenoxyacetic acid (2,4-D); 2,4,5-trichlorophenoxyacetic acid (2,4,5-T); 3,6-dichloroanisic acid (dicamba) and 4-amino-3,5,6-trichloro-picolinic acid (picloram) were tested in the culture medium.

The auxins, 2,4-D, dicamba, 2,4,5-T, and picloram, were inhibitory in the maintenance of the cultures, and developed into a soft brown callus. No somatic embryos could be induced from this callus even after transfer to the hormone free medium. In the presence of IAA, IBA, IPA and cpa, repetitive

*Table 3.* Effect of organic nitrogen compounds on EM cultures.

| Compound | Concentration mg/l | Total embryos/ 200 mg of EM | Mature embryos/ 200 mg of EM |
|---|---|---|---|
| – | – | 42.8 ± 1.06 | 15.0 ± 0.83 |
| CH | 500 | 44.2 ± 0.92 | 16.0 ± 0.31 |
| CH | 1000 | 43.2 ± 0.91 | 17.0 ± 0.44 |
| Glutamine | 500 | 46.6 ± 0.86 | 17.8 ± 0.73 |
| Glutamine | 1000 | 44.8 ± 0.58 | 17.2 ± 0.84 |
| CH + Glutamine | 500 + 500 | 48.2 ± 0.58 | 18.4 ± 0.51 |
| CH + Glutamine | 1000 + 1000 | 47.6 ± 0.39 | 17.6 ± 0.39 |

Basal medium: B5 minerals and vitamins.
Supplements: NAA – 5 mg/l; Sucrose – 3 g/l.

*Table 4.* Standard maintenance medium. B5 salts and vitamins (Gamborg *et al.*, 1968).

| | |
|---|---|
| α-Naphthalene acetic acid: | 5 mg/l |
| Casein Hydrolysate: | 500 mg/l |
| Glutamine: | 500 mg/l |
| Sucrose: | 30,000 mg/l |

embryogenesis continued to occur more or less in a fashion similar to that of the NAA containing medium (Table 2). However, all three NAA concentrations were superior for embryo development. Among the three NAA concentrations, 5 mg/l concentration was most effective with highest EM growth as well as total number of embryos. At the lower concentrations (0.5 and 1 mg/l) NAA was more conducive to embryo maturation. Hence, 5 mg/l of NAA was used for the maintenance of EM cultures.

### 3.4. *Effect of organic nitrogen compounds*

Casein hydrolysate (CH) and glutamine were added at two concentrations (500 and 1000 mg/l) to the induction medium (with NAA increased to 5 mg/l). Combinations of both compounds were also tried (Table 3). The combination of CH and glutamine (500 mg/l each) increased the total number of embryos as well as the number of mature embryos as compared to when they were used individually.

### 3.5. *Standard maintenance medium*

Based on the results obtained from the testing of the above three parameters, a Standard Maintenance Medium (SM) (Table 4) was formulated. This medium was used for further routine maintenance of the EM.

*Table 5.* Effect of abscisic acid on EM cultures.

| ABA mg/l | Total embryos/200 mg of EM | Mature embryos/200 mg of EM |
|---|---|---|
| 0.0 | 42.8 ± 1.21 | 15.4 ± 0.45 |
| 0.1 | 16.2 ± 0.89 | 8.0 ± 0.45 |
| 0.5 | 15.0 ± 0.21 | 12.8 ± 0.26 |
| 1.0 | 12.6 ± 0.34 | 11.6 ± 0.25 |
| 5.0 | 7.4 ± 0.34 | 6.0 ± 1.25 |
| 10.0 | 6.8 ± 0.54 | 6.2 ± 0.89 |

Medium: Standard maintenance medium.

*Table 6.* Effect of storage temperature on EM culture.

| °C | Growth of EM (fold increase) | Period of viability (weeks) |
|---|---|---|
| 0 | 0.0 | 8 |
| 5 | 0.2 | 36 |
| 10 | 4.2 | 12 |
| 27 | 3.6 | 10 |
| 30 | 0.0 | 1 |

Medium: Standard maintenance medium.

### 3.6. *Effect of abscisic acid*

Five different concentrations (0.1, 0.5, 1.0, 5.0 and 10.0 mg/l) of abscisic acid (ABA) were added aseptically to the standard maintenance medium.

The growth of EM and the total number of embryos decreased with an increasing concentration of ABA (Table 5). Cultures turned dark brown as the embryos turned moribund with increasing concentration of ABA. The proportion of mature embryos on ABA-containing media was greater than that on ABA-free media.

### 3.7. *Long term storage*

For the determination of an optimum temperature for growth and maintenance, embryogenic mass cultures, growing on the standard maintenance medium, were incubated at 5°C, 10°C, 22°C, 27°C and 30°C in the dark. The viability of the cultures was checked by their ability to regain the previous growth rates by subculturing at monthly intervals on to the fresh media at 27°C.

The best growth was recorded at 27°C followed by 22°C and 15°C. At 30°C and 10°C, growth was poor (Table 6). The cultures retained their viability over longer periods when they were maintained (wrapped with a plastic film to prevent desiccation) at 10°C and were observed to survive more than 9

*Table 7.* Effect of gibberellic acid on EM cultures.

| $GA_3$ mg/l | Growth of EM (mg) |
|---|---|
| 0.0 | 1080.0 ± 2.10 |
| 0.5 | 799.2 ± 2.16 |
| 1.0 | 851.6 ± 2.16 |
| 2.0 | 852.6 ± 3.04 |
| 5.0 | 855.2 ± 3.03 |

Culture medium: Standard maintenance medium.

months of storage. The next best results were obtained at 22°C followed by 27°C and 30°C.

Because of the ability of EM cultures to retain viability for a longer period, they can be equated to the cultures maintained under minimal growth conditions. Although cryopreservation of somatic embryos is possible in several species, long-term preservation at low temperatures has obvious advantages in terms of capital and maintenance costs over cryopreservation which involves the use of equipment requiring liquid nitrogen as a coolant.

### 3.8. *Formation of an embryogenic callus*

The effect of giberellic acid ($GA_3$) on EM cultures was tested by adding it (filter sterilized) to the standard maintenance medium at different concentrations (0.5, 1.0, 2.0 and 5.0 mg/l) (Table 7).

A soft, friable and translucent embryogenic callus (EC) with red stippling (Fig. 1c) developed in two weeks. The embryogenic nature was evident when the cultures were examined under the microscope. The proembryos (Fig. 1d) and early stages of embryo development had red pigmentation. This was confirmed when EC was cultured onto a hormone free medium, and all stages of embryo development could be followed. The amount of callus formed was proportional to the $GA_3$ concentration up to 5 mg/l (Table 7).

### 3.9. *Suspension cultures*

When cultured in a liquid standard maintenance medium with or without shaking, a viable culture of EM could not be established. The effect of inositol was tested at five concentrations (Table 8). Cultures were maintained in 250 ml Erlenmeyer flasks, each containing 20 ml of medium. At concentrations of 500–5000 mg/l, proliferation of embryo clusters was observed and a suspension culture was established in a week. Cultures consisted of free floating clusters of embryos undergoing repetitive embryogenesis. Inositol concentrations 1000 mg/l were found to be optimal, and growth was inhibited at 10,000 mg/l. Better synchrony in embryo (Fig. 2a) development was observed in cultures grown in the presence of inositol which could be attributed to its role as an osmoticum.

*Table 8.* Effect of inositol on suspension cultures of EM.

| Inositol mg/l | Growth of EM (fold increase) | Total embryos/200 mg of EM | Mature embryos/200 mg of EM |
|---|---|---|---|
| 0.0 | – | – | – |
| 500 | 2.6 | 42.6 ± 0.66 | 15.2 ± 0.64 |
| 1000 | 4.4 | 47.2 ± 0.24 | 18.6 ± 0.82 |
| 5000 | 4.0 | 46.4 ± 0.68 | 17.2 ± 0.84 |
| 10,000 | 0.7 | 16.2 ± 0.18 | 10.2 ± 0.26 |

Culture medium: Standard maintenance medium (liquid).

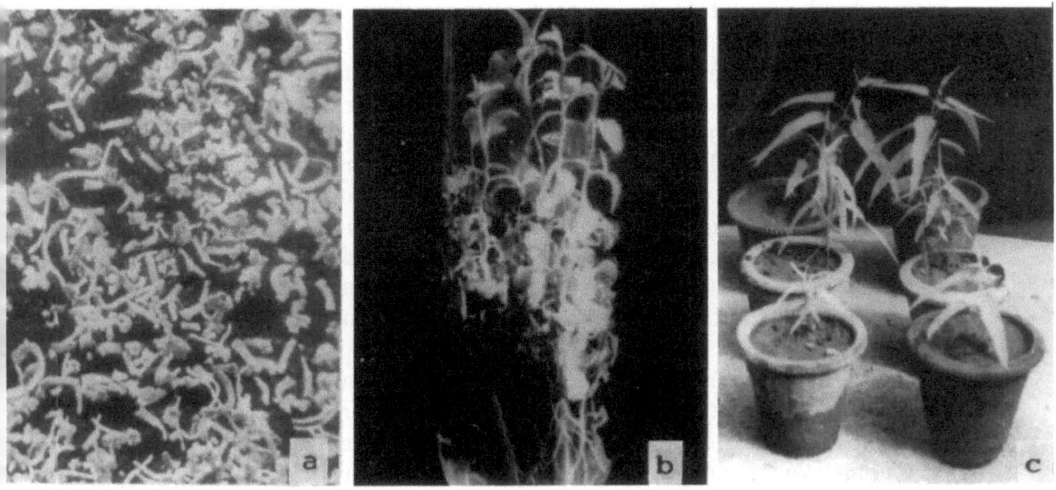


*Figure 2.* (a) Free floating embryos in suspension culture of EM (× 10). (b) Emblings of *E. citriodora* ready for transfer to soil. (c) Emblings after 8 weeks of transfer to soil.

Embryo development continued up to the torpedo stage in the liquid cultures of EM in the presence of auxin. Subculture to the hormone free medium was required for further development and germination of embryos.

## 4. Germination of somatic embryos

In preliminary experiments, it became clear that mature embryos germinated easily when transferred to an auxin-free medium. It was observed that even on an auxin containing media, embryos that were lifted off from the media

*Table 9.* Effect of illumination on germination of somatic embryos.

| Light source (intensity) | Germination (%) | Survival in soil (%) |
|---|---|---|
| Fluorescent (10 $\mu$E m$^{-2}$ s$^{-1}$) | 52 | 62 |
| Fluorescent (25 $\mu$E m$^{-2}$ s$^{-1}$) | 58 | 68 |
| Fluorescent (80 $\mu$E m$^{-2}$ s$^{-1}$) | 54 | 70 |
| Daylight (200 $\mu$E m$^{-2}$ s$^{-1}$) | 54 | 72 |

Culture medium: B5 + sucrose 2 g/l.
Culture period: 3 weeks.
No. of embryos: 50 per treatment.

surface by the growth of the cultures often germinated, presumably due to the lack of auxin.

The germination medium containing B5 minerals and vitamins and 2 g/l sucrose was used for the experiments. The EM cultures were first dispersed in a liquid medium to separate the embryos. Individual mature embryos with normal morphology were transferred to the solid germination media. Approximately, 52% of the embryos germinated and formed well developed healthy shoot and root systems. Although germination did occur in cultures kept in the dark, plantlets were etiolated and normal development of the leaves occurred only when incubated in the light (about 25 $\mu$E m$^{-2}$ s$^{-1}$). On addition of 10% coconut milk to the germination medium, 58% of the embryos germinated.

Various light sources were used to test their effect on embryo germination and plantlet survival (Table 9). Besides artificial illumination at three different light intensities, cultures were also kept at ambient temperature (28–32°C) in daylight (sunlight) which was available at an average of 7 h per day (intensity varied from 0–200 $\mu$E m$^{-2}$ s$^{-1}$). Daylight gave better results than fluorescent illumination. At 25 $\mu$E m$^{-2}$ s$^{-1}$ light intensity where the germination percentage was the highest (58%), survival of the plantlets was low (68%). Plantlets developing under daylight and fluorescent light (80 $\mu$E m$^{-2}$ s$^{-1}$) were distinctly more vigorous in growth and had darker green leaves than those grown at the lower intensities. The survival rate of daylight-grown plantlets was highest (72%), followed by fluorescent light-grown plantlets (70%).

The use of natural daylight as a source of illumination for tissue cultures appears to be neglected. In our results, it is likely that the unique spectral quality and photoperiod of daylight had helped in the repair of the photosynthetic system which is normally impaired under *in vitro* conditions, and thus contributed to the improved vigour of the plantlets.

*Table 10.* Effect of different substrates on survival of emblings.

| Substrate | Survival% |
|---|---|
| Sand | 66.0 |
| Sand:Soil (1:3) | 82.0 |
| Soil | 74.0 |
| Vermiculite | 42.0 |
| Vermiculite:Soil (1:1) | 78.0 |

Transplants: Emblings with 2 cm shoots.
No. of embryos: 50 per treatment.

## 5. Growth in the field

### 5.1. *Transfer to soil*

Plantlets were ready for transplantation in the soil after they had attained a shoot length of 2 cm. At this stage, they had 3–4 fully opened leaves and a well developed tap root system (Fig. 2b). The plantlets from each culture vessel were carefully separated from each other without damaging the roots. Both separation of plantlets and removal of the medium from roots was performed by holding the plantlets under a running water tap. Of the various substrates tested (Table 10), soil and sand (3:1 ratio) gave the best results with the survival of 86% plantlets. The survival of plantlets on vermiculite was lowest (42%), but when mixed with sand the rate improved (78%). Either with soil or with sand alone, the results were slightly less (74 and 70% respectively).

### 5.2. *Encapsulation of somatic embryos*

To test the feasibility of producing synthetic seeds, somatic embryos were encapsulated in a variety of matrices, stored at different temperatures and then tested for their germination ability. Sodium alginate (1 and 2% w/v, complexed with 0.2% calcium nitrate solution), Gelrite and agar (1 and 2% w/v, both gelling at ambient temperature) were tested for the formation of a suitable coat around isolated somatic embryos. The use of sodium alginate gave the best results, both for the ability to form a suitable seed coat around isolated somatic embryos and in viability of embryos. Alginate (1%) formed a soft bead that eased handling of embryos with a forceps and that resulted in 30% germination of encapsulated embryos on a sterile medium, whereas all the other coats were too soft to handle.

A semi-automatic method was adopted to encapsulate the embryos. This method involved the use of a peristaltic pump to control the rate of droplet formation of sodium alginate (1% w/v of Protanal LF, Protan, Norway) and a magnetic stirrer to ensure uniform mixing of the calcium nitrate (0.2%)

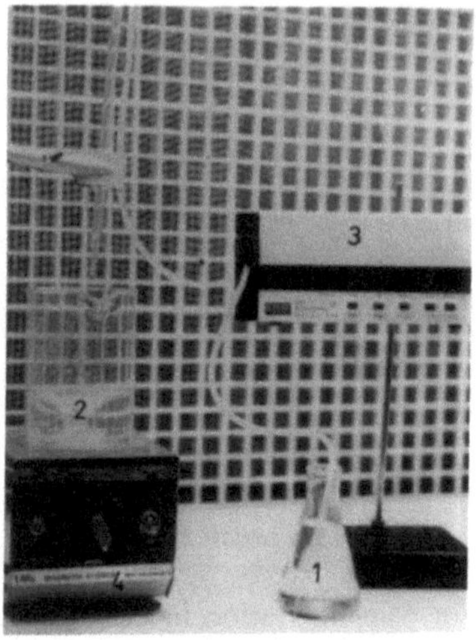


*Figure 3.* Apparatus for encapsulation of Eucalyptus somatic embryos. 1 – Sterile sodium alginate solution (1% w/v); 2 – Calcium nitrate solution (0.2%); 3 – Peristaltic Pump; 4 – Magnetic stirrer.

solution used for complexation (Fig. 3). Individual embryos were grasped with a pair of forceps and introduced into the droplets, which were then allowed to fall into the stirring complexing agent. After 20 min of complexation, all beads were removed, washed with sterile water, and stored in a test tube at room temperature or at 4°C for various periods of time. Encapsulated embryos (Fig. 4a) were then checked for germinability on a variety of media/substrates both under sterile as well as non-sterile conditions.

Germination (Fig. 4b) rates of encapsulated embryos, stored at the room temperature, declined on a sterile germination medium. After 1 day of storage, 38% of the embryos germinated, and the germination rate further declined to 28% after storing for 7 days. No embryo germination was recorded when stored for 10 days. Embryos stored at 4°C decreased their germination rate more rapidly and after 7 days no germination was observed.

Encapsulated embryos stored for 1 day at room temperature were grown on a sterile water agar medium, sterile sand and non-sterile soil. On sterile water-agar the germination rate was 30%, and 98% of the plantlets survived (Table 11). On sterile sand (irrigated with a solution of B5 salts) germination was only 4%, but all the plantlets survived. On non-sterile soil, none of the encapsulated embryos germinated, and they were easily infested with microbial contamination.

*Figure 4.* (a) Individual somatic embryos encapsulated in alginate gel. (b) Magnification of beads with embryos visible through the coat (× 10). (c) A single bead showing emergence of the radicle and the shoot on sterile medium.

*Table 11.* Effect of different substrates on germination and survival of encapsulated embryos.

| Substrate | Germination% | Survival% |
|---|---|---|
| B5 + Agar | 30 | 80 |
| Water + Agar | 30 | 98 |
| Sand + B5 | 4 | 100 |
| Soil + B5 | 0 | 0 |

Duration: 3 weeks.
No. of embryos: 100 per treatment.

### 5.3. *Field studies*

Emblings that survived and continued to grow in a soil:sand mixture were transferred to earthen pots containing garden soil after two weeks and maintained in the greenhouse at 70–80% relative humidity (Fig. 2c). About 200 emblings had thus been transfered. Leaves developed at this stage had the typical hairy appearance of juvenile leaves of this species. After nearly 9 months when smooth adult leaves started to form, plants were shifted to the bigger pots and taken out of the greenhouse. Fifty plants were transferred to pits in the field after they had attained a height of about 120 cm.

Only five emblings survived in the field after 5 years of planting. Extreme drought conditions resulted in heavy mortality of the remaining potted plants transferred to the field conditions. No abnormalities have been observed in morphology and growth rate when compared with the control seedlings. Future studies will be directed at evaluating their variability using RFLP or DNA finger printing techniques. Further field plantings is also envisaged in an area better suited for the trees.

## 6. Conclusion

Tremendous progress has been made in large-scale production of somatic embryos in conifers. Embryogenic cultures of radiata pine (*Pinus radiata*; Smith, 1991), Norway spruce (*Picea abies*), loblolly pine (*Pinus taeda*) and Douglas-fir (*Pseudotsuga menziesii*; Gupta *et al.*, 1993) have been grown in bioreactors or in suspension cultures. The progress has been rapid, however, still it will require more research and development before these methods can be used commercially.

Our system in *E. citriodora* has given a lead and an insight into the problems one will face before it is adopted for large-scale production. The inconsistancy in somatic embryogenesis induction from seeds is a bottleneck, but the characterization of EM culture lines and the requirements for long-term maintenance has given us unprecedented information. Additional research inputs will be necessary to improve maturation and synchrony in embryo development to a higher degree of precision and also to develop an efficient delivery system for synthetic seeds either through fluid delivery systems or containerization. With these refinements, it may be possible to utilize these technologies for commercial application before the turn of this century. Besides having the potential to cope up with the production of millions of plants for reforestation, a good somatic embryogenesis system will also be amenable to gene manipulation, selection and propagation of elite clones, and cryogenic storage for future use.

## 7. References

Boulay, M., 1987. Recherches preliminaires sur l'embryogenese somatique d'Eucalyptus gunnii. Annales de Recherches Silvicoles, Association Foret-Cellulose: 23–37.

Button, J., J. Kochba and C.H. Bornman, 1974. Fine sturcture of embryoid development from embryogenic ovular callus of "Shamouti" orange (*Citrus sinensis*). J. Exp. Bot. 25: 446–457.

Gamborg, O.L., R.A. Miller and K. Ojima, 1968. Nutrient require ments of suspension cultures of soybean root cells. Exp Cell Res. 50: 151–158.

Gupta P.K., A.F. Mascarenhas and V. Jagannathan, 1981. Tissue culture of forest trees – Clonal propagation of mature trees of *Eucalyptus citriodora* Hook, by tissue culture. Plant Sci. Lett. 20: 195–201.

Gupta, P.K., G. Pullman, R. Timmis, M. Kreitinger, W.C. Carlson, J. Grob and E. Welty, 1993. Forestry in the 21 st century. Biotechnology 11: 444–459.

Hartney, V.J., 1980. Vegetative propagation of the eucalypts. Aust. For. Res. 10: 191–211.

Juretic, B. and S. Jelaska, 1991. Plant development in long-term embryogenic callus lines of *Cucurbita pepo*. Plant Cell Rep. 9: 623–626.

Khuspe, S.S., P.K. Gupta, D.K. Kulkarni, U. Mehta and A.F. Mascarenhas, 1987. Increased biomass production by tissue culture of eucalyptus. Can. J. For. Res. 17: 1361–1363.

Lakshmi Sita, G., 1981. Tissue culture of Eucalyptus species. In: A.N. Rao (Ed.), Proc. COSTED Symp. on Tissue Culture of Economically Important Plants, Singapore, pp. 180–184.

Le Roux, J.J. and J. Van Staden, 1991. Micropropagation and tissue culture of Eucalyptus – a review. Tree Physiol. 9: 435–477.

Litz, R.E. and R.A. Conover, 1983. High frequency somatic embryogenesis from Carica suspension cultures. Ann. Bot. 51: 683–686.

McComb, J.A. and I.J. Bennett, 1986. Eucalypts (Eucalyptus ) spp. In: Y.P.S. Bajaj (Ed.), Biotechnology in Agriculture and Forestry, Vol. 1: Trees 1, pp. 340–362. Springer Verlag, Berlin/Heidelberg.

Muralidharan, E.M. and A.F. Mascarenhas, 1987. *In vitro* plantlet formation by organogenesis in *E. camaldulensis* and by somatic embryogenesis in *E. citriodora*. Plant Cell Rep. 6: 256–259.

Muralidharan, E.M., P.K. Gupta and A.F. Mascaranhas, 1989. Plantlet formation through high frequency somatic embryogenesis in long term cultures of *Eucalyptus citriodora*. Plant Cell Rep. 8: 41–43.

Murashige, T and F. Skoog, 1962. A revised medium for rapid growth and bioassays with tobacco tissue culture. Physiol. Plant. 15: 473–497.

Ouyang, Q., H.S. Peng and Q.Q. Li, 1981. Studies on the development of embryoids from Eucalyptus callus. Scientia Silvae Sinicae 17: 1–7.

Parrott, W.A. and M.A. Bailey, 1993. Characterization of recurrent somatic embryogenesis of alfalfa on auxin-free medium. Plant Cell Tiss. Org. Cult. 32: 69–76.

Qin Chang-Le and E.G. Kirby, 1990. Induction of shoots and embryo like structures in cultures derived from juvenile and adult explants of Eucalyptus spp. In: Abstracts, VII International Congress on Plant Tissue and Cell Culture, Amsterdam, Abstract No. A1–74, p. 21.

Rajbhandary, S.B., 1990. Sand rooting of *in vitro* produced shoots of horticulture and forestry species. In: Abstracts, VII International Congress on Plant Tissue and Cell Culture, Amsterdam, Abstract No. A3–178, p. 127.

Raj Bhansali, R., J.A. Driver and D.J. Driver, 1990. Rapid multiplication of adventitious somatic embryos in peach and nectarine by secondary embryogenesis. Plant Cell Rep. 9: 280–284.

Smith, D.R., 1991. An automated bioreactor system for mass propagation of *Pinus radiata*. Agricell Rep. 17: 1–2.

Smith, M.A.L. and B.H. McCown. 1983. A comparision of source tissue for protoplast isolation from three woody species. Plant Sci. Lett. 28: 149–156.

Weissinger II, A.K. and W.A. Parrott, 1993. Repetitive somatic embryogenesis and plant recovery in white clover. Plant Cell Rep. 12: 125–128.

White, P.R., 1954. The Cultivation of Animal and Plant Cells. Ronald Press, New York.

Williams, E.G. and G. Maheshwaran, 1986. Somatic embryogenesis: Factors influencing coordinated behaviour of cell as an embryogenic group. Ann. Bot. 57: 443–462.

Zacharin, R.F., 1978. Emigrant Eucalyptus – Gum Trees as Exotics. Melbourne Univ. Press, Melbourne.

Zobel, B.J., 1993. Clonal forestry in the Eucalyptus. In: M.R. Ahuja and W.J. Libby (Eds.), Clonal Forestry II – Conservation and Application, pp. 139–148. Springer-Verlag, Berlin/Heidelberg.

# 3. Somatic embryogenesis in bamboo

Susan H. Woods, John E. Woods and Glenn B. Collins

**Contents**



## 1. Introduction

Bamboo is one of the most universally useful plant commodities known to man providing food, raw material, shelter, and even medicine for the greater part of the world's population. However, information on the basic biology and reproduction for this complex and diverse group of plants lags behind that for other major agronomic plant species. Even the name bamboo is itself a vernacular term for more than 75 genera and 1250 species of the *Gramineae* (Soderstrom and Ellis, 1988) which are confused taxonomically as a result of their unusual flowering habits. The bamboos occur worldwide and have an extremely wide range of distribution, with some bamboos recorded from a latitude as far north as 46° and as far south as 47°, although the great majority occur in the tropics (Soderstrom and Calderon, 1979).

Although bamboo plays an essential role in the daily life of millions of people in subtropical and tropical regions, it has never been economically important in most of the Western countries. In 1919, the United States Department of Agriculture began introducing and maintaining an extensive collection of bamboo. For a number of years the Department conducted a considerable amount of research on the biology and uses of bamboo (Adamson *et al.*, 1978) and promoted it enthusiastically as an agricultural crop (McIlhenny, 1945). The campaign never really accomplished much, primarily because of a lack of large quantities of bamboo for propagation and because conventional methods of propagation are too labor intensive. The increased awareness of the potential importance of bamboo fiber in the wood and paper industry, as well as a demand for food bamboo created by the proliferation of Asian populations in the U.S., has resulted in a renewed interest in reliable and less labor intensive methods of bamboo propagation.

Once considered to be an inexhaustible source of raw material in the tropics, bamboo is now threatened over a large area of the world. The gregarious flowering habit of bamboo, and the human population pressure

*S. Jain, P. Gupta & R. Newton (eds.), Somatic Embryogenesis in Woody Plants, Vol. 2*, 41–51.

disrupting the natural cycle of reforestation, present an urgent need for developing methods for large scale propagation of bamboo. Propagation by seed is not always possible because of the unavailability of seed and because of their limited period of germination. Vegetative propagation is slow, labor intensive and the degree of failure is still rather high. Furthermore, the vegetatively propagated plants are likely to flower at the same time as the parent clump. Bamboo will become an important new multipurpose tree crop in many areas where it has not been traditionally used, as reliable means of mass propagation become available.

Efficient *in vitro* propagation could prove to be a reliable and useful method for establishment of new bamboo plantations (Rao *et al.*, 1985). Somatic embryogenesis is now recognized as an acceptable method for plant propagation *in vitro* because it enables the rapid production of a large number of uniform plants within a relatively short period of time (Rao *et al.*, 1985). The aseptic nature of regenerants formed *in vitro* is also useful in quarantine situations because such materials can be utilized for safe introduction of crops from one state or country to another.

## 2. Somatic embryogenesis

Somatic embryogenesis can be defined as the process of embryo initiation and development from vegetative or non-gametic cells which differentiate as bipolar structures mimicking zygotic embryos (for review, see Street, 1977). Although much progress has been made since this phenomenon was observed more than thirty years ago in carrot tissue (Stewart, 1958), we are far from a complete understanding of the processes necessary to embryonic initiation and maturation. Grasses in general and bamboos in particular, are among those groups considered recalcitrant to regenerate in culture. In the last ten years important progress has been made in the cell culture of the grasses with extensive evidence now documenting the regeneration of plants via somatic embryogenesis in all the major species of grasses (Vasil, 1987). However, the published reports of efficient protocols for the *in vitro* propagation of bamboos via somatic embryogenesis are few and represent a limited number of species.

Somatic embryogenesis as a method of mass propagation has several advantages. The embryoids are bipolar, germinating to form both root and shoot in the manner of zygotic embryos. Multiplication of embryos is usually very rapid enabling the production of large numbers of propagules and maintaining and manipulating embryogenic cultures is easier and quicker, and hence less labor-intensive and costly than a shoot culture (Rao and Rao, 1988). In addition, the juvenile period and growth habit of plantlets from somatic embryos are similar to those of *in vivo* seedlings. Thus, bamboo plants from somatic embryos can be relied upon to last for their full vegetative life-span (Rao *et al.*, 1990). Disadvantages include the difficulty of initiating embryogenesis from tissue of mature plants and, according to some researchers (Paranjothy *et al.*, 1990), the genetic instability created by an intermediate callus phase. Somaclonal variation is a disadvantage when uniformity of propagules is desired but can be beneficial in isolating valuable variants. Somaclonal variation among bamboo plants regenerated from somatic embryos has not been observed in our laboratory. All the plants of the two species from which large numbers of plants have been produced, *Otatea acuminata aztecorum* and *Lithachne humilis* are remarkably uniform (Woods and Collins Laboratories, unpublished). Rao and Rao (1988) report the isolation of a few somaclonal variants, one of which is a plagiotropic mutant of *Bambusa arundinacea*.

## 3. Species exhibiting somatic embryogenesis

According to published reports (Table 1), somatic embryogenesis has been obtained in *Bambusa arundinacea* (Mehta *et al.*, 1982; Rao and Rao, 1988); *B. balcoa* (Rao *et al.*, 1990); *Dendrocalamus strictus* (Rao *et al.*, 1985, 1988; Rao and Rao, 1988; Mascarenhas *et al.*, 1988; Mukunthakumar and Mathur, 1992); *Sinocalamus latiflora* (Yeh and Chang, 1987; Zamora *et al.*, 1988; Tsay *et al.*, 1990); *B. beecheyana* and *B. oldhamii* (Yeh and Chang, 1986a,b); and an ornamental species, *Otatea acuminata aztecorum* (Woods *et al.*, 1992). Figures 1–3 illustrate somatic embryogenesis in *Otatea*. A report of embryogenesis and plantlet regeneration from vegetative structures of green bamboo, *Phyllostachys viridis* (Hassan and Debergh, 1987), was followed later by a corrigendum indicating that the plant identified as green bamboo was incorrectly identified and was actually *Pogonatherum paniceum*, a grass (Hassan and Debergh, 1988).

Unpublished reports of somatic embryogenesis from our laboratory include plantlet regeneration from *Thysostachys siamensis*, *Dendrocalamus membranaceous*, and *Drapanostachys kasianum*; plants have been established in soil from *Lithachne humilis*, an herbaceous New World bamboo (Figures 4–6); and embryogenic callus and shoot-like structures have been produced from *Phyllostachys aurea* and *P. vivax*.

## 4. Explant sources

The majority of the reports of successful somatic embryogenesis indicate the use of zygotic embryos as the explant source (Mukunthakumar and Mathur, 1992; Woods *et al.*, 1992; Rao and Rao, 1988; Rao *et al.*, 1985, 1988; Mascarenhas *et al.*, 1988; Yeh and Chang, 1987; Mehta *et al.*, 1982). In addition, Tsay *et al.* (1990) and Yeh *et al.* (1991) reported the regeneration of haploid plantlets from the culture of anthers and immature inflorescence explants have been used for the successful regeneration of plants (Rao and

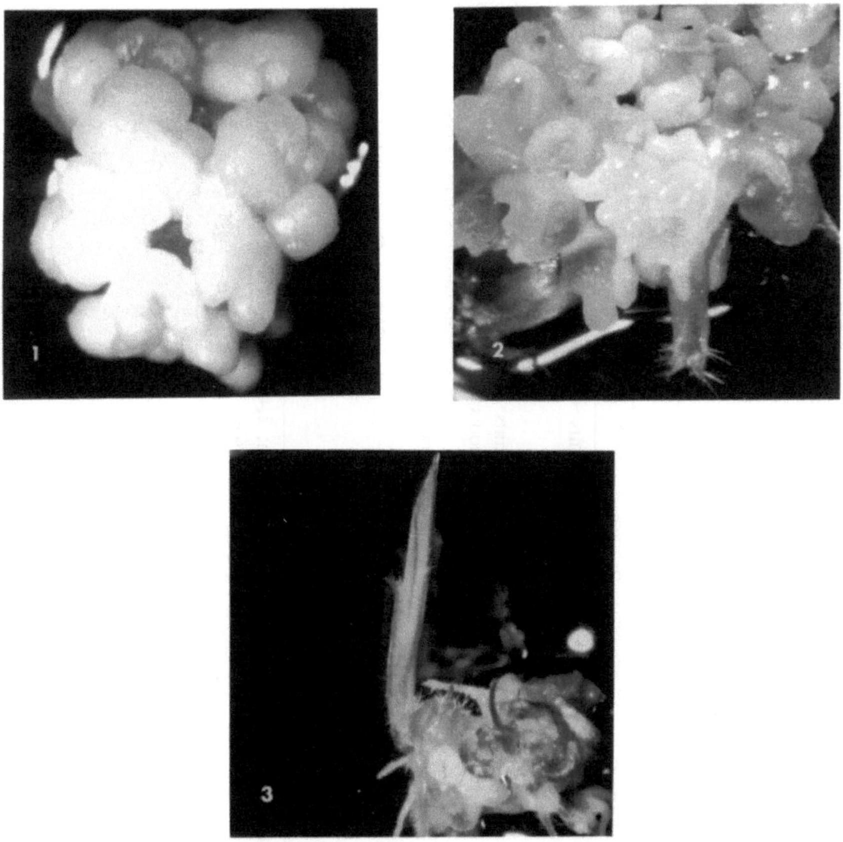


*Figures 1–3. In vitro* somatic embryogenesis and plant regeneration of *Otatea acuminata aztecorum*.
*Figure 1.* Cluster of somatic embryos.
*Figure 2.* Shoots emerging from germinating embryos.
*Figure 3.* Rooted shoot regenerated from somatic embryo.

Rao, 1988; Yeh and Chang, 1986a,b). Yeh and Chang (1986a) used adventitious roots from non-embryogenic callus and Zamora *et al.* (1988) utilized internode tissue for callus induction and plant regeneration. Rao and Rao (1988) reported the use of rhizomes, nodes and leaf sheaths of juvenile plants as explants for somatic embryogenesis and plant production. In our laboratory, all parts (leaf, leaf sheath, node, root and rhizome) of the juvenile axenic plants regenerated from somatic embryos have been used for further production of somatic embryos and plants (Woods *et al.*, 1991). Results from these preliminary studies suggest a definite genotype effect as only fifteen of fifty genotypes tested were embryogenic. We have also been successful with

Table 1. Summary of somatic embryogenesis in bamboo.

| Species | Media | Explant | Results | Reference |
|---|---|---|---|---|
| *D. strictus* | MS + 2,4-D and K | zygotic embryos | encap. SE, plants | Mukunthakumar and Mathur, 1992 |
| *O. acuminata aztecorum* | MS + 2,4-D and BA | seed and embryos | SE, plants | Woods *et al.*, 1992 |
| *B. balcoa* | not specified | not specified | SE | Rao *et al.*, 1990 |
| *S. latiflora* | N6 + 2,4-D and BA | anthers | SE, haploid plants | Tsay *et al.*, 1990 |
| *D. latiflora* | MS + 2,4-D and BA | internode | emb. callus, shoots | Zamora *et al.*, 1988 |
| *D. strictus* | MS + 2,4-D | zygotic embryos | SE, plants | Mascarenhas *et al.*, 1988 |
| *D. strictus*, *B. arundinacea* | B5 + 2,4-D | zygotic embryos | SE, *in vitro* flowering | Rao and Rao, 1988 |
| *S. latiflora* | MS + 2,4-D and K | zygotic embryos | SE, plants | Yeh and Chang, 1987 |
| *B. beecheyana* | MS + 2,4-D and K | florets, ad. roots | SE, plants | Yeh and Chang, 1986a |
| *B. oldhamii* | MS + 2,4-D and K | young inflorescence | SE, plants | Yeh and Chang, 1986b |
| *D. strictus* | B5 + 2,4-D | zygotic embryos | SE, plants | Rao *et al.*, 1985 |
| *B. arundinacea* | N6 + 2,4-D and BA | zygotic embryos | SE, plants | Mehta *et al.*, 1982 |

ad. – adventitious; encap. – encapsulated; SE – somatic embryo; emb. – embryogenic; K – kinetin; 2,4-D – 2,4-dichlorophenoxyacetic acid; BA – benzylaminopurine; MS – Murashige and Skoog, 1962; N6 – Chu *et al.*, 1975; B5 – Gamborg *et al.*, 1968.

*Figures 4–6. In vitro* somatic embryogenesis and plant regeneration of *Lithachne humilis*.
*Figure 4.* Cluster of somatic embryos.
*Figure 5.* Shoot and root emerging from germinating embryo.
*Figure 6.* One year old regenerated plant.

leaf sheath explant material from the mature plant of *Lithachne humilis*. Leaf sheath explants were taken from the youngest unfurled leaf of a branch from a mature plant and were cultured on an MS medium containing BA (0.5 mg/l) and 2,4-D (3 mg/l). The resulting compact callus produced small clusters of embryos from which plants were regenerated. In addition, embryogenic callus has been produced from explant material of young axenic plants of *Phyllostachys aurea* that were generated by organogenesis (Huang *et al.*, 1989).

Current research should be aimed at developing efficient, reliable methods for the induction of somatic embryos from mature tissues of identified elite bamboos. When this is accomplished the objective of mass-propagation of elite bamboo clones will be achieved (Rao *et al.*, 1990).

## 5. Effects of the medium

The basal medium used for the induction of embryogenic callus has mainly been the Murashige and Skoog (1962) medium. However, N6 (Chu *et al.*, 1975) basal medium was found to be more effective than MS medium for anther culture (Tsay *et al.*, 1990). N6 medium was also used by Mehta *et al.* (1982). Although Rao *et al.* (1985) and Rao and Rao (1988) successfully used B5 medium (Gamborg *et al.*, 1968), these reports did not directly compare the effects of different basal media. In our research (Woods *et al.*, 1992) both the B5 and MS basal media were contrasted for their influence on somatic embryogenesis in two direct comparisons: 1) in the presence or absence of 0.5 mg/l BA, and 2) with sucrose concentrations of 5% and 2%. When BA was absent and the sucrose level was 5%, the B5 medium resulted in more embryogenic callus than the MS medium. But, in the presence of BA and 2% sucrose, the MS medium was superior to B5. Moreover, the results of all media variations tested clearly indicated that the presence of BA and 2% sucrose were necessary for optimal production of somatic embryos. Thus, we concluded that the MS medium is superior when other factors in the medium (such as BA and sucrose) are present in optimal combinations. This evidence is supported by observations of callus which has been cultured for more than six passages. Comparison of older cultures grown on all media variations showed that the cultures on the MS media variations consistently maintained embryogenic capacity and continued to produce more somatic embryos than did the cultures on the B5 media variations.

In our laboratory, somatic embryogenesis occurred only on media containing 2,4-D, whereas NAA and picloram were ineffective auxin sources for this response. These results are consistent with a previous report (Yeh and Chang, 1987) which indicated that 2,4-D was essential for bamboo somatic embryogenesis, and NAA and indole-3-acetic acid could not substitute for this effect. All published reports of somatic embryogenesis in bamboo listed 2,4-D as the essential auxin source.

The presence or absence of BA in the culture medium has been tested (Woods *et al.*, 1992) for its influence on the initiation of embryogenic callus and the results provided evidence that BA increases the efficiency of embryo production. Data were not provided in other reports to document the influence of the presence vs absence of a cytokinin on bamboo somatic embryogenesis, although Mehta *et al.* (1982), Rao and Rao (1988), and Zamora *et al.* (1988) used BA and Yeh and Chang (1986a,b, 1987) and Mukunthakumar and Mathur (1992) used kinetin.

## 6. Origin of somatic embryos

Very little has been published documenting the origin of embryogenic callus and somatic embryos in bamboo. Plants regenerated from anther culture revealed their microspore origin as evidenced by the haploid number of chromosomes (Tsay *et al.*, 1990). Rao *et al.* (1991) have examined the callusing zygotic embryo histologically and determined that callusing initiates from the xylem parenchyma cells located towards the base of the vascular bundles. A cambial-like layer forms and the somatic embryos appear to originate from this cambial-like layer. From this point of origin, the embryoid development rapidly spreads to the remainder of the callus resulting in large numbers of somatic embryos.

Secondary somatic embryogenesis from the scutellar area of embryoids has been commonly observed (Rao *et al.*, 1991; Woods *et al.*, 1992; Yeh and Chang, 1986a). Prolonged culture often results in the proliferation of large quantities of secondary embryos, sometimes producing cultures which consist entirely of secondary embryos (Rao *et al.*, 1991; Woods *et al.*, 1992).

## 7. Future trends

The induction of somatic embryogenesis in species where it has not been observed or where it occurs at low frequencies is always a challenge and in species where somatic embryogenesis routinely occurs, the quality of somatic embryos needs to be improved. A systematic evaluation of the minimal needs of bamboo is needed, as far too little effort has been made to study the basic mineral requirements of bamboos *in vitro* (Krikorian and Rao, 1988). Problems such as the control of darkening and the use of antibiotics and fungicides in the culture media need to be addressed along with an increased emphasis on somatic embryo induction from tissues of mature elite plants.

Another critical research goal should be directed toward developing efficient protocols for embryogenic suspension cultures. Some efforts are being made in this area as Rao and Rao (1988) have mentioned the establishment of continuously growing suspension cultures. In our laboratory, plants have been regenerated from somatic embryos multiplied in embryogenic suspension cultures using an adaptation of Finer's system for embryogenic suspension cultures of soybean (Finer and Nagasawa 1988). An efficient system for the large scale production of high quality embryos could provide the material for artificial seed production. Mukunthakumar and Mathur (1992) have reported the successful preparation of artificial seeds from encapsulated somatic embryos obtained from seed derived callus. The induction of embryos from vegetative tissues of mature elite plants accompanied by mass proliferation in liquid suspension culture could provide a low-cost, high-volume propagation system. Embryogenic suspension cultures could also be a source of regenerable protoplasts and possibly useful for genetic transformation.

In spite of the long history of its use, bamboo probably remains the only major agronomic crop that has had little or no improvement by selective breeding. Genetic engineering or biotechnology could deliver genetically improved bamboo into the hands of plant breeders. One such instance that has been reported is the induction of *in vitro* flowering from the somatic embryos of *Dendrocalamus strictus* and *Bambusa arundinacea* (Rao and Rao, 1988). The development of reliable, consistent methods for the initiation of *in vitro* flowering should make possible *in vitro* breeding for better bamboos.

It is also reasonable to expect that gene transfer technology is applicable to bamboo culture. In our laboratory, preliminary experiments with microprojectile bombardment have been encouraging. Clusters of ten somatic embryos, four clusters per plate, and leaf and leaf sheath explants from axenic four-week-old plantlets were subjected to bombardment with 1.1 μl tungsten particles coated with 5 μl of a pMON 530 GUS (β-glucuronidase) construct. Two embryo clusters from each plate were incubated for five hours and tested for GUS activity. The remaining two clusters from each plate were allowed to germinate. These bombarded embryoids survived and regenerated plants although gene expression was not observed either in embryoids or leaf tissue tested for GUS activity.

The tissue culture literature reporting on the successful results of regeneration of bamboo via somatic embryogenesis is still very limited. Additional research is desperately needed for many more species of bamboo in order to fully use these systems for widespread propagation and genetic improvement of bamboo.

## 8. Acknowledgements

This paper (94–3–22) is published with the approval of the Director of the Kentucky Agricultural Experiment Station.

## 9. References

Adamson, W.C., G.A. White, H.T. DeRigo and W.O. Hawley, 1978. Bamboo production research at Savannah, Georgia, 1956–77. Agricultural Research Service, U.S. Department of Agriculture, ARS-S-176, pp. 1–17.

Chu, C.C., C.C. Wang, C.S. Sun, C. Hsu, K.C. Yin and C.Y. Chu, 1975. Establishment of an efficient medium for anther culture of rice through comparative experiments of the nitrogen sources. Sci. Sin. 16: 659–688.

Finer, J.J. and A. Nagasawa, 1988. Development of an embryogenic suspension culture of soybean (*Glycine max* Merrill.) Plant Cell Tiss. Org. Cult. 15: 125–136.

Gamborg, O.L., R.A. Miller and K. Ojima, 1968. Nutrient requirements of suspension cultures of soybean root cells. Exptl. Cell Res. 50: 151–158.

Hassan, A.A. El and P. DeBergh, 1987. Embryogenesis and plantlet development in the bamboo *Phyllostachys veridis* (Young) McClure. Plant Cell Tiss. Org. Cult. 10: 73–77.

Hassan, A.A. El and P. DeBergh, 1988. Corrigendum. Plant Cell Tiss. Org. Cult. 15: 93.

Huang, L.C., B.L. Huang and W.L. Chen, 1989. Tissue culture investigations of bamboo – IV. Organogenesis leading to adventitious shoots and plants in excised shoot apices. Environ. Exp. Bot. 29: 307–315.

Krikorian, A.D. and A.N. Rao, 1988. Bamboo production: imperatives and research strategies. In: I.V.R. Rao, R. Gnanaharan and C.B. Sastry (Eds.), Proceedings of the Int. Bamboo Workshop, pp. 353–370. Kerala Forest Research Institute, India, and I.D.R.C., Canada.

Mascarenhas, A.F., A.L. Nadgir, S.R. Thengane, C.H. Phadke, S.S. Khuspe, M.V. Shirgurkar, V.A Parasharami and R.S. Nadgauda, 1988. Potential application of tissue culture for propagation of *Dendrocalamus strictus*. In: I.V.R. Rao, R. Gnanaharan and C.B. Sastry (Eds.), Proceedings of the Int. Bamboo Workshop, pp. 159–166. Kerala Forest Research Institute, India, and I.D.R.C., Canada.

McIlhenny, E.H., 1945. Bamboo, a must for the South. Am. Hort. Mag. 24: 120–125.

Mehta, U., I.V.R. Rao and H.Y.M. Ram, 1982. Somatic embryogenesis in bamboo. In: A. Fujiwara (Ed.), Proceedings of the 5th Int. Congr. of Plant Tissue and Cell Culture, pp. 109–110. Maruzen Co., Ltd., Tokyo.

Mukunthakumar, S. and J. Mathur, 1992. Artificial seed production in the male bamboo *Dendrocalamus strictus* L. Plant Sci. 87: 109–113.

Murashige, T. and F. Skoog, 1962. A revised medium for rapid growth and bioassays with tobacco tissue culture. Physiol. Plant. 15: 473–497.

Paranjothy, K., S. Saxena, M. Bannerjee, V.S. Jaiswal and S.S. Bhojwani, 1990. Clonal multiplication of woody perennials. In: S.S. Bhojwani (Ed.), Plant Tissue Culture: Applications and Limitations, pp. 190–198. Elsevier, Amsterdam.

Rao, I.V.R., V. Narang and I.U. Rao, 1991. Origin and development of embryogenic callus and somatic embryos and secondary somatic embryogenesis in the bamboo, *Dendrocalamus strictus*. In: Programme for the IV Int. Bamboo Workshop, p. 69 (Abstract). Chiangmai, Thailand.

Rao, I.V.R. and I.U. Rao, 1988. Tissue culture approaches to the mass-propagation and genetic improvement of bamboos. In: I.V.R. Rao, R. Gnanaharan and C.B. Sastry (Eds.), Proceedings of the Int. Bamboo Workshop, pp. 151–158. Kerala Forest Research Institute, India, and I.D.R.C., Canada.

Rao, I.V.R., A.M. Yusoff, A.N. Rao and C.B. Sastry, 1990. Propagation of bamboo and rattan through tissue culture. The I.D.R.C. Bamboo and Rattan Research Network, Canada, pp. 11–60.

Rao, I.U., I.V.R. Rao and V. Narang, 1985. Somatic embryogenesis and regeneration of plants in the bamboo *Dendrocalamus strictus*. Plant Cell Rep. 4: 191–194.

Rao, I.U., I.V.R. Rao, V. Narang, R. Jerath and K.G. Pillai, 1988. Mass-propagation of bamboos from somatic embryos and their successful transfer to the forest. In: I.V.R. Rao, R. Gnanaharan and C.B. Sastry (Eds.), Proceedings of the Int. Bamboo Workshop. Kerala Forest Research Institute, India, and I.D.R.C., Canada, pp. 167–172.

Soderstrom, R. and C.E. Calderon, 1979. V. Ecology and phytosociology and bamboo vegetation. In: M. Numata (Ed.), Ecology of Grasslands and Bamboolands in the World, pp. 223–236. VEB Gustav Fischer Verlag, Jena.

Soderstrom, R. and R.P. Ellis, 1988. The woody bamboos (*Poaceae*: *Bambuseae*) of Sri Lanka: a morphological-anatomical study. Smithsonian Contribution No. 72: 1–74.

Stewart, F.C., 1958. Growth and development of cultivated cells. III. Interpretation of the growth from free cell to carrot plant. Amer. J. Bot. 45: 709–713.

Street, H.E., 1977. Plant Tissue and Cell Culture. Blackwell Scientific Publications, 2nd Edition, Oxford, UK.

Tsay, H.S., C.C. Yeh and J.Y. Hsu, 1990. Embryogenesis and plant regeneration from anther culture of bamboo (*Sinocalamus latiflora* (Munro) McClure). Plant Cell Rep. 9: 349–351.

Vasil, I.K., 1987. Developing cell and tissue culture systems for the improvement of cereal and grass crops. J. Plant Physiol. 128: 193–218.

Woods, S.H., G.C. Phillips, J.E. Woods and G.B. Collins, 1991. Somatic embryogenesis and plant regeneration in Mexican weeping bamboo, *Otatea acuminata aztecorum*. In: Programme for the IV Int. Bamboo Workshop, p. 62 (Abstract). Chiangmai, Thailand.

Woods, S.H., G.C. Phillips, J.E. Woods and G.B. Collins, 1992. Somatic embryogenesis and plant regeneration from zygotic embryo explants in Mexican weeping bamboo, *Otatea acuminata aztecorum*. Plant Cell Rep. 11: 257–261.

Yeh, C.C., J.Y. Hsu and H.S. Tsay, 1991. Effects of microspore developmental stage, sucrose concentration and temperature treatment on embryoid and callus formation of bamboo [*Sinocalamus latiflora* (Monroe) McClure] anther culture. Chinease Agron. J. 1: 47–56.

Yeh, M.L. and W.C. Chang, 1986a. Somatic embryogenesis and subsequent plant regeneration from inflorescence callus of *Bambusa beecheyana* Munro var. beecheyana. Plant Cell Rep. 5: 409–411.

Yeh, M.L. and W.C. Chang, 1986b. Plant regeneration through somatic embryogenesis in callus culture of green bamboo (*Bambusa oldhamii* Munro). Theor. Appl. Genet. 73: 161–163.

Yeh, M.L. and W.C. Chang, 1987. Plant regeneration via somatic embryogenesis in mature embryo-derived callus culture of *Sinocalamus latiflora* (Monro) McClure. Plant Sci. 51: 93–96.

Zamora, A.B., S.Sm. Gruezo and O.P. Damasco, 1988. Callus induction and plant regeneration from internode tissues of bamboo (*Dendrocalamus latiflorus* cv. Machiku). Phil. Agric. 71(1): 76–84.

# 4. Somatic embryogenesis of *Bambusa oldhamii*, *Bambusa beecheyana* and *Sinocalamus latiflora*

Wei-Chin Chang

**Contents**



## 1. Introduction

Bamboo are often giant, long-lived, woody, evergreen grasses, members of the family Gramineae, tribe Bambuseae, mostly of the genera *Arundianaria*, *Bambusa*, *Chimonobambusa*, *Dendrocalamus*, *Phyllostachy*, *Paeudosasa*, *Sasa*, *Semiarundinaria*, and *Shibataea*, known all over the world. These genera are native to every continent (except Europe and Antarctica), mainly Asia but some species of *Arundinaria* and *Bambusa* are native to the western hemisphere.

Bamboo occur mostly as natural vegetation in tropical, subtropical, and temperate regions and are found in great abundance in tropical Asia. Their genera and species are distributed widely. At present, approximately 75 genera and 1250 species cover more than 14 million ha of land. Many of them are indigenous to the monsoon areas of tropical Asia and have a limited distribution; many are found only in cultivation. The distribution of bamboo has been greatly modified by human intervention.

Bamboo are adapted mainly to the warmer parts of the world. The tropical and subtropical clump-forming bamboo, such as *Bambusa* and *Dendrocalamus*, are mostly very sensitive and are killed or severely injured by even a few degrees of frost, although a few can endure temperatures down to 16°F without serious injury.

Due to the long intervals of flowering in bamboo and large-scale consumption of their seeds by wild animals, establishing of large cultivation by seed is extremely difficult. Although bamboo can be propagated vegetatively by offset planting, branch cutting, ground layering of the entire culm, and aerial layering of nodes, the rate of production is not sufficient to meet the demand.

*S. Jain, P. Gupta & R. Newton (eds.), Somatic Embryogenesis in Woody Plants, Vol. 2*, 53–65.

Moreover, the convenient vegetative propagation is beset with problems such as bulkiness of the propagules, their transport over long distances, and low survival rate. Efficient methods of *in vitro* micropropagation would help alleviate the problem (Chang, 1991).

## 2. Brief survey on somatic embryogenesis in bamboo

Micropopagation through nodal segments and somatic embyrogenesis are methods which could produce of large numbers of plants in a relatively short time, and at minimal cost, and with the lowest labor input. Beside our work on bamboo somatic embryogenesis on three bamboo species (Yeh and Chang, 1986a,b, 1987; Chang *et al.*, 1994), there are five referred reports relating to embryoids and their derived plantlets of other 3 species (Mehta *et al.*, 1982; Rao *et al.*, 1985; Tsay *et al.*, 1990; Mukunthakumar and Mathur, 1992; Woods *et al.*, 1992). Before describing our experience with three bamboo spp., we briefly summarize of the work on somatic embryogenesis of bamboo of other workers.

Mehta *et al.* (1982) first reported somatic embryogenesis and plant regeneration of *Bambusa arundinacea*. An embryogenic callus was obtained from the embryonal axis cultured on a medium containing 2,4-dichlorophenoxyacetic acid (2,-4-D), benzyladenine (BA), and polyvinylpyrrolidone(PVP). Embryoids originated in two ways: (1) Isolated densely cytoplasmic cells which give rise to embryoids become distinguished from the surrounding cells. These usually divide to form filamentous embryoids which either mature into plantlets or become dormant and brown. These dormant embryoids can, however, be revived by transference to a fresh medium. (2) Organized structures develop into shoots, roots, or plantlets. Once a shoot, root, or plantlet is organized, lots of secondary embryoids develop in its vicinity, resulting in formation of clusters of embryoids and plantlets. Mehta *et al.* (1982) also stated that the callus mass containing embryoids that develop in either way becomes dark with age and gives a superficial impression that growth has ceased. From within this mass green shoots and root emerge. The callus retains the embryogenic ability through subculture (period not mentioned).

Rao *et al.* (1985) obtained friable, compact, white calli within 10–12 days from cultures of seeds of *Dendrocalamus strictus* on $B_5$ basal medium (Gamberg *et al.*, 1968) plus 2,4-D at $10^{-5}$ M and $3 \times 10^{-5}$ M. The callus had a lobed or nodular appearance; subsequently, localized development of chlorophyll occurred and several small whitish to green embryoids, some with well developed scutella were obtained in the compact calli by 30 days. An average of 8.4 embryoids at the chlorophyllous stage of the callus were observed in a 30-day-old culture. Secondary embryoids were observed proliferating from the scutellar region. Thus, embryoids can be obtained continually from these embryogenic calli. Some of these embryoids germinated while in the callus and gave rise to well-developed plantlets. Rao *et al.* (1985), however, preferred culturing the mature embryoids for germination on filter paper bridges on a $B_5$ liquid medium supplemented with sucrose (2%), indole-3-butyric acid (IBA, $5 \times 10^{-7}$ M), and naphthalene acetic acid (NAA, $10^{-7}$ M). Within 15–20 days, 40% of the embryoids germinated and formed plantlets with roots and shoots. Normal bamboo plantlets with one or two leaves were formed in 20 days. Lateral root growth was observed. Transfer of plantlets to half-strength $B_5$ liquid medium plus sucrose (1%), IBA ($5 \times 10^{-7}$ M), and NAA ($10^{-7}$ M) produced normal plants with four to six leaves. The root system was well developed at this stage, with long roots and a few lateral roots.

Embryogenic callus was initiated from anthers of *Sinocalumus satiflora* (Munro) McClure cultured on N6 medium supplemented with 1 mg/l 2,4-D, 1 mg/l BA, 2 g/l charcoal, 0.8% agar, and 9% sucrose (Tsay *et al.*, 1990). Prolonged culture of these embryogenic calli on the orginal medium or subculture to an auxin-free medium resulted in embryoid formation and subsequent germination to form rooted plantlets. More than 10 plantlets have been produced in a period of 15 months. Several plantlets were successfully transplanted to the field and grew for 6 months. These transplanted haploid plants eventually died due to a lack of vigor. The authors also stated the anther-derived plantlets were haploid as demonstrated by chromosome counting studies on root-tip cells, but no data were included.

Mukunthakumar and Mathur (1992) reported successful artificial seed production through somatic embryogenesis from seeds of *Dendrocalamus strictus*. Surface-sterilized seeds started callusing within 15 days of inoculation on MS medium containing 3 mg/l of 2,4-D and 0.5 mg/l kinetin (Kin) over 45 days, the callus became compact and developed small green-white structures which progressively differentiated into embryoids. Subculture of the embryogenic callus on to fresh medium resulted in further proliferation of the callus, while leaving it on the old medium resulted in the development of 5–7 embryoids per gram fresh wt. of the callus. Balanced growth of roots and shoots was achieved when the embryoids were transferred to the medium containing 1.0 mg/l NAA and 0.5 mg/l Kin, and therefore the matrix (6% sodium alginate and 100 mM $CaCl_2 \cdot 2H_2O$) containing these growth regulators in addition to the usual MS salts was used for encapsulation. A germination frequency of 96% and 45% was achieved *in vitro* and in soil, respectively. The *in vivo* plantlet conversion frequency was increased to 56% following an additional coating of mineral oil on the alginate beads. Germinated artificial seeds could be raised into plants.

A protocol has been developed for the micropropagation of Mexican weeping bamboo (*Otatea acuminata aztecorm*) through embryoid formation from zygotic embryo explants (Woods *et al.*, 1992). Optimal somatic embryogenesis and plant regeneration were obtained by culture in the dark on MS basal medium supplemented with 3 mg/l 2,4-D, 0.5% mg/l BA, and 2% sucrose. The authors claimed more than 95% of the germinating somatic embryos developed shoots, and roots and were transferred to soil with 85% success.

## 3. Protocols for somatic embryogenesis of *B. oldhamii*, *B. beecheyana*, and *S. latiflora*

In our laboratory, somatic embryogenic has been achieved in three bamboo, namely, *Bambusa oldhamii*, *B. beecheyana*, and *Sinocalamus latiflora*. Embryogenic cultures can be initialized mainly from the reproductive tissues, i.e., young florets and mature seeds and the enclosed zygotic embryos. Vegetative tissues, such as roots and leaves as the source for embryogenic callus, have also been demonstrated.

### 3.1. B. oldhamii (*inflorescence tissue*)

Young inflorescence explants of *B. oldhamii* were sterilized by three chemical treatments: 0.01% Antiseptol (China Chemical & Pharmaceutical Co., Taipei) for 3 h, 75% ethanol for 1 min, and 2% sodium hypochlorite for 15 min, followed by thorough washing in sterile distilled water (Yeh and Chang, 1986a). Lower contamination rates (less than 5%) were usually obtained. The basic medium of Murashige and Skoog (1962) supplemented with 2,4-D (3 mg/l), Kin (2 mg/l) and a higher content of sucrose (60%) proved to be effective for callus induction, spontaneous somatic embryogenesis, and subsequently plant regeneration.

Three types of calli developed from the browning floret explants: (1) soft, friable, and nonembryogenic, consisting of filamentous cells which produced adventitious roots and bristles, (2) off-white to pale yellow in color, smooth glossy nodule-like in appearance, and compact, consisting of small, generally rounded cells rich in cytoplasm and having prominent nuclei and starch grains from which embryogenic callus lines were derived. Its occurrence was associated more frequently with, but not restricted to, the slower-growing original brown callus, (3) a kind of gelatinous callus to be found interspersed among between the two types of callus mentioned above.

The surface of the embryogenic callus became increasing convoluted during the 6 months following its appearance, eventually giving rise to clusters of 6 or more embryoids. Embryoids were attached only loosely to the callus mass at maturity. The embryogenic calli could be subcultured and maintained on the same medium used for callus initiation or on auxin-free medium for over 16 months without loss of totipotency. There is no sign of dormancy, and embryoids germinated either on callus initiation medium or on auxin-free medium, and grew into plantlets that could be readily established in soil without hardening.

Segments of adventitious roots excised from rooting callus derived from young inflorescence explants were found to be another source for em-

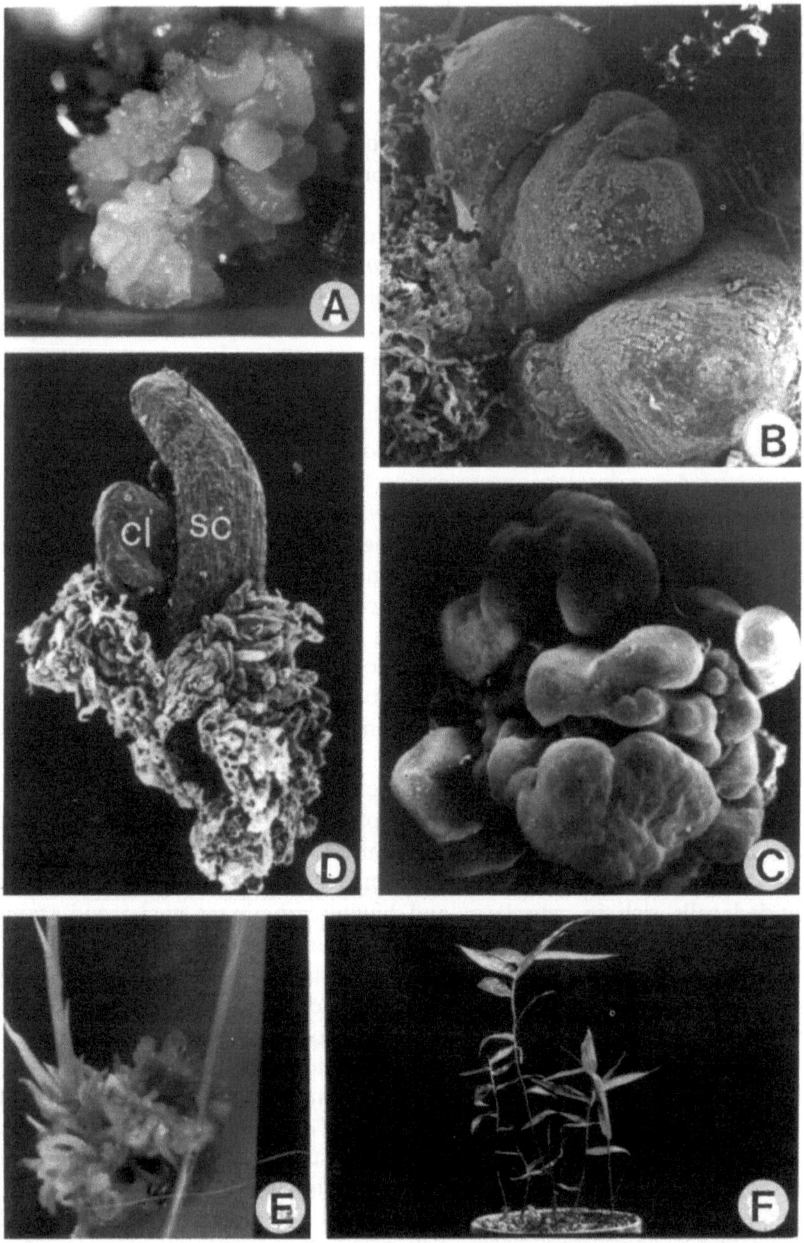


*Figure 1.* Somatic embryogenesis and plant regeneration of *B. oldhamii in vitro* (Yeh and Chang, 1986a). (A) Embryogenic callud derived form cultured young inflorescence segment after 70 days of culture on MS medium supplemented with 3 mg/l 2,4-D, 2 mg/l kinetin and 6% sucrose. (B) Early stage of embryoid formation on the surface of callus. (C) Cluster of embryoids on the surface of callus. (D) Embryoid formed *in vitro*. (E) Cluster of embryoids and plantlets. (F) Green plants derived from embryoids.

bryogenic callus. On the MS medium containing 3 mg/l 2,4-D and 2 mg/l Kin, callus became visible to the unaided eye within one week. Rapid proliferation followed to yield an abundance of callus in about 2 months. The callus first induced was rather gelatinous, usually translucent and smooth at the surface. Subculturing of the callus on a medium of the same composition caused rapid growth. Nodular structures developed within the callus, and after subculturing several times on the same medium, the callus appeared to be composed of more nodular structures. When subcultured for 9 months, albino plantlets occasionally formed from this root-derived callus. Thus, vegetative tissue (at least, in our case, the roots) is a potential source of somatic embryogenesis (Fig. 1).

We have subcultured and maintained the embryogenic callus for more than seven years, and the culture still retains the totipotency for somatic embryogenesis. But more albino plantlets was found among the regenerated plantlets recently.

### 3.2. B. beecheyana (*inflorescence tissue*)

Young inflorescences (0.5–1.0 cm long) taken from local flowering *B. beecheyana* (Munro var. Beecheyana) plants, were excised and disinfected as follows: after being soaked in 0.01% Antiseptol (containing benzethonium chloride 10% w/v and alkyl-arylpolyether alcohol 10% w/v [China Chemical & Pharmaceutical Co. Ltd., Taipei]) for 3 h, inflorescence segments were wetted with ethanol for 1 min, dipped into 2% sodium hypochlorite for 15 min, and then rinsed thoroughly in sterile distilled water. The lemma and paleas of florets were removed, and the remaining floret tissues were cut into 2–5 mm segments for culturing.

A basal medium with macro- and micro-elements of Murashige and Skoog (1962) supplemented with (mg/l): myo-inositol (100), nicotinic acid (0.5), pyridoxine HCl (0.5), thiamine HCl (0.1), glycine (2.0), casein hydrolysate, (1,000), sucrose (30,000–60,000), and Sigma agar (7,000) was satisfactory for callus proliferation and morphogenesis.

The floret explants became swollen and turned brown or black within one week after being placed on MS medium containing 3 mg/l 2,4-D, 2 mg/l Kin and 6% sucrose in darkness at 26 ± 1°C, and callus became visible to the naked eye after about one month. The first callus was rather compact, usually translucent and uneven on the surface. Subculturing of the callus onto the same medium resulted in rapid proliferation. Three distinct callus types were recognized. One callus was soft, friable, non-embryogenic, and consisted of filamentous cells, which produced many adventitious roots and bristles. The second was off-white to pale yellow in color, smooth, glossy, nodule-like in appearance, and compact. It consisted of small, generally spherical cells, usually rich in cytoplasm and starch grains, from which embryogenic callus-lines were derived. The third was a kind of gelatinous callus concurrent with the two types. Nodular structures developed within the compact and organized callus, and after several transfers on the same medium the subcultures appeared to be almost entirely composed of these nodular structures. Further development of these nodular structures eventually gave rise to clusters of embryoids.

The compact and organized callus tissues maintained their embryogenic potential for more than 7 years and through subcultures on the same medium used for callus initiation and embryoid induction. Most embryoids germinated precociously, either on the medium for callus initiation or on the hormone-free medium, and grew into normal plants that could be successfully transplanted to soil without acclimatization.

The non-embryogenic callus derived from floret tissues produced many adventive roots. Segments of these adventitious roots were excised for callus induction. On MS medium containing 3 mg/l 2,4-D and 2 mg/l Kin, callus became visible within one month. Rapid proliferation followed, giving rise to an abundance of callus in about 2 months. The initial callus was rather gelatinous, usually translucent, and smooth at the surface. Subculturing of the callus on the same medium resulted in rapid growth. Nodular structures developed within the root-derived callus, and after several subcultures on the same medium the callus appeared to be composed of more nodular structures. Embryoids eventually formed from these nodular structures and developed into plantlets.

The medium required for whole plant regeneration through somatic embryogenesis in our callus culture of *B. beecheyana* was simpler than that for *D. strictus* (Rao *et al.*, 1985). Rao *et al.* (1985) used $B_5$ (Gamborg *et al.*, 1968) as basic medium. Induction of callus and embryoids was achieved in the presence of 2,4-D. NAA and IBA were employed to promote embryoid germination, and healthy plants were developed in a half-strength mineral liquid medium containing NAA and IBA. In the case of *B. beecheyana*, considerable amounts of 2,4-D (3 mg/l) and Kin (2 mg/l) in an agar-gelled medium were sufficient to induce embryoid formation and for subsequent development of plantlets (Fig. 2).

### 3.3. Sinocalamus latiflora (*young inflorescence*)

Mature seeds of *Sinocalamus latiflorus* (Munro) McClure collected from local accessions were used to establish *in vitro* culture. Mature seeds were wetted with 75% ethanol for one min, dipped into 2% sodium hypochlorite for 10 min, and then rinsed thoroughly in autoclaved distilled water. The mature zygotic embryos were dissected for culture. A basal medium with the macro- and micro-elements of Murashige and Skoog (1962) supplemented with myo-inositol, nicotinic acid, pyridoxine HCl, thiamine HCl, glycine, Kin, 2,4-D, and PVP was sufficient for callus induction and further morphogenesis. Cultures were incubated at 26 ± 1°C in darkness or exposed to artificial light (day-light fluorescent tubes, 40 W) of 15–40 $\mu E\ m^{-2}\ s^{-1}$ with a light-dark cycle of 16/8 h.

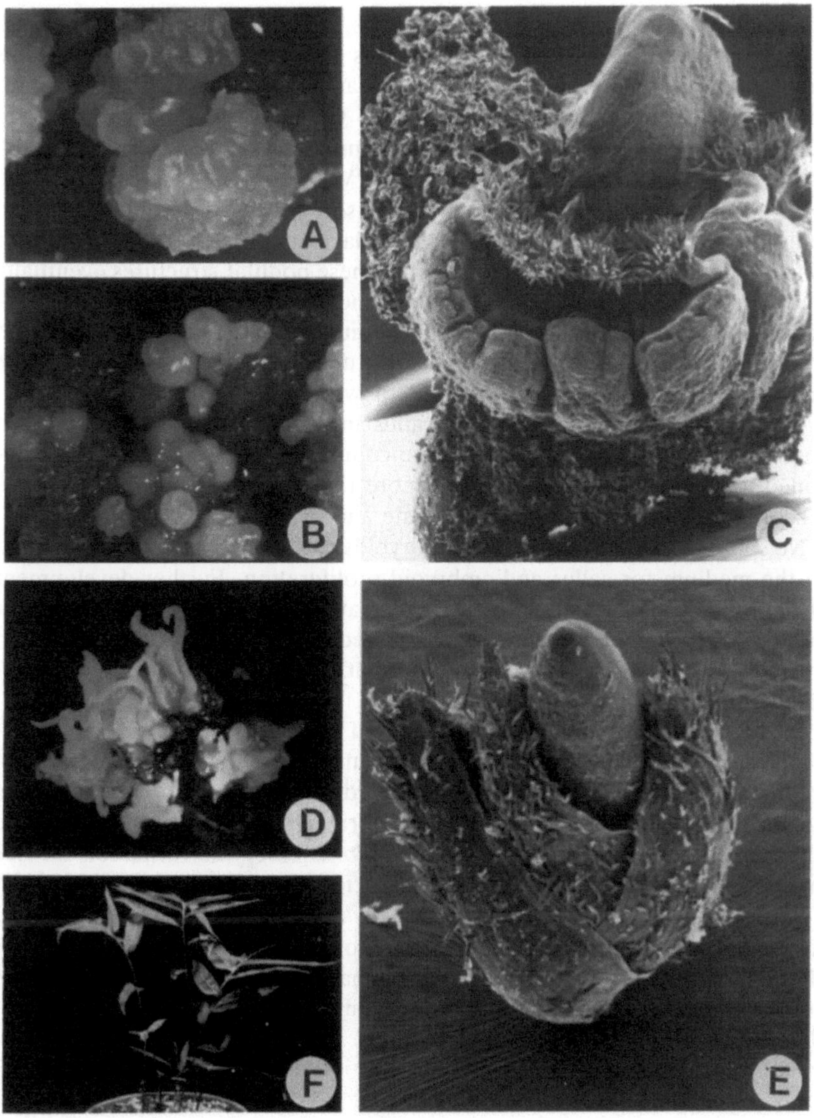


*Figure 2.* Embryoid formation and plant regeneration in *Bambusa beecheyana* (Yeh and Chang, 1986b). (A) Nodular structure proliferation from the subcultured callus derived from floret explants. (B) Clusters of embryoids protruding from the embryogenic callus. (C) SEM micrograph of a row of embryoids at erly developing stage. (D) Germinating embryoids. (E) Single embryoid with well-developed coleoptile and scutellum.

Initially, the explants exuded compounds, probably phenols, into the medium, and some degree of browning was observed in all primary callus proliferation. Thus, PVP (250 mg/l) was added to the medium to retard accumulation of the phenolic compounds. Callus formation from cultured mature zygotic embryos was rapid on MS medium containing 6 mg/l 2,4-D, 3 mg/l Kin, 50 g/l sucrose, and 250 mg/l PVP maintained in darkness. The callus generally grew slowly as dark-brown soft masses. The embryo-derived callus initiated nodular structures after 90 days of culture without re-transfer to fresh medium. Further development of these nodular structures eventually gave rise to clusters of embryoids. Further culture of these brown embryogenic calli for 150 days in the same medium without subculturing resulted in proliferation of white soft callus. Subculture of the white soft callus to the same fresh medium for 3 weeks led to white compact callus formation. Prolonged culture of the compact callus in the same medium resulted in nodular structures and embryoids. The compact and organized callus maintained its embryogenic potential for more than 17 months, through several subcultures, and continued to proliferate and produce embryoids.

No spontaneous germination of embryoids was seen in cultures maintained on the medium for callus and embryoid proliferation. Further development of the embryoids and whole plantlet formation occurred following transfer to an agar (6%) MS medium supplemented with 3 mg/l 2,4-D, 2 mg/l Kin and devoid of PVP. The regenerated plantlets could be successfully transplanted to soil without hardening.

In the presence of high concentrations of Kin (3 mg/l) and sucrose (5%), a high level of 2,4-D (6 mg/l) was found necessary for callus proliferation and for somatic embryogenesis. Only limited callusing and no embryogenesis occurred in the presence of lower concentrations of 2,4–D (1-3 mg/l). The concentration of 2,4-D required for *S. latiflorus* embryoid formation is twice that needed for the two other *Bambusa* species (Yeh and Chang, 1986, 1987). No other auxins (indoleacetic acid (IAA) and NAA) could substitute for 2,4-D. In the presence of such a high level of 2,4-D (6 mg/l), however, parts of embryoids callused soon after their formation. Thus, once the embryoids formed, transfer to the germination medium was necessary to obtain normal plantlets (Fig. 3).

### 3.4. B. beecheyana *Munro var. Beecheyana* (*roots*)

Sterile roots of regenerated green bamboo plants (Yeh and Chang, 1986a) were used for callus induction and subsequent embryogenic callus selection. The regenerated plants derived from young inflorescence were incubated in a hormone-free medium for one month. The roots were excised and washed in sterile distilled water to remove the residual medium, then either directly incubated on the medium or cut into 3-cm segments before culture.

*Figure 3.* Somatic embryogenesis and plant regeneration of *S. latiflorus*. (Yeh and Chang, 1987). (A) A mature seed. (B) Longitudinal section of mature seed showing well-developed zygotic embryos (Arrows). (C) Clusters of embryoids pretruding from the primary callus derived from maturezygotic embryos grown on MS mdeium plus 6 mg/l 2, 4-D, 3 mg Kin, 5% sucrose, and 250 mg/l polyvinylpyrrolidon. (E) Embyoid formation formthe prolonged culture of subcultured callus derived formthe brown primary callus. (F) Embryoid germiantion on MS medium with 3-mg/l 2,4-D and 2 mg/l Kin. (G) A plantlet derived from embryoid, on MS medium plus 3 mg/l 2,4-D and 2 mg/l Kin.

MS media were used as basal media. Cultures were maintained in darkness at 20 ± 1°C or exposed to artificial light of 15–40 $\mu$E $m^{-2}$ $s^{-1}$ with a dark/light cycle of 16/8 h.

Usually, roots turned brown and dark brown after three weeks when cultured on solid MS medium supplemented with 3 mg/l 2,4-D, 2 mg/l Kin, 100 mg/l myo-inositol, and 6% sucrose, then white to pale-yellow-color callus protruded from the culture surfaces. The timing of the callus's emergence was hard to control, and depended on the mass of the roots. Usually, a single cultured root had little chance of survival, and there was no callusing, Aggregated roots, however, were more likely to survive and produced a lot of callus. The longer the cultured roots or the greater the cultured root mass, the sooner and more massive the callus production. In twelve weeks, callus would cover the explants, and brown compounds (presumably phenolic compounds) were released into the medium. Subcultures and refreshment of the media every other 8 weeks were performed to keep the callus viable. The requirements of medium for subculture and germination was the same as the callus induction medium. Dividing the callus during subculture highly accelerated callus proliferation. Initially, the root-inducted callus was somewhat loose and heterogeneous in appearance. Through two subcultures, the newly formed callus became globular, cluster-like, translucent and smooth surfaced. Further development of the translucent callus made some more callus; compact, spherical, but non-translucent, which was composed of oval-like cells with condensed cytoplasm, but no filamentous cells. Somatic embryos spontaneously protruded from these compact, organized and spherical callus. The dormancy period of somatic embryos was uncertain: some of them would germinate once they formed, others would remain in the embryoid stage up to 8 months when maintained in darkness. The regenerated plants were transplanted into soil 6 weeks later without further hardening. We found that embedding the root explants in agarose-gelled media greatly suppressed callus induction. After transferring the root explant onto the medium surface, the callus proliferated rapidly. This demonstrated that the exposure of root explants was critical for callus initiation. Both 2,4-D and Kin were indispensable for callus induction and somatic embryogenesis. NAA could replace up to a half the quantity of 2,4-D in media and BA and $N^6$-($\Delta^2$-isopentenyl)adenine(2iP) could replace half the quantity of Kin, but they could not substitute for 2,4-D and Kin completely. Medium comprising no 2,4-D or Kin resulted in rapid browning and death of callus and somatic embryos. The optimal concentration of 2,4-D in medium was 3 mg/l, and either decreasing or increasing the dosage would lead to callus browning. The dosage of Kin could be varied from 1.5 mg/l to 4 mg/l without provoking callus browning, but increasing the dosage would make the smooth callus surface turn rough and loose. It would not, however, influence the embryogenic potential of the callus. The optimum range of sucrose concentration for callus proliferation and somatic embryogenesis was 3% to 6%. Overdosage made callus susceptible to browning, and under-dosage caused clumsy growing of culture. Addition of GA (1–10 mg/l) had no significant affect on either embryogenesis or embryo germination.

Light has no apparent affect on somatic embryogenesis, but incubating culture under dark conditions reduces the probability of callus browning. On the other hand, light dramatically increases the ratio of embryoid germination. Subculture into light of somatic embryos after incubation under dark conditions resulted in germination within six weeks. Turn-over of the globular, compact callus into somatic embryos was hard to control, and remained the bottleneck through the whole process of somatic embryogenesis. By manipulating the composition of media, including addition of abscisic acid (ABA, 0.1–2 mg/l), coconut milk (0.1%–1%), polyamine (spermidine or spermine, 1–100 mg/l), derivatives of salicylic acid (acetylsalicylic acid and p-aminosalicylic acid, 10–200 mg/l), or changing the medium osmolarity (2%–8% PEG 8000 or mannitol), somatic embryo formation from globular callus has been slightly improved but is still not satisfying. Callus exuding brown-colored compound was another annoying problem which would cause callus senescence and delay the somatic embryogenesis. Supplementing the medium with PVP (250 mg/l) did not improve the situation, but addition of activated charcoal (0.2%) reduced browning.

## 4. Prospects

The limited results achieved thus far by bamboo tissue culture do not reflect an incapacity for *in vitro* development, but rather the lack of sufficient study of these woody Gramineae species. Considering the near nonexistence of conventional breeding methods, cell and tissue culture technologies represent not merely a supplementary mode of improvement in bamboo, but are indispensable tools in overcoming the existing difficulties. It is encouraging that several protocols for bamboo regeneration through somatic embryogenesis have been established. This provides a sound basis for the adaptation of more sophisticated culture methods and novel genetic techniques to bamboo cells.

## 5. References


Chang, W.C., 1991. Bamboo. In: Y.P.S. Bajaj (Ed). Biotechnology in Agriculture and Forestry, Vol. 16, Trees III, pp. 221–237. Springer-Verlag, Berlin/Heidelberg/New York.

Chang, W.C., T.H. Lan, H.R. Chen and T.J. Wen, 1994. Somatic embryogenesis and plant regeneration from roots of bamboo (*Bambusa beecheyana* Munro var. beecheyana). J. Plant Physiol. (in press).

Gamborg, O.L., R.A. Miller and L. Ojima, 1968. Nutrient requirement of suspension culture of soybean root cells. Exp. Cell Res. 50: 151–158.

Mehta, U.I., V.R. Rao and H.Y.M. Ram, 1982. Somatic embryogenesis in bamboo. Proc. 5th Intl. Congr. Plant Tissue Cell Culture, A. Fujiware (ed.), Tokyo, pp. 109–110.

Mukunthakumar, S. and J. Mathur, 1992. Artificial seed production in the male bamboo *Dendrocalamus strictus* L. Plant Sci. 87: 109–113.

Murashige, T and F. Skoog, 1962. A revised medium for rapid growth and bio-assays with tobacco tissue cultures. Physiol. Plant 15: 473–479.

Rao, I.O., I.V.R. Rao and Y. Narang 1985, Somatic embryogenesis and regeneration of plants in the bamboo *Dendrocalamus strictus*. Plant Cell Rep. 4: 191–194.

Tsay, H.S., C.C. Yeh and J.Y. Hsu, 1990. Embryogenesis and plant regeneration from anther culture of bamboo (*Sinocalamus latiflora*) (Munro) McClure) Plant Cell Rep. 9: 349–351.

Woods, S. H., G.C. Phillips, J.E. Woods and G.B. Collins, 1992. Somatic embryogenesis and plant regeneration form zygotic embryo explants in Mexica weeping bamboo, *Otatea acuminata aztecorum*. Plant Cell. Rep. 11: 257–261.

Yeh, M.L. and W.C. Chang, 1986a. Plant regeneration through somatic embryogenesis in callus culture of green bamboo (*Bambusa oldhamii* Munro). Theor. Appl. Genet. 72: 161–163.

Yeh, M.L. and W.C. Chang, 1986b. Somatic embryogenesis and subsequent plant regeneration from infloresence callus of *Bambusa beecheyana* Munro var beecheyana. Plant Cell Rep. 5: 409–411.

Yeh, M.L. and W.C. Chang, 1987. Plant regeneration via somatic embryogenesis in mature embryo-derived callus of *Sinocalamus latiflora* (Munro) McClure. Plant Sci. 50: 93–96.

# 5. Somatic embryogenesis in oak (*Quercus* spp.)

Vladimír Chalupa

**Contents**



## 1. Introduction

The genus *Quercus* includes about 450 species with a large number of varieties and hybrids. The oaks are widely distributed throughout the temperate regions of the northern hemisphere, in Europe, North America, and Asia. Few oak species are found south of equator. Oaks have an important economic value. In Europe, oaks represent about 9% of the growing stock in forests, and cover 25–30% of the forested area in some European countries (France, Greece, Romania, UK, Belgium). Oaks are mostly trees, either deciduous or evergreen. Most of the species are restricted to the warmer temperature regions. Oak species tolerate wide range of soils. Some species (for example *Q. robur*) prefer moist heavy soils, rich in mineral nutrients, the others grow on well-drained soils (for example *Q. petraea*).

Oaks provide the fine hardwood, valued not only because of its great strength and durability, but also for its beauty. The wood of oaks is hard, heavy and strong, and has a pronounced growth-ring figure. Oak wood is used for furniture, panelling, house and ship construction, and often is used as a surface veneer over ordinary timbers. Cork is obtained from the bark of the cork oak (*Q. suber*). Many oak species are grown as ornamental trees, often producing attractive fall colours.

After the virgin forests had been cleared, oaks were often grown as cop-

*S. Jain, P. Gupta & R. Newton (eds.), Somatic Embryogenesis in Woody Plants, Vol. 2, 67–87.*

pice, and trees were cut every 20–30 years to yield firewood, charcoal, fence posts, along with harvest of tanbark. However, this is no longer profitable, and most oak forests are grown from seeds to larger sizes, over a lifetime of 100–150 years and more. Oak species are propagated mainly generatively, by acorns. Geographic races developed in many widely distributed oak species, and these often exhibit marked differences in growth rates and other characteristics. Provenance field tests show variation in phenology, growth rates, adaptability, cold hardiness, stem form, wood quality, susceptibility to insects and diseases (Kriebel, 1993; Kleinschmit, 1993; Jensen, 1993). Oak species with large and continuous distributions exhibit high levels of gene diversity (Kremer and Petit, 1993). Hybridization within the genus *Quercus* appears to be extensive and the occurrence of hybrids is well documented (Rushton, 1993).

Oaks are relatively difficult for breeding. Improvement is constrained by the limited knowledge of genetic variation, the long period to reproductive maturity and long intervals in seed years, impossibility of storing seed for extended periods, and difficulties in vegetative propagation. The goals of tree improvement activities are focused on fast growth, straightness and lack of forking in the stem, self pruning, disease resistance and wood quality traits. Quality timber production for veneer and sawn wood will continue to be the primary goal of breeding of important European oak species (Savill and Kanowski, 1993; Nepveu, 1993). Seed orchards are difficult to use in oak breeding due to long reproductive cycles and low acorn production. Vegetative propagation is an alternative to a breeding system based on seed orchards. Although commercial oak propagation is currently done by seed, asexual propagation systems will enable the establishment of clones with superior characteristics (Chalupa, 1990c, 1993).

Species of the genus *Quercus* are diploid with 24 chromosomes. Johnsson (1946) found three triploid seedlings of *Q. robur*, and Butorina (1993) described two spontaneous triploid mature *Q. robur* trees, which were found in the Voronezh region and were unusually large. In both mature trees 36 chromosomes were predominant in cells, with occasional observations of diploid, hypoaneuploid and hyperaneuploid cells. Meiosis in both trees was abnormal.

As far as it is known, all oak species are anemophilous. The unisexual flowers are produced from lateral buds, with the male flowers grouped into long, stalked, pendulous catkins. The flowers open in central Europe usually in May. Pollen shedding varies from 5 to 10 days, with most anthers discharging all pollen within few days. At the time of pollination the ovules are rudimentary and the embryo-sac undeveloped. Fertilization occurs about 3–4 weeks after pollination (in white oaks). The fruit is a large nut (acorn) and is usually one-seeded, occurring singly or in clusters. The acorns mature either in one year (white oaks) or in two years (red and black oaks). Fruit ripening and seed dispersal occur in the autumn. The embryo has two fleshy cotyledons. Acorns of most species can be stored only for a short time.

### 1.1. *Vegetative propagation*

Vegetative propagation is important for oak reproduction and tree improvement. The long reproductive cycle of oak is a serious obstacle to tree improvement by conventional tree-breeding techniques. As vegetative propagation techniques improve, clonal propagation of oak species is becoming more widely practised. Vegetatively propagated plants have a potential for increasing forest yield. *In vitro* propagation of oak species can be used for the production of plants with desirable genetic traits. A reliable plant regeneration system from meristems and embryogenic cultures is a prerequisite for the application of recombinant DNA technology to oak improvement.

The increasing interest in vegetative propagation stimulated development of new techniques which enable production of plants either by a stem-cutting system or by *in vitro* methods. Experiments with vegetative propagation of oak by cuttings were started a long time ago (Ure, 1937; Thimann and Delisle, 1939; Muhle Larsen, 1946). The rooting of various oak species proved to be difficult. In subsequent years, Cornu *et al.* (1975, 1977), Kleinschmit *et al.* (1975), Garbaye *et al.* (1977), Chalupa (1982, 1990d, 1993), Spethmann (1982, 1986), and Spethmann and Harms (1993) described the production of rooted cuttings of important European oak species.

### 1.2. In vitro *propagation*

Experiments with tissue culture propagation of oak were not successful for a long time. The main objective of older experiments was to establish callus cultures and to regenerate shoots and plants from callus cultures. Despite numerous experiments, the induction of shoots in oak callus cultures has not been achieved. Recently a system based on *in vitro* multiplication of shoots from axillary buds has been developed (Chalupa, 1979, 1981, 1983, 1984; Bellarosa, 1981; Pardos, 1981; Vieitez *et al.*, 1985). Micropropagated plantlets were transplanted into soil and later were planted in the field. The system of axillary shoot multiplication was used for micropropagation of various oak species: *Q. robur* and *Q. petraea* (Chalupa, 1979, 1981, 1983, 1984, 1985b, 1987b, 1988, 1990c,d, 1993; Vieitez *et al.*, 1985; Pevalek-Kozlina and Jelaska, 1986; Civinová and Sladký, 1987; Favre and Juncker, 1987; Meier-Dinkel, 1987; Meier-Dinkel *et al.*, 1993; San-José *et al.*, 1988, 1990; Juncker and Favre, 1989; Volkaert *et al.*, 1990; Gebhardt *et al.*, 1993; Evers *et al.*, 1993), *Q. suber* (Bellarosa, 1981, 1989; Pardos, 1981; Manzanera and Pardos, 1990), *Q. shumardii* (Bennett and Davies, 1986), *Q. acutissima* (Ide and Yamamoto, 1986; Sato *et al.*, 1987), *Q. serrata* (Ide and Yamamoto, 1987); *Q. lobata* (Johnson and Walker, 1990), and *Q. alba* and *Q. rubra* (Schwarz and Schlarbaum, 1993).

In the past several years, rapid progress in plant regeneration by somatic embryogenesis has demonstrated the promising prospects of this method for clonal propagation and tree improvement. Somatic embryogenesis has been

*Table 1.* Somatic embryogenesis in *Quercus* spp.

| Species | Explant | Response | Reference |
|---|---|---|---|
| *Q. lebani* | Zygotic embryo segments | Somatic embryoids, roots, shoots | Srivastava and Steinhauer, 1982 |
| *Q. robur* | Immature zygotic embryos, | Somatic embryos | Chalupa, 1985b |
| | anthers | Embryos | Chalupa, 1985b |
| | immature zygotic embryos | Somatic embryos, plantlets, plants in soil | Chalupa, 1987a, 1990a, 1993 |
| *Q. petraea* | Immature zygotic embryos | Somatic embryos, plantlets, plants in soil | Chalupa, 1987a, 1989, 1990a,b, 1993 |
| | anthers | embryos, plantlets | Jörgensen, 1988, 1993 |
| *Q. suber* | Internode stem fragments of seedlings | Somatic embryos | El Maataoui and Espagnac, 1987 |
| | Internode stem fragments of seedlings | Histology of somatic embryogenesis | El Maataoui *et al.*, 1990 |
| | Immature zygotic embryos | Somatic embryos, plantlets, plants in soil | Bueno *et al.*, 1992; Manzanera *et al.*, 1993 |
| *Q. acutissima* | Cotyledons of zygotic embryos | Somatic embryoids, plantlets | Haraguchi, 1987, 1988 |
| | Embryonic axes of zygotic embryos | Somatic embryoids, plantlets, plants in soil | Sasaki *et al.*, 1988, Shoyama *et al.*, 1992 |
| *Q. rubra* | Immature zygotic embryos | Somatic embryos, plantlets, plants in soil | Gingas and Lineberger, 1989 |
| | Immature zygotic embryos | somatic embryos, plantlets, plants in soil | Chalupa, 1991, 1992 |
| *Q. alba* | Immature zygotic embryos | Somatic embryos | Gingas and Lineberger, 1989 |
| *Q. ilex* | Leaf segments from adult trees (50 year-old) | Somatic embryos | Féraud-Keller and Espagnac, 1989 |
| *Q. serrata* | Immature zygotic embryos | Somatic embryos | Sasamoto and Hosoi, 1989 |
| | Immature zygotic embryos | Callus proliferation from protoplasts | Sasamoto and Hosoi, 1992 |
| *Q. cerris* | Immature zygotic embryos | Somatic embryos | Ostrolucká and Pretová, 1991 |
| *Q. bicolor* | Male catkins | Somatic embryos, plantlets, plants in soil | Gingas, 1991 |

reported in many important forest tree species (Radojević, 1979; Sommer and Brown, 1980; Tulecke and McGranahan, 1985; Chalupa, 1985a,b; Hakman *et al.*, 1985; Nagmani and Bonga, 1985; Gupta and Durzan, 1986; Merkle and Sommer, 1986; Durzan and Gupta, 1987). Somatic embryogenesis has a potential to produce a great number of complete plantlets in a short time. It may play an important role in tree breeding programs and in production of transgenic trees obtained by recombinant DNA technology. High-velocity microprojectils carrying new genes can penetrate the cell wall and deliver nucleic acids in a biologically active form to a large number of cells. Furthermore, embryogenic cultures can be used for cryogenic storage, when plant cells are exposed to ultra-low temperature and cellular metabolic activities are arrested. They can be used as a source of protoplasts that have the capacity to form somatic embryos, and after protoplast fusion, to form somatic hybrids.

Recently, somatic embryogenesis has been induced in several oak species (Table 1). Immature or mature embryos, male catkins, anthers, stem seedling segments or leaf fragments were used as initial explants for induction of somatic embryogenesis in the genus *Quercus*. Somatic embryogenesis was reported in *Q. lebani* (Srivastava and Steinhauer, 1982), *Q. robur* and *Q. petraea* (Chalupa, 1985b, 1987a, 1989, 1990a,b, 1992, 1993; Jörgensen, 1988, 1993), *Q. suber* (El Maataoui and Espagnac, 1987; El Maataoui *et al.*, 1990; Bueno *et al.*, 1992; Manzanera *et al.*, 1993), *Q. acutissima* (Haraguchi, 1987, 1988; Sasaki *et al.*, 1988; Shoyama *et al.*, 1992), *Q. rubra* (Gingas and Lineberger, 1989; Chalupa, 1991, 1992), *Q. alba* (Gingas and Lineberger, 1989), *Q. ilex* (Féraud-Keller and Espagnac, 1989), *Q. serrata* (Sasamoto and Hosoi, 1989, 1992), *Q. bicolor* (Gingas, 1991), and *Q. cerris* (Ostrolucká and Pretová, 1991). Plant regeneration from oak somatic embryos has proved to be difficult. The conversion of somatic embryos into plantlets has been achieved only in some species. Although somatic embryogenesis was obtained in many *Quercus* species, low efficiency of embryo conversion into plantlets remains a major problem. Initial source of explant, genotype, nutrient medium, and phytohormones greatly influence the induction of embryogenic cultures and plant regeneration.

## 2. Initiation of embryogenic cultures

The initiation of embryogenic cultures in *Quercus* is greatly dependent on the type of explant used. Due to difficulties in culturing tissues from adult specimens, oak embryogenic cultures are usually initiated from young tissues, most frequently from embryonic tissue. Immature zygotic embryos are often used as explants and shown to be highly embryogenic. The embryogenic potential of immature zygotic embryos usually varies and is dependent on the genotype. Somatic embryos can develop either from subcultured embryogenic tissue or directly from explants without involving any intermediate callus stage. *Quercus* species were induced to form embryogenic cultures and somatic embryos from immature and mature zygotic embryos, anthers and male catkins, and leaf fragment and stem internode segments.

*Table 2.* Effect of collection date and nutrient medium on the formation of embryogenic tissues from cultured immature embryos of *Quercus robur* L.

| Collection date | Embryogenic cultures (%) | |
|---|---|---|
| | MS | WPM |
| July 5 | 62 | 54 |
| July 12 | 80 | 76 |
| July 19 | 82 | 74 |
| July 26 | 76 | 70 |
| August 2 | 75 | 72 |
| August 9 | 61 | 65 |

### 2.1. *Initiation of cultures from immature and mature zygotic embryos*

Embryogenic cultures of *Quercus robur* and *Quercus petraea* were initiated from immature zygotic embryos (Chalupa, 1985b, 1987a, 1989, 1990a,b, 1992, 1993). The acorns were harvested from open pollinated trees at weekly intervals during July and August. Following their removal from the trees, fruits were kept moist in a plastic bag stored at 4°C for 1 to 8 days. Acorn sterilization consisted of 70% ethanol dip followed by a 20 min immersion in calcium hypochlorite (7.5% w/v) solution. Seeds were rinsed twice in sterile distilled water. Following the surface sterilization, immature zygotic embryos were removed using aseptic procedures. Explants were placed in 100 ml flasks containing 20 ml of agar nutrient medium. The embryogenic cultures were initiated on MS medium (Murashige and Skoog, 1962), SH medium (Schenk and Hildebrandt, 1972), and WPM medium (Lloyd and McCown, 1980). The basal media were supplemented with glutamine (1.37 mM) or casein hydrolysate (500 $mg \cdot l^{-1}$), cytokinin 6-benzylaminopurine (BA 4.44 μM), or cytokinin plus auxin indole-3-butyric acid (IBA 0.49 μM or 4.90 μM) or cytokinin plus gibberellic acid ($GA_3$ 2.89 μM). Embryogenic cultures were initiated in the dark and were grown at 24°C. Explants from immature acorns collected during July and in the beginning of August gave rise to embryogenic tissue in high frequency (61–82%; Table 2). Immature zygotic embryos cultured in the dark produced white embryogenic tissues within 5 to 7 weeks. Embryogenic cultures were subcultured to a fresh medium every 4–5 weeks.

Embryogenic cultures of *Quercus rubra* were initiated from immature zygotic embryos (Gingas and Lineberger, 1989). Immature leaf and ovular explants failed to produce embryogenic tissue. After 2 months in culture, all foliar tissue became necrotic. Developing acorns were collected at weekly intervals from June to August. Acorns were sterilized in a 0.5% Alconox solution for 20 min, transferred to a 0.5% hypochlorite solution for 20 min, and rinsed twice with sterile distilled water. Late heart and early cotyledonary explants of immature zygotic embryos, cultured in light (80–100 $\mu E\ m^{-2}\ s^{-1}$) on modified MS medium containing various combinations of 2,4-

dichlorophenoxyacetic acid (2,4-D) and BA, proved to be highly embryogenic. The highest embryoid numbers were generated from tissues explanted 4 to 7 weeks post-fertilization and grown in light. Embryogenic cultures arose from explants cultured on a modified MS medium containing 200 mg · $l^{-1}$ casein hydrolysate and various combinations of BA (0.04–4.44 μM) and 2,4-D (0.05–4.52 μM). Chalupa (1991, 1992) established embryogenic cultures of *Quercus rubra* from immature embryos on modified SH medium containing glutamine (1.37 mM), casein hydrolysate (500 mg · $l^{-1}$), and cytokinin BA (4.44 μM), or cytokinin plus auxin (IBA 0.49 μM). Rapidly growing white embryogenic tissue developed from immature embryos within 3–5 weeks.

Embryogenic liquid culture lines were induced from immature zygotic embryos of *Quercus serrata* (Sasamoto and Hosoi, 1989, 1992). After sterilization (2% NaClO for 30 min), immature zygotic embryos including cotyledons were excised and cultured in 50 ml tubes containing liquid MS medium supplemented with BA (0.1 μM) and 2,4-D (0.1 μM). Small cell clumps were transferred to fresh medium with the same composition at monthly intervals.

Embryogenic cultures of *Quercus cerris* were initiated from cotyledon segments of immature zygotic embryos (Ostrolucká and Pretová, 1991). The induction of embryogenic tissue was stimulated on WPM containing 4.44 μM BA, 2.32 μM kinetin, and 4.90 μM IBA. The same combination of phytohormones added to MS medium failed to produce embryogenic tissue. The process of somatic embryogenesis in *Q. cerris* was stimulated on WPM medium containing BA, kinetin, and IBA. Subculture of callus on WPM medium containing 2,4-D (2.26–9.05 μM) stimulated growth of non-embryogenic, watery callus.

Bueno *et al.* (1992) and Manzanera *et al.* (1993) initiated embryogenic callus of *Quercus suber* from immature embryos. Acorns were collected during embryo development, from June to September. Embryos were cultured on modified Sommer *et al.* (1975) medium, supplemented with 2,4-D (2.26 μM to 45.24 μM). Zygotic embryos formed embryogenic callus on all media containing 2,4-D. Higher concentrations of 2,4-D (22.6 μM and 45.2 μM) provided a lower percentage of embryogenic tissue. Globular structures and embryoids were formed directly on the zygotic embryos or in the callus during the second and third week after the beginning of the 2,4-D treatment.

Embryogenic structures were initiated in *Quercus lebani* (Srivastava and Steinhauer, 1982) from segments of mature embryos cultured on MS medium. Middle segments of mature embryos cultured on modified MS medium containing casein hydrolysate (1 g · $l^{-1}$), $N^6$-(2-isopentenyl) adenine (2iP, 24.61 μM), indole-3-acetic acid (IAA, 11.42 μM) callused profusely and produced a subculturable mass of tissue within 6 weeks. After transfer to a fresh medium, the callus became compact and green, and differentiated into dicotyledonous embryoids.

Embryogenic cultures of *Quercus acutissima* were established from cotyledons and embryonic axes of zygotic embryos (Haraguchi, 1987, 1988; Sasaki *et al.*, 1988; Shoyama *et al.*, 1992). Sasaki *et al.* (1988) sterilized seeds with 3% NaClO for 10 min, then with 70% ethanol for 30 s. Embryonic axes were cultured on half-strength MS medium containing BA (4.44 μM) and $GA_3$(2.89 μM). Somatic embryogenesis occurred during an 11-week culture of embryonic axes on half-strength MS medium supplemented with BA (4.44 μM) and $GA_3$(2.89 μM). The addition of BA and $GA_3$ stimulated the production of secondary somatic embryos. On the other hand, the addition of 2,4-D or naphthaleneacetic acid (NAA) stimulated production of non-embryogenic callus, and somatic embryogenesis did not occur.

### 2.2. *Initiation of cultures from anthers and male catkins*

Embryogenic cultures of *Quercus robur* were established from anthers and male catkins on modified MS media and WPM medium supplemented with 2,4-D (4.52–9.04 μM) and BA (0.89–4.44 μM) (Chalupa, 1985b). Only a few anthers and male catkins produced embryogenic tissues containing small globular multicellular structures. After transfer of embryogenic tissues with induced globular structures to medium with reduced concentration of hormone (BA 0.89 μM), embryoids developed.

Embryogenic cultures from anthers, excised from flower buds of *Quercus petraea*, were established by Jörgensen (1988, 1993). The inflorescences were surface sterilized in 70% ethanol for 1 min followed by 5–7 min 6.5% NaClO treatment. After sterilization, anthers were excised from flower buds and placed on agar nutrient media. Calli developed from the interior of anthers cultured on modified MS media containing 2.5 to 5.0 μM 2,4-D and 0.0 to 10 μM BA, or on modified WPM containing 0.5 to 10 μM BA and 0.5 to 10 μM 2,4-D. Anthers in the early uninucleate stage produced embryogenic calli after 2 to 3 months of cultivation.

Gingas (1991) initiated embryogenic cultures of *Quercus bicolor* from partially expanded male catkins. Staminate catkins were collected at the semipendent stage of development. Whole catkins were disinfected with a 0.5% NaClO and 0.5% Alconox solution for 15 min followed by two rinses in sterile water. Catkins were cultured on a modified MS medium containing $200\ mg \cdot l^{-1}$ casein hydrolysate and supplemented with either BA (4.44 μM) or 2,4-D (4.52 μM). Cultures were subjected to darkness for 48 h and then placed under $30\ \mu mol\ m^{-2}\ s^{-1}$ of continuous illumination provided by cool-white fluorescent lamps. Explants on medium containing 2,4-D produced a yellow embryogenic callus originating from the junction of the pedicel and peduncle. Subsequent transfers to MS medium with BA (4.44 μM) and then to MS without growth regulators resulted in callus proliferation. After 6 to 8 weeks on a medium with BA, callus transferred to MS without growth regulators produced opaque, white embryoids.

### 2.3. *Initiation of cultures from stem and leaf segments*

Embryogenic cultures of *Quercus suber* were initiated from internode stem segments taken from 6–8 month-old greenhouse-grown seedlings (El Maataoui and Espagnac, 1987; El Maataoui *et al.*, 1990). Internode fragments (5–8 mm long) were sterilized for 1 min in 70% ethanol and then for 20 min in 4.5% calcium hypochlorite solution. Embryogenic calli were initiated on MS medium containing 111 mM glucose, 10 μM IBA, 8.88 μM BA, and 1 g $\cdot$ $1^{-1}$ casein hydrolysate. The first divisions leading to the callus formation were observed in the cortical parenchyma after 5–6 days. Calli developed within 2 weeks, and were separated and transferred on the fresh medium where the proliferation continued. Embryogenic calli were friable and translucent, and contained spherical nodules which varied in size.

Féraud-Keller and Espagnac (1989) initiated embryogenic tissue of *Quercus ilex* from newly formed leaves which were collected from adult trees (about 50 years-old). Leaf fragments were sterilized for 20 min in calcium hypochlorite solution and then rinsed three times in sterile distilled water. The leaf fragments were cultured on modified MS medium supplemented with BA (17.76 μM) and NAA (2.69 μM). Only leaves formed in the current year, taken from trees in October, produced embryogenic tissues. The neoformations appeared only on calli that had not been subcultured for more than 7 months.

## 3. Development of somatic embryos

Reports on somatic embryogenesis in *Quercus* provide evidence that oak tissue is highly embryogenic. Presently, conditions are being optimized in order to increase the frequency of somatic embryo formation and to establish the best conditions for embryogenesis. The efficient production of oak somatic embryos and recovery of plantlets from embryos require better understanding of the principles of embryogenesis. Oak somatic embryos originate in embryogenic tissue through a series of developmental stages. Somatic embryos develop in embryogenic cultures as bipolar structures bearing both root and shoot meristem necessary for complete plant development. Initiation and maturation of the embryos occurs in one or more steps involving various media composition. Somatic embryos of *Quercus* often developed in embryogenic tissue maintained on nutrient medium used for the initiation of cultures, or embryogenic tissues were transferred to different media where the development or maturation of somatic embryos was stimulated. Formation of oak somatic embryos can occur in the absence of exogenous hormones, however, in most cases, the development of somatic embryos was stimulated by manipulation of hormonal balance in the nutrient medium.

*Figure 1.* Somatic embryoids arising from the embryogenic culture of *Quercus robur*.

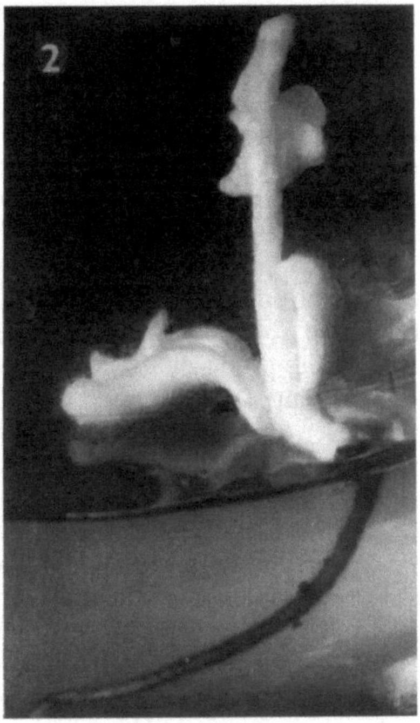



*Figure 2.* Germinating mature somatic embryo of *Quercus robur*.

### 3.1. *Development of somatic embryos in cultures initiated from immature and mature zygotic embryos*

Development of somatic embryos in *Quercus robur* cultures was stimulated on MS, WPM, and SH media (Chalupa, 1985b, 1987a, 1990a, 1992, 1993). Within 6–10 weeks, globular structures developed in 70% of the cultures grown on modified MS or SH medium, or WPM containing BA (4.44 μM), or BA plus $GA_3$ (2.89 μM), or BA plus IBA (4.90 μM). The development of globular and torpedo-shaped structures continued in subcultured tissues and the frequency of embryoid formation often increased. Embryoids were only loosely attached to tissue at the root pole. Gradually somatic embryos showing bipolar structure developed. Repetitive somatic embryogenesis was frequent and led to the formation of many small embryoids. Embryogenic tissues maintained on a medium containing cytokinin, have retained their embryogenic potential for more than two years (Figs. 1 and 2).

The first globular and torpedo-shaped structures developed in tissues of *Quercus petraea* cultured on modified SH or MS medium (containing BA 4.44 μM, or BA plus IBA 4.90 μM) within 5–8 weeks (Chalupa 1987a, 1989,

1990a,b, 1993). The first somatic embryos developed two months after the initiation of cultures. Embryogenic tissues cultured on SH or MS medium containing cytokinin (BA) maintained their embryogenic potential for more than three years. Secondary somatic embryogenesis was frequent and led to the formation of new somatic embryos.

High embryoid numbers were obtained in *Quercus rubra* immature zygotic embryo explants cultured on modified MS medium supplemented with 4.44 μM BA and 4.52 μM 2,4-D (Gingas and Lineberger, 1989). Embryoids developed from explants cultured on media containing various combinations of BA and 2,4-D. However, the highest percentages of normal polar embryoids were produced by explants cultured on growth regulator-free media. Forty percent of the normal polar oak embryoids resulted from explants that were never exposed to growth regulators. This includes *Quercus* in the list of species that do not require auxin for the induction of somatic embryogenesis. *Q. rubra* embryoids were produced via direct and indirect embryogenesis. Initially, embryoids were produced directly from cotyledon surfaces of the explant and began to appear after 3 to 4 weeks in culture. Callus could easily be maintained and globular structures developed from it, later giving rise to polar embryoids. Repetitive asexual embryogenesis was common. Secondary and tertiary embryoids formed on the cotyledons of other embryoids. Embryogenic tissues and somatic embryos of *Q. rubra* were also produced on modified SH medium supplemented with 4.44 μM BA (Chalupa 1991, 1992). Numerous somatic embryos developed and enlarged within 7–10 weeks. Embryogenic tissues cultured on modified SH medium containing cytokinin (BA) retained their embryogenic potential for more than three years. Secondary embryoids formed frequently on somatic embryos.

Somatic embryogenesis in *Quercus alba* was initiated from immature zygotic embryos cultured on modified MS medium (Gingas and Lineberger, 1989). Embryoids developed directly from the cotyledon explants. Repetitive embryogenesis was both direct (primary from radicles) and indirect. Less than 1% of the white oak embryoids reached the heart developmental stage. None exhibited root or shoot development.

Embryogenic cultures and embryoids of *Quercus cerris* were induced on WPM containing BA (4.44 μM), kinetin (2.32 μM), and IBA (4.90 μM) (Ostrolucká and Pretová, 1991). After 7–8 weeks, white callus was formed with nodular structures, which proliferated intensively. After transfer to a fresh medium, the nodular structures produced bipolar embryoids. Increased embryogenic activity was observed after addition of activated charcoal ($1.5 \text{ g} \cdot 1^{-1}$) in the medium. The cotyledon-derived cultures maintained their embryogenic activity for two years. Germination of somatic embryos and production of plants were not achieved.

Somatic embryos of *Quercus suber* developed in embryogenic cultures initiated from immature zygotic embryos (Bueno *et al.*, 1992; Manzanera *et al.*, 1993). Globular structures were initiated on modified Sommer *et al.* (1975) medium containing 2,4-D. Torpedo-shaped somatic embryos, loosely attached to the callus, developed on the same medium. The development of somatic embryos was stimulated after transfer of embryogenic cultures on the basal medium supplemented with $GA_3$ (2.88 μM), or with $GA_3$ plus BA (0.44 μM) and NAA (0.05 μM), or on growth regulator-free medium.

Embryogenic cultures and somatic embryos in *Quercus acutissima* were initiated from embryonic axes of zygotic embryos (Sasaki *et al.*, 1988; Shoyama *et al.*, 1992). Somatic embryogenesis occasionally occurred during the 11-week culture of embryonic axes on the half-strength MS medium supplemented with BA (4.44 μM) and $GA_3$ (2.89 μM). On the other hand, the addition of 2,4-D or NAA to the MS medium, stimulated production of non-embryogenic callus. Somatic embryogenesis did not occur on media containing auxin. BA and $GA_3$ stimulated production of secondary somatic embryos, and this ability was maintained at least for 3.5 years.

### 3.2. *Development of embryos in cultures initiated from anthers and male catkins*

Embryogenic tissues were initiated from anthers of *Quercus robur* cultured on modified MS medium and WPM supplemented with 2,4-D and BA (Chalupa, 1985b). Embryogenic tissues with developing globular structures were transferred to media with reduced concentrations of growth regulators (to WPM containing 0.89 μM BA), where embryoids and embryos developed.

The initiation of embryogenic calli from anthers of *Quercus petraea* and development of globular structures were stimulated on modified WPM supplemented with 2,4-D and BA (Jörgensen, 1988, 1993). To enhance the embryogenic potential, the calli were transferred to media lacking auxin and containing only BA. Embryos developed on modified WPM supplemented with 2.5 μM BA. Secondary embryos developed on the surface of the primary embryos on the same medium.

Embryogenic tissue in *Quercus bicolor* was initiated from partially expanded male catkins cultured on a modified MS medium containing 2,4-D (Gingas, 1991). Embryoids were produced after callus was transferred to MS medium containing BA (4.44 μM) and then to MS medium without growth regulators. After 6 to 8 weeks on MS medium with BA, callus transferred to MS without hormones produced white embryoids. Separated and individually cultured embryoids underwent repetitive embryogenesis. Repetitive embryogenesis directly from the surface of the embryoids has continued for more than 1.5 years.

### 3.3. *Development of somatic embryos in cultures initiated from stem and leaf segments*

Embryogenic calli produced from stem fragments excised from greenhouse-grown seedlings of *Quercus suber*, were able to form somatic embryos (El Maataoui and Espagnac, 1987; El Maataoui *et al.*, 1990). Embryoids were produced in calli cultured on MS medium supplemented with BA (8.88 μM) and IBA (10 μM). Globular structures differentiated an apical meristem flanked with two cotyledons and a radicular meristem on the opposite end. In most cases, embryoid development stopped at this stage. Somatic embryos were able to produce secondary embryoids in large numbers. This multiplication capacity has been maintained for a long time. A histological study showed that calli consisted of parenchyma cells surrounded by a few layers of meristematic cells. Proembryos formed in groups around the edge of some calli. They were produced by the evolution of two categories of cells: one category had the appearance of embryogenic cells with a small vacuole and a well-developed nucleus. The morphology of other cells was similar to that of suspensor cells. The ontogenetic stages were similar to those described in zygotic embryos of the genus *Quercus*.

Embryogenic tissues and somatic embryos were produced from leaf fragments of adult trees of *Quercus ilex* (Féraud-Keller and Espagnac, 1989). Leaf fragments were cultured on modified MS medium containing BA (17.76 μM) and NAA (2.69 μM). Primary nodules produced on calli were removed and subcultured on the same medium and grown either in the dark or in the light. In the dark only, they produced secondary nodules, which were the source of somatic embryos both in light and dark. It seems, that the somatic embryos had regulated their structure in the dark, however, they failed to develop in such a manner that leads to germination.

## 4. Conversion of somatic embryos into plantlets and growth of regenerated plants

A serious problem of oak regeneration via somatic embryogenesis is either an absence, or a low frequency of somatic embryo conversion into plantlets. Even somatic embryos that appear normal often fail to complete germination and conversion into plantlets. The development of somatic embryos is often blocked after the formation of cotyledons. Therefore, for further development of somatic embryos into plants, various treatments were tested. Maturation and conversion of embryos involve a sequence of medium changes and alternation of physical conditions.

Germination of *Quercus robur* somatic embryos and formation of plantlets occurred rarely on media containing a high cytokinin concentration (Chalupa, 1987a, 1990a, 1992). The transfer of oak embryogenic tissues with induced embryoids to WPM containing a reduced cytokinin concentration (BA 0.88–2.66 μM) and culturing them in light, led to greening and further development of embryoids. Embryos formed either roots only, or both shoots and roots. The frequency of somatic embryo conversion into plantlets was low. The majority of the cultures produced somatic embryos which did not germinate. Desiccation improved germination and conversion of embryos into plantlets. Embryogenic tissues with induced embryoids, continuously cul-

*Figure 3.* Plantlet regenerated from somatic embryo of *Quercus robur*.

*Figure 4.* Plants of *Quercus robur* regenerated from somatic embryos, growing in a greenhouse.

tured on the same medium without transfer for 3–4 months, produced more plantlets after transfer to fresh WPM, when compared with tissues subcultured to a fresh medium every month. Similarly, embryogenic tissues with induced embryoids, cultured for 3 weeks on MS medium supplemented with cytokinin and sorbitol (6%) and then transferred to WPM containing a low concentration of BA (0.88 μM), produced more plantlets than tissues cultured on a medium lacking sorbitol. Dehydration of somatic embryos inside sterile sealed dishes for 2–3 weeks increased the frequency of somatic embryo conversion into plantlets (Fig. 3). More than 70 plantlets of *Q. robur*, regenerated from somatic embryos, were transplanted into potting mixture (peat and perlite, 1:1 v/v). Plantlets were grown under high air humidity and continuous light (100 $\mu E\ m^{-2}\ s^{-1}$) in growth cabinets for 2–3 months to achieve a fast stem elongation and formation of new leaves. High humidity was gradually reduced to normal values. Seventy eight percent of the plantlets survived and after hardening-off, they were transferred outdoors and later planted in the field (Fig. 4). The survival of oaks planted in the field after hardening-off was high (81–92%). Planted oaks withstood winter frosts (−10° to −15°C) without significant losses. The growth in height of trees produced from embryogenic cultures was comparable with the growth of trees produced by *in vitro* organ cultures. At the end of the third growing season, oaks regenerated by somatic embryogenesis were 35–50 cm high. Trees produced from somatic embryos exhibited normal growth and appearance.

The germination and conversion of *Quercus petraea* somatic embryos were stimulated by alternation of physical conditions and medium changes (Chalupa, 1987a, 1989, 1990b, 1993). The embryo conversion was stimulated by desiccation (dehydration of somatic embryos inside sterile sealed dishes for 2–3 weeks) and by exposure to cold (2° to 3°C for 3–4 weeks). After the treatments, somatic embryos were cultured on WPM containing low concentration of cytokinin (BA 0.44 μM) under continuous light (60 $\mu E\ m^{-2}\ s^{-1}$). Well-developed plantlets were transplanted into the potting mixture and were grown under high air humidity and continuous light. After acclimatization, 62 plants regenerated from somatic embryos were planted in the nursery. Jörgensen (1993) regenerated *Quercus petraea* plantlets from somatic embryos on a modified WPM without hormones.

Somatic embryoids of *Quercus rubra* exhibited root elongation on MS medium (3–24% embryoids) (Gingas and Lineberger, 1989). As no further development beyond root elongation occurred, embryoids were subjected to chilling. However, chilling did not improve epicotyl development in embryoids, and neither gibberellic acid treatments were effective in breaking the dormancy. Epicotyl dormancy of embryoids was overcome by desiccation (air drying or use of an osmoticum) and rehydration treatments. Plantlets with well-developed epicotyls and roots were produced after rehydration on half-strength basal medium. Increased size and opaqueness of the embryoids entering desiccation positively influenced survival. Three plantlets were successfully acclimated to non-sterile conditions. A high relative humidity was maintained around the above-soil plant parts for several weeks. Healthy and viable plants have been growing in the artificial medium (a mix of peat, perlite and shredded bark, 1:1:1 v/v) for 9 months under a 16 h photoperiod (30 μE $m^{-2}$ $s^{-1}$). Chalupa (1991, 1992) stimulated conversion of somatic embryos of *Q. rubra* by medium changes and alternation of physical conditions. The desiccation treatment (air drying in sealed sterile Petri dishes for 2 weeks) improved the conversion of somatic embryos. After the desiccation, somatic embryos were transferred on WPM containing 0.44 μM BA and were cultured under continuous light (40 μE $m^{-2}$ $s^{-1}$). Plantlets regenerated from somatic embryos were transplanted into potting mixture and grown under high air humidity and continuous light. After hardening off, the acclimated plants were planted in the nursery.

The conversion of *Quercus acutissima* somatic embryos into plantlets was stimulated on WPM containing a low concentration of BA (Sasaki *et al.*, 1988; Shoyama *et al.*, 1992). Mature somatic embryos developed shoots when transferred on this medium. Root formation occurred upon transfer of the shoots on half-strength WPM supplemented with 2.46 μM or 4.90 μM IBA. Plants obtained by this process had the normal chromosome number (2n = 24) and did not show any phenotypic abnormalities.

The conversion of *Quercus suber* somatic embryos into plantlets was promoted by cold treatment (Bueno *et al.*, 1992; Manzanera *et al.*, 1993). Air drying had little effect on embryo germination. In contrast, cold storage treatment (10 weeks at 5°C and 2 weeks at 2°C) stimulated radicle and epicotyl growth. Epicotyl dormancy was overcome by placing the somatic embryos that had developed only a radicle, on paper bridges, in test tubes containing Sommer *et al.* (1975) medium supplemented with 0.4 μM BA. Thirty plantlets with normally developed shoots were transferred to soil and acclimated in a mist tunnel at high relative humidity inside the greenhouse.

The germination and conversion of *Quercus bicolor* embryoids into plantlets was stimulated on a modified MS medium (Gingas, 1991). Polar somatic embryoids were separated and cultured on half-strength MS medium without growth regulators. About 20% of the embryoids germinated and developed roots and shoots. Neither a chilling (4°C for 30 days) nor a desiccation treatment enhanced germination. Two plantlets have been transplanted into a soilless potting medium. However, successful acclimation did not occur after 6 months.

## 5. Conclusions

The long reproductive cycle of oaks is a serious obstacle to effective tree improvement by conventional methods. The present situation calls for the development of new methods in oak propagation and breeding. Recently, significant progress was achieved in oak vegetative propagation by a stem-cutting system and by *in vitro* methods. Organogenesis and somatic embryogenesis are the most promising methods of oak micropropagation.

Somatic embryogenesis has been reported in various oak species. Embryogenic cultures were initiated from immature and mature zygotic embryos, male catkins, anthers, leaf fragments, and stem internode segments. Reports on somatic embryogenesis in *Quercus* indicate that oak tissue is highly embryogenic. Oak somatic embryos originate in embryogenic tissue through a series of developmental stages. Initiation of embryogenic tissue and formation of somatic embryos can occur in the absence of exogenous hormones, however, in most cases, the initiation and development of somatic embryos were stimulated by manipulation of hormonal balance in nutrient media. Repetitive somatic embryogenesis in oak cultures was frequent and led to the formation of many secondary embryoids. Embryogenic tissues have often retained their embryogenic potential for several years.

A serious problem of oak regeneration by somatic embryogenesis is the low frequency of somatic embryo conversion into plantlets. The development of somatic embryos is often blocked after the formation of cotyledons. It is expected that a high percentage of oak somatic embryos with a well structured and functional shoot apex and radicle meristem will be able to develop into plants. Desiccation treatment or exposure to cold improved conversion of somatic embryos into plantlets in some species. Plantlets of some species regenerated from somatic embryos were successfully transferred to soil and continued to grow. More research is needed to improve the process of maturation and conversion of somatic embryos into plants.

## 6. References


Bellarosa, R., 1981. *In vitro* culture of *Quercus suber* L. embryos. In: Colloque Int. Cult. *In Vitro* des Essen. Forest. AFOCEL, Nangis, pp. 119–126.

Bellarosa, R., 1989. Oak (*Quercus* spp.) In: Y.P.S. Bajaj (Ed.), Biotechnology in Agriculture and Forestry, Vol. 5, Tress II, pp. 387–401. Springer-Verlag, Berlin.

Bennett, L.K. and F.T. Davies, 1986. *In vitro* propagation of *Quercus shumardii* seedlings. HortSci. 21: 1045–1047.

Bueno, M.A., R. Astorga and J.A. Manzanera, 1992. Plant regeneration through somatic embryogenesis in *Quercus* suber. Physiol. Plant 85: 30–34.

Butorina, A.K., 1993. Cytogenetic study of diploid and spontaneous triploid oaks, *Quercus robur* L. Ann. Sci. For. 50, Suppl. 1: 144s–150s.

Chalupa, V., 1979. *In vitro* propagation of some broad-leaved forest trees. Commun. Inst. For. Cech. 11: 159–170.

Chalupa, V., 1981. Clonal propagation of broadleaved forest trees *in vitro*. Commun. Inst. For. Cech. 12: 255–271.

Chalupa, V., 1982. Vegetative propagation of broadleaved trees by cuttings. Lesnictví (Forestry) 28: 21–30.

Chalupa, V., 1983. Micropropagation of conifer and broadleaved forest trees. Commun. Inst. For. Cech 13: 7–39.

Chalupa, V., 1984. *In vitro* propagation of oak (*Quercus robur* L.) and linden (*Tilia cordata* Mill.). Biol. Plant 26: 374–377.

Chalupa, V., 1985a. Somatic embryogenesis and plantlet regeneration from cultured immature and mature embryos of *Picea abies* (L.) Karst. Commun. Inst. For. Cech. 14: 57–63.

Chalupa, V., 1985b. *In vitro* propagation of *Larix*, *Picea*, *Pinus*, *Quercus*, *Fagus*, and other species using adenine-type cytokinins and thidiazuron. Commun. Inst. For. Cech. 14: 65–90.

Chalupa, V., 1987a. Somatic embryogenesis and plant regeneration in *Picea*, *Quercus*, *Betula*, *Tilia*, *Robinia*, *Fagus* and *Aesculus*. Commun. Inst. For. Cech. 15: 133–148.

Chalupa, V., 1987b. European hardwoods. In: J.M. Bonga and D.J. Durzan (Eds.), Cell and Tissue Culture in Forestry, Vol. 3, pp. 224–246. Martinus Nijhoff Publishers, Dordrecht.

Chalupa, V., 1988. Large scale micropropagation of *Quercus robur* L. using adenine-type cytokinins and thidiazuron to stimulate shoot proliferation. Biol. Plant 30: 414–421.

Chalupa, V., 1989. Plant regeneration by somatic embryogenesis in Norway spruce (*Picea abies*/L./Karst.) and sessile oak (*Quercus petraea*/Matt./Liebl.). Commun. Inst. For. Cech. 16: 135–143.

Chalupa, V., 1990a. Plant regeneration by somatic embryogenesis from cultured immature embryos of oak (*Quercus robur* L.) and linden (*Tilia cordata* Mill.). Plant Cell Rep. 9: 398–401.

Chalupa, V., 1990b. Somatic embryogenesis and plant regeneration in *Quercus petraea* (Matt.) Liebl., *Tilia platyphyllos* Scop., and *Aesculus hippocastanum* L. Lesnictví (Forestry) 36: 599–604.

Chalupa, V., 1990c. Biotechnology in forest tree improvement: trees of the future. In: R. Rodrigues, R. Sánchez Tamés and D.J. Durzan (Eds.), Plant Aging, pp. 311–318. Plenum Press, New York.

Chalupa, V., 1990d. Vegetative propagation of oak (*Quercus robur* L.), beech (*Fagus sylvatica* L.) and linden (*Tilia cordata* Mill.) by means of cuttings and explant culture. Lesnictví (Forestry) 36: 589–598.

Chalupa, V., 1991. Somatic embryogenesis and plant regeneration in European silver fir (*Abies alba* Mill.) and red oak (*Quercus rubra* L.). Commun. Inst. For. Cech. 17: 51–58.

Chalupa, V., 1992. Somatic embryogenesis and plant regeneration in *Quercus robur* L. and *Quercus rubra* L. Lesnictví-Forestry 38: 475–481.

Chalupa, V., 1993. Vegetative propagation of oak (*Quercus robur* and *Q. petraea*) by cutting and tissue culture. Ann. Sci. For. 50, Suppl. 1: 295s–307s.

Civinová, B. and Z. Sladký, 1987. A study of the regeneration capacity of oak (*Quercus robur* L.). Scr. Fac. Sci. Nat. Univ. Purk. 17: 103–110.

Cornu, D., J. Garbaye and F. Le Tacon, 1975. Résultat d'un essai préliminaire sur le bouturage du chene et du hetre. Tech. For. 27: 139–140.

Cornu, D., S. Delran, J. Garbaye and F. Le Tacon, 1977. Recherche des meilleures conditions d'enracinement des boutures herbacées du chene rouvre (*Q. petraea* /M./ Liebl) et du hetre (*Fagus silvatica* L.). Ann. Sci. For. 34: 1–16.

Durzan, D.J. and P.K. Gupta, 1987. Somatic embryogenesis and polyembryogenesis in Douglas-fir cell suspension cultures. Plant Sci. 52: 229–235.

El Maataoui, M. and H. Espagnac, 1987. Néoformation de structures de type embryons somatique sur des cultures de tissue de chene liege (*Quercus suber* fL.). C.R. Acad. Sc. Paris 304, Sér. III: 83–88.

El Maataoui, M., H. Espagnac and N. Michaux-Ferriere, 1990. Histology of callogenesis and somatic embryogenesis induced in stem fragments of cork oak (*Quercus suber*) cultured *in vitro*. Ann. Bot. 66: 183–190.

Evers, P., E. Vermeer and S. van Eeden, 1993. Rejuvenation of *Quercus robur*. Ann. Sci. For. 50, Suppl. 1: 330s–335s.

Favre, J.M. and B. Juncker, 1987. In vitro growth of buds taken from seedlings and adult plant material in *Quercus robur* L. Plant Cell Tiss. Org. Cult. 8: 49–60.

Féraud-Keller, C. and H. Espagnac, 1989. Condition d'apparition d'une embryogénese somatique sur des cals issus de la culture de tissus foliaires du chene vert (*Quercus ilex*). Can. J. Bot. 67: 1066–1070.

Garbaye, J., B. Kazandjian and F. Le Tacon, 1977. Développement de boutures racinées de chene rouvre (*Quercus petraea*/M/Liebl.). Ann. Sci. For. 34: 245–260.
Gebhardt, K., U. Frühwacht-Wilms and H. Weisgerber, 1993. Micropropagation and restricted-growth storage of adult oak genotypes. Ann. Sci. For. 50, Suppl. 1: 323s–329s.
Gingas, V.M., 1991. Asexual embryogenesis and plant regeneration from male catkins of *Quercus*. HortSci. 26: 1217–1218.
Gingas, V.M. and R.D. Lineberger, 1989. Asexual embryogenesis and plant regeneration in *Quercus*. Plant Cell Tiss. Org. Cult. 17: 191–203.
Gupta, P.K. and D.J. Durzan, 1986. Plantlet regeneration via somatic embryogenesis from subcultured callus of mature embryos of *Picea abies* (Norway spruce). In Vitro 22: 685–688.
Hakman, I., L.C. Fowke, S. von Arnold and T. Eriksson, 1985. The development of somatic embryos in tissue cultures initiated from immature embryos of *Picea abies* (Norway spruce). Plant Sci. 38: 53–59.
Haraguchi, M., 1987. Induction of embryoids by cotyledon culture in *Quercus acutissima*. Bull. Kanto Branch Japan For. Soc. 39: 99–100.
Haraguchi, M., 1988. Plantlet regeneration from adventitious embryos induced by cotyledon culture in *Quercus acutissima*. J. Jap. For. Soc. 70: 411–416.
Ide, Y. and S. Yamamoto, 1986. *In vitro* plantlet regeneration from axillary buds of juvenile seedlings of kunugi (*Quercus acutissima*). J. Jap. For. Soc. 68: 472–474.
Ide, Y. and S. Yamamoto, 1987. *In vitro* plantlet regeneration from axillary buds of juvenile seedlings of konara (*Quercus serrata*). J. Jap. For. Soc. 69: 109–112.
Jensen, J.S., 1993. Variation of growth in Danish provenance trials with oak (*Quercus robur* L. and *Quercus petraea* Mattuschka Liebl). Ann. Sci. For. 50, Suppl. 1: 203s–207s.
Johnson, K.R. and R.F. Walker, 1990. Micropropagation of valley oak shoots. Tree Planters' Notes 41: 27–30.
Johnsson, H., 1946. Chromosome numbers of twin plants of *Quercus robur* and *Fagus silvatica*. Hereditas 32: 469–472.
Jörgensen, J., 1988. Embryogenesis in *Quercus petraea* and *Fagus sylvatica*. J. Plant Physiol. 132: 638–640.
Jörgensen, J., 1993. Embryogenesis in *Quercus petraea*. Ann. Sci. For. 50, Suppl. 1: 344s–350s.
Juncker, B. and J.M. Favre, 1989. Clonal effects in propagating oak trees via *in vitro* culture. Plant Cell Tiss. Org. Cult. 19: 267–276.
Kleinschmit, J., 1993. Intraspecific variation of growth and adaptive traits in European oak species. Ann. Sci. For. 50, Suppl. 1: 166s–185s.
Kleinschmit, J., R. Witte and A. Sauer, 1975. Möglichkeiten der züchterischen Verbesserung von Stiel-und Traubeneichen (*Quercus robur* and *Quercus petraea*). II. Versuche zur Stecklingsvermehrung von Eiche. Allg. Forst. u. Jagdzt. 146: 179–185.
Kremer, A. and R.J. Petit, 1993. Gene diversity in natural populations of oak species. Ann. Sci. For. 50, Suppl. 1: 186s–202s.
Kriebel, H.B., 1993. Intraspecific variation of growth and adaptive traits in North American oak species. Ann. Sci. For. 50, Suppl. 1: 153s–165s.
Lloyd, G. and B. McCown, 1980. Commercially-feasible micropropagation of mountain laurel, *Kalmia latifolia*, by use of shoot-tip culture. Comb. Proc. Int. Plant Prop. Soc. 30: 421–427.
Manzanera, J.A. and J.A. Pardos, 1990. Micropropagation of juvenile and adult *Quercus suber* L. Plant Cell Tiss. Org. Cult. 21: 1–8.
Manzanera, J.A., R. Astorga and M.A. Bueno, 1993. Somatic embryo induction and germination in *Quercus suber* L. Silvae Gen. 42: 90–93.
Meier-Dinkel, A., 1987. In vitro Vermehrung und Weiterkultur von Stieleiche (*Quercus robur* L.) und Traubeneiche (*Quercus petraea*/Matt./Liebl). Allg. Forst u. Jagdzt. 158: 199–204.
Meier-Dinkel, A., B. Becker and D. Duckstein, 1993. Micropropagation of several clones of late-flushing *Quercus robur* L. Ann. Sci. For. 50, Suppl. 1: 319s–322s.
Merkle, S.A. and H.E. Sommer, 1986. Somatic embryogenesis in tissue cultures of *Liriodendron tulipifera*. Can. J. For. Res. 16: 420–422.

Muhle Larsen, C., 1946. Experiments with softwood cuttings of forest trees. Forstl. Forsøgsv. Danmark 17: 289–443.

Murashige, T. and F. Skoog, 1962. A revised medium for rapid growth and bioassays with tobacco tissue cultures. Physiol. Plant 15: 473–497.

Nagmani, R. and J.M. Bonga, 1985. Embryogenesis in subcultured callus of *Larix decidua*. Can. J. For. Res. 15: 1088–1091.

Nepveu, G., 1993. The possible status of wood quality in oak breeding programs (*Quercus petraea* Liebl. and *Quercus robur* L.). Ann. Sci. For. 50, Suppl. 1: 388s–394s.

Ostrolucká, M.G. and A. Pretová, 1991. The occurrence of somatic embryogenesis in the species *Quercus cerris* L. Biologia 46: 9–14.

Pardos, J.A., 1981. *In vitro* plant formation from stem pieces of *Quercus suber* L. In: Coll. Int. Cult. *In Vitro* des Essen. Forest. AFOCEL, Nangis, pp. 186–190.

Pevalek-Kozlina, B. and S. Jelaska, 1986. *In vitro* growth and development of oaks (*Quercus robur* and *Q petraea*). Acta Bot. Croat. 45: 55–61.

Radojević, L., 1979. Somatic embryogenesis and plantlets from callus cultures of *Paulownia tomentosa* Stued. Z. Pflanzenphysiol. 91: 57–62.

Rushton, B.S., 1933. Natural hybridization within the genus *Quercus* L. Ann. Sci. For. 50, Suppl. 1: 73s–90s.

San-José, M.C., A. Ballester and A.M. Vieitez, 1988. Factors affecting *in vitro* propagation of *Quercus robur* L. Tree Physiol. 4: 281–290.

San-José, M.C., A.M. Vieitez and A. Ballester, 1990. Clonal propagation of juvenile and adult trees of sessile oak by tissue culture techniques. Silvae Gen. 39: 50–55.

Sasaki, Y., Y. Shoyama, I. Nishioka and T. Suzaki, 1988. Clonal propagation of *Quercus acutissima* Caruth. by somatic embryogenesis from embryonic axes. J. Fac. Agr., Kyushu Univ. 33: 95–101.

Sasamoto, H. and Y. Hosoi, 1989. Somatic embryogenesis in suspension cultures of *Quercus serrata* Thunb. J. Jap. For. Soc. 71: 20–22.

Sasamoto, H. and Y. Hosoi, 1992. Callus proliferation from the protoplasts of embryogenic cells of *Quercus serrata*. Plant Cell Tiss. Org. Cult. 29: 241–245.

Sato, T., N. Mori and A. Saito, 1987. In vitro plantlet propagation from epicotyl segments of young seedlings of kunugi (*Quercus acutissima*). J. Jap. For. Soc. 69: 113–117.

Savill, P.S. and P.J. Kanowski, 1993. Tree improvement programs for European oaks: goals and strategies. Ann. Sci. For. 50, Suppl. 1: 368s–383s.

Schenk, R.H. and A.C. Hildebrandt, 1972. Medium and technique for induction and growth of monocotyledonous and dicotyledonous plant cell culture. Can. J. Bot. 50: 199–204.

Schwarz, O.J. and S.E. Schlarbaum, 1993. Axillary bud proliferation of 2 North American oak species: *Quercus alba* and *Quercus rubra*. Ann. Sci. For. 50, Suppl. 1: 340s–343s.

Shoyama, Y., Y. Sasaki, I. Nishioka and T. Suzaki, 1992. Clonal propagation of oak (*Quercus acutissima* Carruth). In: Y.P.S. Bajaj (Ed.), Biotechnology in Agriculture and Forestry, Vol. 18, High-Tech and Micropropagation II, pp. 179–192. Springer-Verlag, Berlin.

Sommer, H.E. and C.L. Brown, 1980. Embryogenesis in tissue cultures of sweetgum. For. Sci. 26: 257–260.

Sommer, H.E., C.L. Brown and P.P. Kormanik, 1975. Differentiation of plantlets in longleaf pine (*Pinus palustris* Mill.) tissue cultured in vitro. Bot. Gaz. 136: 196–200.

Spethmann, W., 1982. Stecklingsvermehrung von Laubbaumarten. I. Versuche mit Ahorn, Esche, Eiche, Buche, Kirsche, Linde, Birke. Allg. Forst u Jagdzt. 153: 13–24.

Spethmann, W., 1986. Stecklingsvermehrung von Stiel – und Traubeneiche (*Quercus robur* L. und *Quercus petraea*/Matt./Liebl). Schriften Forstl. Fak. Univ. Göttingen 86: 1–99.

Spethmann, W. and P. Harms, 1993. Influence of fertilized substrate on rooting and growth of oak cuttings. Ann. Sci. For. 50, Suppl. 1: 308s–313s.

Srivastava, P.S. and A. Steinhauer, 1982. In vitro culture of embryo segments of *Quercus lebani*: Organogenesis and callus growth as a differential response to experimental conditions. Z. Pflanzenphysiol. 106: 93–96.

Thimann, K.V. and A.L. Delisle, 1939. The vegetative propagation of difficult plants. J. Arnold Arb. 20: 116–136.

Tulecke, W. and G. McGranahan, 1985. Somatic embryogenesis and plant regeneration from cotyledon tissue of walnut, *Juglans regia* L. Plant Sci 40: 53–67.

Ure, C.R., 1937. Preliminary tests with softwood cuttings of trees and shrubs in Alberta. Sci. Agr. 17: 720–726.

Vieitez, A.M., M.C. San-José and E. Vieitez, 1985. *In vitro* plantlet regeneration from juvenile and mature *Quercus robur* L. J. Hort. Sci. 60: 99–106.

Volkaert, H., J. Schoofs, A. Pieters and E. De Langhe, 1990. Influence of explant source on in vitro axillary shoot formation in oak seedlings. Tree Physiol. 6: 87–93.

# 6. Somatic embryogenesis in *Populus* spp.

Charles H. Michler

**Contents**



## 1. Introduction

Forest trees in the genus *Populus* are utilized for timber, pulp, and components of plywood and flakeboard manufacturing, and more recently, as whole tree fuelwood for bioenergy production. The more commercially important species are from two of five sections in the genus *Populus*, Leuce and Aigeiros. Poplars are widely dispersed throughout the northern hemisphere and occur along waterways and bottom lands, and in colder regions of temperate climactic zones including those regions in the USA, Belgium, Canada, UK, China, India and Japan (Ahuja, 1993). As a pioneer and coppicing species, poplars adapt to a wide variety of sites and the main limitation for growth is water availability. In general, members of the genus respond remarkably to cultural inputs such as weed control, irrigation, and fertilization which result in phenomenal growth rates for biomass production (Chun, 1993). Despite the ease of hybridization between many *Populus* species, traits for biotic and abiotic resistances and adequate adventitious rooting are still lacking in many important commercial clones (Michler, 1993; Ostry and Michler, 1993).

Poplar tree improvement requires the identification, isolation, description, and multiplication of useful genes or gene combinations that result in desirable growth and yield characteristics while maintaining resistance to important insect pests and pathogens. Biotechnology has become an important component of poplar tree improvement because classical breeding methods require long generation times for selection, breeding, and testing. Emerging technologies enable desirable genes to be introduced into important germplasm that results in the expression of useful traits in a short timeframe (Ostry and Michler, 1993). The ability to genetically modify important clones is limited by lack of adequate *in vitro* vegetation regeneration methods for all species and the lack of known resistance genes.

Shoot organogenesis is one method of *in vitro* vegetation propagation and

*S. Jain, P. Gupta & R. Newton (eds.), Somatic Embryogenesis in Woody Plants, Vol. 2*, 89–97.

has previously been reviewed in depth by others (Ahuja, 1993; Chun, 1993; Ernst, 1993). In general, explants can be successfully initiated *in vitro* from most plant tissues including shoot and root tips, stem nodes and internodes, axillary buds, flower primordia, anthers, leaf and petiole segments, and cambial tissue (Ernst, 1993). High rates of shoot regeneration are species- and genotype-dependent and shoot production can be maximized through hormonal and nutritional alterations using a wide variety of culture media formulations. For some species and interspecific hybrids, *in vitro* rooting remains a limitation to successful propagation and *ex vitro* transition to field tests.

The ability to regenerate *Populus* trees from somatic embryogenic cells will be the focus of the remaining discussion because of its utility for clonal propagation and genetic modification systems. Use of somatic embryogenesis may overcome difficulties with clonal propagation of difficult-to-root species, speed the introduction of improved clones into commercial production (Evers *et al.*, 1988), and provide a regeneration system amenable to genetic modification techniques for alteration of tree genomes.

Poplar somatic embryogenesis is not limited by physiological barriers that impede progress with some woody angiosperms (Parrott *et al.*, 1991). These limitations can include the inability to initiate embryogenic callus from non-embryonic tissues, low frequency of somatic embryo formation and germination, lack of control of aberrant morphology, and difficulty in plantlet acclimatization to *ex vitro* environments.

Poplar somatic embryogenesis has been achieved with *P. alba* × *P. grandidentata* (Michler and Bauer, 1987, 1991), *P. ciliata* (Cheema, 1989), *P. deltoides* (Michler, unpublished data), *P. euramericana* (Chung and Chun, 1991), and *P. nigra* × *P. maximowiczii* (Chung and Chun, 1991; Park and Son, 1988). In addition, Stoehr and Zsuffa (1990) have reported regeneration of embryoids from anther-derived embryogenic callus of *P. maximowiczii*. Furthermore, poplar somatic embryogenesis has been coupled with *in vitro* genetic selection to produce commercially important, genetically modified hybrid poplars with increased tolerance to the herbicides, glyphosate and sulfometuron methyl (Michler *et al.*, 1989).

## 2. Approaches

### 2.1. *Culture initiation and maintenance*

For generation of embryogenic callus, a variety of non-embryonic explants have been employed including punctured leaf sections collected from 20 year old trees (Park and Son, 1988), and leaf (Cheema, 1989; Michler and Bauer, 1991) and internode (Michler and Bauer, 1991) sections of *in vitro* leaf and shoot cultures that were derived from mature hardwood cuttings from trees up to 40 years of age. Thus, it appears that *Populus* tissue has the unique

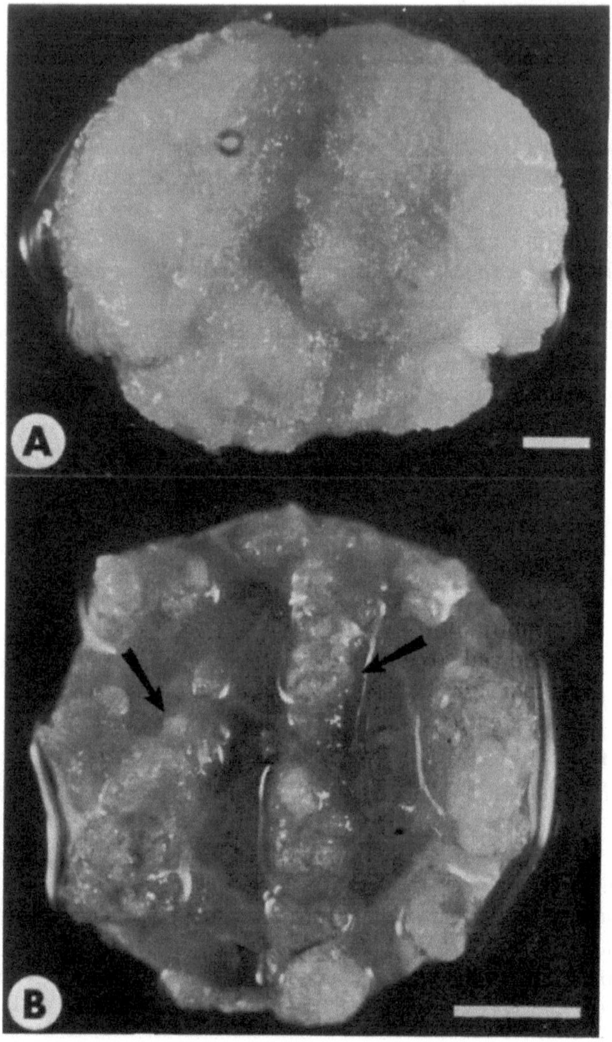


*Figure 1.* Green interveinal, non-embryogenic callus (A) and clear, nodular, embryogenic veinal callus (B) produced on *P. alba* × *P. grandidentata* "Crandon" leaf disks (Michler and Bauer, 1991). Scale bar = 0.25 mm.

ability, compared to other angiosperm woody species tested to date (Gronroos *et al.*, 1989; Radojević, 1989; Tulecke and McGranahan, 1985), to redifferentiate embryogenic cells from non-embryonic rejuvenated tissue. This will allow approaches that have been developed for *Populus* somatic embryogenesis to be used for mass propagation of proven elite species.

Somatic embryos have been regenerated directly on leaf veins (Michler and Bauer, 1991; Park and Son, 1988) and secondarily on hypocotyls and cotyledons of somatic embyos (Cheema, 1989). Leaf-derived cell suspensions (Cheema, 1989; Michler and Bauer, 1991), callus (Cheema, 1989), and liquid nodule culture (Chung and Chun, 1991) have been used for indirect somatic embryo initiation. In most reports on poplar somatic embryo formation, authors have described a similar non-chlorophyllous, densely cytoplasmic, nodular embryogenic callus type that arises in association with stem or leaf veinal tissue (Fig. 1B) with a possible origin from cells within vascular parenchyma. Embryogenic callus can be separated from green organogenic callus that arises from interveinal tissue (Fig. 1A). Failure to do so can result in shoot organogenesis and somatic embryogenesis occuring in close proximity (Park and Son, 1988). Furthermore, Michler and Bauer (1991) have described the importance of leaf explant maturity and found it necessary to use developing leaves prior to full expansion.

### 2.2. *Embryo development*

The use of MS (Murashige and Skoog, 1962) culture medium has been successful for all published approaches although the testing of other plant culture media has not been described. The synthetic auxin, 2,4-dichlorophenoxy acetic acid (2,4-D) was a media addendum used in all cases to initiate embryogenic callus. A 5:1 ratio of auxin (2,4-D) to cytokinin (6-benzylamino-purine BA) stimulated embryogenic callus initiation (Park and Son, 1988), and the same auxin to cytokinin ratio resulted in the greatest number of globular somatic embryos using BA and 2,4-D as plant growth regulators (Michler and Bauer, 1991). A 100:1 ratio of 2,4-D to zeatin was optimal for globular embryo production when different cytokinins were tested. In the latter tests, kinetin and thidiazuron were found to be the least effective growth regulators when used in combination with 2,4-D.

The addition of glutamine in the embryogenic initiation medium has been hypothesized to enhance differentiation of predetermined direct embryogenic cells (Kamada and Harada, 1979). In our tests, this was the case, but L-glutamine inhibited embryogenic callus formation. Exogenous glutamine provided at 40 $\mu$M resulted in the regeneration of 9.4 somatic embryos per leaf disk which was twice the production rate at the 20 $\mu$M treatment level. Inorganic nitrogen provided by MS medium may be sufficient in quantity to stimulate embryogenic callus formation and organic nitrogen sources may have a more specialized role in cellular differentiation.

As with most angiosperm species, sucrose is an important osmoticum and carbon source (Ammirato, 1983). Altering the culture media carbohydrate concentration and type can result in strong developmental control of somatic embryogenesis. Optimization can result in control of precocious germination and abnormal morphology (Ammirato and Steward, 1971), induction of competence for non-embryogenic cells (Kochba and Button, 1974), and promotion of secondary embryogenesis (Steward *et al.*, 1975). From our factorial studies, we found that 3% sucrose was optimal for promotion of early somatic embryo development through enhancement of regeneration frequency up to 66 somatic embryos per petri plate. Sucrose concentration of 2 and 4% resulted in a significant reduction in embryo production. Without optimization studies, other protocols have utilized similar sucrose levels in the culture medium.

Light quality and fluence rates have not been varied in previous tests. In general, embryogenic callus and cell suspensions were grown in darkness (Cheema, 1989; Michler and Bauer, 1991), and embryo maturation and germination seemed to have a requirement for light. Further study is warranted on optimization of the culture environment because others (Michler and Lineberger, 1987) have reported a critical role for photomorphogenic control of endogenous plant growth regulator balance and resulting effects on somatic embryo morphology.

The loss of embryogenic potential in rapidly dividing cell cultures is not fully understood. Loss of embryogenic ability has been attributed to the presence of 2,4-D in the medium (Halperin, 1966) and eventual loss of embryogenic diploid cells (Fujimura and Komamine, 1975). With poplar cell suspensions cultures, reports are conflicting. Embryogenic potential of cell lines was not maintained by cell suspensions after repeated subcultures by Cheema (1989), but was maintained with cultures grown on solid agar-based media. In our laboratory, with different genotypes and different plant growth regulator ratios in the culture medium, high frequency regeneration has been maintained for embryogenic cell suspension and callus cultures for over 2 years.

Michler and Bauer (1991) performed a comparative microscopic and anatomical study of the morphologies of zygotic and somatic embryos of a particular clone, *P. alba* × *P. grandidentata* "Crandon" (Fig. 2). All main stages of development were compared including globular, heart, torpedo, and mature embryos. The authors found that somatic embryos throughout their development closely resembled their zygotic counterparts in shape, but embryo size was significantly different. Globular stage somatic embryos were twice the diameter of zygotic embryos, and at the heart-shaped stage, embryos were twice the length measured from tip of the radicle end to cotyledon apex. Torpedo stage somatic embryos were 10% greater in length. Anatomically, the difference in size was attributed to cell enlargement rather than additional cell division. The size differences did not affect the ability to acclimatize germinated embryos to *ex vitro* environments.

### 2.3. *Embryo maturation*

Exogenous abscisic acid (ABA) is not found to play a necessary developmental role with embryo maturation and morphology in any published studies on poplar somatic embyrogenesis. This finding is not surprising since seed

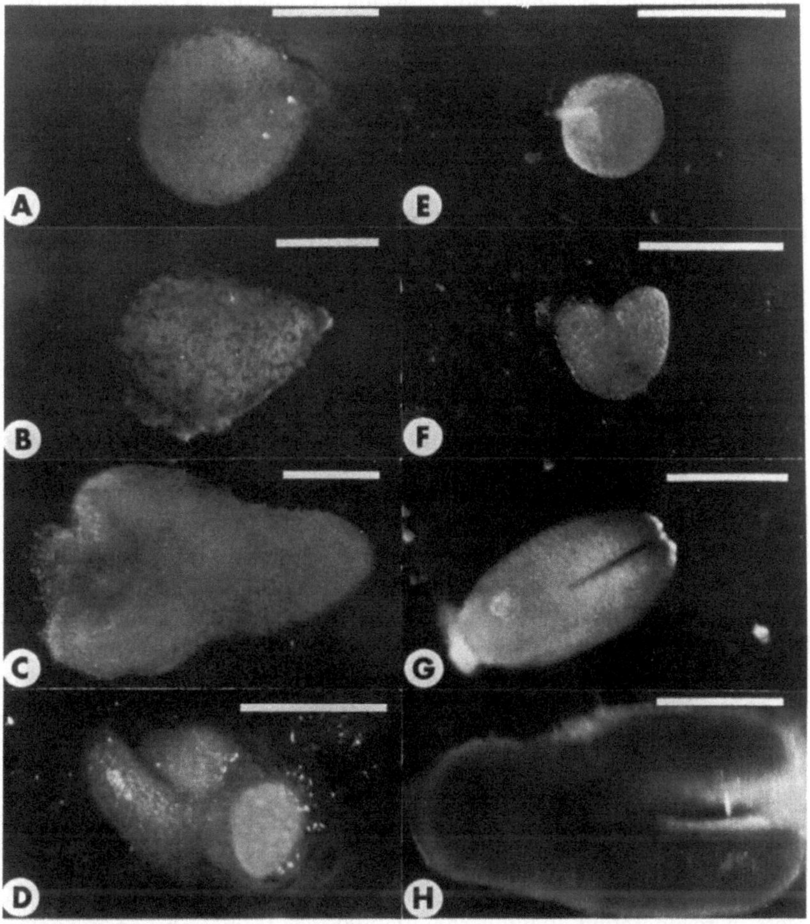


*Figure 2.* Developmental stages of *P. alba* × *P. grandidentata* "Crandon" somatic (A-D) and zygotic embryos (E-H) (Michler and Bauer, 1991). (A, E) globular-shaped, (B, F) heart-shaped, (G) torpedo-shaped, (C, H) mature, and (D) germinating embryo. Scale bar = 0.25 mm.

from many *Populus* species mature with high water content and germination occurs immediately following dispersal (P.R. Larsen, personal communication). In our studies, germination was not delayed by dormancy factors nor was aberrant morphology apparent if somatic embryos matured on solid medium. Teratomous and abnormal multiple embryo clumps were reported by Cheema (1989), but the cytokinin to auxin ratio for developmental culture media was considerably different from that of the other authors, and embryogenic callus was initiated from semi-organized leaf cultures that were >8 years old. Endogenous abscisic acid (ABA) levels are unknown for poplar somatic embryogenic systems, and knowledge of this would be an important aspect for development of artificial seed delivery systems where control of germination is critical to success.

### 2.4. *Embryo germination*

Poplar somatic embryos mature and germinate with minimal cultural manipulations. Cheema (1989) and Michler and Bauer (1991) reported the necessity for transfer of globular-shaped embryos from liquid to solid medium to promote somatic embryo maturation and germination. In addition, it was necessary to provide pulse treatments of either indole-3-acetic acid (IAA) or naphthaleneacetic acid (NAA) in the culture medium to stimulate radical elongation. Similar requirements for auxin exposure have been found for *Carya* (Wetzstein *et al.*, 1989) and *Robinia* (Merkle and Wiecko, 1989).

### 2.5. Ex vitro *fitness*

From our studies, greenhouse grown plantlets did not differ in growth rate and phenotype compared with germinated seedlings of the identical clone, *P. alba* × *P. grandidentata* "Crandon". Field fitness and productivity of various hybrid poplar clones derived from somatic embryos has been evaluated for 3 years, and preliminary data suggests a significant increase in branch number. Data is currently being analyzed from measurements of tree height, diameter, branch length, and disease incidence.

## 3. Conclusions

Success in regenerating *Populus* somatic embryos from embryogenic callus has depended upon modification of protocols developed for other woody angiosperms.

The significant gain in knowledge comes from approaches developed that allow mature trees to be propagated by these methods since embryonic tissues are not needed for source explants. Critical steps include proper rejuvenation of plant material used for explants, proper selection of an embryogenic callus type, determination of optimal hormonal balance in the embryogenic development media to maintain highly morphogenic cultures and insure production of aberrant-free embryos, and stimulation of radical elongation in arrested embryos by auxin treatment. To date, there is little basic knowledge of molecular and physiological controls of embryogenic callus initiation and somatic embryo development. Through further study to optimize culture protocols, poplar somatic embryogenesis could become an important model system to gain further understanding of embryogenesis.

## 4. Acknowledgements

The author is grateful for technical support in this research from Ms. Therese Voelker and Edmund Bauer. Financial support was received in part from the US Department of Energy, Interagency Agreement No. DE-A105-800R20763 and Drs. D. Skilling and M. Ostry, North Central Forest Experiment Station, St. Paul, MN.

## 5. References

Ahuja, M.R., 1993. Regeneration and germplasm preservation in aspen-*Populus*. In: M.R. Ahuja (Ed.), Micropropagation of Woody Plants, pp. 187–194. Kluwer Academic Pubishers, Dordrecht.

Ammirato, P.V., 1983. Embryogenesis. In: D.A. Evans, W.R. Sharp, P.V. Ammirato and Y. Yoshida (Eds.), Handbook of Plant Cell Culture, pp. 82–123. MacMillan, New York.

Ammirato, P.V. and F.C. Steward, 1971. Some effects of the environment on the development of embryos from cultured free cells. Bot. Gaz. 132: 149–158.

Cheema, G.S., 1989. Somatic embryogenesis and plant regeneration from cell suspension and tissue cultures of mature himalayan poplar (*Populus ciliata*). Plant Cell Rep. 8: 124–127.

Chun, Y.W., 1993. Clonal propagation in non-aspen poplar hybrids. In: M.R. Ahuja (Ed.), Micropropagation of Woody Plants, pp. 209–222. Kluwer Academic Publishers, Dordrecht.

Chung, K.H. and Y.W. Chun, 1991. Plant regeneration of hybrid poplar using nodule culture system. J. Korean For. Soc. 80: 1–8.

Ernst, S.G., 1993. *In vitro* culture of pure species non-aspen poplars. In: M.R. Ahuja (Ed.), Micropropagation of Woody Plants, pp. 195–207. Kluwer Academic Publishers, Dordrecht.

Evers, P.W., J. Donkers, A. Prat and E. Vermeer, 1988. Micropropagation of Forest Trees Through Tissue Culture. Pudoc, Wageningen.

Fujimura, T. and A. Komamine, 1975. Effects of various growth regulators on the embryogenesis in a carrot cell suspension culture. Plant Sci. Lett. 5: 359–364.

Gronroos, L., S. von Arnold and T. Eriksson, 1989. Callus production and somatic embryogenesis from floral explants of basket willow (*Salix viminalis* L.). J. Plant Physiol. 134: 558–566.

Halperin, W., 1966. Alternative morphogenetic events in cell suspensions. Amer. J. Bot. 53: 443–453.

Kamada, H. and H. Harada, 1979. Studies on the organogenesis in carrot tissue cultures. I. Effects of growth regulators on somatic embryogenesis and root formation. Z. Pflanzenphysiol. 91: 255–266.

Kochba, J. and J. Button, 1974. The stimulation of embryogenesis and embryoid development in habituated ovular callus from the "Shamouti" orange (*Citrus sinensis*) as affected by tissue age and sucrose concentration. Z. Pflanzenphysiol. 73: 415–421.

Merkle, S.A. and A.T. Wiecko, 1989. Regeneration of *Robinia pseudoacacia* via somatic embryogenesis. Can. J. For. Res. 19: 285–288.

Michler, C.H., 1993. *In vitro* genetic selection for woody plant improvement. In: M.R. Ahuja (Ed.), Micropropagation of Woody Plants, pp. 443–455. Kluwer Academic Publishers, Dordrecht.

Michler, C. and E. Bauer, 1987. Somatic embryogenesis in plant cell cultures of *Populus*. In Vitro Cell. Dev. Biol. 23(3) Part II: 46A.

Michler, C.H. and E.O. Bauer, 1991. High frequency somatic embryogenesis from leaf tissue of *Populus* spp. Plant Sci. 77: 111–118.

Michler, C.H. and R.D. Lineberger, 1987. Effects of light on somatic embryo development and abscisic levels in carrot suspension cultures. Plant Cell Tiss. Org. Cult. 11: 189–207.

Michler, C.H., T.M. Voelker and E.O. Bauer, 1989. Production of putative tolerant plants from somatic embryos regenerated from *in vitro* selected callus. In Vitro Cell. Dev. Biol. 25(3) Part II: 62A.

Murashige, T. and F. Skoog, 1962. A revised medium for rapid growth and bioassays with tobacco tissue cultures. Physiol. Plant 15: 473–497.

Ostry, M.E. and C.H. Michler, 1993. Use of biotechnology for tree improvement in *Populus* model systems. In: M.R. Ahuja (Ed.), Micropropagation of Woody Plants, pp. 471–483. Kluwer Academic Publishers, Dordrecht.

Park, Y.G. and S.H. Son, 1988. *In vitro* organogenesis and somatic embryogenesis from punctured leaf of *Populus nigra* × *P. maximowiczii*. Plant Cell Tiss. Org. Cult. 15: 95–105.

Parrott, W.A., S.A. Merkle and E.G. Williams, 1991. Somatic embryogenesis: potential for use in propagation and gene transfer systems. In: D.R. Murray (Ed.), Advanced Methods in Plant Breeding and Biotechnology, pp. 158–200. CAB International, Wallingford, Oxon.

Radojević, L., 1989. Plant regeneration of *Aesculus hippocastanum* L. (Horsechestnut) through somatic embryogenesis. J. Plant Physiol. 132: 322–326.

Steward, F.C., H.W. Israel, R.L. Mott, H.J. Wilson and A.D. Krikorian, 1975. Observations on growth and morphogenesis in cultured cells of carrot (*Daucus carota* L.). Phil. Trans. R. Soc. London, Sec. B 273: 33–53.

Stoehr, M.U. and L. Zsuffa, 1990. Induction of haploids in *Populus maximowiczii* via embryogenic callus. Plant Cell Tiss. Org. Cult. 23: 49–58.

Tulecke, W. and G. McGranahan, 1985. Somatic embryogenesis and plant regeneration from cotyledons of walnut, *Juglans regia* L. Plant Sci. 40: 57–63.

Wetzstein, H.Y., J.R. Ault and S.A. Merkle, 1989. Further characterization of somatic embryogenesis and plantlet regeneration in pecan (*Carya illinoensis*). Plant Sci. 64: 193–201.

# 7. Somatic embryogenesis in walnut (*Juglans regia*)

John E. Preece, Gale H. McGranahan, Lynn M. Long
and Charles A. Leslie

**Contents**



## 1. Introduction

There are about 20 *Juglans* (walnut) species distributed through Asia, Europe, North America, and South America (Liberty Hyde Bailey Hortorium 1976). Of these species, *J. regia* (English or Persian walnut, primarily grown for nut production) and *J. nigra* (Eastern black walnut, valued mainly for its wood) are the most important commercially (McGranahan and Catlin, 1987; McGranahan and Leslie, 1990). Within the United States, the annual farm gate value of English walnuts is about $ 150,000,000 (McGranahan, 1987). In 1987, the export value of Eastern black walnut logs from the United States was $ 13,174,000; this represented only 4.6% of the volume, but 14.9% of the total value of log exports across all species (McCurdy, 1989).

Traditional breeding and selection of walnuts is expensive and time consuming (McGranahan, 1987). Genetic improvement of Eastern black walnut is hampered because in the wild, trees are distributed irregularly, making comparisons among trees difficult (Beineke, 1989). Clonal testing of progeny and propagation of cultivars is primarily limited to grafted plants on seedling rootstocks (Beineke, 1989; McGranahan *et al.*, 1988a).

Because of the high value of both English walnut and Eastern black walnut, most of the focus on walnut tissue culture has been on these two species (Preece *et al.*, 1989). The primary method of adventitious regeneration of walnut has been via somatic embryogenesis. This method has not been used for commercial propagation because only juvenile explants have been responsive (Table 1) and possible somaclonal variation has not been adequately tested. Repetitive somatic embryogenesis has been used primarily for genetic manipulations of walnut.

*S. Jain, P. Gupta & R. Newton (eds.), Somatic Embryogenesis in Woody Plants, Vol. 2, 99–116.*

*Table 1.* Chronological summary within species of successful treatments from published studies that achieved somatic embryogenesis in walnut[1].

| Species | Nutrient salts | Plant growth regulators (μM) | | Percent embryogenesis | Embryo germination | Acclimatized plants in soil | Reference |
|---|---|---|---|---|---|---|---|
| | | Primary medium | Secondary medium | | | | |
| *Juglans hindsii* | DKW | 4.4 BA, 9.3 kinetin + 0.05 IBA | none | 10 | yes | yes | Tulecke and McGranahan, 1985 |
| *J. major* | DKW | 4.4 BA, 9.3 kinetin + 0.05 IBA | none | not reported | yes | not reported | Cornu, 1988 |
| *J. nigra* | DKW | 4.4 BA, 9.3 kinetin + 0.05 IBA | none | not reported | yes | not reported | Cornu, 1988 |
| *J. nigra* | WPM | 0.05–5.0 TDZ + 0.1–10.0 2,4-D | none | 0–75 | yes | no | Neuman *et al.*, 1993 |
| *J. regia* | DKW | 4.4 BA, 9.3 kinetin + 0.05 IBA | none | 17–44 | yes | yes | Tulecke and McGranahan, 1985 |
| *J. regia*[2] | DKW | 4.4 BA, 9.3 kinetin + 0.05 IBA | none | not reported | yes | yes | Tulecke *et al.*, 1988 |
| *J. regia* | DKW | 0.45 TDZ, 4.6 zeatin + 17.0 IAA, or 0.05 TDZ, 5.7 IAA + 10 $AgNO_3$ or 0.44 BA, 0.93 kinetin + 5.0 IBA | none or 0.05 TDZ, 5.7 IAA + 10 $AgNO_3$ | not reported | not attempted | not attempted | Aly *et al.*, 1992 |
| *J. nigra* hybrid | DKW | 4.4 BA, 9.3 kinetin + 0.05 IBA | none | not reported | yes | not reported | Cornu, 1989 |
| *J. nigra* × *J. regia* | DKW | 4.4 BA, 9.3 kinetin + 0.05 IBA | none | not reported | yes | not reported | Cornu, 1988 |
| *Pterocarya* sp. × *J. regia* | DKW and DKW with half $K_2SO_4$ | none | none | 9 | yes | yes | McGranahan *et al.*, 1986 |

[1] Cotyledons or zygotic embryos from immature seeds were used as the explant, unless otherwise noted.
[2] Endosperm from immature seeds used as the explant and triploid embryos were produced.

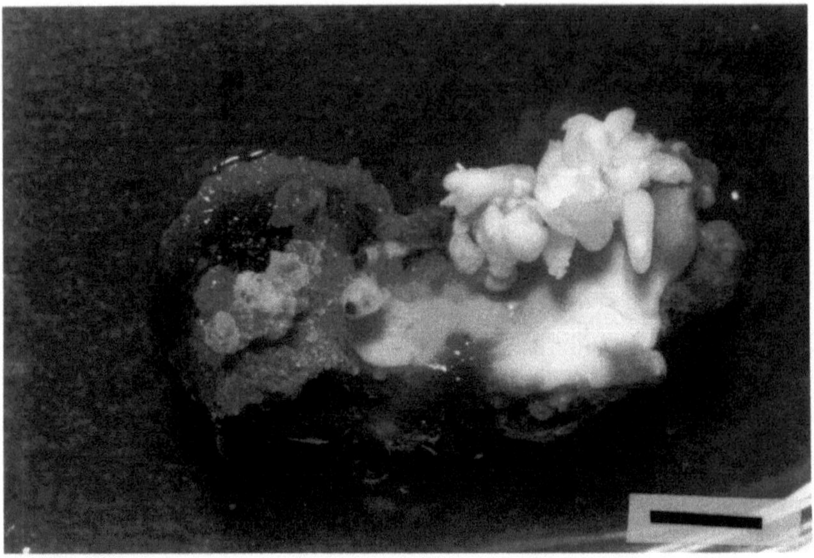

*Figure 1.* Somatic embryos forming from the surface of a *J. nigra* cotyledonary explant. Black scale bar = 5 mm.

Clonal micropropagation has been primarily limited to axillary shoot proliferation from juvenile explants (for reviews, see Leslie and McGranahan, 1992; Preece *et al.*, 1989). There are no reports of successful micropropagation of adult Eastern black walnut. Both reviews noted that rooting of microshoots is an area that requires further research to be applicable to more walnut genotypes. Because somatic embryos have radicles, rooting has been less problematic than with microshoots.

## 2. Primary somatic embryogenesis

We are defining primary somatic embryogenesis as the initiation of somatic embryos from an explant or from callus that forms on an explant (Fig. 1). The growth and maturation of these somatic embryos is also included in this section.

### 2.1. *Explants*

*Disinfestation.* Immature fruit, collected during the growing season, have been disinfested in 50–95% ethanol for 10 s to five min, followed by 0.1% NaClO for five min, 1% NaClO for 20 min, or 5.25% NaClO for 5–30 min (Aly *et al.*, 1992; Neuman *et al.*, 1993; Rodriguez *et al.*, 1989; Tulecke and McGranahan, 1985). Tulecke and McGranahan (1985) were the only researchers to report the use of a wetting agent with the bleach. Rodriquez *et al.* (1989) recommended a commercial sterilant in combination with the NaClO if disinfestation was a problem. Neuman *et al.* (1993) used a commercial sterilant prior to immersion in NaClO and a 0.01 N HCl rinse following the NaClO soak.

Immature fruit generally have been free from internal microorganisms. Immature seeds are covered by more intact fruit layers than mature seeds. Mature seeds abscise from the tree and fall to the ground. The combination of ground contact, age, and deterioration of fruit coverings may have contributed to greater contamination rates with explants from mature seeds. Neuman *et al.* (1993) found it necessary to soak cotyledon pieces from mature seeds in a mixture of four antibiotics for four hours prior to surface disinfestation. This was because of bacterial contamination of mature cotyledon explants.

Immature fruit have been opened under aseptic conditions to extract explants. When sufficiently immature, fruit have been cut in half with a sterile knife because the endocarp was soft and non-lignified. At this time the endosperm was liquid or gelatinous and the cotyledons were not visible or were beginning to fill the seed cavity. Gradually, the endocarp hardened making cutting with a knife difficult. Fruit with hardened endocarps have been cracked under aseptic conditions with a vise.

*Timing.* There has been a window of time between anthesis and fruit maturity in which somatic embryogenesis was possible with walnut. Tulecke and McGranahan (1985) reported that the optimum time for obtaining somatic embryos from walnut was 6–11 weeks after pollination. Cornu (1989) obtained somatic embryos from seeds collected from black walnut hybrids during August (when seeds had gelatinous endosperm with expanding cotyledons) and early September (when the seeds had filled and there was no visible endosperm). Neuman *et al.* (1993) were able to expand the window of time when somatic embryos could be induced from Eastern black walnut by using increasing concentrations of thidiazuron (TDZ). With a concentration as high as 5.0 $\mu$M TDZ, explants from fruit collected 7–17 weeks after anthesis were embryogenic. However, with seeds collected 16 weeks after anthesis, adventitious shoots formed on cotyledon explants from Eastern black walnut cultured on medium containing TDZ and 2,4-dichlorophenoxyacetic acid (2,4-D) (Long *et al.*, 1992).

*Types.* Most researchers have reported somatic embryogenesis from immature walnut cotyledons (Cornu, 1988, 1989; Deng and Cornu, 1992; McGranahan *et al.*, 1987; Neuman *et al.*, 1993; Tulecke and McGranahan, 1985). However, some have used intact zygotic embryos to produce somatic embryos (Aly *et al.*, 1992; Cornu, 1988; McGranahan *et al.*, 1986). Walnuts also have been regenerated from endosperm tissue. Cheema and Mehra

(1982) reported the formation of callus, roots, and shoot structures from gelatinous endosperm. The callus and roots were 3n = 48. Tulecke *et al.* (1988) regenerated triploid *J. regia* from gelatinous endosperm via somatic embryogenesis. Because triploid plants are probably sterile, it is likely that these plants cannot be used as pollinizers or for production of nuts. Triploid *J. regia* may be useful as rootstocks or for their wood.

Maternal tissue has not been shown to be embryogenic. Aly *et al.* (1992) analyzed somatic embryos that appeared to form from maternal tissues attached to fertilized ovules. Using restriction fragment length polymorphism (RFLP) and isozyme analyses, they showed that these somatic embryos were of zygotic and not maternal origin. It would be a major breakthrough for walnut biotechnology if adult tissue could be regenerated via somatic embryogenesis. This would allow researchers to manipulate genotypes with known phenotypic performance rather than untested juvenile plants.

### 2.2. *Media*

*Salts*. Most researchers have used DKW medium (Driver and Kuniyuki, 1984) for walnut somatic embryogenesis (see McGranahan *et al.*, 1987; Tulecke and McGranahan, 1985; Tulecke *et al.*, 1988, for corrected recipe of DKW) (Table 2). Neuman *et al.* (1993) cultured *J. nigra* cotyledon explants on both DKW and woody plant medium (WPM; Lloyd and McCown, 1980). WPM had been shown to be successful for somatic embryogenesis from pecan, another member of the Juglandaceae (Merkle *et al.*, 1987). Neuman *et al.* (1993) obtained only adventitious roots and no somatic embryos from cotyledons on DKW, but did obtain somatic embryos from explants on WPM. Likewise, Long *et al.* (1992) found that only 38% of *J. nigra* cotyledon explants were embryogenic when cultured on DKW, and produced a mean of only 1 somatic embryo per explant while 78% of explants cultured on WPM produced a mean of 22 somatic embryos per explant.

It may be that DKW is a better salt formulation for most walnut species, including *J. hindsii*, *J. regia*, and interspecific hybrids and WPM is better for *J. nigra*. However, a thorough study comparing response of walnut genotypes on the two media with the same plant growth regulator combinations has not been reported. Such a study could elucidate medium and walnut genotype combinations that would maximize somatic embryogenesis from explants.

*Plant growth regulators*. Most studies with walnut have repeated the combination of 4.4 μM benzyladenine (BA), 9.3 μM kinetin, and 0.05 μM indolebutyric acid (IBA) first reported by Tulecke and McGranahan (1985) (Tables 1 and 2). Tulecke and McGranahan (1985) placed cotyledon explants on DKW containing this combination of BA, kinetin, and IBA for 2–4 weeks, then transferred explants to basal medium lacking plant growth regulators. Because Neuman *et al.* (1993) obtained no embryogenesis from *J. nigra* with this approach, they developed a procedure, in part following the pecan

*Table 2.* Components of media used by Tulecke and McGranahan (1985), Neuman *et al.* (1993) and Long and Preece (LP) for somatic embryogenesis of walnut.

| | Concentration (mg/l) | | |
|---|---|---|---|
| | DKW[1] | WPM[2] | LP |
| Nitrates | | | |
| $NH_4NO_3$ | 1416.0 | 400.0 | 908.0 |
| $Ca(NO_3)_2 \cdot 4HOH$ | 1968.0 | 556.0 | 1262.0 |
| $Zn(NO_3)_2 \cdot 6HOH$ | 17.0 | – | 8.5 |
| Sulfates | | | |
| $K_2SO_4$ | 1559.0 | 990.0 | 1274.5 |
| $MgSO_4 \cdot 7HOH$ | 740.0 | 370.0 | 555.0 |
| $MnSO_4 \cdot HOH$ | 33.5 | 22.3 | 27.9 |
| $ZnSO_4 \cdot 7HOH$ | – | 8.6 | 4.3 |
| $CuSO_4 \cdot 5HOH$ | 0.25 | 0.25 | 0.25 |
| Phosphates | | | |
| $KH_2PO_4$ | 265.0 | 170.0 | 217.5 |
| $H_3BO_3$ | 4.8 | 6.2 | 5.5 |
| $NaMoO_4 \cdot 2HOH$ | 0.39 | 0.25 | 0.32 |
| Calcium | | | |
| $CaCl_2 \cdot 2HOH$ | 149.0 | 96.0 | 122.5 |
| Iron | | | |
| $Na_2$ EDTA | 45.4 | 37.3 | 41.35 |
| $FeSO_4 \cdot 7HOH$ | 33.8 | 27.8 | 30.8 |
| Organics | | | |
| Thiamine · HCl | 2.0 | 1.0 | 1.5 |
| Nicotinic Acid | 1.0 | 0.5 | 0.75 |
| Glycine | 2.0 | 2.0 | 2.0 |
| Pyridoxine · HCl | – | 0.5 | 0.25 |
| MyoInositol | 100.0 | 100.0 | 100.0 |
| Sucrose | 30000.0 | 30000.0 | 30000.0 |
| Gelrite | 2000.0 | – | 2000.0 |
| Agar | – | 7000.0 | – |
| L-glutamine | 250.0 | – | – |
| Casein hydrolysate | – | 1000.0 | 500.0 |
| Growth Regulators | | | |
| BA | 4.4 μM | – | – |
| IBA | 0.05 μM | – | – |
| Kinetin | 9.3 μM | – | – |
| 2,4-D | – | 0.1 μM | – |
| TDZ | – | 5.0 μM | – |

[1] Driver and Kuniyuki, 1984.
[2] Lloyd and McCown, 1980.

protocol of Merkle *et al.* (1987), using WPM supplemented with 1–10 μM 2,4-dichlorophenoxyacetic acid (2,4-D) plus 0.05–5.0 μM TDZ (Table 2) and obtained somatic embryos from up to 78% of the explants. Aly *et al.* (1992) obtained somatic embryogenesis from *J. regia* explants using DKW supplemented with 4.6 μM zeatin, 0.45 μM TDZ, and 17.0 μM indoleacetic acid (IAA); 0.05 μM TDZ and 5.7 μM IAA; or 0.44 μM BA, 0.93 μM kinetin and 5.0 μM IBA. Therefore, there is a variety of plant growth regulator combinations in the primary medium that will stimulate (or possibly not inhibit) somatic embryogenesis in walnut. In a direct comparison between DKW with 4.4 μM BA, 9.3 μM kinetin, and 0.05 μM IBA and WPM with 5.0 μM TDZ and 0.1 μM 2,4-D, Long *et al.* (1992) reported 1 somatic embryo per *J. nigra* explant and 38% embryogenesis with DKW, BA, kinetin, plus IBA compared to 22 somatic embryos per explant and 78% embryogenesis with WPM, TDZ, and 2,4-D.

Aly *et al.* (1992) used 10 μM silver nitrate and 500 mg/liter cefotaxime and obtained somatic embryogenesis from *J. regia*. However, using a different combination of plant growth regulators, they obtained somatic embryogenesis without either of these additives. The value of using these additives was not discussed by the authors. Silver nitrate inhibits ethylene binding and the antibiotic cefotaxime stimulates walnut somatic embryo production.

Not only may there be different plant growth regulator requirements among the different walnut genotypes, some may require no plant growth regulators to form somatic embryos. McGranahan *et al.* (1986) obtained somatic embryos from fertilized ovules of the intergeneric hybrid *Pterocarya* sp. × *J. regia* placed on DKW without plant growth regulators. Similar results have been obtained using genetically pure *J. regia* (McGranahan and Leslie, unpublished observation).

*Organic nitrogen*. Tulecke and McGranahan (1985) used 250 mg/liter L-glutamine in their conditioning medium and obtained somatic embryogenesis from *J. hindsii* and *J. regia*. Most studies have had success following this protocol and have used this same amino acid. Because Merkle *et al.* (1987) had better success with pecan using casein hydrolysate rather than with glutamine, Neuman *et al.* (1993) substituted casein hydrolysate for L-glutamine and successfully obtained somatic embryogenesis of *J. nigra*. Because L-glutamine and casein hydrolysate were not systematically compared for any walnut species, it is not clear whether one is superior, or even if either is necessary. Using fertilized ovules from the intergeneric hybrid *Pterocarya* sp. × *J. regia*, McGranahan *et al.* (1986) obtained somatic embryogenesis on basal medium without either supplemental form of reduced nitrogen. Similar results have been obtained using genetically pure *J. regia* (McGranahan and Leslie, unpublished observation).

### 2.3. *Secondary media*

Secondary media have been used to allow somatic embryos to develop. Typically, after 2–4 weeks on primary medium walnut explants have been transferred to secondary medium containing the same salts, but lacking plant growth regulators and the supplemental organic nitrogen.

Most researchers have used DKW with no plant growth regulators for the secondary medium (Table 1). Neuman *et al.* (1993), used WPM with no plant growth regulators for *J. nigra*. Tulecke *et al.* (1988) obtained somatic embryogenesis by culturing gelatinous endosperm, first on DKW with 4.4 μM BA, 9.3 μM kinetin, and 0.05 μM IBA for 2–3 weeks; then on basal DKW with 6% sucrose for two weeks; then on 2.2 μM BA and 0.5 μM IBA and 10% sucrose for three weeks; and then on basal DKW with 3% sucrose.

### 2.4. *Transfers*

Walnut tissue cultures have produced a toxic exudate *in vitro* (Compton and Preece, 1988). Cultures of walnut species have declined (growth slowed and tissues became brown or black) if not transferred to fresh medium at regular intervals. It is our observation that the most healthy walnut tissue cultures were those that were transferred to fresh medium weekly. However, most reports have indicated transfer intervals of every 2–3 weeks. Waiting three weeks or longer between transfers has been less successful than with more frequent transfers. Once walnut explants have begun to decline it has been difficult or impossible to rescue them and obtain viable cultures again. If there is one critical and special procedure for success with walnut tissue cultures, it is the absolute necessity for frequent transfers to new medium. If this is not anticipated and planned for, it is not worth the time and effort to place walnut species *in vitro*.

### 2.5. *Origin of primary somatic embryos*

In their initial paper on walnut somatic embryogenesis, Tulecke and McGranahan (1985) reported that all somatic embryos that they observed on *J. hindsii* and *J. regia* formed directly from the explant without an intervening callus (direct embryogenesis). However, using a black walnut hybrid, Cornu (1989) reported that clusters of somatic embryos formed from dense calluses (indirect embryogenesis). Neuman *et al.* (1993) reported that *J. nigra* somatic embryos formed both directly from the original cotyledon explant and indirectly via callus. There were either differences among walnut genotypes regarding direct and indirect somatic embryogenesis, or similar pathways were followed and have not been sufficiently characterized histologically.

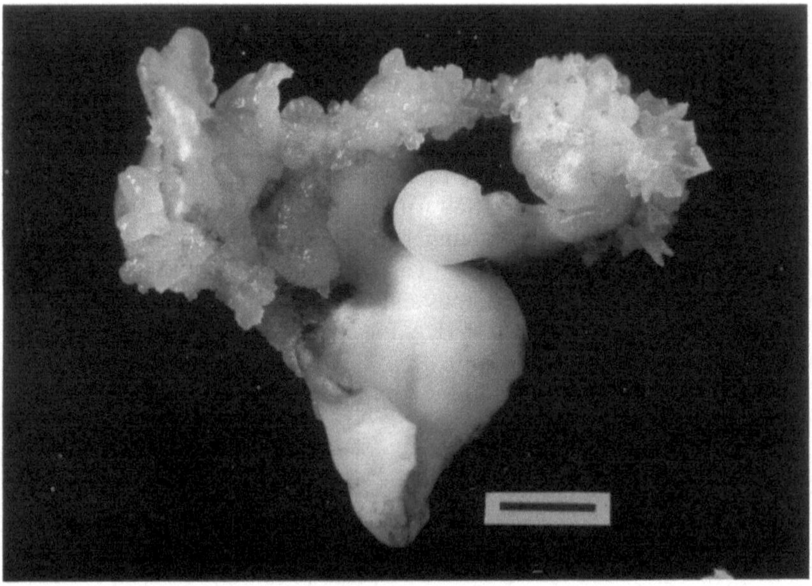

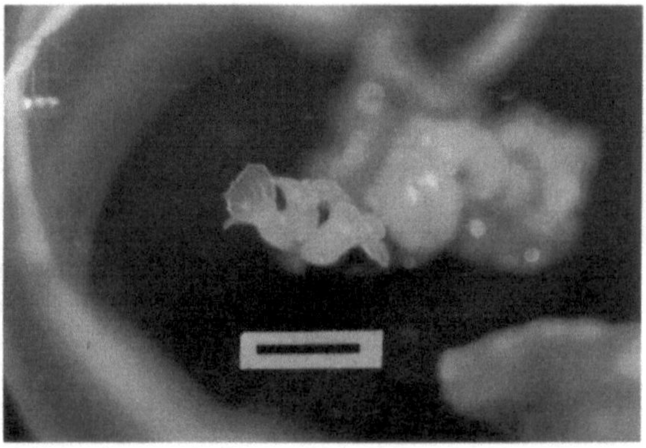

*Figure 2.* (A) Secondary somatic embryos forming from a *J. nigra* × *J. hindsii* somatic embryo. (B) Close up of secondary somatic embryos of the same genotype. Black scale bar = 5 mm.

## 3. Secondary (repetitive) somatic embryogenesis

Secondary somatic embryos are those that form from other somatic embryos (repetitive embryogenesis, Fig. 2). If transfers are made with sufficient regularity, secondary somatic embryogenesis should continue indefinitely and is now routine with walnut.

*Figure 3.* (A) Normal, high quality secondary somatic embryos from *J. regia*. (B) Well-formed secondary somatic embryos of *J. regia* from the surface of an older somatic embryo that had turned black. Black scale bar = 5 mm.

## 3.1. *Explants*

The best explants for secondary somatic embryogenesis have been 3–5 mm long somatic embryos. Well-formed somatic embryos, such as those pictured in Figs. 2B and 3A have been separated from a mass of somatic embryos with the tips of forceps. This has been done in the Petri dish or other vessel in which the somatic embryos were growing. Intact somatic embryos have been placed individually upon the surface of a basal medium where they have expanded and given rise to other somatic embryos (Fig. 2A). Generally, within one week the new secondary walnut embryos have become visible.

It has been possible to transfer pieces of cotyledons from somatic embryos, masses of incompletely-formed somatic embryos, or larger somatic embryos with small somatic embryos on the surface to fresh medium and achieve additional somatic embryogenesis. Older and larger somatic embryos generally have produced dark exudates or have become black, but often they have produced large numbers of somatic embryos for a period of weeks (Fig. 3B).

The rate of production of secondary somatic embryos has varied considerably with walnut species, embryo clone within species, or even individual embryos within an established clone. This may determine the availability of suitable, well-formed somatic embryos for further subculturing or other manipulations.

It has been our experience that rapidly-growing clustered masses including apparently globular structures (Fig. 4A) and embryos exhibiting a "cabbage-type" growth, characterized by many fused cotyledons emerging (Fig. 4B), have not been suitable to use to multiply secondary somatic embryos. Although such lines may be plentiful and multiply rapidly, it generally has been difficult or impossible to stimulate them to grow into well formed somatic embryos that are capable of germination. This barrier could be related to genotype of a specific cell line, or to the medium that is used (see Salts and plant growth regulators section below). Another phenotype that should be avoided is the solid, smooth, opaque white type because they have produced very few secondary somatic embryos (Fig. 5). We have been able to select for or against this type of growth.

## 3.2. *Media*

*Salts and plant growth regulators.* Most studies reporting secondary embryogenesis from walnut somatic embryos have used DKW with no plant growth regulators as the medium (Aly *et al.*, 1992; Deng and Cornu, 1992; McGranahan *et al.*, 1986, 1987, 1988, 1990; Polito *et al.*, 1989; Tulecke and McGranahan, 1985; Tulecke *et al.*, 1988). We had different results with *J. nigra*. After three months on DKW with no plant growth regulators, most somatic embryos failed to develop beyond the globular stage (Fig. 6). All cultures were then placed on WPM with 1 g/l casein hydrolysate based on use of this medium as a secondary medium for *J. nigra* (Neuman *et al.*, 1993). Somatic embryos developed beyond the globular stage with this medium, but after 3 additional months new embryos were mainly globular. Rotating every 2–3 months between DKW without plant growth regulators or supplemental organic nitrogen and WPM as described above (Table 2) allowed some

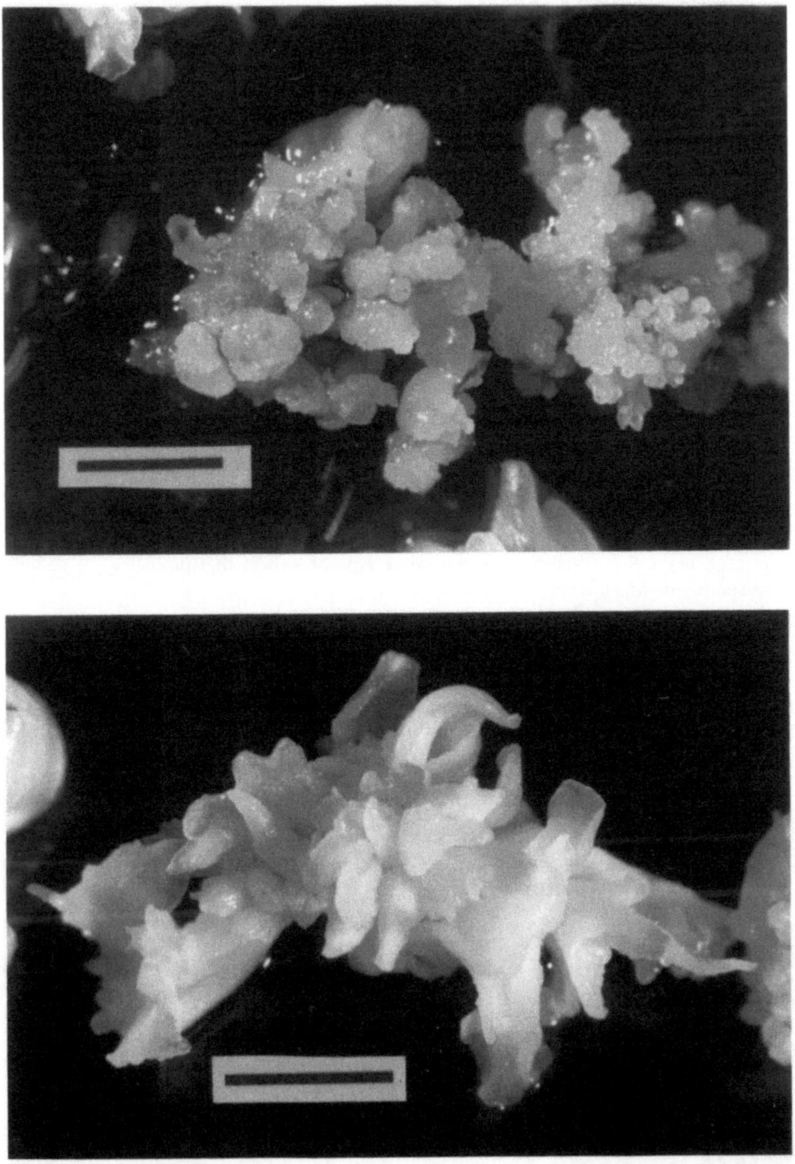

*Figure 4. J. regia.* (A) Rounded, globular type that grows and multiplies rapidly, but tends to form callus. (B) "Cabbage-type" growth with fused cotyledons emerging. Although this multiplies rapidly, normal embryos that are capable of germination are seldom obtained. Black scale bar = 5 mm.

*Figure 5. J. regia.* Solid, smooth, opaque white type of growth that produces few secondary somatic embryos. Black scale bar = 5 mm.

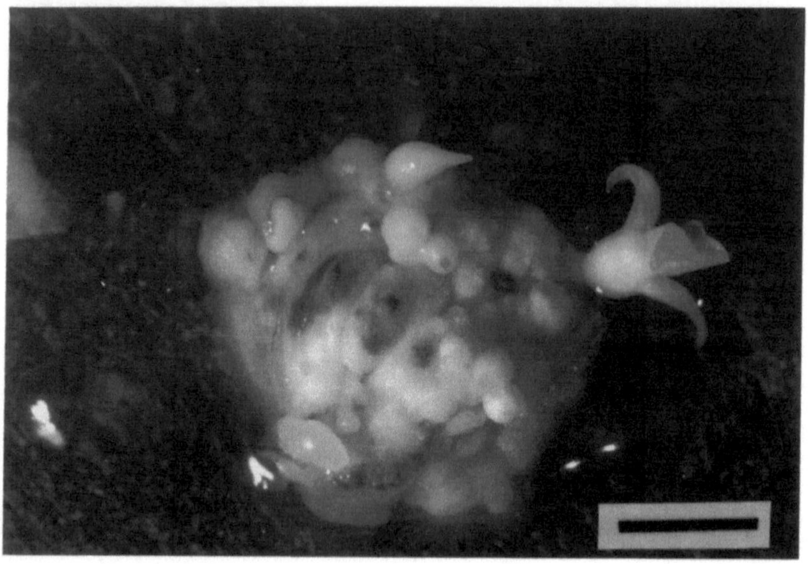

*Figure 6.* One well-formed secondary somatic embryo and several arrested secondary globular embryos on *J. nigra* cultured on DKW. These arrested secondary globular embryos grew when placed onto LP medium. Black scale bar = 5 mm.

globular embryos to develop further. However, toward the end of a cycle on one medium the somatic embryos would mainly be globular.

Therefore, we developed a new medium that is a combination of half WPM and half DKW. This medium has been named Long and Preece (LP) medium (Table 2). Within 3 months after placing the explants on LP medium, many more embryos were produced that matured beyond the cotyledonary stage than with any previous treatment. *J. nigra* repetitive somatic embryos have been cultured on this medium for nearly one year and have continued to develop normally.

### 3.3. *Origin of secondary somatic embryos*

The most complete study on origin of secondary somatic embryos was with *J. regia* (Polito *et al.*, 1989). Somatic embryos formed from surface cells on the cotyledons and hypocotyl without an intervening callus. Secondary somatic embryos were of single cell origin. This has direct implications for genetic transformation (see below).

## 4. Genetic transformation

Individual, well-formed somatic embryos have been inoculated with *Agrobacterium tumefaciens* to produce genetically transformed secondary embryos (Table 3; Dandekar *et al.*, 1989; McGranahan *et al.*, 1988b, 1990). Young and intermediate aged somatic embryos were superior for inoculation compared to older embryos (McGranahan *et al.*, 1988b). It was not necessary to wound walnut somatic embryos to transform them (McGranahan *et al.*, 1988b). Inoculation of somatic embryos for 5–10 minutes in *Agrobacterium* suspension followed by co-cultivation with the bacteria for 48 h on DKW with 100 μM acetosyringone (added to activate the virulence genes of the bacterial vector) has been sufficient for genetic transformation of walnut (Dandekar *et al.*, 1989).

Following inoculation and co-cultivation, somatic embryos have been placed on basal DKW with antibiotics to inhibit the growth of the *Agrobacterium* (Dandekar *et al.*, 1989). On this medium, additional secondary somatic embryos have been produced. The single epidermal cell origin has made walnut somatic embryos ideal for genetic transformation. Some members of the first generation of secondary somatic embryos following inoculation may have been chimeras because they may have been past the single cell stage when genetic transformation occurred. Therefore, subsequent generations of somatic embryos have been tested to confirm transformation or germinated to produce transgenic plants. The materials and protocol for transformation of walnut somatic embryos with *Agrobacterium* have been outlined by Dandekar *et al.* (1989).

*Table 3.* Summary of published papers on genetic transformation of *Juglans* spp. by inoculating with *Agrobacterium tumefaciens*.

| Species | Bacterial strains | Tissue inoculated | Tumor (callus) formation | Genes inserted | Transgenic embryo germination | Acclimatized transgenic plants in soil | Reference |
|---|---|---|---|---|---|---|---|
| *Juglans hindsii*, *J. regia*, and *J. hindsiix J. regia* | pTiB6S3, pMON200, pTiA6, K12 × 562E, K12 × 167, and K12 | *in vitro* microshoots | yes | opine APH(3′)II hormone | NA | NA | Dandekar *et al.*, 1988 |
| *J. regia* | K12 × 562E and K12 × 167 | somatic embryos | yes | opine, hormone, APH(3′)II, | yes | yes | McGranahan *et al.*, 1988b |
| *J. regia* | EHA101, EHA101/pCGN7001 EHA101/pCGN7314 | somatic embryos | no | APH(3′)II, GUS | yes | no | McGranahan *et al.*, 1990 |
| Hybrid walnut | not reported | somatic embryos | no | chalcone synthase antisense, NPTII, GUS | yes | no | Jay-Allemand *et al.*, 1991 |

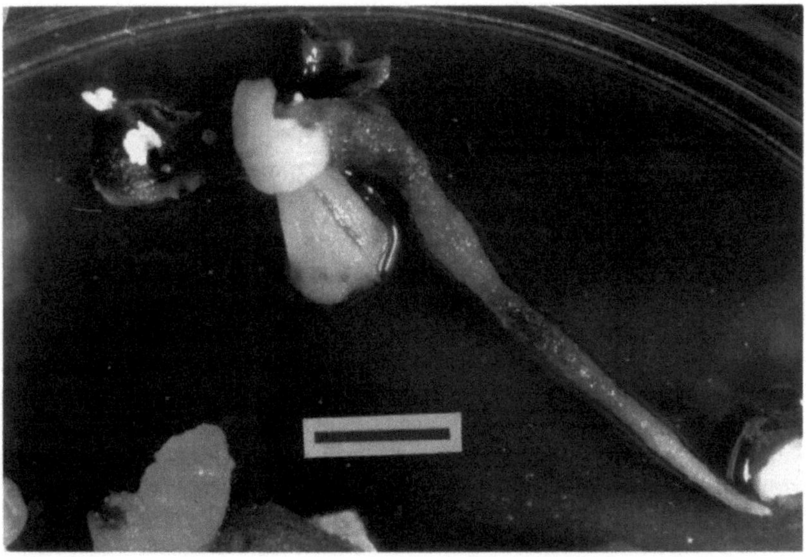

*Figure 7.* Germinated *J. regia* somatic embryo with elongated radicle, but without a well-developed epicotyl. Black scale bar = 5 mm.

## 5. Germination

Tulecke and McGranahan (1985) reported that radicles would elongate on *J. hindsii* and *J. regia* somatic embryos when incubated in light, but epicotyl development generally was lacking (Fig. 7). They found that if somatic embryos were given a 8–10 week cold (2–4°C) treatment, epicotyls elongated normally. Two to four weeks in the cold resulted in plants with short internodes and generally weak growth.

Deng and Cornu (1992) improved germination of *J. nigra* × *J. regia* somatic embryos from 0–5% (non-chilled) to 10% when large, developed embryos were chilled for two months then placed in light. Germination was improved to 45–46% when non-chilled somatic embryos were desiccated for three or five days, then placed onto cotton plugs with liquid DKW. Large embryos (5 × 10 mm, length × width) germinated better than small embryos (2 × 1 mm, length × width). This is the best reported result with germination of walnut somatic embryos.

When walnut somatic embryos have germinated, they have been transplanted to soil, or they have been clonally micropropagated via axillary shoot proliferation (Dandekar *et al.*, 1989; McGranahan *et al.*, 1988). When epicotyls were sufficiently long (up to 1 cm), they have been excised and placed onto axillary shoot proliferation medium. The proliferation medium for walnut was DKW with 4.4 μM BA and 0.05 μM IBA (McGranahan *et al.*, 1987).

Epicotyls from germinated somatic embryos have been clonally micropropagated if the somatic embryos were desirable genotypes and more than one plant was needed for testing. The primary use for this has been with genetically transformed walnut (Dandekar *et al.*, 1989; McGranahan *et al.*, 1988).

## 6. Acclimatization

About one month after germination, somatic embryos were transplanted into preformed peat plugs soaked in half-strength DKW (Tulecke and McGranahan, 1985; McGranahan *et al.*, 1986). When roots and shoots had grown sufficiently (in 2–4 weeks) they were transplanted into plastic cups, then pots covered with polyethylene bags. Over a 2–4 week period, the bags were perforated, then removed. Plants then grew normally.

## 7. References

Aly, M.A.M., R.G. Fjellstrom, G.H. McGranahan and D.E. Parfitt, 1992. Origin of walnut somatic embryos determined by RFLP and isozyme analysis. HortSci. 27: 61–63.

Beineke, W.F., 1989. Selection and evaluation for better varieties. In: J.E. Phelps (Ed.), The Continuing Quest for Quality: Proceedings of the 4th Black Walnut Symposium, July 30 – August 2, 1989, Carbondale, IL, pp. 145–152. Walnut Council, Indianapolis, IN.

Cheema, G.S. and P.N. Mehra, 1982. Morphogenesis in endosperm cultures. Proceedings of the 5th Intl. Cong. Plant Tissue and Cell Culture. pp. 111–112.

Compton, M.E. and J.E. Preece, 1988. Response of tobacco callus to shoot tip exudation from five species. HortSci. 23: 208–210.

Cornu D., 1988. Somatic embryogenesis in tissue cultures of walnut (*Juglans nigra*, *J. major* and hybrids *J. nigra* × *J. regia* In: M.R. Ahuja (Ed.), Somatic Cell Genetics of Woody Plants, pp. 45–49. Kluwer Academic Publishers, Dordrecht.

Cornu D., 1989. Walnut somatic embryogenesis: physiological and histological aspects. In: E. Dreyer, G. Assenac, M. Bonnet-Masimbert, P. Dizengremel, J.M. Favre, J.P. Garrec, F. Le Tacon and F. Martin (Eds.), Ann Sci. For. 46, Suppl. Forest Tree Physiology. pp. 133s–135s. Elsevier, Paris.

Dandekar, A.M., L.A. Martin and G. McGranahan, 1988. Genetic transformation and foreign gene expression in walnut tissue. J. Amer. Soc. Hort. Sci. 113: 945–949.

Dandekar, A.M., G.H. McGranahan, C.A. Leslie and S.L. Uratsu, 1989. *Agrobacterium*-mediated transformation of somatic embryos as a method for the production of transgenic plants. J. Tissue Culture Methods 12: 145–150.

Deng, M.-D. and D. Cornu, 1992. Maturation and germination of walnut somatic embryos. Plant Cell Tiss. Org. Cult. 28: 195–202.

Driver, J.A. and A.H. Kuniyuki, 1984. *In vitro* propagation of paradox walnut rootstock. HortSci. 19: 507–509.

Jay-Allemand, C., L. Jouanin, M.D. Deng, A.C. Claudot, A. Drouet and D. Cornu, 1991. Transfer of chalcone synthase antisense gene: new strategy for studying polyphenols involved in walnut rhizogenesis. Plant Sci. Today 59: 305.

Leslie, C. and G. McGranahan, 1992. Micropropagation of persian walnut. In: Y.P.S. Bajaj

(Ed.), Biotechnology in Agriculture and Forestry, Vol. 18, High-Tech and Micropropagation II, pp. 136–150. Springer-Verlag, Berlin.

Liberty Hyde Bailey Hortorium, 1976. Hortus Third: a Concise Dictionary of Plants Cultivated in the United States and Canada, 3rd. Ed. MacMillan, New York.

Lloyd, G. and B. McCown, 1980. Commercially-feasible micropropagation of mountain laurel, *Kalmia latifolia*, by use of shoot tip culture. Comb. Proc. Intl. Plant Prop. Soc. 30: 421–427.

Long, L.M., J.E. Preece, G.R. Gaffney and J.W. Van Sambeek, 1992. Somatic embryogenesis and organogenesis of eastern black walnut (*Juglans nigra*). HortSci. 27: 107 (Abstr.).

McCurdy, D.R., 1989. The United States walnut log export market. In: J.E. Phelps (Ed.), The Continuing Quest for Quality: Proceedings of the 4th Black Walnut Symposium, July 30 – August 2, 1989, Carbondale, IL, pp. 300–310. Walnut Council, Indianapolis, IN.

McGranahan, G.H., 1987. Research priorities in genetic improvement of Persian walnuts and their rootstocks. 78th Annual Report, Northern Nut Growers Association, pp. 73–77.

McGranahan, G.H. and P.B. Catlin, 1987. *Juglans* rootstocks. In: R.C. Rom and R.F. Carlson (Eds.), Rootstocks for Fruit Crops, pp. 411–450. Wiley, New York.

McGranahan, G. and C. Leslie, 1990. Walnut (*Juglans* L.), In: J.N. Moore and J.R. Ballington (Eds.), Genetic Resources of Fruit and Nut Crops, Vol. 2, pp. 907–951. International Society for Horticultural Science, Wageningen.

McGranahan, G.H., J.A. Driver and W. Tulecke, 1987. Tissue culture of *Juglans*. In: J.M. Bonga and D.H. Durzan (Eds.), Cell and Tissue Culture in Forestry 3, pp. 261–271. Martinus Nijhoff, Boston.

McGranahan, G.H., C.A. Leslie and J.A. Driver, 1988a. *In vitro* propagation of mature persian walnut cultivars. HortSci. 23: 220.

McGranahan, G.H., C.A. Leslie, S.L. Uratsu, L.A. Martin and A.M. Dandekar, 1988b. Agrobacterium-mediated transformation of walnut somatic embryos and regeneration of transgenic plants. Bio/Technol. 6: 800–804.

McGranahan, G.H., W. Tulecke, S. Arulsekar and J.J. Hansen, 1986. Intergeneric hybridization in the *Juglandaceae*: *Pterocarya* sp. × *Juglans regia*. J. Amer. Soc. Hort. Sci. 111: 627–630.

McGranahan, G.H., C.A. Leslie, S.L. Uratsu and A.M. Dandekar, 1990. Improved efficiency of the walnut somatic embryo gene transfer system. Plant Cell Rep. 8: 512–516.

Merkle, S.A., H.Y. Wetzstein and H.E. Sommer, 1987. Somatic embryogenesis in tissue cultures of pecan. HortSci. 22: 128–130.

Neuman, M.C., J.E. Preece, J.W. Van Sambeek and G.R. Gaffney, 1993. Somatic embryogenesis and callus production from cotyledon explants of Eastern black walnut (*Juglans nigra* L.). Plant Cell Tiss. Org. Cult. 32: 9–18.

Polito V.S., G. McGranahan, K. Pinney and C. Leslie, 1989. Origin of somatic embryos from repetitively embryogenic cultures of walnut (*Juglans regia* L.): implications for *Agrobacterium*-mediated transformation. Plant Cell Rep. 8: 219–221.

Preece, J.E., J.W. Van Sambeek, C.A. Huetteman and G.R. Gaffney, 1989. Biotechnology: *In vitro* studies with walnut (*Juglans*) species. In: J.E. Phelps (Ed.), The Continuing Quest for Quality: Proceedings of the 4th Black Walnut Symposium, July 30 – August 2, 1989, Carbondale, IL, pp. 159–180. Walnut Council, Indianapolis, IN.

Rodriguez, R., A. Revilla, M. Albuerne and C. Perez, 1989. Walnut (*Juglans* spp.). In: Y.P.S. Bajaj (Ed.), Biotechnology in Agriculture and Forestry, Vol. 5, Trees II, pp. 99–126. Springer-Verlag, Berlin.

Tulecke, W. and G. McGranahan, 1985. Somatic embryogenesis and plant regeneration from cotyledons of walnut, *Juglans regia* L. Plant Sci. 40: 57–63.

Tulecke, W., G. McGranahan and H. Ahmadi, 1988. Regeneration by somatic embryogenesis of triploid plants from endosperm of walnut, *Juglans regia* L. cv Manregian. Plant Cell Rep. 7: 301–304.

# 8. Somatic embryogenesis in rubber (*Hevea brasiliensis* Müll. Arg.)

M.P. Carron, H. Etienne, L. Lardet, S. Campagna, Y. Perrin, A. Leconte and C. Chaine

**Contents**



## 1. Introduction

### 1.1. *Origin, cultivation and propagation*

Natural rubber, produced by *Hevea brasiliensis* (Müll. Arg.) has been known since ancient times and the Mayas and Aztecs used it in their daily life and religious ceremonies (Serier, 1993). In forests, this tropical tree can reach 40 m in height and 5 m in circumference. It is much smaller in plantations, where twice-weekly tapping of the tree bark considerably reduces its growth. Paradoxically, cultivation of this tree is very recent, beginning in Southeast Asia at the turn of the century, i.e., outside its original zone in the Amazonian basin, where "South American Leaf Blight", caused by the fungus *Microcyclus ulei* is rife. The normal cultivation zone lies between 20° N and 20° S, apart from plantations in China and northern India (Compagnon, 1986). The areas planted cover around 7 million hectares, primarily in Thailand, Indonesia and Malaysia, but also in China, India, Africa and Latin America.

Natural rubber production has been steadily increasing since the turn of the century and currently stands at 5,360,000 t (Rubber Statistical Bulletin, 1992). The increase in demand should persist in the coming decades due to

*S. Jain, P. Gupta & R. Newton (eds.), Somatic Embryogenesis in Woody Plants, Vol. 2*, 117–136.

the foreseeable increase in consumption in many countries (Livonière, 1992). The possibilities of extending the planted areas seem limited nowadays and it is primarily through yield increases that growers will be able to satisfy demand. The quality of the planting material used is an essential component of yield.

True-to-type multiplication of selected planting material has long been faced with the difficulty of developing a cutting technique, due to early loss of rhizogenesis capacity during young tree growth, then the observation of deficient root systems on cuttings (Wiersum, 1955; Muzik and Crusado, 1958; Levandowsky, 1959). Budgrafting onto plants obtained from seed was gradually generalized from the 1920s onwards. Nevertheless, a certain heterogeneity is seen in budgrafted clones even today, which is particularly attributable to the non-selected stocks. Moreover, mean yields are also significantly lower than those of the best plants obtained from seed (McIndoe, 1958), probably linked to the mature state of the scions used. These two major handicaps would be potentially removed by the *in vitro* propagation of plants on their own roots (Carron *et al.*, 1989).

### 1.2. *Merits of somatic embryogenesis for rubber*

Research to develop somatic embryogenesis as an *in vitro* propagation method for rubber (*Hevea brasiliensis*, Müll. Arg.) began in the 1970s. Like most woody species, *Hevea brasiliensis* initially seemed recalcitrant to *in vitro* culture. The first plantlets obtained by somatic embryogenesis were reported by Wang *et al.* (1980). Others were then obtained by the Rubber Research Institute of Malaysia (Wan *et al.*, 1982) and by CIRAD-IRCA (Carron and Enjalric, 1982, 1985).

After several years of in-depth research, this technique can now be used to obtain plantlets from several clones of industrial interest, in a reliable and reproducible manner (Carron *et al.*, 1994). After producing a few hundred plantlets by somatic embryogenesis, and before considering using this procedure to supply planting material, it was necessary to analyze the *in vitro* performance of different *Hevea* clones with regard to the established technique, and assess their biological yield at the different stages of *in vitro* culture. It is very difficult to find any bibliographical references on biological yields in *in vitro* culture, whether it involves micropropagation by bud development or somatic embryogenesis. Extrapolation from yields reported from one or other steps is virtually always vitiated because certain handling parameters, such as the selection of plant material at different stages of the procedure, are not taken into account. However, an analysis of yield factors seems essential, to compare the effectiveness of the different procedures, but also to reveal truly limiting phases and adapt techniques accordingly.

The juvenility and vigour of the somaplants produced were also examined from two angles. Firstly successful acclimatization and early growth in the field, and secondly, their micropropagation capacity. Whilst the micropropagation of juvenile individuals has been relatively well mastered in each of the phases of the process (establishment, multiplication, rooting, weaning), application of this technique to elite trees has come up against low reactivity in clonal material, as in the case of cuttings (Carron *et al.*, 1989). In the absence of juvenile controls for these genotypes, the hypothesis of an insufficient juvenility level in this material has been put forward to explain why they remain recalcitrant. However, somatic embryogenesis is, theoretically, the most effective way of rejuvenating mature trees (Nozeran *et al.*, 1977; Bonga, 1982). This effect can be seen in the characteristics of the somaplants during field growth, but also in their micropropagation capacity which, as we have seen, is a morphogenetic characteristic of juvenile individuals.

### 1.3. *Plant material and culturing conditions for somatic embryogenesis*

The explant was a thin slice of internal seed integument taken from immature fruits (8–10 weeks after anthesis). It should be noted that the internal integument is a maternal tissue and, therefore, possesses the same genotype as the mother-tree from which the fruits were taken. This naturally sterile tissue does not require surface disinfection before culturing. There were six different culturing phases: the first five were carried out in the dark.

1) Initiation of callogenesis (day 0–25) on MH1 medium (Carron and Enjalric, 1985) containing 234 mM sucrose, 3,4-dichlorophenoxyacetic acid (3,4-D) and kinetin (Kin) (4.44 μM each), 10 μM $AgNO_3$, and gelified with 2 g/l gelrite.
2) Embryogenesis expression (day 26–50). The calli were transferred to the modified MH1 medium amended with 3,4-D and benzyladenine (BA) (1.35 μM each), $5 \times 10^{-3}$ μM abscisic acid (ABA), and 50 μM spermidine.
3) Proembryo development (day 51–80). The embryogenic calli were subcultured to a modified MH1 medium supplemented with 1.8 μM 3,4-D and 0.9 μM BA, without $AgNO_3$.
4) Maturation of somatic embryos (day 81–105). The isolated embryos were placed in multi-well dishes containing a medium having 30% of the macroelements and 200% of the microelements of MS mineral medium (Murashige and Skoog, 1962), the vitamins of MH1 medium, 351 mM sucrose, 500 mg/l activated charcoal, and gelified with 2 g/l gelrite.
5) Germination of somatic embryos (day 106–130). The embryos were cultured in test tubes poured with a modified maturation medium having 146 mM sucrose and 8.7 μM $GA_3$.
6) Plantlet development (day 131–156). The embryos were brought into the light (50 $\mu E/m^2/s$) and transferred to an identical medium to the previous one (step 5), except for 73 mM sucrose and 28.9 μM $GA_3$.

The experiment described involved four clones of industrial interest wich have different parental origin and different agronomical features: PB 260, PB 235 (created in Prang Besar, Malaysia), PR 107 (created in Proefstation,

Indonesia) and RRIM 600 (created in Rubber Research Institute of Malaysia). The immature fruits were sent from the IDEFOR-DPL station in the Ivory Coast. For each clone, 600 explants were cultured, representing 20 seeds, or 7 fruits per clone.

During the culture, each clone was characterized *in vitro* performance by studying: a) *histogenesis* – All the percentages were calculated in relation to the initial number of explants cultured: (1) callogenesis rate, (2) embryogenesis rate (a callus is defined as embryogenic if it bears one or more somatic embryos clearly identifiable with the naked eye, whatever their stage of development), (3) the mean number of embryos per embryogenic callus, (4) the percentage of total embryos, (5) the percentage of normal embryos (only white embryos with a "pear" shape or two clearly visible cotyledons were isolated and moved to the maturation stage), and (6) the rate of conversion into plantlets; b) *callus morphology* – The appearance of calli can be characterized by colour, reflecting the level of browning, and structure. The structure is either compact or friable/soft meaning that the different parts of the calli can easily be disected; and c) *handling* – The transfer times were recorded at each stage of the culturing process. The timing is not really significant in absolute terms, given the particular conditions of the experiment; nevertheless, in relation to the quantity of plant material (embryogenic calli, normal embryos, etc.) they shed a new light, in a more applied manner, on the biological yield of each clone, in terms of labor necessary for the production of a plantlet.

### 1.4. *Conditions of acclimatization and field trial for plants reared* in vitro

When somaplants are removed from *in vitro* conditions, they are comparable to miniature seedlings with a 2–3 cm tap root, "spider leg" type roots and a stem with one or two fully developed leaves. The plants are transferred from the *in vitro* tube to a peat/leaf mould mixture (vol/vol). Three successive stages can then be distinguished, which take a total of 4–6 months to develop, before the plants can be planted in the field: (1) Weaning – this is designed to adapt the young plantlets to *in vivo* conditions where they have to resume apex growth, to form of a new growth unit, and to become autotrophic. During the first phase, which lasts 4 to 6 weeks, periodic water spray is applied to avoid water stress by maintaining the relative humidity close to saturation at 20–30°C, and fortnightly fungicide (e.g., 0.5 g/l of Benomyl) treatments; a polyethylene curtain produce 50% shading, (2) Hardening – the climatic protection, periodic water spray against dryness and curtain against sunlight, is gradually removed. Once apex growth has resumed, fertilizers should be applied to compensate for the low reserves of somaplants compared to seedlings; the applications should be low and repeated each week: 1 g/l of N-P-K 10–52–10 in the first 4 weeks, then 1 g/l of N-P-K 20–20–20 in the following 4 weeks and 2 g/l of N-P-K 28–14–14 during the nursery phase, (3) Nursery – comparable to a conventional nursery. The plants develop until they have acquired sufficient vigour to be planted in the field. Generally speaking, 20 cm tall plants can be transferred to the field.

The somaplants of three clones (PR 107, PB 260, and RRIM 600), acclimatized in greenhouses in Montpellier, were transferred to the IDEFOR-DPL Station in the Ivory Coast on May, 1992. They were immediately transferred to 3 l – polyethylene bags filled with soil and hardened under 50% shading for three weeks then taken out into the full-sunlight nursery until planting on June, 1992. The 13–20 cm tall somaplants were compared to seedling plants green grafted at 7–8 months, and cut just above the graft (zero growth) at the time of planting, 10 months after sowing in polyethylene bag. In an agronomical trial, growers usually consider planting as the absolute starting point. During field growth, the height and girth of the somaplants have been compared to those of the scions of the grafted clone. Planting was carried out in rows of $7 \times 2.80$ m, i.e., a density of 510 trees/ha (normal planting density).

In view of the variable number of somaplants available (66 PR 107, 18 PB 260 and 2 RRIM 600), the trial was divided into two parts with two different planting designs: Fisher blocks with 2 treatments × 6 replicates for PR 107, and total randomization for PB 260 and RRIM 600.

### 1.5. *Somaplant micropropagation capacity tests*

#### 1.5.1. *Plant material preparation*

*Mother-trees from somaplants.* The response of a primary explant in micropropagation is directly correlated to the juvenility and vigour of the mother-tree. Somaplants are theoretically juvenile, developped from embryonic cells; nevertheless their vigour is low up to the end of acclimatization, compared to initial growth of the seedlings. Consequently, vigorous 6–10 months old somaplants of clones PB 260 and RRIM 600 are selected. Dormant axillary buds are then taken from the base of the somaplants and grafted onto the green branches of young plants obtained from seed. Approximately after one month of grafting, the stocks are cut back and the graft develops a "graft branch" from which primary explants will be taken. This operation therefore multiplies the ortet in order to have a large number of primary explants available.

*Mother-trees of mature genotypes.* The mature mother-trees were prepared in the Ivory Coast. They are also grafted plants of clones PB 260 and RRIM 600; the scions were taken from a conventional budwood garden of woody shoots. These plants are being maintained in containers in tropicalized greenhouses at the CIRAD centre in Montpellier.

#### 1.5.2. *Culturing conditions for micropropagation*

*Primary culture.* Primary explants are taken from graft branches of a growth unit. They include a portion of about 2 cm long stem with a closed axillary

*Table 1.* Quantification of the output from the different embryogenesis stages in four *Hevea* clones. The values given are the results of an experiment carried out with an initial 600 explants per clone.

| Success rate of the different steps in somatic embryogenesis | Clones | | | |
|---|---|---|---|---|
| | PB 260 | PR 107 | RRIM 600 | PB 235 |
| Explants | 100 | 100 | 100 | 100 |
| Callus (% explants) | 89 | 87 | 94 | 99 |
| Embryogenic callus (% explants) | 51 | 46 | 36 | 37 |
| Mean number of embryos per embryogenic callus | 4.1 | 2.5 | 1.2 | 1.9 |
| Total embryos (% explants) | 209 | 115 | 43 | 70 |
| Normal embryos (% explants) | 61 | 95 | 34 | 24 |
| Plantlets (% explants) | 7.7 | 31 | 17 | 2 |

bud. After soaking (2 h) in a hormone solution containing 44 μM BA and 25 μM IBA, the explants are transferred to a basic micropropagation medium (Enjalric and Carron, 1982) without growth regulators, and containing 175 mM sucrose. After 4 weeks of culture, the bud develops a primary shoot, which will be used for the multiplication phase.

*Multiplication.* Amplification is carried out by culturing the upper part from the primary shoot (shoot tip culture); the primary explant is also subcultured. The multiplication medium contained 4.4 μM BA, 1.25 μM IBA, and 175 mM sucrose. The primary explant and the shoot tip were then subcultured every 4 weeks.

*Rooting.* Root induction in regenerated shoots is done in two phases: (1) the base of the shoots is soaked for three days in an auxin solution containing 25 μM IBA and 27 μM NAA, (2) Root initiation occurs on a solid medium containing the macroelements of Quoirin and Lepoivre (1977) diluted 8

*Figure 1.* Embryogenic callus from clone PB 260 at the end of the "development phase" (Day 80). Note the compact callus and the more or less developped somatic embryos. Bar = 0.55 mm.
*Figure 2.* Embryogenic callus from clone PR 107 at the end of the "development phase" (Day 80). Note the soft, brown or beige callus and the cluster of somatic embryos. Bar = 0.55 mm.
*Figure 3.* Somatic embryos from the clone PB 260 at the end of the "maturation time" (Day 105). Cotyledons are poorly developed and produced an unusual shape. Note right up the coalescent embryos which possibly will form two distinct plantlets. Bar = 0.55 mm.
*Figure 4.* Somatic embryos from the clone PR 107 at the beginning of the "maturation phase" (Day 90). The embryos are well pear-shaped and the cotyledons begin to elongate. Bar = 1.1 mm.
*Figure 6.* Young somaplant from clone PB 260 at the beginning of the "conversion phase" (Day 140). Note the analogy with the seedling development. Bar = 1.37 mm.
*Figure 7.* Young tree PB 260 from somatic embryogenesis, 6 months after planting. Bar = 8.3 cm.

1
4
2
3
6

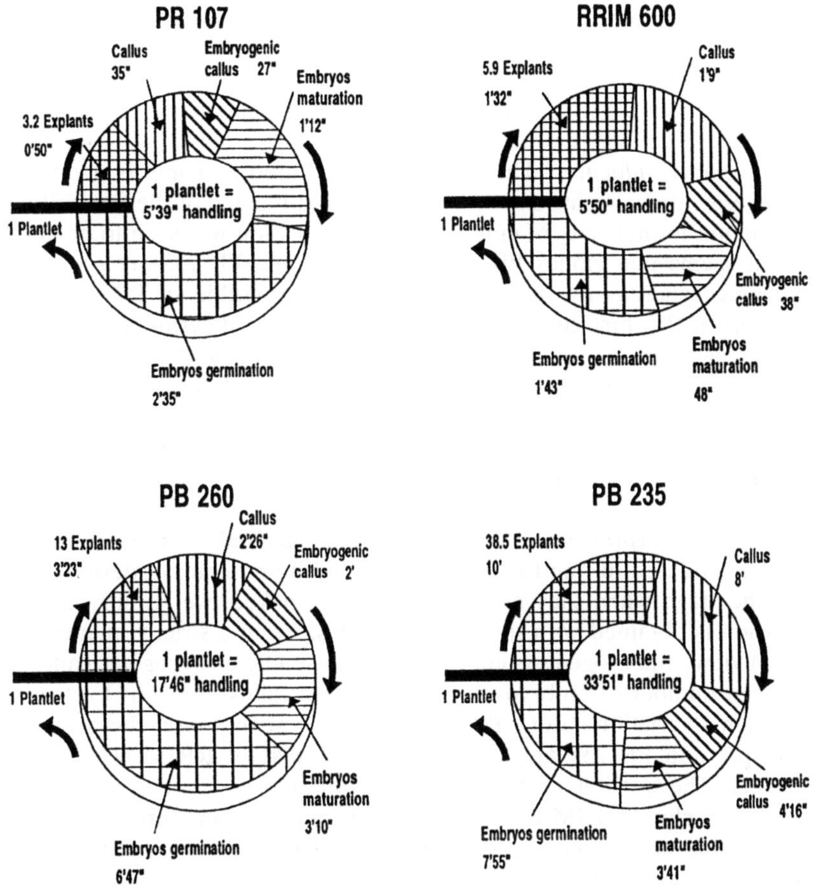


*Figure 5.* Handling time for obtaining one developed plantlet from four clones in rubber.

times, and 117 mM sucrose. After 10 days' culturing on the rooting medium, the shoots, bearing one or several root tips, are then moved on to acclimatization. Root growth, therefore, takes place in a horticultural substrate into green house.

## 2. Conditions for the expression of somatic embryogenesis in different clones

### 2.1. *Embryogenesis development*

*Embryogenesis induction*. The response of cultured explants is good, the callogenesis rate being around 90% (Table 1). At the end of the first culture, the variation in precocity of embryogenesis expression was seen to be genotypic. It seems that clones PB 260, RRIM 600 and PB 235 are somewhat precocious, whereas the phenomenon is not yet visible in PR 107. The conditions of this first culture appear to have a clear determining effect on the later performance of the plant material; among the different medium components, the humidity parameters (Etienne *et al.*, 1991), auxin/cytokinin content (Etienne *et al.*, 1993a) and the type of cytokinin (Montoro *et al.*, 1992) have a dominant effect on both the frequency of embryogenesis expression and the quality of the embryo development.

*Embryogenesis expression*. This phase is characterized by proliferation of the callus initiated during the first culture, and by proembryo development up to the stage when they are visible with the naked eye. Embryogenesis expression is obtained with the four genotypes. From a strictly quantitative point of view, embryogenesis intensity is expressed by embryogenic calli rate and mean number of embryos per embryogenic callus; it seems to demonstrate clonal variation. Clone PB 260 gives the best response; the responses of PR 107 and PB 235 are average, whereas that of RRIM 600 is weak (Table 1). The qualitative development of proembryos is seen to be more advanced in clone RRIM 600, which look similar to immature zygotic embryos. Callus development and embryogenesis expression on calli are dependant on the genotype. The embryos of PB 260 occur in dense clumps on a very compact, light beige callus, to which they are firmly attached by their base (Fig. 1). In clone PR 107, the embryos are usually in a cluster and develop in any part of a large soft brown callus: on the top, in the centre, and underneath at the interface with the medium (Fig. 2). It should be noted in this case that addition of ABA in the medium favours the standard development and singulation of the embryos (Etienne *et al.*, 1993b,c).

*Somatic embryo development*. Callus growth is halted during this phase and towards the end of the culturing period, the calli start to turn brown and necrotic. Depending on the cases, a proportion of the proembryos, visible at the end of the second culture, revert to calli and disappear; these are usually embryos at the globular stage with less advanced development. At the same time, some other embryos continue to develop. The more developped embryos became arrested at the cotyledonary stage. From their original translucent yellow appearance, they turn white. There is little or no more new proembryo formation at this stage. The embryos existing on a callus are heterogeneous at the end of culturing, in terms of morphology and developmental stage. In the case of white embryos, which are the most advanced, 3 morphological types can be seen:

a) Nodular type with many structural aberrations (budding, multiple abnormal procambial bundles, absence of polarization). When placed under conditions conducive to germination, most of the embryos swell or reform a non-embryogenic callus; a small percentage of the embryos elongate and 3% root. None of these embryos germinated.
b) Coalescent type, which in fact represents several fused somatic embryos. This type has a higher degree of organization than the previous type and a very small percentage of them can singulate and germinate normally.
c) Pear shaped type, which has normal morphology: the root pole and the cauline pole are clearly distinct and the embryos are well singulated. However, as with the other two types, the cauline meristem is not usually formed, the starch reserves are very low and protein reserves nil. These characteristics and a very turgid state reflect their immature state (Michaux-Ferrière *et al.*, 1992; Etienne *et al.*, 1993b,c).

The relative frequency of these different morphological types, and the level of maturity of normal, pear shaped type embryos vary considerably depending on the genotype. In clone RRIM 600, almost all the somatic embryos formed correspond to the pear shaped type. They can be distinguished from those of the other clones by better developed cotyledons and a greater vigour in size. Although the root and cauline meristems are not completely singulated this stage, substantial starch and protein reserves are already seen building up. This relatively "normal" development, in relation to the zygotic ontogenesis, leads to substantial germination and conversion frequencies. Clone PR 107 also has embryos that are morphologically well-developed (Fig. 3), but the cauline meristem at the anlage stage and the absence of protein reserves are responsible for low viability at this stage. In clones PB 260 and PB 235, only a small percentage of embryos are of the "normal" type (around 25%); they are small, their dry weight being around 60 times less than that of zygotic embryos at the same stage, and they are not completely differentiated as compared with clone PR 107 (Fig. 4). Hence, in these two clones, teratogenesis is the main factor responsible for the low conversion rate. But, whatever the clone, insufficient embryo maturity characterized this stage.

*Somatic embryo maturation.* In rubber, and just like in other plants, a study of the zygotic model has revealed the need to induce a maturation phase prior to embryo germination, primarily for the completion of cauline meristem formation, accumulation of starch and protein reserves and desiccation. This is partially achieved by adding ABA, activated charcoal and a high sucrose molarity in the medium. The increase in starch content and osmotic potential in embryos is linked to sucrose concentration, whereas ABA favours somatic

*Table 2.* Effect of the maturation phase on the performance of somaplants from two clones during acclimatization. This performance is expressed as the survival rate and the time required for the resumption of growth (striking time). The values given are the means ± s.e. for three replicates.

| Clone | Maturation step | Number of somaplants | Survival rate (%) | Time for growth resumption (weeks) |
|---|---|---|---|---|
| PB 260 | no | 50 | 3 ± 3 | 22 ± 3 |
| | yes | 50 | 60 ± 10 | 2 ± 1 |
| PR 107 | no | 50 | 15 ± 5 | 15 ± 2 |
| | yes | 150 | 70 ± 12 | 1 ± 0.5 |

embryo structure formation (cotyledon development, organization of the meristem, procambial bundles, the epidermis and the latex-bearing vessels), protein reserve accumulation and a maximum drop in water and osmotic potentials (Etienne *et al.*, 1993b). However, as regards the water level in particular, a perfect correspondence has still to be obtained with the characteristics of mature zygotic embryos, which can be linked to the differences recorded in the level of conversion rates.

*Embryo germination.* The germination medium is characterized first of all by reducing the sucrose concentration (146 mM), hence the osmotic potential, which encourages embryo hydration. It proved necessary to add 8.7 μM $GA_3$ gibberellin to stimulate germination through elongation of the hypocotyl and stem emergence in somatic embryos, just like in zygotic embryos. In particular, the germination precocity limits callogenesis problems. In addition, browning of the plantlet base and tap root is prevented by adding 500 mg/l activated charcoal. The latter element does not interfere with gibberellin activity, indeed it increases conversion efficacy, greatly improves plantlet vigour and ensures their survival. Complete plantlet development is obtained by a second culture on a similar medium, but with a lower sucrose content (73 mM) and, especially, a transfer to light (50 $\mu E/m^2/s^1$), gradually preparing the plants for autotrophy, and a slightly higher gibberellin concentration (28.9 μM).

*Plantlet acclimatization.* It appeared that induction of a maturation phase into the procedure had repercussions on acclimatization, where it keeps the time taken for plant striking, resumption of growth and development to a minimum, one to two weeks after leaving the *in vitro* tube (Table 2). The acclimatization rates and vigour of young plantlets is thus improved.

### 2.2. *Biological yields of different clones*

The "yield" concept is usually lacking in scientific reports or is reduced to parameters such as the "success frequency" of a phenomenon or the "multiplication coefficient". It is, therefore, often difficult to distinguish between a one-off phenomenon and a regular phenomenon, between reliability on a research level and possible routine use.

The somatic embryogenesis yield in different rubber clones was evaluated at two levels: firstly on the basis of success rates at each stage of the process (Table 1), and secondly based on the plant material *in vitro* handling time required to obtain a plantlet (Fig. 5).

*Clonal* in vitro *performances.* Embryogenesis is expressed well in the four clones studied, but under conditions highly specific to each of them. Clone PB 260 has very high embryogenic efficiency in terms of embryogenic callus rate and the number of embryos per callus. On the other hand, it reveals high teratogenesis, cumulated with a relatively low conversion rate, which greatly affects overall embryogenesis yield. On the other hand, clone RRIM 600 has rather low embryogenic efficiency, but the embryos formed on the calli are almost all normal and have a high conversion rate. The performance of PR 107 falls somewhere between the previous two, with better embryogenic efficiency than RRIM 600 and better conversion rate than PB 260. Conversely, clone PB 235 combines the low embryogenic efficiency of RRIM 600 and the low conversion rate of PB 260.

*Clonal handling charges.* The overall plant material handling time in relation to the number of plants produced is commonly accepted as being the major cost price parameter for plants grown *in vitro*. We attempted to evaluate the value of this parameter in this small-scale production experiment, with the sole aim of examining the respective incidence of the characteristics of the different embryogenesis steps.

The transfer times were measured under normal research laboratory conditions, with no attempts at optimization, i.e., 26 min to culture 100 explants – 21 min to transfer 100 calli – 30 min to transfer 100 embryogenic calli – 40 min to isolate and transfer 100 maturing embryos – 43 min for each of the following two transfers of 100 germinating embryos, then conversion. Fig. 5 was obtained by crossing these data with those in Table 1.

Taking into account the overall handling time reveals the paramount importance of the embryo conversion rate and minimizes that of embryogenesis efficiency, i.e., it is more important to master a high conversion rate with an average embryogenic intensity than the other way round. Thus, clone RRIM 600 becomes equivalent to PR 107 on this level, despite its handicap as regards embryogenic efficiency. On the other hand, clone PB 260 proves to be much "heavier" to handle, despite its high embryogenic efficiency. It is normal that here too PB 235 has the poorest performance.

## 3. Somaplant performance *ex vitro*

*Acclimatization.* Somaplant acclimatization to natural conditions is a particularly delicate stage in all *in vitro* culture processes. The losses recorded during the move *ex vitro* can drastically reduce the effectiveness of micropropagation procedures. Such losses usually depend on two types of factors: a) external, the fact that somaplants are unadapted to low relative humidity conditions, high light intensity and temperature fluctuations, and/or b) internal, impeded development following hormone treatment *in vitro*, or the inability of somaplants to function autotrophically.

Under our culture conditions, the relative humidity shock was attenuated by maintaining high relative humidity in the acclimatization tunnel in which the somaplants spent two months (80% < RH < 100%). Partial somaplant autotrophy was stimulated as early as the *in vitro* stage through the gradual reduction in sucrose content (351 mM to 73 mM) during the final three cultures; low lighting (440 $\mu E/m^2/s^1$) at the start of acclimatization prolonged this effect. In the absence of a maturation phase, somaplant survival during acclimatization was very low (0 to 15%) and growth resumption only occurred after several months. Integrating such a maturation phase into the procedure greatly improved these results, with survival frequencies varying from 50 to 80% for clones PR 107 and PB 260, and a rapid growth resumption in the first two weeks of transfer in the natural conditions (Table 2). Two months after leaving the tube, somaplant morphology was very similar to that of a miniaturized seedling (Fig. 6): the root system developed from the base of the hypocotyl and was initially composed of a tap rootlet and a more or less complete bundle of fine "spider leg" roots. At 5–10 cm depth in the soil, the tap root formed secondary roots. In most cases, the cotyledons remained very small. The hypocotyl was surmounted by the epicotyl with one or two leaves, with two latent cotyledonary buds at its base.

*Field trial.* The survival rate of the somaplant was very high (99%) relative to the budded clones (81%). Growth resumption of the somaplants in the field was excellent during the first year, despite the initial small size of the planting material relative to the rootstocks of the budded clones. Growth was vigorous from the outset, as reflected in the height and girth measurements (Table 3). It is still too early to reach any conclusion as to the relative growth rate of the two somaplant/budded plant treatments, particularly since the grafted plants, theoretically, have an initial advantage due to the age of the budgrafted stock. At this stage, the two treatments were roughly equal for clone PR 107, whereas the somaplants showed markedly better growth in PB 260 and RRIM 600 (Fig. 7).

*Table 3.* Field growth of somaplants (S) and grafted plants (G) of three clones of industrial interest: PB 260, PR 107 and RRIM 600. Plant height and girth were measured at the time of planting, at six months and at one year. The measurements marked with the same letters are not significantly different according to the Duncan test at 5%.

| Clone | Plants | Planting | | Growth at 6 months | | Growth at 12 months | |
|---|---|---|---|---|---|---|---|
| | | Height (cm) | Survival (%) | Height (cm) | Girth (mm) | Height (cm) | Girth (mm) |
| PR 107 | 66 S | 13 ± 1.6 | 98 | 64[a] | 8.2[a] | 177[a] | 23.3[a] |
| | 66 G | – | 80 | 55[a] | 9.2[a] | 186[a] | 24.2[a] |
| PB 260 | 18 S | 20 ± 3.2 | 100 | 83[a] | 11.3[a] | 262[a] | 34.7[a] |
| | 23 G | – | 87 | 54[b] | 9.5[a] | 206[b] | 27.6[b] |
| RRIM 600 | 2 S | 18 | 100 | 110 | 13.9 | 351 | 37.1 |
| | 23 G | – | 65 | 76 | 11.1 | 245 | 29.3 |

*Table 4.* Clones RRIM 600 and PB 260 – *in vitro* response of primary explants: effect of mother-tree origin.

| | Origin of mother-plants | | | |
|---|---|---|---|---|
| | Clone RRIM 600 | | Clone PB 260 | |
| | Mature | Somatic embryo | Mature | Somatic embryo |
| Primary culture | | | | |
| Shoot forming explant (%) | 100 | 80 | 78.1 | 82 |
| Shoots length (mm) | 7.6 ± 1.6 | 20.5 ± 2.7 | 14.9 ± 2.0 | 14.1 ± 3.0 |
| Multiplication step (subculture of primary explants) | | | | |
| % explants forming shoots (mean of 4 subcultures) | 43.6 ± 18 | 79.2 ± 5 | 4.8* | 82.8 ± 6.7 |
| number of shoots/explant (mean of 4 subcultures) | 1.2 ± 0.2 | 2.4 ± 0.2 | 1.0 ± 0.0 | 2.4 ± 0.5 |
| Shoots length (mm) (mean of 4 subcultures) | 6.9 ± 1.8 | 9.2 ± 0.9 | 10.5 | 9.5 ± 0.9 |

* Value from the first subculture; reactivity was nul during the subsequent subcultures.

## 4. Rejuvenation and micropropagation capacity from somaplant

The experiment consisted in comparing the micropropagation capacities of clones introduced *in vitro*, either from mature plants or from somaplants, hence theoretically juvenile mother-trees. The study involved two genotypes: RRIM 600 and PB 260 clones.

The response of each type of plant material was assessed during primary

*Table 5.* Clones RRIM 600 and PB 260 – *in vitro* response of stem tips taken from primary shoots: effect of mother-tree origin.

| | Origin of mother-plants | | | |
|---|---|---|---|---|
| | Clone RRIM 600 | | Clone PB 260 | |
| | Mature | Somatic embryo | Mature | Somatic embryo |
| % shoot-tips forming callus (1 transfert on fresh medium) | 14.3 | 97.4 | 50 | 94.4 |
| % active growing shoot-tips (2 transferts on fresh medium) | 0 | 94.5 | 0 | 94.4 |
| % decapited shoot-tips forming axillary buds (1 transfert on fresh medium) | 28.6 | 92.0 | 20 | 88.9 |
| % decapited shoot-tips forming axillary shoots (1 transfert on fresh medium) | 0 | 73.0 | 0 | 83.3 |

culture, the multiplication phase (subcultures of the primary explant and stem tips) and the rooting phase.

*Primary culture and subcultures of primary explants* (*Table 4*). In clone RRIM 600, the primary explants from somaplants produced *in vitro* primary shoots greater than that of the primary shoots obtained from primary explants taken from mature mother-trees. This phenomenon was not seen in clone PB 260. It is, therefore, seem that the state of maturity does not affect morphogenetic capacities in the same way in all genotypes. This genotypic variability was also seen during subculturing of primary explants. Mature clone RRIM 600 retained a certain capacity to produce shoots during subculturing, whereas in clone PB 260 this capacity is lost right from the first subculture and disappeared completely in the second subculture. For both genotypes, the use of mother-trees obtained from somaplants very significantly increased the capacity of primary explants to produce axillary shoots during subculturing.

*Stem tip cultures* (*Table 5*). For this micropropagation stage, the responses of the mature state in the two genotypes were similar. A small proportion of the stem tips were capable of rapidly establishing a basal callus, and there was no resumption of apical growth. Topping the stem tips led to the reactivation of axillary buds on a small number of explants. None of the buds were capable of continued growth to form a shoot. On the other hand, the stem tips of primary shoots obtained from primary explants, taken from mother-trees established from somatic embryos, had morphogenetic possibili-

*Table 6.* Clones RRIM 600 and PB 260 – rooting of shoots obtained during primary explant subculturing: effect of mother-tree origin. The values in brackets correspond to the number of shoots treated.

| | Origin of mother-plants | | | |
|---|---|---|---|---|
| | Clone RRIM 600 | | Clone PB 260 | |
| | Mature | Somatic embryo | Mature | Somatic embryo |
| % rooting of shoots from | | | | |
| 1st subculture | 0 (10) | 44 (25) | not enough shoots produced | 41 (56) |
| 2nd subculture | – | 77 (35) | not enough shoots produced | 46 (28) |
| 3rd subculture | – | 63 (41) | not enough shoots produced | – |
| 4th subculture | 0 (8) | 46 (37) | not enough shoots produced | – |
| 5th subculture | – | 41 (41) | not enough shoots produced | – |
| Mean | 0 (18) | 55 (119) | | 43 (84) |

ties enabling indefinite multiplication by cascade culturing the stems formed *in vitro*.

*Shoot rooting* (*Table 6*). With clone RRIM 600, it was impossible to initiate rooting in shoots obtained from mature plant material. Only the shoots regenerated from explants taken from somatic embryo-derived mother-trees rooted. With clone PB 260, the incapacity of the mature material to develop shoots during primary explant subculturing prevented us significantly testing the rooting capacity of the shoots obtained from this material. A trial on a not statistically significant sample did however show that this material was incapable of rooting. As with clone RRIM 600, rooting is only obtained in shoots regenerated from explants taken from somatic embryo-derived mother-trees.

## 5. Discussion

The experiments carried out on different clones enable us to determine the characteristics of somatic embryogenesis under the culture conditions in question, and the specific responses of each clone. It is already clear that despite the small number of genotypes tested, these responses were quite distinct at both qualitative and quantitative levels.

However, the genotypic effect seem to be less marked than in *Oryza sativa* (Ozawa and Komamine, 1989), *Medicago sativa* (Chen *et al.*, 1987) and *Gossypium* sp. (Trolinder and Xhixian, 1989), for which any modification to culture medium has positive effects on ontogenesis in certain clones and negative effects in others. For *Hevea*, we recently showed that optimizing the hormone supply based on a single clone leads to a significant improvement in embryogenic expression and ontogenesis in the four clones tested (Montoro *et al.*, 1992).

However, the structure of calli and their morphogenetic capacities are not strictly specific to the genotype but rather to a genotype/medium interaction (Montoro *et al.*, 1993); we in fact showed that for each genotype, callus structure and embryogenic expression could be modified by changing the auxin/cytokinin balance, or by increasing the sucrose or calcium concentration of the medium. Under the culture conditions described above in clone PB 260, histological studies revealed two types of embryogenesis: a) firstly, single-cell origin that aborts early, and b) the second one from highly meristematic clumps that change into globular proembryos (Michaux-Ferrière *et al.*, 1992). The abundance of embryogenic cells in the friable calli of PB 260, obtained on a medium with a high calcium content, and the formation of highly singulated embryos are similar to those obtained with PR 107 and RRIM 600, which are soft under normal culture conditions, prompt the hypothesis that embryogenesis from single-cell origin is successful in friable calli. Analyzing the *in vitro* performance of a species, and even a genotype, is, therefore, *a priori* a very complex business, and can only be carried out if closely linked to culture conditions.

The output of a technique results from the success rates for the different successive culture stages. The studies carried out make it possible to accurately define the importance of each stage and identify the most limiting stages. Hence, the work yield is the same for PR 107 and RRIM 600 (around 6 min per plantlet), although the strictly biological output for the one is twice that of the other: 3.2 explants/plantlet for PR 107 as against 5.9 explants/plantlet for RRIM 600. Despite its high embryogenic efficiency, clone PB 260 gives considerably lower yields due to its poor germination and embryo conversion rates. For clone PB 235, the very low yields suggest that culturing conditions were inappropriate, unless the cause is of genetic origin, but this is difficult to determine with any certainty.

The regeneration of plants obtained by somatic embryogenesis confirms the theory of complete rejuvenation insofar as the young plantlets reproduce the morphological characteristics of seedlings and the plants have very vigorous initial growth in the field. Furthermore, the results obtained show that the micropropagation capacity of explants taken from somatic embryo-derived mother-trees is markedly greater than that of explants taken from mature mother-trees. Logically, these results can be put down to a difference in the state of maturity/juvenility between the two types of plant material.

Rejuvenation by the somatic embryo technique is, therefore, reflected in the re-acquisition of micropropagation capacity. This confirms the juvenile character of plants obtained by somatic embryogenesis.

Somaplants will play a key role as juvenile controls of the genotypes selected in fundamental studies on understanding the physiological mechanisms involved in the change from a juvenile to a mature state or more simply in the search for biochemical markers of the juvenile state.

## 6. Conclusion

The results quoted here clearly show that although reproducible in several clones, this technique will not however enable a direct shift to mass production. The main limiting factors would seem to be, in order of importance: a) the development of an embryogenic material amplification phase, and b) for certain genotypes, the improvement of somatic embryo germination/conversion rates. The former is currently being studied; the conditions for obtaining calli that are both friable and embryogenic have been defined (Montoro *et al.*, 1993) and have enabled the establishment of sustained embryogenic cultures either on a gel medium or by cellular aggregate suspension in a liquid medium (Montoro, 1993). The optimization of the conditions for embryogenesis expression on these types of material is currently being looked into.

However, the technique developed is already sufficiently effective for small-scale production of a few hundred trees. These experimental productions could make it possible to set up field trials to evaluate the growth and latex production characteristics of this new planting material, i.e., to quantify the initial hypotheses on expected gains. Furthermore, somatic embryogenesis makes it possible to obtain juvenile mother-trees suitable for use in micropropagation. This will provide an opportunity of determining whether the recalcitrance observed on selected genotypes as regards micropropagation is of physiological origin, linked to the state of maturity of the plant material, or of genetic origin. Secondly, small-scale production by micropropagation could also be launched to set up field trials of the genotypes whose recalcitrance is physiological and whose embryogenic intensity is very weak. These trials will provide the first hints of the agronomic merits of the selected *Hevea* clones grown on their own roots.

## 7. References

Bonga, J.M., 1982. Vegetative propagation in relation to juvenility, maturity and rejuvenation. In: J.M. Bonga and D.J. Durzan (Eds.), Tissue Culture in Forestry, pp. 387–412. Martinus Nijhoff/Dr. W. Junk Publishers, The Hague.

Carron, M.P. and F. Enjalric, 1982. Studies on vegetative micropropagation of *Hevea brasiliensis* by somatic embryogenesis and *in vitro* microcutting. In: IAPTC (Eds.), Proc. 5th Intl. Cong. Plant Tissue and Cell Culture, 11–16 July, 1982, Tokyo, pp. 751–752.

Carron, M.P. and F. Enjalric, 1985. Somatic embryogenesis from inner integument of the seed of *Hevea brasiliensis* (Kunth) Müll. Arg. C.R. Acad. Sc. Paris 300(III): 653–658.

Carron, M.P., F. Enjalric, L. Lardet and A. Deschamps, 1989. Rubber (*Hevea brasiliensis* Müll. Arg.). In: Y.P.S Bajaj (Ed.), Biotechnology in Agriculture and Forestry, Vol. 5, Trees II, pp. 222–245. Springer Verlag, Berlin/Heidelberg.

Carron, M.P., H. Etienne, N. Michaux-Ferriere and P. Montoro, 1994. Somatic embryogenesis in rubber tree (*Hevea brasiliensis* Müll.Arg.). In: Y.P.S. Bajaj (Ed.), Biotechnology in Agriculture and Forestry, Vol. 30: Somatic Embryogenesis and Synthetic Seed I. Springer-Verlag, Berlin/Heidelberg (in press).

Chen T.H.H., J. Marowitch and B.G. Thompson, 1987. Genotypic effects on somatic embryogenesis and plant regeneration from callus cultures of *alfalfa*. Plant Cell Tiss. Org. Cult. 8: 73–81.

Compagnon, P., 1986. Le Caoutchouc Naturel. Biologie, Culture, Production. Maisonneuve and Larose, Paris.

Enjalric, F. and M.P. Carron, 1982. *In vitro* microcutting of *Hevea brasiliensis* (Kunth, Müll. Arg.) young plants. C.R. Acad. Sci. Paris 295: 259–264.

Etienne, H., A. Berger and M.P. Carron, 1991. Water status of callus from *Hevea brasiliensis* during induction of somatic embryogenesis. Physiol. Plant 82: 213–218.

Etienne, H., B. Sotta, P. Montoro, E. Miginiac and M.P. Carron, 1993a. Relations between exogenous growth regulators and endogenous indole-3-acetic acid and abscisic acid in the expression of somatic embryogenesis in *Hevea brasiliensis* (Müll. Arg.). Plant Sci. 88: 91–96.

Etienne, H., P. Montoro, N. Michaux-Ferrière and M.P. Carron, 1993b. Effects of desiccation, medium osmolarity and abscisic acid on the maturation of *Hevea brasiliensis* somatic embryos. J. Exp. Bot. 44(267): 1613–19.

Etienne H., B. Sotta, P. Montoro, E. Miginiac and M.P. Carron, 1993c. Comparison of endogenous ABA and IAA contents in somatic and zygotic embryos of *Hevea brasiliensis* (Müll. Arg.) during ontogenesis. Plant Sci. 92: 111–119.

Levandowsky, D.W., 1959. Multiplication de l'*Hevea brasiliensis* par bouture. Rev. Gen. Caoutch. Plast. 36(9): 1132–1141.

Livonière, H. de, 1992. Caoutchouc naturel: actualité et avenir. Caoutch. Plast. 719: 66–72.

McIndoe, K.G., 1958. The development of clonal rootstocks in Hevea. Quart. Circ., Ceylan Rubber Research Institute 34: 39–57.

Michaux-Ferrière, N., H. Grout and M.P. Carron, 1992. Origin and ontogenesis of somatic embryos in *Hevea brasiliensis*. Amer. J. Bot. 79(2): 174–180.

Montoro, P., H. Etienne, M.P. Carron and A. Nougarède, 1992. Effect of cytokinins on the induction of embryogenesis and the quality of somatic embryos in *Hevea brasiliensis* Müll. Arg. C.R. Acad. Sci. Paris 315: 567–574.

Montoro, P. 1993. Establishment of embryogenic suspension cultures in *Hevea brasiliensis*: effects of culture medium on callus histogenesis, friability and expression of embryogenesis. Thesis, University Montpellier II, Montpellier.

Montoro P., H. Etienne, N. Michaux-Ferrière and M.P. Carron, 1993. Callus friability and somatic embryogenesis in *Hevea brasiliensis*. Plant Cell Tiss. Org. Cult. 33: 331–338.

Murashige, T. and F. Skoog, 1962. A revised medium for rapid growth on bioassays with tobacco tissue cultures. Physiol. Plant 15: 473–497.

Muzik, T.J. and M.J. Crusado, 1958. Transmission of juvenile rooting ability from seedling to adults of *Hevea brasiliensis*. Nature 181: 1288.

Nozeran, R., G. Ducreux and L. Rossignol-Bancilhon, 1977. Réflexions sur les problèmes de rajeunissement chez les végétaux. Bull. Soc. Bot. Fr. Lett. Bot. 129(2): 107–130.

Ozawa, K. and A. Komamine, 1989. Establishment of a high frequency embryogenesis from long-term cell suspension cultures of rice (*Oriza sativa* L.). Theor. Appl. Genet. 77: 205–211.

Quoirin, M. and P. Lepoivre, 1977. Étude de milieux adaptés aux cultures *in vitro* de *Prunus*. Acta Hort. 78: 437–442.

Rubber Statistical Bulletin, 1992. International Study Group, London.

Serier, J.B., 1993. Histoire du Caoutchouc. Desjonquères, Paris.

Trolinder, N.L. and C. Xhixian, 1989. Genotype specificity of the somatic embryogenesis response in cotton. Plant Cell Rep. 8: 133–136.

Wan, A.R., H. Ghandimathi, O. Rohani and K. Paranjothy, 1982. Recents developments in tissue culture of Hevea. In: A.N. Rao (Ed.), Tissue Culture of Economically Important Plants, pp. 152–158. COSTED, Singapore.

Wang, Z., X. Zeng, C. Chen, H. Wu, Q. Li, G. Fan and W. Lu, 1980. Induction of rubber plantlets from anther of *Hevea brasiliensis* Müll. Arg. *in vitro*. Chinese J. Trop. Crops 1: 25–26.

Wiersum, N., 1955. Observations of the rooting of Hevea cuttings. Arch. Rubber Cultivation 32: 213–243.

# 9. Somatic embryogenesis in birch (*Betula pendula* Roth.)

Vladimír Chalupa

**Contents**



## 1. Introduction

The genus *Betula* consists of 40–60 species of deciduous trees and shrubs, distributed in the cooler parts of Europe, Asia, and North America. Most species are extremely hardy, some grow at the far north around the Arctic Circle, where they form the northern treeline. Birches vary in size from tiny shrubs to large forest trees reaching up to 30 m in height. Birches are fast growing and pioneer species, able to establish a cover on bare and devastated or burned lands. They are often the first trees which appear in the sequence of natural rejuvenation. However, birches are not long-lived trees, and they are gradually replaced by other species, such as oaks, beeches or spruces that have grown up in their shelter. Foresters, frequently, use the light shelter of birches to establish forest stands of shade tolerant trees such as beech or spruce. Birches are often used for ornamental planting because of their graceful growth habit, handsome foliage and white bark.

Silver birch (*Betula pendula* Roth.) is one of the most important birch species. The natural range of silver birch is throughout Europe and central parts of Asia. It merges into another closely related species, *B. platyphylla* Sukacz., in central Siberia. Silver birch is commonly grown in the lowland areas as well as at high altitudes on the mountains. It can be seen growing at maximum elevation of 3000 m in eastern Turkey. Silver birch is an economically important fast growing species and has a great resistance to unfavourable external conditions like air pollution. For these reasons, it is often used for reforestation in regions with high air pollution hazards. Silver birch grows on a wide range of sites, from rich moist loams to dry sandy soils.

Birch wood is diffuse-porous, light-coloured, and too soft to be used as

*S. Jain, P. Gupta & R. Newton (eds.), Somatic Embryogenesis in Woody Plants, Vol. 2,* 137–151.

building material. However, its excellent grain quality is very valuable for making furniture. A special form of silver birch, the curly-birch (*Betula pendula* var. *carelica*) exhibits the unusual structure of wood that is very valuable. In Scandinavia, silver birch trees of a good form and large size are common. The wood is widely used for furniture, pulpwood, and veneer and plywood production. It makes also an excellent firewood and provides a considerable quantity of charcoal. Birch wood decays quickly when exposed to wetness and dryness alternatively. It is not durable outside, but lasts well if kept dry. The bark can be used to make a variety of containers. Birch twigs and bark also produce an oil which is used as a leather preservative giving it fragrance.

Genetic polymorphism occurs in birch and different forms have been recognized. Provenance studies have shown a considerable variability within provenances. Silver birch with large natural distribution includes forms and geographic races which exhibit marked differences in growth rates, stem form, wood quality, and resistance to diseases. Hybridization within the genus *Betula* is well-documented, and natural and artificial hybrids between different species have been described (Johnsson, 1945; Eifler, 1964).

Silver birch possesses many desirable traits and biological characteristics which facilitate its breeding. High natural variability, fast growth, early flowering, and high seed production favour the fast improvement of birches by various breeding techniques. In Finland, large quantities of improved seed are produced using plastic greenhouses where selected trees are grown and cross-pollinated naturally or artificially. Twelve silver birch seed orchards (in polythene greenhouses) are used for the production of genetically improved seed. There are five two-clone silver birch seed orchards producing tested crossing combinations (Lepistö, 1973; Hagquist, 1991).

*Betula pendula* is a diploid species with 2n = 28 chromosomes (some other birch species are tetraploids, for example *B. pubescens* 2n = 56, some species are hexaploids 2n = 84). Polyploidy has been induced by colchicine treatment in *B. pendula* (Valanne, 1972). The rate of growth of colchicine-induced *Betula* polyploids was inferior to that of the chromosomally normal, diploid parental trees (Johnsson, 1956; Pieninkeroinen and Valanne, 1989).

Silver birch trees are anemophilous. The pendulous staminate catkins are formed in late summer and open after considerable elongation in early spring. Catkin discharges all pollen within few days (Sarvas, 1952). The pistillate catkins, terminal on shoot spurlike branchlets, appear with leaves in the spring. The female flower has a single ovary with two styles and a three-lobed bract. The silver birch produces seed in small catkin-like cones. The fruit is a nutlet with a membranous wing on each side for wind dispersal. The cones mature in one season and break up on the trees, when the seed, a tiny winged nut, is released. Seed dispersal is usually done by wind, sometimes by water. Natural germination of seed occurs in the spring following dispersal. Germination is epigeous. Most seeds complete germination within four weeks without a preliminary period of stratification.

### 1.1. *Vegetative propagation*

Foresters are interested in birch vegetative propagation because of its potential to multiply economically important varieties in large numbers. Forest yield can be enhanced significantly by large-scale multiplication of selected genotypes with improved growth rates, valuable quality of wood, and high stress and disease tolerance. Vegetative propagation is an important tool of preserving unique characteristics of some of the selected trees. Much of the fundamental biology of birch vegetative propagation is still poorly understood. Many different factors, both internal and environmental, influence the success of vegetative propagation. Currently, birch is vegetatively propagated by grafting, or rooting of cuttings, and by *in vitro* methods. Grafting techniques are laborious. Graft failure can occur due to delayed incompatibility.

The success of rooting in birch cuttings is largely dependent on the age of the mother plant. The rooting capacity of cuttings decreases with the increasing age of a mother tree. Vegetative propagation by cuttings is restricted mainly to young material because aging reduces the rooting ability. Cuttings collected from young birch plants exhibited high rooting potential. Aldén *et al.* (1977) found that the rooting ability was high in 3–5 year-old birch ortets. In our experiments (Chalupa, 1982), cuttings of *B. pendula* were taken from 2–4 year-old plants. The rooting percentage was high (60–80%) when softwood cuttings were used and treated with growth regulators (IBA, indole-3-butyric acid). Auxin treatment of cuttings promoted significantly adventitious root formation. The success in rooting of cuttings was greatly dependent on the maintenance of a satisfactory water balance in plants. Clonal differences in rooting ability of birch cuttings were observed. There is a substantial evidence which indicates that the rooting is genetically controlled (Chalupa, 1982).

### 1.2. In vitro *propagation*

Birch can be propagated *in vitro* by three different type of cultures: a) callus, b) organ, and c) embryogenic.

#### 1.2.1. *Callus cultures*

Callus cultures are easily initiated from various type of explants. Experiments with birch callus cultures started a long time ago. Jacquiot (1955) described formation of unorganized callus and observed differentiation of organs in birch callus cultures. Huhtinen and Yahyaoglu (1974) regenerated early flowering trees of *B. pendula* from calli initiated on MS medium (Murashige and Skoog, 1962) containing kinetin (2.32 $\mu$M) and indole-3-acetic acid (IAA, 142.70 $\mu$M). The regenerated trees produced male flowers at the young age. Srivastava and Steinhauer (1981a) regenerated birch plants from callus derived from catkins. Callus maintained its organogenic capacity for more than 10 months. Histological studies of the callus revealed cells of different shape. Cell clusters simulating early stages of embryogeny were also observed. New plants were also regenerated from callus initiated from shoot buds (Srivastava and Steinhauer, 1981b). Simola (1985) produced plants from subculturable callus derived from young leaf blades of *B. pendula* f. *purpurea*. Shoot differentiation was stimulated on a modified N7 medium (Chu *et al.*, 1975) containing zeatin or zeatin riboside. Excised shoots were rooted on media containing a low level of auxin or lacking auxin. Cameron (1990) found autotetraploids (2n = 56) in a population of plants regenerated from callus cultures derived from stem internodes of *B. pendula*. These plants exhibited twisted foliage and slow rate of growth. The frequency of autotetraploidy varied with the auxin used in the medium. The frequency was 3.8% when IBA was used, and increased to 11% using $\alpha$-naphthalene-acetic acid (NAA). Most regenerated plants were morphologically normal with 2n = 28 chromosomes. The autotetraploids have a potential to produce triploid birches (2n = 42).

### 1.2.2. *Organ cultures*

Rapid clonal micropropagation of *B. pendula* was achieved by organ culture, using shoot tips, nodal segments, and isolated buds as initial explants (Chalupa, 1981a,b, 1983, 1985b). The main advantages of using this method are: a) the multiplication rate is high, and b) cultures are genetically stable. This method was successfully used for micropropagation of juvenile seedlings and 20–30 year-old mature trees. Rapid shoot multiplication was achieved on a modified MS medium, WPM (Lloyd and McCown, 1980), and BTM (Chalupa, 1981b, 1983) supplemented with 6-benzylaminopurine (BA, 0.88–4.44 μM) and 0.1 μM IBA. Shoot tips and axillary buds started to grow in few days and produced new shoots within 4–5 weeks. At the basal end of explants touching the culture medium, new tissue with induced adventitious buds developed and later on multiple adventitious shoots were produced. Axillary and adventitious shoots were harvested and used either for rooting or multiplication. After removing long shoots, the remaining basal tissues with adventitious buds and short shoots were divided into 3–8 pieces and placed on a fresh multiplication medium where shoot proliferation continued. Tissue with adventitious buds also formed on leaves if developing leaves were in contact with the medium (Chalupa, 1981a). By using organ culture, birch was multiplied either by axillary shoots or by a combination of adventitious and axillary shoots. Regenerated microshoots were rooted on WPM or GD (Gresshoff and Doy, 1972) medium (half strength) containing low level of auxin (IBA 0.49 μM), or were rooted *ex vitro* in a potting mixture (peat and perlite 1:1, v/v). Roots developed within 2–3 weeks. The rooting percentage was high (85–95%). Experiments with micropropagation of selected mature trees of *B. pendula* (Chalupa, 1989) showed that mature trees can be reliably propagated by organ cultures using apical and axillary buds, shoot tips, nodal segments, and leaf segments as initial explants. Shoot multiplication was stimulated on WPM containing low levels of BA (2.22–4.44 μM) or BA plus IBA (0.25 μM). Adventitious shoots were also induced on leaf segments cultured on WPM containing 4.44–8.87 μM BA. Each leaf segment with induced adventitious buds produced 2–11 shoots. Mature birch trees were multiplied using adventitious and axillary shoots which were rooted in WPM or in a potting mixture. Regenerated trees were planted in the field. At the end of the fifth growing season, micropropagated trees were more than 3 m tall. No abnormalities were observed in the growth and morphology of micropropagated trees. Trees regenerated from the same mother tree exhibited uniformity in stem form and leaf morphology (Chalupa, 1989).

Ryynänen and Ryynänen (1986) produced plants from adult curly-birch trees (*B. pendula* var. *carelica* Mercklin). Curly-birch wood is sold according to weight. It costs much more than the ordinary birch. Cultures were initiated on a modified MS medium containing 4.44 μM BA. Adventitious buds were induced after explants were transferred on the bud inducing MS medium containing 44.38 μM BA and 1.07 μM NAA. Shoot elongation was stimulated on a modified MS medium containing 2.22 μM BA and 2.85 μM IAA. Shoots were rooted on a medium lacking growth regulators. Särkilahti (1988) regenerated tetraploid plantlets from cultured apical and axillary buds of a mature polyploid *B. pendula* tree. Bud explants were grown on a modified MS medium supplemented with 8.87 μM BA and 0.05 μM NAA. On this medium, initially, adventitious buds were induced, and later on developed into shoots. The regenerated plantlets had a tetraploid chromosome set (n = 56), and an altered leaf morphology, typical of colchicine-polyploid birches. In Finland, selected mature *B. pendula* trees were micropropagated commercially using axillary shoot proliferation system (Jokinen and Törmälä, 1991). Initiation and multiplication of shoot cultures occurred on WPM supplemented with 4 μM BA. Regenerated shoots were rooted on WPM containing 1 μM IBA. The clonal fidelity of micropropagated plants was high. Genotype of the tree affected the multiplication rate more than the age of mother plant. Meier-Dinkel (1992) studied internal and external factors affecting micropropagation of juvenile and mature trees of *B. pendula*, *B. pubescens*, and a hybrid *B. platyphylla* var. *japonica* × *B. pendula*. Trees were propagated by organ cultures. Explants were grown on MS medium or WPM. The culture establishment and shoot multiplication was best stimulated by 8.87 μM BA. High rooting percentage was obtained on half strength MS medium containing 0.49 μM IBA or 0.54 μM NAA. In Sweden, Welander (1993) investigated the factors affecting micropropagation of juvenile and adult trees of *B. pendula*. The modified N6 medium containing 4.4 μM BA and 0.005 μM NAA was most suitable for culture initiation (Chu *et al.*, 1975). The best medium for shoot multiplication was WPM containing 2.2 μM BA plus 0.005 μM NAA. A rooting medium was developed on the basis of WPM amended with one fifth strength of macronutrients, and 0.49 μM IBA. On this medium, rooting efficiency was 95–100%. In field experiments, the number of surviving plants correlated with the initial height. The regenerated plants produced from adventitious or axillary buds were uniform.

### 1.2.3. *Embryogenic cultures*

Somatic embryogenesis has a number of advantages when compared with other *in vitro* propagation systems. It has a potential for: a) fast clonal propagation of superior genotypes, and b) the production of large number of plantlets in a short time. Somatic embryogenesis has been described in various species of conifers and broadleaves (Radojević, 1979; Tulecke and McGranahan, 1985; Chalupa, 1985a,b; Hakman *et al.*, 1985; Nagmani and Bonga, 1985; Gupta and Durzan, 1986; Merkle and Sommer, 1986; Durzan and Gupta, 1987). Embryogenic regeneration system is highly desirable in tree breeding and gene transfer applications. Transformed embryogenic tissue can produce large number of somatic embryos and plantlets. Embryogenic cultures can also be used for cryogenic storage, and as a source of protoplasts that possess the capacity to form somatic embryos. Recently, somatic embryos have been produced in embryogenic cultures of *B. pendula* (Chalupa, 1987; Kurtén *et al.*, 1990; Nuutila *et al.*, 1991; Nuutila and Kauppinen, 1992; Hämäläinen *et al.*, 1992, 1993), and plants have been regenerated from birch somatic embryos.

## 2. Initiation of embryogenic cultures

Formation of birch embryogenic cultures has been achieved using different type of explants such as immature and mature seeds, shoot apices, and leaf segments. The ability to produce embryogenic cultures differed markedly between genotypes. Low numbers of embryogenic cultures were often initiated in tested genotypes. Our experiments suggested a marked genetic influence on the potential to produce embryogenic cultures. Initiation of birch embryogenic cultures was also dependent on the medium composition, and the type and concentration of auxin in the medium. The data suggest that an optimal induction medium has to be often determined empirically for genotypes used in the experiment. So far birch embryogenic cultures have been initiated only from juvenile explants. A reliable system needs to be developed using explants from mature and genetically proven superior trees.

Embryogenic cultures of *B. pendula* were initiated from immature and mature seeds, and shoot apices and leaf segments of 1–3 month-old seedlings (Chalupa, 1987). Seeds were harvested from eleven open pollinated trees during June – September. Seeds were surface sterilized by immersion in 70% ethanol for 3 min followed by a 20 min immersion in 7.5% calcium hypochlorite solution and then washed three times with sterile distilled water (shoot apices were surface sterilized in 0.1% mercuric chloride solution for 20 min). Explants were placed in 100 ml flasks containing 20 ml of agar

(0.6%) medium. Explants were cultured on a modified MS medium containing 1.37 mM glutamine, and 800 mg · $l^{-1}$ casein hydrolysate. The medium was supplemented with 2,4-dichlorophenoxyacetic acid (2,4-D, 9.0 or 13.6 μM), BA (2.22 μM) and kinetin (2.32 μM). Sucrose (20 g · $l^{-1}$) was used as a carbon source. The pH of the medium was adjusted to 5.7 before autoclaving for 20 min at 121°C. Cultures were incubated in growth cabinets at 23°C with a 16 h photoperiod (light intensity 50 μE $m^{-2}$ $s^{-1}$) under cool fluorescent lamps. The medium supported callus formation from all type of explants. Calli were subcultured on the fresh medium every 4–5 weeks. Within 5–7 weeks, the frequency of callus production was 4–73%. After 6–9 weeks, the globular structures developed in some cultures. The development of embryogenic structures was stimulated after the transfer of callus tissue on a medium supplemented with reduced concentration of growth regulators (0.67 μM 2,4-D and 0.23 μM kinetin). Globular and torpedo shaped structures developed in embryogenic cultures within 4–6 weeks. Embryogenic tissue consisted of small meristematic cells, which were thin-walled and highly cytoplasmic. Only a small portion of growing callus cultures (2–8%) exhibited embryogenic potential. The potential to produce embryogenic cultures varied greatly due to genotypic differences.

Kurtén *et al.* (1990) induced somatic embryogenesis in *B. pendula* using eight different families, obtained by crossing parent plants with different herbivore resistance. Seeds, seedlings, and leaves from 1-year-old plants were used for the establishment of tissue culture systems. The plant material was surface sterilized with 70% ethanol (5 min) and sodium hypochlorite (leaves for 10 min, and seeds for 30 min). The explants were cultured in petri dishes poured with agar-solidified N7 basal medium (Simola, 1985) supplemented with 9.05 μM 2,4-D, 2.32 μM kinetin, and 1 g · $l^{-1}$ casein hydrolysate. Media were solidified with 0.8% agar, and autoclaved at 121°C for 20 min. Cultures were grown under 16 h photoperiod (light intensity 125 μE $m^{-2}$ $s^{-1}$) at 23°C. Callus cultures were initiated from all type of tested explants. Embryogenic calli were initiated after the transfer of calli on N7 basal medium. White globular structures were observed on the callus surface 3 weeks after transfer. Embryogenic calli were induced in three birch families out of eight. The family had a more pronounced effect on somatic embryogenesis than the type of explant. Embryogenic cultures were derived from all type of explants. The embryogenic calli were friable and anthocyanin was produced in cultures.

Nuutila *et al.* (1991) studied the effect of different concentrations of sucrose and inorganic nitrogen on somatic embryogenesis in birch cultures. Calli were initiated from leaf segments cultured on agar N7 basal medium supplemented with 9.05 μM 2,4-D, 2.32 μM kinetin and 1 g · $l^{-1}$ casein hydrolysate. The calli produced on agar medium were suspended and cultured in the liquid medium. Later the cells were passed through the sieve (343 μm) and collected by centrifugation and they were used as inoculum onto the embryo production medium. Embryo production medium consisted of N7 basal medium supplemented with 0.45 μM 2,4-D, 0.09 μM kinetin, 50 mg · $l^{-1}$ casein hydrolysate, and was solidified with 0.3% gelrite. Somatic embryos developed on this medium.

### 3. Development of somatic embryos

The development of somatic embryos was stimulated after transfer of embryogenic tissue from a nutrient medium containing a higher concentration of growth regulators onto the medium containing either reduced concentration or devoid of growth regulators. Chalupa (1987) observed the formation of birch somatic embryos after transfer of embryogenic tissue on to a modified MS medium supplemented with reduced concentration of 2,4-D (0.67 μM) and kinetin (0.23 μM). After transfer to this medium, somatic embryos developed in embryogenic tissue within 4–6 weeks (Fig. 1). Embryogenic

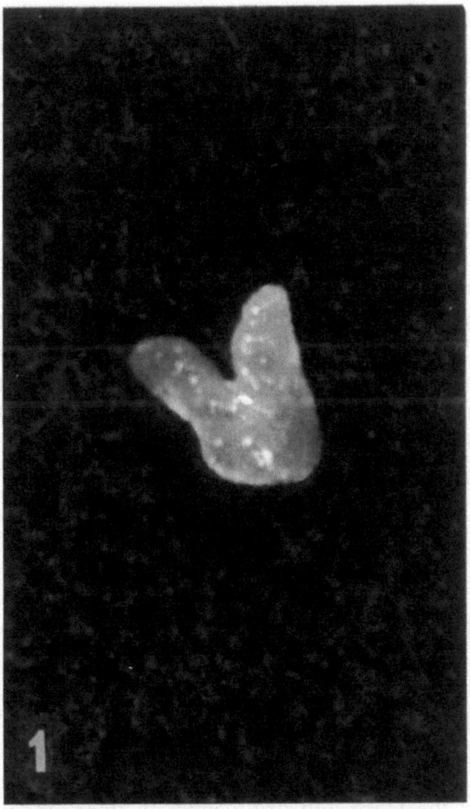


*Figure 1.* Somatic embryo of birch (*Betula pendula*).

tissue was characterized by regions of meristematic cells, which were small, thin-walled and highly cytoplasmic. Globular somatic embryos developed at the periphery of the meristematic regions and consisted of densely cytoplasmic cells. At later stages, embryos were surrounded by epidermic cells. Late heart-shaped embryos had developed well-defined apical and root meristems. Somatic embryos developed only in small part of calli initiated from explants. On the other hand, somatic embryos were produced in cultures derived from all used different explants. This indicates the importance of genotypic factors in the ability of producing birch embryogenic cultures. The influence of the developmental stage of the initial explants was not clearly documented in birch embryogenic cultures. Somatic embryos were produced in cultures derived from immature seeds and also in cultures derived from leaf segments.

Kurtén *et al.* (1990) achieved the formation of somatic embryos after calli were transferred from N7 basal medium containing 2,4-D and kinetin onto N7 basal medium lacking growth regulators and solidified with 0.3% gelrite (N 70 medium). Embryogenic calli formed somatic embryos on the callus surface 3–4 weeks after transfer. In two cell lines demonstrating a strong embryogenic potential, globular shaped embryos appeared on callus surface even without transferring embryogenic tissue on N 70 medium. However, the somatic embryos did not develop beyond the heart shaped stage. Somatic embryos also developed in the suspension cultures (N7 growth medium) without transferring for 2–3 weeks. These results indicate that birch cultures have an ability to form embryos even on media containing higher concentrations of auxin (2,4-D).

Nuutila *et al.* (1991) described the formation of somatic embryos on N7 basal medium supplemented with 0.45 μM 2,4-D, 0.09 μM kinetin, and 50 mg $\cdot\, l^{-1}$ casein hydrolysate and solidified with 0.3% gelrite (embryo production medium). They developed a model on the basis of experiments and showed that maximum embryo production can be obtained in the medium amended with 35 mM inorganic nitrogen and 20.8 g $\cdot\, l^{-1}$ sucrose. The results of eight experiments supported this model. Somatic embryo production ranged from 113 to 150 embryos/100 mg inoculum. Experimental design and mathematical modelling were found quite useful in embryogenic studies.

The development of somatic embryos, cell growth and nutrient uptake of an embryogenic and non-embryogenic cell line were studied by Nuutila and Kauppinen (1992). Embryogenic and non-embryogenic cell lines were cultured in liquid N7 basal medium supplemented with 0.45 μM 2,4-D, 0.09 μM kinetin, and 50 mg $\cdot\, l^{-1}$ casein hydrolysate. The embryogenic and non-embryogenic cell suspensions had different growth rate in the same medium. The non-embryogenic cells grew well in the medium, and the highest dry weight was achieved in 21 days. Embryogenic suspension culture exhibited a small increase in the fresh weight during the first two weeks. After two weeks the fresh and dry weight of embryogenic cultures started to increase. At the end of the experiment (42 days), the dry weight of embryogenic and non-embryogenic cultures was almost the same. The highest increase of somatic embryo formation was observed in the third and fourth week. During this period, a number of somatic embryos increased from 30 to 500 embryos/flask. The non-embryogenic cell lines rapidly hydrolyzed sucrose and consumed the glucose as a carbon source. The embryogenic cell line also rapidly hydrolyzed sucrose, but the monosacharides were consumed after the embryos started to germinate. It was suggested that high concentrations of sucrose enhanced the initiation of embryo maturation. In embryogenic suspensions, ammonium and nitrate were taken up only when the embryos were visibly germinating and ammonium was used faster than nitrate. According to Durzan (1987) ammonia assimilation contributes to cell organization.

### 3.1. *Classification of somatic embryos with computer vision*

Hämäläinen *et al.* (1992, 1993) classified birch somatic embryos by computer vision. Embryogenic cells were cultured in liquid N7 basal medium supplemented with 0.45 μM 2,4-D, 0.09 μM kinetin, and 50 mg · $1^{-1}$ casein hydrolysate. After two weeks, globular-shaped proembryos were visible, and within four weeks the suspension contained embryos of all developmental stages, undifferentiated cells, and cell clumps. The developing somatic embryos were collected, resuspended in fresh medium, and poured into petri dishes for embryo classification. The specific classification algorithm developed for birch somatic embryos showed that automated recognition of embryos at different developmental stages is possible without a significant number of false selections. The probability of classifying a non-embryo as an embryo was less than 1%. These results suggest that a reliable computer vision system suitable for automated monitoring of samples from the bioreactor has been developed.

## 4. Conversion of somatic embryos into plantlets and growth of regenerated plants

Germination and conversion of birch somatic embryos into plantlets was achieved either on the same medium used to develop somatic embryos, or germination was stimulated by medium changes and alternation of physical conditions. Chalupa (1987) observed the germination of birch somatic embryos after transfer of embryogenic tissue with developing embryos on WPM containing 0.25 μM IBA or lacking growth regulators. Somatic embryos germinated and leafy plantlets developed within 4–6 weeks. Germinating somatic embryos produced mostly normal plantlets with a leafy shoot and radicle. Development of abnormal plantlets was also observed. Plantlets with multiple shoots or lacking radicle were found. Well-developed plantlets were

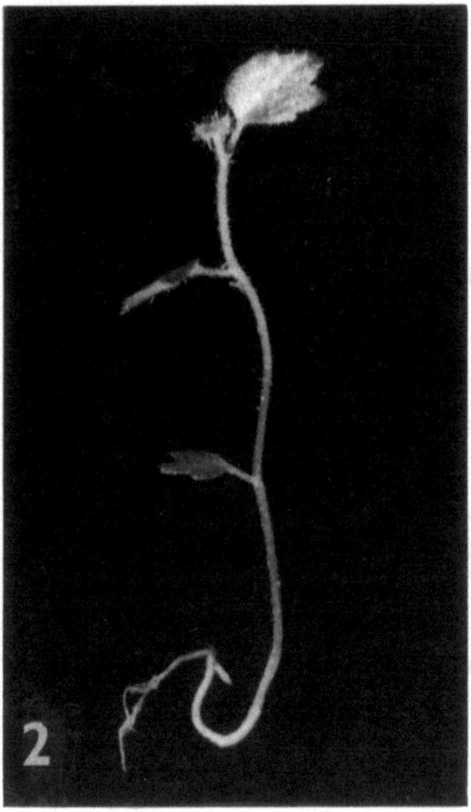


*Figure 2.* Birch plantlet derived from a somatic embyo.

transplanted into a potting mixture (peat and perlite, 1:1 v/v) (Fig. 2). Plantlets were grown under high air humidity and continuous light in growth cabinets for 2–3 months to achieve stem elongation and formation of new leaves. High humidity was gradually reduced to normal values. Most potted plants survived. After acclimatization, 18 plants regenerated from somatic embryos were planted in a nursery. The survival of birch plants regenerated from somatic embryos and planted in the nursery was high, only two plants died. The height of trees produced from somatic embryos was comparable with those produced *in vitro* by organ cultures. Trees produced from somatic embryos exhibited normal growth and appearance (Fig. 3).

Kurtén *et al.* (1990) found that somatic embryos either formed on N 70 medium or later transferred onto it, developed and germinated normally. Some abnormal germination was also observed. Embryos with fused cotyledons, with multiple shoots or swollen hypocotyl and non-functioning shoot apices were found. So far, 49 somatic embryo plantlets (emblings) of one

*Figure 3.* Plant of *Betula pendula*, regenereted from a somatic embryo, growing in the field.

birch family have been potted in the soil and transferred to greenhouse conditions.

## 5. Conclusions

Silver birch is a tree species which demonstrates a considerable amount of variability including significant differences in growth rates, stem form, wood quality, and resistance to diseases. *In vitro* propagated valuable forms and selected genotypes have a potential for increasing forest yield. Somatic embryogenesis has a number of advantages when compared with other *in vitro* propagation systems.

Birch embryogenic cultures were initiated from different type of explants, such as immature and mature seeds, shoot apices, and leaf segments. The genotypic effect exhibited marked differences in the ability to produce embryogenic cultures. Exogenous auxin was essential to initiate birch embryogenic tissue. Embryogenic cultures have an ability to form embryos on media containing auxin, however, the development of somatic embryos was stimulated after transfer of embryogenic tissue onto media lacking growth regulators, or containing reduced concentration of auxin and cytokinin. The germination of somatic embryos was achieved on media lacking growth regulators and produced mostly normal plantlets with leafy shoot and radicle. However, abnormal germination of somatic embryos was also observed. Well-developed plantlets were transplanted into potting mixture and after acclimatization, regenerated plants were planted in the nursery.

At present, the efficiency of plant production from birch embryogenic cultures is low. The development of an efficient system to increase the number of emblings will be helpful in commercial applications of birch somatic embryogenesis. So far birch embryogenic cultures have been initiated from juvenile explants. In the future, a reliable efficient plant regeneration system using explants from adult trees needs to develop.

## 6. References


Aldén, T., I. Dormling, C. Ehrenberg, H. Kellerstam and S. Persson, 1977. Some methods for vegetative propagation. In: Vegetative Propagation of Forest Trees-Physiology and Practice, Symposium, Uppsala, pp. 137–147.

Cameron, A.D., 1990. Autotetraploid plants from callus cultures of *Betula pendula* Roth. Tree Physiol. 6: 229–234.

Chalupa, V., 1981a. *In vitro* propagation of birch (*Betula verrucosa* Ehrh.). Biol. Plant 23: 472–474.

Chalupa, V., 1981b. Clonal propagation of broadleaved forest trees *in vitro*. Commun. Inst. For. Cech. 12: 255–271.

Chalupa, V., 1982. Vegetative propagation of broadleaved trees by cuttings. Lesnictví (Forestry) 28: 21–30.

Chalupa, V., 1983. Micropropagation of conifer and broadleaved forest trees. Commun. Inst. For. Cech. 13: 7–39.

Chalupa, V., 1985a. Somatic embryogenesis and plantlet regeneration from cultured immature and mature embryos of *Picea abies* (L.) Karst. Commun. Inst. For. Cech. 14: 57–63.

Chalupa, V., 1985b. *In vitro* propagation of *Larix*, *Picea*, *Pinus*, *Quercus*, *Fagus*, and other species using adenine-type cytokinins and thidiazuron. Commun. Inst. For. Cech. 14: 65–90.

Chalupa, V., 1987. Somatic embryogenesis and plant regeneration in *Picea*, *Quercus*, *Betula*, *Tilia*, *Robinia*, *Fagus* and *Aesculus*. Commun. Inst. For. Cech. 15: 133–148.

Chalupa, V., 1989. Micropropagation of mature trees of birch (*Betula pendula* Roth.) and aspen (*Populus tremula* L.). Lesnictví (Forestry) 35: 983–993.

Chu, C.C., C.C. Wang, C.S. Sun, C. Hsü, K.C. Yin, C.Y. Chu and F.Y. Bi, 1975. Establishment of an efficient medium for anther culture of rice through comparative experiments on the nitrogen sources. Sci. Sin. 18: 659–668.

Durzan, D.J., 1987. Ammonia: Its analogues, metabolic products and site of action in somatic embryogenesis. In: J.M. Bonga and D.J. Durzan (Eds.), Cell and Tissue Culture in Forestry, Vol. 2, pp. 92–136. Martinus Nijhoff Publishers, Dordrecht.

Durzan, D.J. and P.K. Gupta, 1987. Somatic embryogenesis and polyembryogenesis in Douglas-fir cell suspension cultures. Plant Sci. 52: 229–235.

Eifler, I., 1964. Untersuchungen zum Bestäubungsvorgang und der Samenentwicklung bei Birkenartkreuzungen. Züchter 34: 305–312.

Gresshoff, P.M. and C.H. Doy, 1972. Development and differentiation of haploid *Lycopersicon esculentum* (tomato). Planta 107: 161–170.

Gupta, P.K. and D.J. Durzan, 1986. Somatic polyembryogenesis from callus of mature sugar pine embryos. Bio/Techn. 4: 643–645.

Hagquist, R., 1991. Production of genetically improved birch seed and micropropagated seedlings. In: Foundation for Forest Tree Breeding in Finland, Annu. Rep., Helsinki, p. 31.

Hakman, I., L.C. Fowke, S. von Arnold and T. Eriksson, 1985. The development of somatic embryos in tissue cultures initiated from immature embryos of *Picea abies* (Norway spruce). Plant Sci. 38: 53–59.

Hämäläinen, J.J., U. Kurtén and V. Kauppinen, 1993. Classification of plant somatic embryos by computer vision. Biotechn. Bioeng. 41: 35–42.

Hämäläinen, J.J., U. Kurtén, V. Kauppinen and J. Heilala, 1992. Automated classification of somatic plant embryos. Acta Hort. 319: 601–606.

Huhtinen, O. and Z. Yahyaoglu, 1974. Das frühe Blühen von aus Kalluskulturen herangezogenen Pflänzchen bei der Birke (*Betula pendula* Roth.). Sil. Gen. 23: 32–34.

Jacquiot, C., 1955. Formation d'organes par le tissu cambial d'*Ulmus campestris* L. et de *Betula verrucosa* Gaert. cultivé *in vitro*. C.R. Acad. Sci. Paris 240: 557–558.

Johnsson, H., 1945. Interspecific hybridization within the genus *Betula*. Hereditas 31: 163–176.

Johnsson, H., 1956. Auto- and allotriploid *Betula* families, derived from colchicine treatment. Z. Forstgen. Forstpflanzenzücht. 5: 65–70.

Jokinen, K. and T. Törmälä, 1991. Micropropagation of silver birch (*Betula pendula* Roth.) and clonal fidelity of mass propagated birch plants. In: M.R. Ahuja (Ed.), Woody Plant Biotechnology, pp. 31–36. Plenum Press, New York.

Kurtén, U., A.M. Nuutila, V. Kauppinen and M. Rousi, 1990. Somatic embryogenesis in cell cultures of birch (*Betula pendula* Roth.). Plant Cell Tiss. Org. Cult. 23: 101–105.

Lepistö, M., 1973. Accelerated birch breeding in plastic greenhouses. For. Chron. 49: 172–173.

Lloyd, G. and B. McCown, 1980. Commercially-feasible micropropagation of mountain laurel, *Kalmia latifolia*, by use of shoot-tip culture. Comb. Proc. Int. Plant Prop. Soc. 30: 421–427.

Meier-Dinkel, A., 1992. Micropropagation of birches (*Betula* spp.). In: Y.P.S. Bajaj (Ed.), Biotechnology in Agriculture and Forestry, Vol. 18, pp. 40–81. Springer-Verlag, Berlin.

Merkle, S.A. and H.E. Sommer, 1986. Somatic embryogenesis in tissue culture of *Liriodendron tulipifera*. Can. J. For. Res. 16: 420–422.

Murashige, T. and F. Skoog, 1962. A revised medium for rapid growth and bioassays with tobacco tissue cultures. Physiol. Plant 15: 473–497.

Nagmani, R. and J.M. Bonga, 1985. Embryogenesis in subcultured callus of *Larix decidua*. Can. J. For. Res. 15: 1088–1091.

Nuutila, A.M. and V. Kauppinen, 1992. Nutrient uptake and growth of an embryogenic and a non-embryogenic cell line of birch (*Betula pendula* Roth.) in suspension culture. Plant Cell Tiss. Org. Cult. 30: 7–13.

Nuutila, A.M., U. Kurtén and V. Kauppinen, 1991. Optimization of sucrose and inorganic nitrogen concentrations for somatic embryogenesis of birch (*Betula pendula* Roth.) callus cultures: A statistical approach. Plant Cell Tiss. Org. Cult. 24: 73–77.

Pieninkeroinen, K. and T. Valanne, 1989. Old colchicine-induced polyploid materials of *Betula pendula* Roth. and *Betula pubescens* Ehrh. Ann. Sci. For. 46, Suppl: 264s–266s.

Radojević, L., 1979. Somatic embryogenesis and plantlets from callus cultures of *Paulownia tomentosa* Stued. Z. Pflanzenphysiol. 91: 57–62.

Ryynänen, L. and M. Ryynänen, 1986. Propagation of adult curly-birch succeeds with tissue culture. Silva Fenn. 20: 139–147.

Sarvas, R., 1952. On the flowering of birch and the quality of seed crop. Commun. Inst. For. Fenn. 40.7: 1–38.

Särkilahti, E., 1988. Micropropagation of a mature colchicine-polyploid and irradiation-mutant of *Betula pendula* Roth. Tree Physiol. 4: 173–179.

Simola, L.K., 1985. Propagation of plantlets from leaf callus of *Betula pendula* f. *purpurea*. Sci. Hort. 26: 77–85.

Srivastava, P.S. and A. Steinhauer, 1981a, Regeneration of birch plants from catkin tissue cultures. Plant Sci. Lett. 22: 379–386.

Srivastava, P.S. and A. Steinhauer, 1981b. Isozymes in differentiating shoot bud cultures of *Betula pendula* Roth. Z. Pflanzenphysiol. 103: 341–346.

Tulecke, W. and G. McGranahan, 1985. Somatic embryogenesis and plant regeneration from cotyledon tissue of walnut, *Juglans regia* L. Plant Sci. 40: 53–67.

Valanne, T., 1972. Colchicine effects and colchicine-induced polyploidy in *Betula*. Ann. Acad. Sci. Fenn. Ser. A., IV Biol. 191: 1–28.

Welander, M., 1993. Micropropagation of birch. In: M.R. Ahuja (Ed.), Micropropagation of Woody Plants, pp. 223–246. Kluwer Academic Publishers, Dordrecht.

# 10. Somatic embryogenesis in sandalwood (*Santalum album* L.)

P.S. Rao and V.A. Bapat

**Contents**



## 1. Introduction

### 1.1. *Brief description of the plant*

*Santalum album* L., popularly called sandalwood is one of the most important forest trees of India. It is the source of the world famous "East Indian sandalwood oil". The oil stands aloft among all known natural essential oils and is a universal fixative par excellence used in perfumary (Chadha, 1972).The scented oil is present in heartwood and roots and hence the tree is invariably harvested by uprooting. The heartwood formation generally begins after 20 years of plantation. The best heartwood formation takes place when the trees are from 30 to 60 year old and attain a girth of 40–60 cm (Sen-Sarma, 1977). The fragrant wood is extensively used in furniture and related products. The oil has a great demand in cosmetic, soap and perfume industries. Sapwood of sandalwood is a good raw material for toys. The bark of the tree contains juvenile hormonal activity which disturbs the physiological processes and acts as an inhibitor for growth of insect larvae (Shankaranarayana *et al.*, 1980).

*S. Jain, P. Gupta & R. Newton (eds.), Somatic Embryogenesis in Woody Plants, Vol. 2*, 153–170.

### 1.2. *Problems in sandalwood propagation*

Multiple uses of *Santalum* puts this plant in great demand and prices of the oil and wood have shot up owing to the acute shortage of good material. The major constraints for multiplication of sandalwood have been diseases and pests. The plant is plagued by a range of pathogens including bacteria, fungi and mycoplasma like organisms (MLO) (Parthasarathy and Venkatesan, 1982). Spike disease caused by an MLO has threatened the extinction of *Santalum*. In addition to this, there is uncontrolled cutting of the trees because of its great economic importance.

The routine method for sandalwood propagation is only through seeds. Owing to cross pollination the progenies are variable. Seeds also lose viability after 8 months. The other methods such as rooting of cuttings and grafting are not successful. The plant is a semi-root parasite and generally requires the host from seedling stage for its survival. The host parasite connection is continued even in adult trees. Hence, sandalwood seeds are sown always with a host in polybags in forest nurseries. All these factors make this plant an ideal candidate for *in vitro* multiplication.

The *in vitro* multiplication of forest trees either through direct organogenesis or somatic embryogenesis has acquired unlimited importance and substantial progress has been made in many species such as Poplar (*Populus*), birch (*Betula*), teak (*Tectona*), *Eucalyptus*, several *Pinus* sps., Douglas fir (*Pseudotsuga*) and in many others (Ahuja, 1993). Somatic embryogenesis has been reported in many forest tree species (details in Ahuja, 1993) and many parameters such as growth regulator combinations, type of carbohydrate source, light, temperature, nature of the explant and genotype controlling the induction of somatic embryos and their subsequent regeneration into plantlets have been elucidated for many species (Rangaswamy, 1986; Gray and Purohit, 1991; Tautorus *et al.*, 1991). The occurence of somatic embryos in tissue cultures of *Santalum* is one of the earliest reports on somatic embryogenesis in forest trees (Bapat and Rao, 1979, 1984). The work carried out in our laboratory has been reviewed earlier (Bapat and Rao, 1989; Rao and Bapat, 1992). In the present chapter, we have documented our continuing efforts and recent results.

## 2. Summary of the past work

### 2.1. *Organ culture*

*Hypocotyl.* Shoots were regenerated directly on hypocotyl explants on a basal nutrient medium containing auxins such as indoleacetic acid (IAA, 5.71 μM), napthaleneacetic acid (NAA, 10.74 μM), napthoxyacetic acid (NOA, 4,95 μM) or indolebutyric acid (IBA, 4.90 μM). Shoots did not develop on 2,4-dichlorophenoxy acetic acid (2,4-D, 4.52 μM) or *para* cholorophenoxy acetic acid (pCPA, 5.36 μM) supplemented medium. However, prolific regeneration of shoot buds occurred on hypocotyl explants placed on a basal medium amended with individual cytokinin. Kinetin (4.60 μM), zeatin (4.56 μM), benzylaminopurine (BA, 4.40 μM), 6-benzyl 9-tetrahydropyrane adenine (PBA, 4.90 μM) and adenine (74.02 μM) induced a high incidence of bud initiation between 8–10 buds per explant. Bud induction became optimum with BA in the medium and 15–20 buds were formed on a single explant. Green, leafy shoots (15 cm high) were isolated and subcultured on a variety of root inducing media but rooting was poor and was observed only in a few cultures (Rao and Bapat, 1978).

*Nodal segments.* Nodal segments containing axillary buds from mature trees also produced numerous shoot buds on BA supplemented medium. These shoot buds on transfer to the BA free medium developed further and elongated. However, such shoot buds did not develop roots on various root inducing media.

### 2.2. *Callus cultures from mature explants*

Young shoots from a 20-year-old "plus" tree were isolated and 5 mm long stem internodes excised from shoots were cultured on MS basal medium (Murashige and Skoog, 1962) containing 2,4-D (4.52 μM) and sucrose (87.6 mM). Twenty percent of the cultured explants developed callus. Callus was separated from the original explant and was subcultured on the fresh medium with the same composition. From the 2nd subculture onwards the callus mass proliferated rapidly and produced a sufficient mass of tissue on a variety of media combinations. On MS basal medium supplemented with IAA (2.85 μM) and BA (2.22 μM), callus tissue developed into a highly regenerative embryogenic tissue that contained somatic embryos with all stages from globular to torpedo. In several cultures the young embryos redifferentiated again and produced a callus mass which again showed prolific differentiation of embryos. In an actively growing callus, embryogenesis was initiated in small aggregates of highly cytoplasmic cells. Following this, embryos passed through the successive ontogenetic development showing globular, heart and torpedo shaped stages (Plate 2A-D) (Rao and Raghava Ram, unpublished data). A sequential change in the media helped the development of somatic embryos into viable plantlets. Callus tissues bearing regenerated embryos were first subcultured on hormone free MS medium and afterwards on White's medium containing IAA (5.71 μM) with a lower sucrose concentration. Under these conditions the embryos developed into normal plantlets with a good root and shoot system.

A
B
C
D
E

### 2.3. *Callus cultures from endosperm tissues*

Callus was initiated from endosperm excised from mature plants on MS medium containing 2,4-D (4.52 μM)) and kinetin (0.93 μM). The tissue was friable and was subcultured every four weeks (Rao and Raghava Ram, 1983). Various factorial experiments were conducted using several plant growth regulators and somatic embryos developed on MS supplemented with IAA (5.71 μM) and BA (4.44 μM)). The results have been presented in Plate 3A-F (Rao and Raghava Ram, unpublished data).

### 2.4. *Cell suspension cultures*

The cell suspension cultures were initiated by transferring friable callus to MS medium supplemented with 2,4-D (2.26, 4.52 and 11.31 μM). It was noticed that the nature of the suspension had an effect on the efficiency of embryo formation. Upon plating on various media, cells from suspensions developed embryos. The hormone concentrations tested did not appear to be critical for embryo formation, although slightly more organized embryos were observed on media with low BA (2.22 and 4.40 μM) and 2,4-D. In liquid media, the embryos did not develop past the early cotyledonary stage; however, on agar media many pluricotyledonous structures were observed (Rao and Ozias-Akins, 1985).

### 2.5. *Protoplast culture*

Protoplasts were isolated from a variety of sources such as stem callus, cell suspensions and leaf mesophylls (Bapat *et al.*, 1985; Rao and Ozias-Akins, 1985). Protoplasts derived from callus and cell suspensions were more amenable for further studies since leaf mesophyll protoplasts failed to undergo repeated divisions. High yields of protoplasts ($8.73 \times 10^{-6}$) were obtained from callus with a mixture of cellulase (1%), macerozyme (0.5%) and 0.5 M sorbitol or mannitol. For obtaining a high yield of cell suspension protoplasts, cellulase (1%), macerozyme (1%) and 0.5% driselase mixture in 0.55 m mannitol was essential. Various basal media with BA and 2,4-D in different concentrations and combinations were tested for their ability to support cell wall formation and divisions. V-47 medium (Binding, 1974) was suitable for protoplasts isolated from suspensions whereas MS was found to be favourable for protoplasts from stem callus. Sustained and repeated divisions leading to

*Plate 1.* Somatic embryogenesis in sandalwood. (A) Regeneration of plantlets from callus established from stem segments. (B-C) Regenerated somatic embryos isolated from callus tissue. (D) Synthetic seeds obtained after encapsulation of somatic embryos. (E) Somatic embryos from cell suspensions.

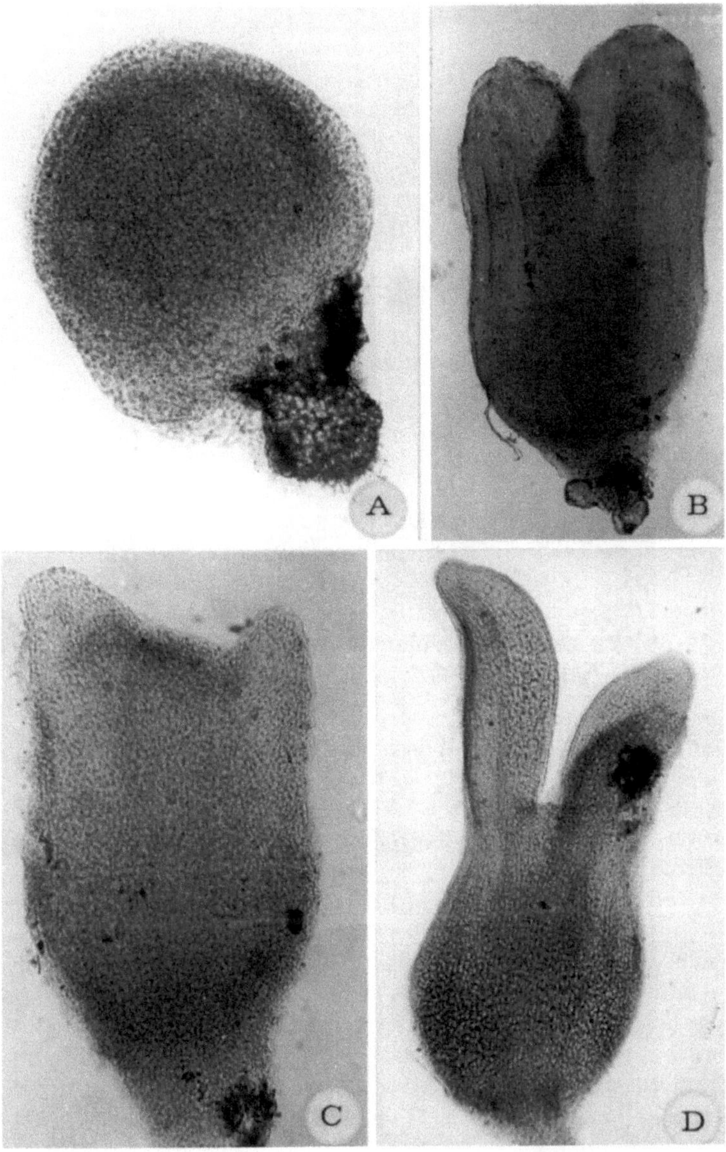


*Plate 2.* Somatic embryogenesis in sandalwood. (A-D) Whole mounts of somatic embryos of various developmental stages.

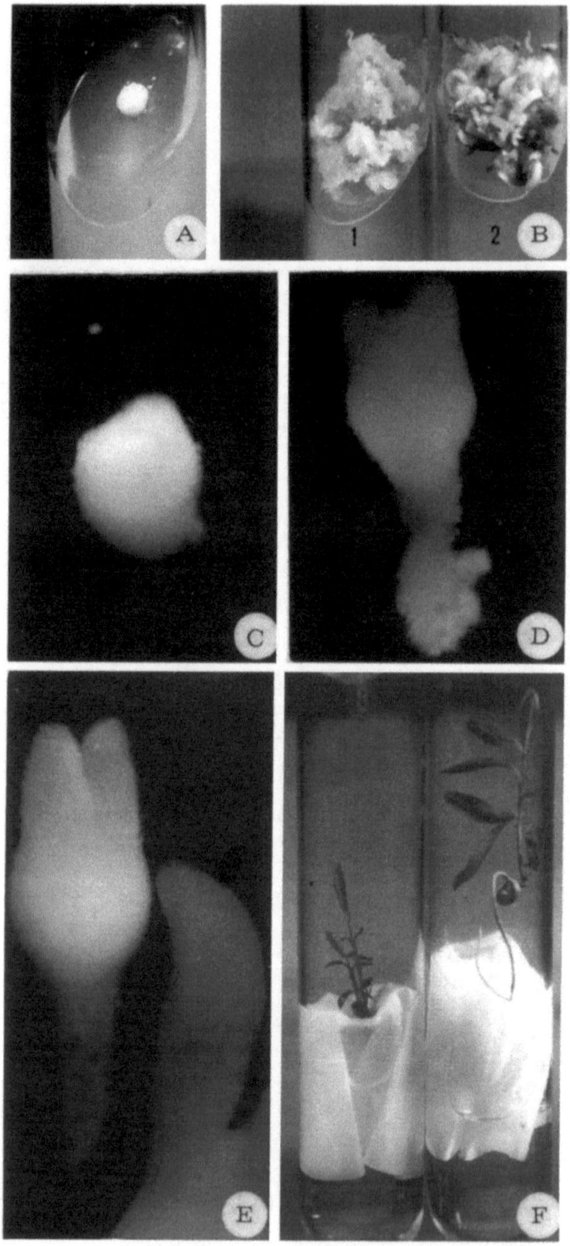


*Plate 3.* Somatic embryogenesis in sandalwood. (A) Endosperm at the time of culture. (B) Callus development from endosperm. (C-E) Somatic embryos regenerated from endosperm callus. (F) Plantlets regenerated from somatic embryos.

large cell colonies and callus formation was obtained from protoplasts isolated from both sources. The macroscopic colonies and globular embryos resulting from protoplast cultures proliferated further and began to form somatic embryos on MS medium supplemented with IAA (5.71 μM), BA (4.4 μM), and casaminoacids (400 mg/l). No significant differences were observed between two levels of osmolality (400 and 700 m Os/kg $H_2O$). Although embryos of all developmental stages could be recognized, there was a predominance of globular embryos which had a clearly discernible suspensor giving it a club-shaped appearance. Plantlet development was observed from somatic embryos regenerated from protoplasts on basal medium containing IAA (5.71 μM), IBA (2.46 μM) and GA (2.89 and 8.66 μM) (Rao and Ozias-Akins,1985).

### 2.6. *Production of somatic embryos in a bioreactor*

In recent years, bioreactors have been designed to scale up the production of somatic embryos. In the case of *Santalum* cell suspension cultures established in flasks were used as the inoculum for the two types of bioreactors used. Firstly, a seven litre stirrer tank bioreactor was used for the conversion of non-embryogenic cells into embryogenic cells. Secondly, a bell jar bioreactor of one litre capacity was used for the development of preglobular embryos to mature embryos. Conversion of non-embryogenic cells to proglobular embryos occurred in 7 l capacity bioreactor after 3 weeks. In one litre capacity bioreactor mature embryos with well-developed cotyledons became visible after 4 weeks. Normal plantlets were obtained from such embryos. The detailed experimental protocol has been described earlier (Bapat *et al.*, 1990).

## 3. Current studies

### 3.1. *Maturation and desiccation of somatic embryos*

#### 3.1.1. *Experimental*

Embryogenic cultures at late torpedo stage somatic embryos were used. Several cultures with uniform age were pooled together and somatic embryos were manually picked and transferred to White's medium (White, 1954) containing abscisic acid (ABA, 1.89, 3.78 and 18.92 μM, Sigma) and sucrose (58.4 mM). The ABA solution was prepared in 1 N NaOH, filter-sterilized and added to cooled autoclaved White's medium (WM). Since sucrose alone at high concentration is known to induce maturation and desiccation, somatic embryos were cultured on various concentrations of sucrose (87.6 to 350.4 mM).

### 3.1.2. *Heat shock treatment*

In one experiment, somatic embryos were cultured on WM having different concentrations of ABA (1.89, 3.78 and 18.92 μM) and were kept at 37°C for 30,60 and 90 min and then incubated in darkness for 30 days at 25 ± 2°C. The pH of the medium was adjusted to 5.8 before autoclaving and the medium was gelled with 0.8% agar (RM 134, HIMEDIA). The medium was autoclaved for 20 min at 121°C under the steam pressure of 1.06 kg $cm^2$. All the experiments were conducted twice with 100 embryos in each treatment.

### 3.1.3. *Procedure for drying of somatic embryos*

Somatic embryos were rinsed in sterile distilled water in order to remove the traces of agar. The washed embryos were placed in 6 cm dia petri dishes and were exposed to laminar flow air for 24 h in light (950 to 1000 lux) at 25°C. The petri dishes were sealed with parafilm and were kept in dark at 28 ± 2°C for 10, 20 or 30 days according to the experimental design. After the desiccation period the embryos were removed and cultured on WM supplemented with sucrose (58.4 mM). The embryogenic tissue and rehydrated embryos were maintained at 25 ± 2°C, 1000 lux light intensity and 55% relative humidity. The germination response of the embryos was scored after 60 days.

### 3.1.4. *Encapsulation of somatic embryos*

The desiccated somatic embryos were dipped in 3% w/v sodium alginate (Sigma) and the mixture of somatic embryos and alginate was dropped into $CaCl_2$ $2H_20$ (1.036 g/150 ml of distilled water in 250 ml flask) and was kept for 1 h in this solution. The flasks were placed on a gyratory shaker (80 rpm) under light (1000 lux). Following the encapsulation, beads were recovered by decanting the $CaCl_2$ solution and were washed with sterile distilled water 3–4 times. Encapsulated beads were cultured on WM supplemented with 58.4 mM sucrose.

In addition to sodium alginate, another type of gel known as composite gel was tested. The composite gel was prepared by mixing 25% (w/v) silica gel (H) and 4% (w/v) sodium alginate in MS medium. After mixing with composite gel, the selected embryos were dropped in 5% $CaCl_2$ $2H_2O$ solution and were allowed to remain in this solution for 30 min. The developed beads were thoroughly washed in sterile distilled water before culturing them on an appropriate nutrient medium. The encapsulation procedure was same for both nondesiccated and desiccated somatic embryos (details in Bapat, 1992).

### 3.1.5. *Drying of embryogenic tissue*

The embryogenic tissue, having a heterogenous composition (cells,cell clusters and individual globular embryos), was transferred to 6 cm dia petri dishes on sterile filter paper and exposed to laminar flow air for 8 h at 28°C in light (1000 lux). Six pieces of calli (1 g fresh weight) were used for each treatment. The petri dishes were sealed with parafilm and kept in darkness

*Table 1.* Effect of ABA on survival and germination of desiccated somatic embryos of *Santalum album* L.

| ABA (μM) | No. of embryos which survived desiccation (days) | | | | | | No. of embryos which converted to plants (days) | | | | | |
|---|---|---|---|---|---|---|---|---|---|---|---|---|
| | 10 | | 20 | | 30 | | 10 | | 20 | | 30 | |
| | a | b | a | b | a | b | a | b | a | b | a | b |
| 0 | 16 | 14 | 12 | 10 | 7 | 4 | 3 | 0 | 2 | 0 | 0 | 0 |
| 1.89 | 22 | 19 | 21 | 24 | 21 | 18 | 6 | 3 | 4 | 1 | 4 | 2 |
| 3.78 | 36 | 37 | 31 | 29 | 29 | 26 | 8 | 6 | 5 | 3 | 3 | 2 |
| 18.92 | 49 | 51 | 46 | 40 | 45 | 39 | 11 | 8 | 10 | 7 | 7 | 3 |

100 somatic embryos per treatment, a: nonencapsulated, b: encapsulated.
Data scored after 60 days.
From Bapat & Rao (1992) with permission.

for 10, 20 or 30 days at 28 ± 2°C. The tissue was rehydrated by placing it on the fresh MS basal medium containing IAA (2.85 μM), BA (2.22 μM) and sucrose (87.6 mM). The growth was measured on a both fresh and dry weight basis by weighing the individual calli. For measuring dry weight, the tissue was pressed on a filter paper to remove excessive water and then was dried in an oven at 60 ± 2°C for 24 h.

## 4. Results and discussion

### 4.1. *Influence of abscisic acid*

Embryos of late torpedo stage showed arrested development on WM supplemented with ABA (1.89, 3.78 or 18.92 μM) in darkness. Size of embryos was reduced and many embryos turned brown or developed only rudimentry cotyledons. Embryos treated with ABA for 30 days showed an increase in weight. On an average, they weighed 4 mg when compared with weight (1.2 to 2 mg) of untreated somatic embryos. The frequency of plant regeneration also varied according to the concentration of ABA in the medium. The conversion frequency of somatic embryos was also dependent upon the length of the ABA treatment period. Abscisic acid (18.92 μM) caused a better desiccation tolerance and even after 30 days of desiccation, embryos remained viable and could be regenerated into plantlets. Nevertheless, all the embryos that survived desiccation did not convert into plantlets when rehydrated and cultured on WM. Some of the embryos continued to remain in an arrested state of growth and subsequently necrosed (Table 1).

The development pattern of desiccated encapsulated embryos was similar to nonencapsulated desiccated embryos except for the difference in the number of embryos responding to a particular concentration of ABA. The

*Table 2.* Effect of low temperatures on ABA treated somatic embryos of *Santalum album* L.

| ABA (μM) | 28°C (Control) | | −4°C | | −10°C | |
|---|---|---|---|---|---|---|
| | a | b | a | b | a | b |
| 0 | 12 | 4 | 0 | 0 | 0 | 0 |
| 1.89 | 17 | 9 | 8 | 2 | 7 | 0 |
| 3.78 | 24 | 17 | 11 | 5 | 9 | 0 |
| 18.92 | 39 | 24 | 4 | 1 | 4 | 0 |

100 somatic embryos per treatment, 30 days of storage time for treated embryos, a: No. of embyros survived, b: No. of embryos converted to plants, desiccation period: 30 days after ABA treatment.
From Bapat & Rao (1992) with permission.

*Table 3.* Effect of high temperature (37°C) on survival and germination of somatic embryos of *Santalum album* L.

| Treatment time (minutes) | No. of embryos which survived desiccation (days) | | | | | | No. of embryos which converted to plants (days) | | | | | |
|---|---|---|---|---|---|---|---|---|---|---|---|---|
| | 10 | | 20 | | 30 | | 10 | | 20 | | 30 | |
| | a | b | a | b | a | b | a | b | a | b | a | b |
| 0 | 14 | 12 | 11 | 6 | 7 | 5 | 4 | 1 | 2 | 0 | 0 | 0 |
| 30 | 26 | 24 | 22 | 19 | 19 | 14 | 8 | 6 | 1 | 2 | 1 | 1 |
| 60 | 27 | 26 | 18 | 16 | 15 | 8 | 9 | 6 | 6 | 2 | 6 | 2 |
| 90 | 20 | 18 | 17 | 11 | 15 | 8 | 5 | 2 | 3 | 0 | 2 | 1 |

100 somatic embryos per treatment, a: nonencapsulated, b: encapsulated.
Data scored after 60 days.
From Bapat & Rao (1992) with permission.

emergence of roots by breaking of matrix followed by shoot primordia was observed in most of the beads.

### 4.2. *Effect of temperature treatment*

Nonencapsulated and desiccated somatic embryos treated with ABA were subjected to low temperature treatment (−4 and −10°C) for 30 days. A drastic reduction in the number of embryos retaining the capacity to form plantlets was observed (Table 2). However, both nonencapsulated and encapsulated somatic embryos treated with ABA (18.92 μM) for 30 days and then given a heat treatment (37°C) showed a varied germination response which depended upon the time of exposure to temperature (Table 3). The heat treatment did not increase the desiccation tolerance. A 90 min exposure of embryos to 37°C adversely affected the conversion of somatic embryos into plantlets (Table 3).

*Table 4.* Effect of sucrose on survival and germination of somatic embryos of *Santalum album* L.

| Sucrose (Conc.) | No. of embryos which survived desiccation (days) | | | | | | No. of embryos which converted to plants (days) | | | | | |
|---|---|---|---|---|---|---|---|---|---|---|---|---|
| | 15 | | 30 | | 45 | | 10 | | 30 | | 45 | |
| | a | b | a | b | a | b | a | b | a | b | a | b |
| 87.6 mM | 18 | 16 | 15 | 12 | 8 | 6 | 4 | 2 | 3 | 1 | 1 | 0 |
| 175.2 mM | 33 | 31 | 28 | 27 | 6 | 6 | 7 | 5 | 3 | 1 | 2 | 1 |
| 233 mM | 24 | 20 | 15 | 16 | 11 | 8 | 6 | 4 | 6 | 0 | 3 | 1 |
| 292 mM | 21 | 20 | 18 | 19 | 14 | 12 | 4 | 1 | 1 | 0 | 1 | 0 |
| 350.4 mM | 19 | 18 | 16 | 17 | 11 | 9 | 2 | 1 | 1 | 0 | 1 | 0 |

100 somatic embryos per treatment, a: nonencapsulated, b: encapsulated.
Data scored after 60 days.
From Bapat & Rao (1992) with permission.

### 4.3. *Effect of sucrose concentrations*

Induction of desiccation tolerance by high concentrations of sucrose has been reported (Anandarajah and Mckersie, 1990). However, our results indicate that sucrose (87.6 to 350.4 mM) was less effective in inducing desiccation tolerance as compared with ABA. Embryo growth was arrested with different levels (175.2 to 233.6 mM) of sucrose. Most of the embryos produced long roots which penetrated the medium, however, shoot growth was completely inhibited (Table 4). At higher levels of sucrose (242 to 350.4 mM), embryos developed red pigmentation indicating subtle changes in anthocyanin biosynthetic pathways. Radicle elongation in embryos commenced immediately following their removal from the high sucrose medium. The primary root elongated rapidly and was accompanied by the develpment of laterals. Epicotyl elongation became apparent after 2–3 weeks of rehydration. In some cultures, adventitious embryos developed at the cotyledonary tip of the embryos.

### 4.4. *Somatic embryogenesis from dried callus*

Dried callus did not show any visible growth during the first three weeks of culture after transfer to a fresh MS containing IAA (4.52 μM) + BA (4.40 μM) for rehydration. In the 4th and 5th week, the survival and recovery of viable embryos became apparent. The hydrated callus showed revival by the appearance of faint yellow and white patches on the surface of the callus mass. This was followed by the emergence of tiny and bright white somatic embryos. Upon isolation and culture of such embryos on White's medium, complete plantlets were regenerated from them. A decrease in fresh and dry weight was recorded in dried callus on rehydration when compared with control tissues (Table 5).

*Table 5.* Growth and revival of viability of dried callus of *Santalum album* L.

| Days of dry state | Fresh weight (gms) | Dry weight (mg) | No. of callus cultures showing viability |
|---|---|---|---|
| 0 | 5.645 | 650 | 100% |
| 10 | 3.780 | 515 | 60% |
| 20 | 2.110 | 410 | 60% |
| 30 | 1.800 | 370 | 40% |

Growth after 60 days after transfer on the fresh medium. 6 callus replicates for each treatment. From Bapat & Rao (1992) with permission.

*Table 6.* Effect of different concentrations of sodium carbonate on *in vivo* germination of somatic embryos of *Santalum album* L. encapsulated in sodium alginate and composite gel.

| Conc. of Sodium bicarbonate mg/l | No. of seeds germinated | | Mean germination | | % germination | |
|---|---|---|---|---|---|---|
| | SA | CG | SA | CG | SA | CG |
| 00 | 0 | 0 | 0 | 0 | 0 | 0 |
| 10 | 4 | 0 | 2 + 0 | 0 | 8 | 0 |
| 15 | 6 | 1 | 3 + 1.41 | 1 + 0.7 | 13 | 4.16 |
| 25 | 8 | 0 | 4 + 2.82 | 0 | 17 | 0 |
| 50 | 7 | 0 | 3.5 + 0.7 | 0 | 15 | 0 |
| 100 | 5 | 0 | 2.5 + 0.7 | 0 | 10 | 0 |

A total of 16 seeds were kept for germination.
SA – sodium alginate; CG – composite gel. (From Fernandes *et al.*, 1992, with permission.)

The process of maturation and desiccation has been shown to be of vital importance in the development of zygotic embryos. Recent examples have indicated the importance of maturation and desiccation of somatic embryos of many plant species for obtaining large numbers of good quality plantlets (Deng and Cornue, 1992; Gray, 1989; Janick *et al.*, 1992). In several plant species zygotic embryos undergo a process of maturation followed by a dry phase called desiccation (Blackman *et al.*, 1992). This phase has also been shown to be present in somatic embryos. It has been emphasised that prior to development into plantlets somatic embryos need a "resting phase" or a maturation phase (Gray and Purohit, 1991).

The process of maturation and desiccation of zygotic and somatic embryo is controlled by a variety of external and internal factors (Kermode, 1990). Abscisic acid, high levels of proteins and carbohydrates, and less water content are the main factors controlling the process of maturation and desiccation of zygotic embryos (Mayer and Poljakoff-Mayber, 1982), and these factors are also known to influence the maturation and desiccation of somatic embryos as well (Kim and Janick, 1991; Janick *et al.*, 1992; Attree *et al.*, 1991). This has been demonstrated in alfalfa (*Medicago*) (Mckersie *et al.*, 1989), grapes (*Vitis*) (Gray, 1989), white spruce (*Picea*) (Attree *et al.*, 1991) and soybean (*Glycine*) (Hammit and Davey, 1987). It appears that ABA and sucrose in combination with other factors play a vital role in maturation and desiccation of somatic embryos.

Abscisic acid as a plant hormone is known to have multiple roles during the life cycle of a plant other than as a growth inhibitor (Zeevart and Creelman, 1988). The prevention of precocious germination of immature embryos by exogenous ABA has been demonstrated in cotton (*Gossypium*), maize (*Zea*), rapeseed (*Brassica*), soybean (*Glycine*) and wheat (*Triticale*). Abscisic acid is also implicated in the regulation of storage proteins and in the synthesis mRNAs (Gomez *et al.*, 1988). In our work on *Santalum*, hardening treatments with ABA and high sucrose imposed on somatic embryos influenced growth and resulted in embryo maturation following dehydration. Plantlet regeneration from such embryos which underwent a hardening treatment indicated that the growth-retarding effect of these treatments was reversible. In *Santalum*, encapsulation was not necessary for the revival of germination from desiccated embryos. In contrast, carrot somatic embryos did not survive desiccation unless they were encapsulated in a matrix (Kim and Janick, 1991).

The retention of totipotency of dried callus of *Santalum* even after 30 days in a dry state without ABA treatment is quite significant. In similar studies on *Oryza*, Tsukahara and Hirosawa (1992) reported a simple dehydration treatment for plantlet regeneration and suggested that not only the water content of callus is reduced during dehydration, but also a certain period of time is needed for the cells to adjust physiologically to low water content for efficient regeneration.

Although sucrose has been used to enhance desiccation tolerance in alfalfa (*Medicago*) (Anandarajah and Mckersie, 1990), it was ineffective in *Santalum*. This difference is possibly due to the biochemical nature of alfalfa seed which is a legume and *Santalum* being a forest tree.

The present study has demonstrated two important aspects: (1) the ability of the desiccated sandalwood embryos to survive, and (2) retention of the potentiality of desiccated embryos to regenerate into plantlets. These results would significantly expand research on the synthetic seeds of sandalwood.

*Santalum* seeds remain viable for a period of 8 months. Desiccated somatic embryos retained viability even after 30 days. With further modifications in desiccation techniques it should be possible to extend the viability period similar to the zygotic embryos. The final success however, would lie in the germination of desiccated embryos in soil in a nonsterile environment.

### 4.5. *Synthetic seeds*

Under *in vitro* conditions, germination of synthetic seeds and their conversion into plantlets depends upon several factors including the medium composition and stage of the coated embryo. The conversion of somatic embryos to plantlets benefits from the presence of nutrients in the encapsulating matrix acting as a synthetic endosperm. Previously we have reported on various parameters controlling germination of encapsulated somatic embryos of *Santalum* (Bapat and Rao, 1988) and in this article we report some of our recent results. The critical stage in synthetic seed research is *in vivo* germination of encapsulated somatic embryos. Natural seeds have their own inherent mechanism to protect the zygotic embryos from microbial and fungal attack, and also from physical damage in the external environment. To fulfil these requirements in synthetic seeds, the composition of the nutrient medium has to be modified in order to provide nourishment and protection to the encapsulated embryos and enable them to germinate under nonsterile conditions. Hence, the gel matrix was mixed either with a fungicide or with a food preservative.

The fungicides bavastin, carbendazim or diethane M-45 (10, 15, 25, 50 and 100 mg/l) were mixed separately and were added to the matrix. In other experiments, only a food preservative (sodium bicarbonate, 10, 15, 25, 50 and 100 mg/l)) was added to the matrix.

None of the encapsulated somatic embryos germinated when fungicide was included in the matrix. However, there was no contamination of the encapsulated embryos even under nonsterile conditions for 4 weeks.

Among the different concentrations of sodium bicarbonate used, the optimum concentration for germination (17%) of encapsulated embryos was 25 mg/l. The encapsulated somatic embryos in matrix devoid of food preservative did not germinate and eventually died. Germination occurred with the emergence of shoot primordia, but plantlet development was inhibited because of drying of the alginate coat (Table 6).

Encapsulated somatic embryos in the composite gel provided with food preservative showed 4.2% germination which is low. As in the case of sodium alginate, although many beads showed signs of germination, this was not followed by the emergence of shoot primordia. Although the *in vivo* germination percentage was very low, it is still significant that somatic embryos could germinate in soil. The major hurdle for the *in vivo* germination of somatic embryos was not contamination, but was the drying up of the beads before the somatic embryos could develop into plantlet. Further modifications in techniques would make it possible to achieve higher rates of germination comparable with natural seeds. It is interesting to note that the food preservative was effective in checking the contamination and helping the embryos to germinate *in vivo*. This would help in eliminating the use of fungicides in the matrix and its possible harmful effects.

## 5. Conclusion

Sandalwood (*Santalum album* L.) is an evergreen tropical forest tree famous for its scented wood and oil. Since the plant is propagated only through seeds, there is a need for tissue culture methods which provide for faster multiplication of selected genetically superior clones. Somatic embryogenesis will probably be the most appropriate mode for rapid micropropagation of sandalwood. Bioreactors have been employed to maximize the production of somatic embryos, and several designs have been made to fabricate the bioreactors suitable for somatic embryo production (Ammirato and Styer, 1985; Nishimura *et al.*, 1992). The production of somatic embryos in cell suspensions and bioreactor could be achieved in millions. However, the major problem of concern is the low number of conversion of somatic embryos into plantlets. The metabolic processes leading to efficient germination of somatic embryos have to be elucidated. The situation is often complicated by the poor understanding of the factors controlling the process of maturation and desiccation in zygotic as well as in somatic embryos. These reported studies clearly demonstrate that the transition of somatic embryos between induction and germination involves the process of maturation. More intensive investigations are needed to improve the conversion rate of somatic embryos to plantlets before large scale utilizations are realized. The prospects of synthetic seeds as a delivery system for micropropagation are growing (Redenbaugh *et al.*, 1992; Gupta and Kreitinger, 1993) and there is no doubt that the synthetic seed technology will be a viable option for making clonal forestry a success. The use of electroporation with somatic embryos to obtain transgenic plants has been reported (Tautorus *et al.*, 1991). Further, investigations have been carried out with the use of molecular markers to assess the embryogenic potential as well as to know the developmental stage of developed embryos. This approach could be useful for the induction of somatic embryos in many recalcitrant species (Sterk and De Vries, 1992).

## 6. References


Ahuja, M.R, 1993. Micropropagation a la carte. In: M.R. Ahuja (Ed), Micropropagation of Woody Plants, pp. 3–9. Kluwer Academic Publishers, Dordrecht.

Ammirato, P.V. and D.J. Styer, 1985. Strategies of somatic embryogenesis in suspension cultures. In: M. Zaitlin, P. Day and A. Hollaender, pp. 161–178. Academic Press, New York.

Anandarajah, K. and B.D. Mckersie, 1990. Enhanced vigor of dry somatic embryos of *Medicago sativa* L. with increased sucrose. Plant Sci. 71: 261–266.

Attree, S.M., D. Moore, V.K. Swahney and L.C. Fowke, 1991. Enhanced maturation and desiccation tolerance in white spruce (*Picea glauca* (Moench Voss) somatic embryogenesis: Effects of nonplasmolysing water stress and abscisic acid. Ann. Bot. 68: 519– 525.

Bapat, V.A. and P.S. Rao, 1979. Somatic embryogenesis and plantlet formation in tissue cultures of sandalwood (*Santalum album* L.). Ann. Bot. 44: 629–630.

Bapat, V.A. and P.S. Rao, 1984. Regulatory factors for *in vitro* multiplication of sandalwood tree (*Santalum album* L.). I. Shoot bud regeneration and somatic embryogenesis in hypocotyl cultures. Proc. Indian Acad. Sci. (Plant Sci.) 93: 19–27.

Bapat, V.A., R. Gill and P.S. Rao, 1985. Regeneration of somatic embryos and plantlets from stem callus protoplasts of sandalwood tree (*Santalum album* L.). Curr. Sci. 54: 978–982.

Bapat, V.A. and P.S. Rao, 1988. Sandalwood plantlets from "synthetic seeds". Plant Cell Rep. 7: 434–436.

Bapat, V.A. and P.S. Rao, 1989. *In vitro* stategies for sandalwood micropropagation. In: V. Dhawan (Ed.), Application of Biotechnology in Forestry and Horticulture, pp. 145–156. Plenum Press, New York.
Bapat, V.A., D.P. Fulzele, P.S. Rao and M.R. Heble, 1990. Production of somatic embryos of sandalwood (*Santalum album* L.) in bioreactors. Curr. Sci. 59: 746–748.
Bapat, V.A., 1992. Studies on synthetic seeds of sandalwood (*Santalum album* L.) and mulberry (*Morus indica*). In: K. Redenbaugh (Ed.), Synseeds, pp. 381–407. CRC Press, Boca Raton.
Bapat, V.A. and P.S. Rao, 1992. Plantlet regeneration from encapsulated and nonencapsulated desiccated somatic embryos of a forest tree: Sandalwood (*Santalum album* L.). J. Plant Biochem. Biotechnol. 1: 109–113.
Binding, H., 1974. Regeneration von haploiden undiploiden Pflanzen aus Protoplasten von *Petunia hybrida* L. Z. Pflanzenphysiol. 74: 327–356.
Blackman, S.A., R.L. Obendorfand, A.C. Leopold, 1992. Maturation proteins and sugars in desiccation tolerance of developing soybean seeds. Plant Physiol. 100: 225–230.
Chadha, Y.R., 1972. In: The wealth of India 9, pp. 208–224. Pub. Inf. Dir. CSIR, New Delhi.
Deng, M.D. and D. Cornue, 1992. Maturation and germination of walnut somatic embryos. Plant Cell Tiss. Org. Cult. 28: 195–202.
Fernandes, P.C., V.A. Bapat and P.S. Rao, 1992. *In vivo* germination of encapsulated somatic embryos of *Santalum album* L. (Sandalwood). Ind. J. Exp. Biol. 30: 839–841.
Gomez, J., M.D. Sanchez, V. Sttlefel, J. Rlgau, P. Pugamonenechand and M. Pages, 1988. A gene induced by the plant hormone abscisic acid in response to water stress encodes a glycine rich protein. Nature 334: 262–264.
Gray, D.J. 1989. Effects of dehydration and exogenous growth regulators on dormancy, quiescence and germination of grape somatic embryos. In Vitro Cell Dev. Biol. 25: 1173–1178.
Gray, D.J. and A. Purohit, 1991. Somatic embryogenesis and development of synthetic seed technology. Crit. Rev. Plant Sci. 10: 33–61.
Gupta, P.K. and M. Kreitinger, 1993. Synthetic seeds in forest trees. In: M.R. Ahuja (Eds.), Micropropagation of Woody Plants, pp. 107–119. Kluwer Academic Pubishers, Dordrecht.
Hammit, N. and M.R. Davey, 1987. Somatic embryogenesis and plant regeneration from cultured zygotic embryos of soybean (*Glycine max* L. Merr). J. Plant Physiol. 128: 219–226.
Janick, J., Y. Kim, S. Kitto and Y. Saranga, 1992. Desiccated synthetic seeds. In: K. Redenbaugh (Ed.), Synseeds, pp. 11–33. CRC Press, Boca Raton.
Kermode, A.R., 1990. Regulatory mechanisms involved in the transition from seed development to germination. Crit. Rev. Plant Sci. 9: 155–194.
Kim, Y.H. and J. Janick, 1991. Abscisic acid and proline improve desiccation tolerance and increase fatty acid content of celery somatic embryogenesis. Plant Cell Tiss. Org. Cult. 24: 83–89.
Mayer, A.M. and A. Poljakoff-Mayber, 1982. The Germination of Seeds. Pergamon Press, Oxford.
Mckersie, B.D., T. Senaratna, S.R. Bowley, D.C.W. Brown, J.E. Krochko and J.D. Bewley, 1989. Application of artificial seed technology in the production of hybrid alfalfa (*Medicago sativa* L.). In Vitro Cell Dev. Biol. 25: 1183–1188.
Murashige, T. and F. Skoog, 1962. A revised method for rapid growth and bioassays with tobacco tissue cultures. Physiol. Plant. 15: 473–493.
Nishimura,S,T.Terashima,K.Higashi and H.Kamada,1992. Bioreactor culture of somatic embryos for mass propagation of plants. In: K. Redenbaugh (Ed.), Synseeds, pp. 175–181. CRC Press, Boca Raton.
Parthasarathy, K. and K.R. Venkatesan, 1982. Sandal spike disease. Curr. Sci. 51: 225–230.
Rangaswamy, N.S, 1986. Somatic embryogenesis in angiosperm cell, tissue and organ culture. Proc. Indian Acad. Sci. (Plant Sci.) 96: 247–271.
Rao., P.S. and V.A. Bapat, 1978. Vegetative propagation of sandalwood plants through tissue culture. Can. J. Bot. 56: 1153–1156.
Rao, P.S. and N.V. Raghava Ram, 1983. Propagation of sandalwood (*Santalum album* L.)

using tissue and organ culture technique. In: S.K. Sen and K.L. Giles (Eds.), Plant Cell Culture in Crop Improvement, pp. 119–124. Plenum Press, New York.

Rao, P.S. and P. Ozias-Akins, 1985. Plant regeneration through somatic embryogenesis in protoplast cultures of sandalwood (*Santalum album* L.). Protoplasma 124: 80–86.

Rao., P.S. and V.A. Bapat, 1992. Micropropagation of sandalwood (*Santalum album* L.). In: Y.P.S. Bajaj (Ed.), Biotechnology in Agriculture and Forestry, pp. 193–210. Springer-Verlag, Berlin.

Redenbaugh, K., J. Fujii, D. Slade, P. Viss and M. Kossler, 1992. Artificial seeds-encapsulated somatic embryos. In: Y.P.S. Bajaj (Ed.), Biotechnology in Agriculture and Forestry, pp. 2–46. Springer-Verlag, Berlin.

Sen-Sarma, P.K, 1977. Sandalwood – its cultivation and utilization. In: C.K. Atal and B.M. Kapur (Eds.), Cultivation and Utilization of Medicinal and Aromatic Plants, pp. 287–297. Reg. Res. Lab., Jammu Tawi.

Shankaranarayana, K.H., K.S. Aiyer and G.S. Krishna Rao, 1980. Insect growth inhibitor from the bark of *Santalum album* L. Phytochem. 19: 1239–1240.

Sterk, P. and S.C. de Vries, 1992. Molecular markers for plant embryos. In: K. Redenbaugh (Ed.), Synseeds, pp. 115–132. CRC Press, Boca Raton.

Tautorus, T.E, L.C. Fowke and D.I. Dunstan, 1991. Somatic embryogenesis in conifers. Can. J. Bot. 69: 1873–1899.

Tsukahara, M. and T. Hirosawa, 1992. Simple dehydration treatment promotes plantlet regeneration of rice (*Oryza sativa*) callus. Plant Cell Rep. 11: 550–553.

White, P.R., 1954. The cultivation of animal and plant cells. The Ronald Press, New York.

Zeevart, J.A.D. and R.A. Creelman, 1988. Metabolism and physiology of abscisic acid. Ann. Rev. Plant Physiol. 39: 439–473.

# 11. Somatic embryogenesis in olive (*Olea europaea* L.)

Eddo Rugini

**Contents**



## 1. Introduction

### 1.1. *History, classification and distribution*

The olive, which originated in the eastern Mediterranean basin, belongs to the *Oleaceae* family and comprises about 30 genera including various garden and economical plants (Knoblauch, 1895; Ciferri, 1942; Taylor, 1945; Johnson, 1957; Mazzolani and Altamura Betti, 1979, 1980; Altamura Betti *et al.*, 1985). Sax and Abbe (1932) found that six genera of two tribes belonging to the sub-family *Oleoideae* had the basic haploid chromosome number of 23 (2n = 2x = 46). Other than in Mediterranean countries, subspecies are also found in western Africa, Tanzania, the Canary and Azore Islands, South Africa, Mauritius, and Mascarene Comore. During the last 30–40 years, cultivated olives have been introduced into the USA, Australia, South Africa, China, Korea and other Asiatic countries. Most cultivated olives belong to *Olea europaea* L. which comprises many groups of forms; up to 2600 different cultivars have been recorded, although many of these might be ecotypes. These groups may have originated by mutation and hybridization (Chevalier, 1948) and in addition, as indicated by palaeontological studies,

*S. Jain, P. Gupta & R. Newton (eds.), Somatic Embryogenesis in Woody Plants, Vol. 2,* 171–189.

*Table 1.* Summary of work carried out in somatic embryogenesis on olive. The best basal media and hormone concentration are reported in brackets, both in induction () and regeneration [ ] media.

| Initial explant | Media (in mg/l) (of induction) [of regeneration] | Reference |
|---|---|---|
| Root seedlings | (MS + NAA 5 + BA 0.1) [MS + NAA 0.1 + BA 0.1] | Rugini and Tarini, 1986 |
| Immature zygotic embryos | [(1/2MS + BA 0.5 ± NAA 0.5)] | Rugini, 1988 |
| Mature zygotic embryos (*Oleaster*) and (*Sativa*) | (OMc + IBA 5 + 2iP 0.4) [OMc + 0 to 0.5 IBA] | Orinos and Mitrakos, 1991; Mitrakos *et al.*, 1992 |
| Immature zygotic embryos | (SH + IBA 0.25 to 0.5 + 2iP 0 to 0.1) [SH + no hormones] | Leva *et al.*, 1993 |
| Petioles of cvs. (Canino and Moraiolo) | (sol. OMc + 2iP 0.1 + BA 0.1 + IBA 0.05) [liq. OMc + 2iP 0.1 + BA 0.1 + IBA 0.05] | Rugini and Caricato, 1995 |

MS – Murashige and Skoog (1962); NAA – naphthaleneacetic acid; BA – 6-benzylaminopurine; IBA – indolebutyric acid; SH – Schenk and Hildebrand (1972); 2iP – 6-dimethylaminopurine; OMc – olive medium (OM; Rugini, 1984) in which macroelements were replaced with those of Bourgin and Nitsch (1967) with addition of 1 gm/l casein hydrolysate and omitting glutamine.

the tropical and subtropical Afro-Asiatic species, such as *Olea chrysophylla* Lam., *Olea excelsa* Ait., *Olea ferruginea*, *Olea leperrinei* and *Olea europaea* subsp. *cerasiformis*, probably contributed to the evolution of the cultivated olive (Mazzolani and Altamura-Betti, 1976/77). Among the Euro-Mediterranean olive varieties, two subspecies: 1) *oleaster* (wild olive), and 2) *sativa* (cultivated olive) are usually recognized. The *Oleaster* shows thorny shoots, often quadrangular in cross section, small, oval, almost round leaves on young plants and small, elliptical, black-reddish fruit with a low oil content. The *Sativa* has cylindrical shoots, lanceolate leaves, and big, oval or elliptical fruits that are purplish-blue, black-reddish or, rarely, white in color, with a high oil content. The *Oleaster* is found in relatively few locations. In many places, they are often mistaken for *Olevaster* which are cultivated ecotypes of *Olea europaea sativa* kept in a continuous juvenile stage by grazing. These plants, although they have never produced flowers for hundreds of years, undergo a morphogenetic change and lose their similarity to *Oleaster* if transferred into normal cultivation and could be induced to flower within 2–3 years (Rugini and Lavee, 1992).

### 1.2. *Economic importance*

Ninety-seven percent of the olive production (750 million plants) is accounted for in the countries of the Mediterranean basin, about 92% for oil extraction; the remaining percentage belongs to table olives. The annual average world production of olive oil during the 1980s was 1,666,650 tons (FAO Yearbook, 1991). The main producer countries are Spain, Italy, Greece and Tunisia. The average world production of olive oil from 1970 to 1990 increased in tons as did the table olive production, thus the relative increase in table olives produced during the last 20 years was considerably greater than that of oil. The use of olive oil in the world has greatly increased in recent years, about 7.2% from the first five years of the 1980s when compared with the last five years, due to the olive oil's property of reducing the incidence of some cardiovascular diseases for its high oleic acid content. Libya and Greece have the highest pro-capita consumption with about 65 kg per person per year followed by Italy and Spain with 25–30 kg. Eighty-nine percent of the oil consumption belongs to the producer countries and consequently international trade is low. Commercially, olive oil is sold at a number of different grades but, generally, shows quite low quality; overall it is not commercially competitive enough with other vegetable oils, although it is the best for its taste and the only fluid oil which can be used in the crude state. Unfortunately, the irrational techniques of cultivation, such as the lack of harvesting mechanization, irrigation and suitable cultivars, and the scarce research in defining the oil quality characteristics contribute to making olive culture problematic.

### 1.3. *State of the art of genetic improvement*

The olive species comprises a great number of polyclonal cultivars whose identification is difficult due to a lack of cultivar standard and problems of nomenclature (Roselli, 1990; Scaramuzzi and Roselli, 1986; Prevost *et al.*, 1993). Several attempts using pollen grains, leaves and seeds have been made to distinguish olive clones through morphological (Roselli, 1977; Bartolini *et al.*, 1991) and biochemical (Pontikis *et al.*, 1980; Loukas and Krimbas, 1983; Roselli *et al.*, 1992; Perri *et al.*, 1992; Durante *et al.*, 1992) markers and with highly accurate techniques such as DNA fingerprinting (Bogani *et al.*, 1993). The olive has yet to yield to improvement: conventional cross breeding, clonal selection, and gamma irradiation gave poor results.

Breeding of olive cultivars was rather limited until recent years. The long juvenile phase, which exceeds 10 years, has discouraged the establishment of research projects with olives. At present, this has been simplified because a way to shorten the juvenile period has been found. The method is based on a method of single shoot upright training together with controlled thinning and shortening of the lateral juvenile branches. For good response, intensive growing conditions with irrigation and nutrition are obligatory [(Lavee, unpublished data, and Fontanazza (personal communication)]. However, juvenility is still not short enough to speed up the breeding programs. Two new cultivars, one for the table olive "Kadesh" and one for oil extraction "Barnea", and some new potential cultivars with specific characteristics such as vigour, tree shape, big drupes and resistance to *Spilocaea oleagina*, are currently being tested through this method (Lavee *et al.*, 1986; Bellini and Giordani, 1992). The preliminary observations indicated that the similar wide distribution of types in the F1 progenies of most cultivars indicate that the existing cultivars are only a few generations away from the wild type, and many probably are direct clone selections of wild types. It should be noted, however, that some of the properties in those selected clones such as vigor, leaf size, and fruit shape are dominant characters in their F1 progenies (Rugini and Fedeli, 1990; Rugini and Lavee, 1992; Bellini and Giordani, 1992). The hereditability of the most important traits in olive is still very difficult to identify due to the great variability in fruit size, leaf shape, and oil content of the existing olive cultivars.

Crop improvement is mainly based on clonal selection and occasionally on promising seedlings, and is observed in the wild or in orchards, after scion decline. Up to now, this has been almost unsuccessful, except in a few cases (Morettini, 1972; Berenguer, 1978; Fontanazza, 1987; Parlati *et al.*, 1986). However, many accessions in the world are under evaluation (Pannelli *et al.*, 1992b) for the selection in modern intensive cultivation. Clonal selection and germplasm preservation are essential before starting serious breeding or biotechnological programs.

Mutagenesis has also been attempted with some success. Two dwarf mutants, the first an unstable one from cv Ascolana tenera (Roselli and Donini, 1982), and a second very stable one from cv Leccino which blossoms 15 days later than the mother plants (Pannelli *et al.*, 1992a) were selected from irradiated plantlets. In addition, two compact mixoploid mutants from cvs Frantoio and Leccino, have been isolated from the same irradiated plantlets (Pannelli *et al.*, 1992a). These are characterized by large and thick leaves, a great resistance to *Spilocaea oleagina* and they produce, along with the normal ones, very large fruits, almost double the normal size (derived from tetraploid cells). From the mixoploid Frantoio plants, several tetraploid plantlets were isolated by using *in vitro* culture methods and hardening in the greenhouse; the isolation of tetraploid plantlets is underway also from mixoploid tissues of cv. Leccino by using *in vitro* somatic embryogenesis. (Rugini *et al.*, 1993).

### 1.4. *Aims of genetic crop improvement*

Most of the olive cultivars grown are fairly drought resistant and have a wide range of adaptability. Presently, the general aim in olive genetic improvement is to develop cultivars suitable for more intensive, often irrigated agriculture. Thus, a high production potential under non limiting environmental and cultural conditions is being searched for. This also involves the present need for quality products and adaptability to mechanization. Major genetic improvement is directed toward a reduction in alternate bearing, disease resistance, [*Spilocaea oleagina* (Cast.) Hugh, *Verticillium dahliae* Reinke et Berth], and pest resistance [*Dacus oleae* Gmel, *Saissetia oleae* Bern and

*Prays oleae* Bern.]. Furthermore, small sized plants with a low chilling requirement, high frost resistance, self-rooting ability and reduced apical dominance with a high number of lateral shoots, are required. As for drupes, a high pulp/stone ratio, uniform fruit ripening, high oil content and quality are needed. Cultivars with total self-incompatibility are of special importance in the breeding programs as they enable relatively easy and confident final cultivar crossings. In olive, self-incompatibility is in many cases greatly dependent on environmental conditions and unstable characters, which causes, together with the scarce vegetative-reproductive equilibrium, a low percentage of fruit set. At present this equilibrium can be induced by spraying the fungicide Hexaconazole (ICI-Solplant) or abscisic acid at blossoming time to temporarily reduce shoot growth by closing the stomata (Rugini and Pannelli, 1992). For many of the desired characters only a few clones are presently available. Finally, new cultivars are required to extend olive cultivation beyond Mediterranean environments. The non-Mediterranean wild olive, probably incompatible with the Mediterranean ones, might contain useful genes for future breeding.

Considering the importance of this crop for the economy of many countries and the present limits of both existing cultivars and the conventional breeding methods, bioengineering approaches must be speeded up without neglecting the other methods, which should allow for the selection of good varieties. Thus, the development of *in vitro* techniques, such as micropropagation, callus and cell culture and an efficient regeneration method are fundamental.

## 2. Culture initiation

### 2.1. *Brief review of* in vitro *work done*

The details on *in vitro* olive culture have already been reported in previous reviews (Rugini, 1986; Rugini and Fedeli, 1990; Rugini and Lavee, 1992). Following a short review of *in vitro* research, with particular emphasis on the recent reports, somatic embryogenesis will be described in detail.

*Micropropagation by axillary bud stimulation. In vitro* mass propagation is not yet economically convenient because it gives rise to costly and often long juvenile period plants. Currently, most cultivars are easily propagated by semihardwood cutting, and some cultivars require grafting because they are difficult-to-root. Unfortunately for these cultivars, micropropagation has not yet been reported, due mainly to various difficulties encountered in adapting them to *in vitro* growth conditions. Several problems are involved with the use of these techniques: difficulties in obtaining sterile shoots, because shoots do not develop from meristems and require a relatively high concentration (4–5 mg/l) of zeatin to grow; in addition 10–20 mg/l $GA_3$ is advised for faster elongation without affecting the subsequent rooting ability of explants.

Furthermore, the shoots grow with a strong apical dominance (from one bud, only one shoot). Vigorous apical herbaceous twigs, one year node explants, possibly the bifurcation of two shoots in which the bud is protected under bark, and inflorescences from mixed buds, are recommended for culture initiation. A mixture of $HgCl_2$ and sodium hypocloride under vacuum is required for satisfactory disinfection (Rugini and Fedeli, 1990). Considerable progress on olive micropropagation has been accomplished through the development of a medium through measurement of the ratio among the mineral elements in growing shoots in the field and in mature embryos (Rugini, 1984). The resulting olive medium (OM) contains a high concentration of Ca, Mg, S, Zn, B and Cu and glutamine as the main nitrogen source. The ratio between Ca/N should not exceed 1:11 to get good quality shoots. Modified MS media (Rugini, 1981; Fiorino and Leva, 1986) also supports growth for some cultivars. A low agar concentration (0.5–0.6%) and well-aerated conditions are also essential for rapid shoot growth. Among woody fruit trees, olive shows some peculiarities: it can metabolize mannitol (Leva *et al.*, 1992) and shows a low concentration of polyamine content in its tissues compared to other woody plants (Rugini *et al.*, 1992). In fact, the addition of mannitol or sucrose induces a similar growth, but mannitol has the advantage of reducing basal callus formation. Consequently the culture requires less zeatin for the same shoot mass production compared to the use of sucrose (Rugini and Caricato, 1994). In the presence of auxin, the addition to the rooting medium of the diamine putrescine at 160 mg/l, or a rapid basal treatment with one product of its catabolism, i.e., $H_2O_2$, increases and promotes early rooting (Rugini *et al.*, 1991a). Putrescine increases and promotes earlier rooting also when the *Agrobacterium rhizogenes* wild type is used in place of auxin (Rugini, 1992).

*Callus culture, cell suspensions of juvenile and mature tissues.* Callus cultures of olives were first established by Lavee and Messer (1969). Callus formations were induced from parts of juvenile tissues (Rugini, 1986; Rugini and Fedeli, 1990) and maintained for a long time. Significant progress has also been made in the callus culture of mature cultivars. At present it seems that the petioles from *in vitro* grown shoots are the best source material. MS medium with 7 mg/l thidiazuron (TDZ), 0.1 mg/l NAA and 4% sucrose in the dark at 28°C (Rugini and Caricato, 1994) give rise to large whitish, soft callus, while the use of 2,4-dichlorophenoxy acetic acid (2,4-D) produces semi-soft callus that soon turns brown (Rugini, 1986). Chlorogenic acid and cinnamic acid, at concentrations of $10^{-3}$ M and $10^{-4}$ M respectively can replace the auxins. Abscisic acid at a low concentration has a weak effect on the initiation and maintenance of bark callus (Lavee and Adiri, 1974; Lavee and Avidan, 1982).

Small and overfed cell suspensions were produced in a modified OM medium (Rugini and Fedeli, 1990) in an autophyton, in flasks or in multiwells on the shaker. In the beginning, growth was slow and required a replenishment of the medium rather than subculturing. Furthermore, high cell density, more than $10^5$/ml is required for growth.

*Anther and ovary cultures.* Preliminary research on anther culture produced callus of uncertain origin failed to develop any further in the subsequent subcultures on a modified MS medium although amino acids, antioxidants, growth regulators and vitamins were added to the medium (Mulè *et al.*, 1993). Unfertilized immature ovules were able to produce callus, while fertilized ones failed (Fodale *et al.*, 1993).

*Protoplast culture of zygotic and mature tissues of cultivars.* Viable protoplasts from hypocotyls, cotyledons and leaves of micropropagated seedling shoots were obtained by overnight incubation with 1.5% Driselase, 110 g/l mannitol, 500 mg/l of $CaCl_2 \cdot 2H_2O$, and 500 mg/l KCl. In BN (Bourgin and Nitsch, 1967) basal medium supplemented with 1 mg/l NAA, 0.5 mg/l 2,4-D and 0.5 mg zeatin riboside, the protoplasts formed the cell wall within 3–5 days and were able to divide in 15 days after progressively decreasing the osmotic concentration (Rugini, 1986). Cañas *et al.* (1987) used a similar protocol for protoplast isolation and subsequently they added 5 mg/l ornitine to the OMc (modified OM in which macro elements were replaced by those of BN, plus 1 g/l caseine hydrolysate) for protoplast culture. They obtained the formation of some microcalli derived from protoplast cells of mature cotyledons, but not from protoplasts isolated from seedling leaves. Mencuccini and Lucchetti (1992) succeeded in isolating protoplasts from leaves of micropropagated cultivar shoots but with no cell division. Perri *et al.* (1993) found that petiole callus of cv. Dolce Agogia and Canino yielded a high number of viable protoplasts. Subsequently, the derived cells were able to divide, unlike leaf lamina protoplasts, in MS liquid medium with 10 mg/l TDZ, 0.01 mg/l NAA, 0.75% sucrose, and 9% mannitol. However, somatic embryos of "Canino" cultivar tissues were the best protoplast source both in protoplast population density per ml and in their subsequent growth (Perri and Rugini, unpubished data).

*Shoot organogenesis from zygotic tissues.* Bao *et al.* (1980) and Wang *et al.* (1981) regenerated shoots from hypocotyl sections on White in a medium (1963) supplemented with 0.01 mg/l NAA, and 0.5 mg/l BA. Mature embryo fragments were also found to be a suitable material for shoot regeneration (Rugini, 1986). Cañas and Benbadis (1988) regenerated shoots from embryo fragment callus in OMc supplemented with 0.5 to 4 mg/l 2iP, after the callus was induced with 5 mg/l IBA and 0.2 to 0.5 mg/l zeatin riboside. The calli that originated from cotyledon segments proximal to the embryo axes were found more responsive than those from the other parts.

*Shoot organogenesis from mature tissues of cultivars.* Shoot organogenesis and plant recovery from leaf petioles of *in vitro*-grown mature cultivars was

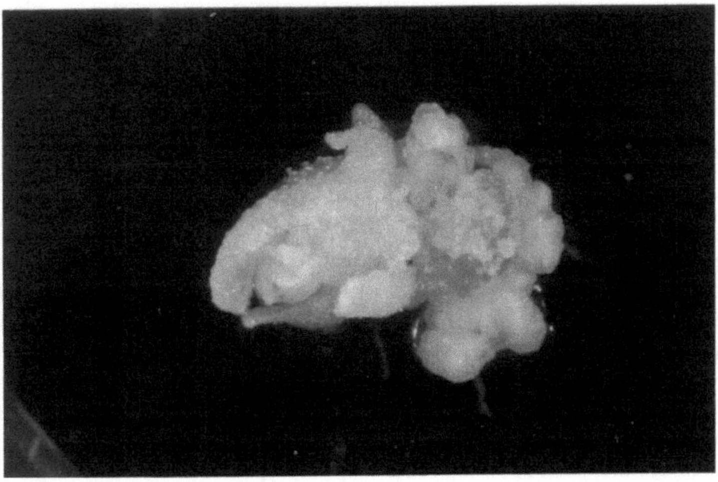

*Figure 1.* Shoot organogenesis from cv. Canino petiole. The petioles of this shoots were used to induce cyclic somatic embryogenesis.

obtained by Mencuccini and Rugini (1993). The efficiency of regenerating shoots, directly from petiole tissues or mediated by callus, ranged from 10 to 20% with only one shoot per explant (Fig. 1). The efficiency varied according to leaf position on the shoot, quality of shoot derived petioles and medium composition, while light photoperiod inhibited regeneration. The best results were obtained in MS medium supplemented with 6.6 mg/l of TDZ or in OMc supplemented with a combination of 1.5–2 mg/l 2iP, 0.5 mg/l BA and caseine hydrolysate. In the latter medium, direct shoot regeneration was more frequent. In both media, shoot regeneration was observed also from callus produced in the presence of cytokinin and a low concentration of auxin (0.5 mg/l IBA), except for medium containing 2,4-D, which inhibited morphogenesis at any concentration. Although mannitol is metabolized by the tissues, it was less efficient than sucrose in callus induction, growth and shoot regeneration (Rugini and Caricato, 1994). Regeneration from leaf lamina, stem and roots of mature tissues has not yet been reported.

## 3. Somatic embryogenesis

### 3.1. *Significance of somatic embryogenesis*

Other than an efficient method of propagation for some difficult-to-root cultivars and some new genotypes, somatic embryogenesis may become an important tool for plant regeneration from genetically manipulated cells and also from anthers or pollen, since the production of homozygous plants by traditional methods is still not possible. Because somatic embryogenesis avoids chimerism in regenerated plants, it can be used to obtain plants with different ploidy level from mixoploid plants. The continuous production of long-term cycles of embryos from epidermal cells, without a loss in capacity, could be of great advantage both in mass propagation and in genetic improvement through the insertion of exogenous DNA. In fact, this technique avoids callus production, which could be a wide source of nondesiderable variability, and thus the transformation of the epidermal embryo cells via *Agrobacterium* is easier. This avoids the wounding of the tissues which could be a cause of their browning, and the use of more sophisticated and less efficient transformation procedures.

In olive, somatic embryogenesis has succeeded and plants have been recovered from 1) immature, and 2) mature zygotic embryo tissues, 3) seedling tissues, 4) normal, and 5) mature transgenic tissues of cultivars (Table 1).

#### 3.1.1. *Somatic embryogenesis from immature zygotic embryos*

Somatic embryogenesis from immature zygotic embryos of cvs Dolce Agogia, Leccino, Frantoio and Moraiolo was first obtained by Rugini (1988). The age of embryos, genotype, basal media, growth regulator combinations (auxins and cytokinins) and light, were found to be critical factors in embryo induction. Somatic embryogenesis was only induced in embryos aged 75 days after ovule fertilization. Before and after this date, they only gave rise to callus and occasionally, in some treatments, some globular structures, but were unable to develop into distinct embryos. Basal MS medium supported more embryogenesis than modified OM (half strength salts, with 800 mg/l $Ca(NO_3)_2 \cdot 4H_2O$ replacing glutamine). However, embryos developed into plantlets quicker in the latter medium. Somatic embryos usually began to appear directly in the radicle area after six weeks in culture and in a low percentage from callus. The highest percentage of somatic embryogenesis was obtained in the absence or in low concentration of auxins combined with 0.1 to 0.5 mg/l BA. At a higher concentration, (3 mg/l) BA alone inhibited embryogenesis, but when it was combined with a low concentration of NAA (0.1 to 1 mg/l), embryogenesis occurred again. Any concentration of 2,4-D inhibited somatic embryogenesis either alone or combined with BA, and produced only yellow callus; when the concentration of 2,4-D was higher than 3 mg/l, callus formation was reduced. The number of somatic embryos derived from each zygotic embryo was 2 to 15, often arranged in a crown-like position. The increase in osmolarity by sorbitol did not affect embryogenesis, but it inhibited germination. Light photoperiod completely inhibited somatic embryogenesis. Leva *et al.* (1993), reached a similar conclusion while using three cultivars (Frangivento, Frantoio, and Picholine). They induced embryogenic callus in SH medium supplemented with 0.25 to 0.5 mg/l NAA and either with 0.1 mg/l 2iP or without it in the dark. Somatic embryos developed on the derived callus in a hormone free medium under 16 h light photoperiod. In one case they obtained a continuous cycle of embryogenesis and were able to maintain this capacity for over one year. The capacity of the immature zygotic embryos to differentiate somatic embryos could be maintained for over 2 months if the fruitlets were collected 75 days after full bloom (best time is when they express the maximum embryogenic capacity) and then stored at 12° to 15° C. Under these conditions the small embryos continued to grow at the expense of the endosperm but they did not lose their embryogenic capacity. On the contrary, if the fruits were left on the plant, the zygotic embryos were unable to differentiate via somatic embryogenesis (Rugini, unpublished data).

3.1.2. *Somatic embryogenesis from mature zygotic embryo segment callus*
By using segments of nongerminated mature zygotic embryos of both wild (Orinos and Mitrakos, 1991) and cultivated olives (Mitrakos *et al.*, 1992), somatic embryogenesis was induced from the callus. The embryos were dissected into three segments (radicle, proximal, and distal cotyledon) and then placed on OMc basal medium (Cañas and Benbadis, 1988) containing 5 mg/l IBA and 0.5 mg/l 2iP for inducing callus formation under 16 h photoperiod at 24°C. Subsequently, the derived calli were transferred to the same basal medium with or without 0.5 mg/l IBA at 7-day intervals (0, 7, 14, 21). Both wild and cultivated olive zygotic embryo callus produced somatic embryogenesis after 35 days in this medium. The difference in the percentage of calli forming embryos was correlated with the origin of the explant and the time of their previous permanence on callus induction medium. Calli derived from the radicle segment produced the highest somatic embryogenesis (40%); in contrast, the other two kinds of segments produced embryos at a low percentage. Both cotyledon segments after dissection and 7-day-old calli placed directly on the regeneration medium did not produce somatic embryos, while 14 and 21-day-old calli did in a percentage ranging from 1.7 to 6.7 dependent upon the growth regulators used in regeneration medium. In fact, in the regeneration medium, cytokinin (2iP) inhibited somatic embryogenesis in all concentrations tested while IBA inhibited it only at a concentration higher than 1 mg/l. Somatic embryogenesis was not influenced by the salt concentration or by light or dark conditions.

3.1.3. *Somatic embryogenesis from seedling tap root callus*
In this technique embryos are dissected aseptically from mature seeds after refrigeration at 4° to 5° for several months, and then placed to germinate *in vitro* as previously reported (Rugini, 1986). When seedlings developed into 2 to 3 nodes, several wounds were made with the knife on the tap root and then the whole seedlings were cultured on the surface of the solid MS basal medium, contained in 100 ml jars supplemented with 5 mg/l NAA, 0.1 mg/l BA and 3% sucrose. The callus developed from the wounds under a 16 h photoperiod at 23°C; the whole seedlings were subcultured in half-strength MS medium contained 0.1 mg/l NAA, 0.1 mg/l BA, 3% sucrose and 0.6%

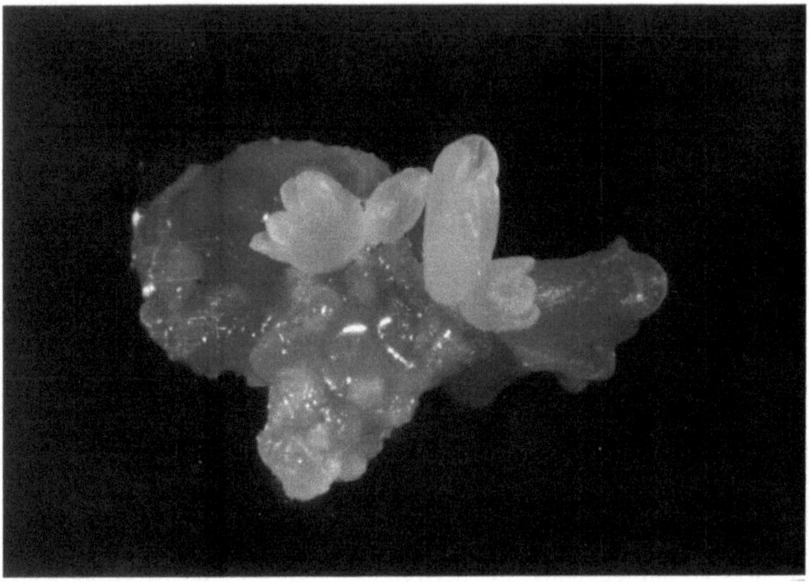

*Figure 2.* Embryos from enlarged somatic embryos which was derived from petioles of cv. Canino.

agar, and placed in the dark. After 35 days the calli started to turn brown, and simultaneously 1 or 2 embryos appeared in 10% of calli (Rugini and Tarini, 1986).

3.1.4. *Somatic embryogenesis from mature tissues of cultivars*

Recently, Rugini and Caricato (1995) developed a method to produce a continuous cycle of somatic embryogenesis from mature tissues of cultivar Canino. The embryo masses were induced in petioles collected from regenerated shoots. The use of neoformed shoots through organogenesis from micropropagated olive cultures seems to be an essential condition in achieving proembryo masses. Cyclic embryogenesis, both from normal embryos or teratomas, could be obtained on both a liquid and a solid OMc medium supplemented with 0.1 2iP, 0.1 BA, 0.05 IBA, 3% sucrose, and 200 mg/l filter-sterilised cefotaxime for over one year (Fig. 2). However, the production of normal embryos was higher and faster on filter paper embedded with a liquid medium. The addition of cefotaxime avoided friable callus formation and increased the number of normal neoformed embryos. The role of cefotaxime in producing more normal embryos and in maintaining the long-term embryogenetic capacity is not yet clear, however it seems that the reduction of the growth rate induced by the antibiotic allowed better cell differentiation.

Proembryo masses were obtained both on solid OMc from petioles of

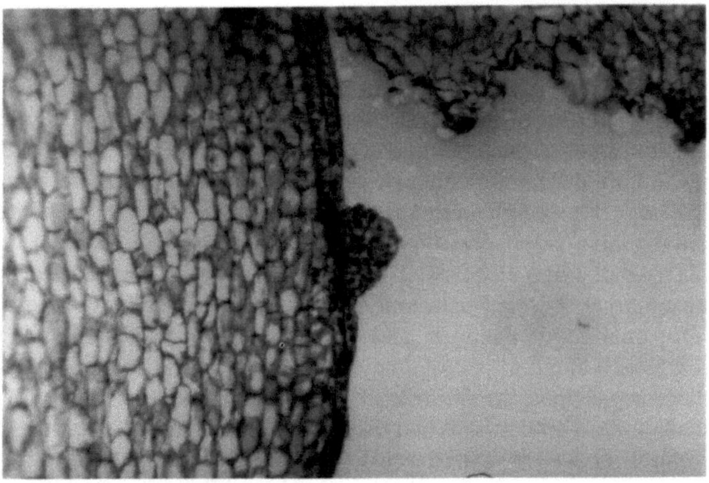

*Figure 3.* Initial embryo formation from epidermid of other embryos.

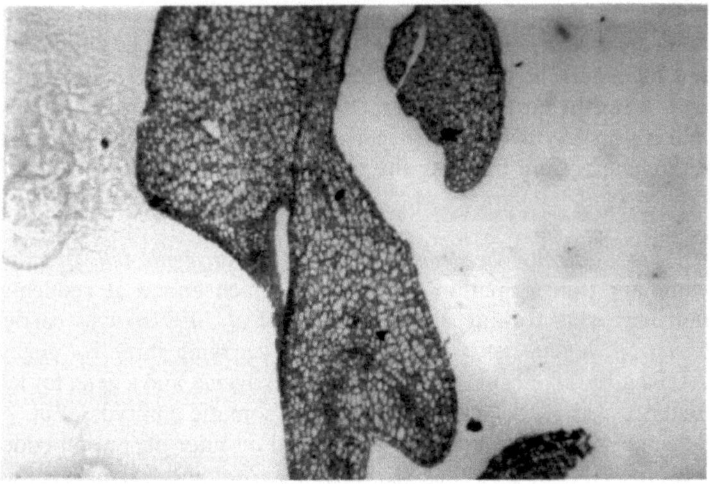

*Figure 4.* Longitudinal section of somatic embryo of cv. Canino, still attached to tissue.

whole leaflets with lamina still not expanded, or regenerated directly from the petiole tissues and from the detached petioles collected from expanded leaves of shoots regenerated from callus. These masses, when transferred to a filter paper with the liquid OMc, started to produce embryos on their secondary surface after about 15 days.

Generally the successive embryos or embryo-like structures originated from the surface of the normal and abnormal embryos but only in the hypocotyls and rootlet zone, not from their cotyledons. The normal embryos, white in color, arose from tissues either singularly or in a cluster of about 2 to 5 and continued growing until the torpedo stage with a round basal pole, but they did not germinate under these conditions. Some of them were easily detached and others remained strongly attached to the tissues. In addition to normal embryos, several types of teratoma were observed: 1) fusions between axes of 2 or 3 embryos, 2) formations of claviform-shaped structures with an enlarged base of attachment, 3) formations of only enlarged and twisted or enlarged cotyledons, and 4) formations of hardened and swollen hypocotyls with longitudinal pointed ribs. Except for the enlarged cotyledons and claviform-shaped structures, all the other teratoma were capable of producing normal embryos. When the normal single embryos or the morphogenetic teratomas were transferred to a 16 h light photoperiod they turned green and, within 2 weeks the white successive embryos continued to appear on their surface before they stopped to form new embryos. It seems that light hampered the differentiation of new embryos. Those that developed after exposure to light had probably already been induced in the dark.

Histological observations of teratoma or normal producing embryos at different stages of development showed that the somatic embryos originated from epidermal cells (Figs. 3 and 4). It is known that the juvenility of tissues acquired by recurrent regeneration increased the regeneration capacity; in this case, however, it is interesting to note that through recurrent regeneration it is possible to switch the type of morphogenesis from organogenesis to embryogenesis even by using the same medium and conditions.

3.1.5. *Somatic embryogenesis from putative transgenic cells of cultivars*

A preliminary transformation experiment, which aimed at reducing plant size and increasing rooting ability (Rugini *et al.*, 1991b) was carried out. *Agrobacterium tumefaciens* strain LBA 4404 carrying three *rol* genes A, B and C (ORFs 10, 11, 12) of *Agrobacterium rhizogenes* and a gene for kanamycin resistance was used. Whole or wounded somatic embryos of cv. Canino were infected with *Agrobacterium* and placed on filter paper embedded with OMc plus 0.1 2iP, 0.1 BA, 0.05 IBA, 3% sucrose, and 200 mg/l cefotaxime, in the dark. The regenerated somatic embryos were tested for kanamycin resistance. The technique of selection consisted of placing the neoformed embryos in a medium containing 200 mg/l kanamycin in the dark for 3 days in order to give the antibiotic the possibility to act inside the cells, then to expose the embryos to light. Among 100 embryos tested, 7–8 of them demonstrated resistance to kanamycin and turned green and developed in presence of the antibiotic (Fig. 6), while the others remained white and died, as the apical shoots of the developed embryos (Rugini and Caricato, unpublished data).

*Figure 5.* Plantlets in Jiffy pots derived from somatic embryos of cv. Canino.

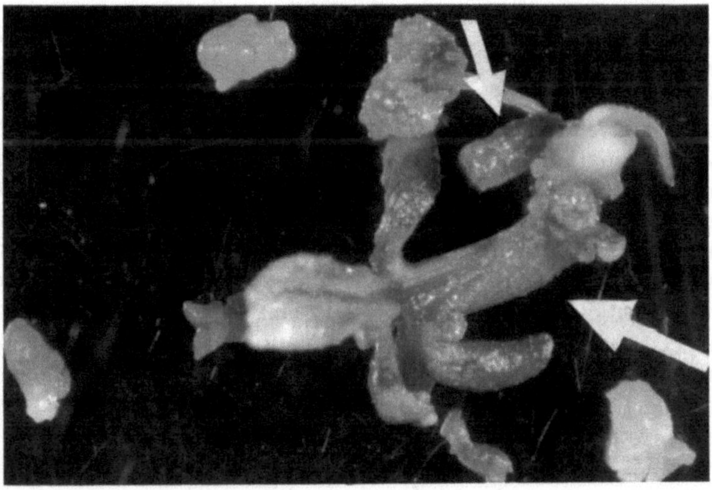

*Figure 6.* White somatic embryos of cv. Canino originated from infected embryos with *Agrobacterium tumefaciens* carrying *rol* A,B,C genes of *Agrobacterium rhizogenes* in kanamycin selection. Some of them became green and germinated (see arrows) in presence of kanamycin when transferred to light and give rise to plantlets which were hardened to greenhouse.

The GUS gene was not very useful in studying transformation because olive itself possess GUS activity quit high.

### 3.2. *Embryo germination and transfer of plantlets to soil*

There is little data on olive embryo germination, however the well formed embryos both from zygotic or mature tissues can germinate in a low mineral salt concentration (1/4 MS) with 2% sucrose, 0.2 to 0.5 mg/l zeatin and 0.6% agar. The addition of 0.02% activated charcoal, 160 mg/l putrescine and an increase of 2 mg/l zeatin to the above medium, induced an earlier greening and germination of about one week than those germinated in medium lacking charcoal and putrescine and when transferred to Jiffy-7 pots, under 50 μmol $s^{-1}$ $m^{-2}$ photon flux density. They were also germinated in liquid OMc (the same used to induce somatic embryos) in a gyrator shaker for 4 to 6 days in the dark and 2 days in the light. Under these conditions germination occurred quickly, but the embryos were subjected to vitrification. This anomaly, however, disappeared within a week after transplanting them to pots; they then started to grow regularly (Rugini and Caricato, 1995). Survival was more than 90% and they formed 2–3 nodes within a month. The derived plantlets appeared to be morphologically similar to those micropropagated by axillary bud stimulation (Fig. 5).

## 4. Conclusions and framework

In the last decade significant progress has been made in developing *in vitro* techniques for cloning some olive cultivars: a) a new medium has been developed, b) mannitol has been found a better carbon source than sucrose in some cultivars, c) polyamines increased rooting percentage, promoted early root initiation, and delayed senescence, and d) hardening of plantlets in the greenhouse was allowed. However, some problems still remain to be solved: a) apical dominance is strong, b) sterile shoots still cannot be initiated from meristems, c) zeatin cannot be substituted by less costly cytokinins, d) several cultivars still do not easily adapt to growing *in vitro*, and e) rejuvenation of plantlets is still too high. Significant progress has been made with callus, cell suspension and protoplast isolation and culture: a new medium together with growth regulator combinations was developed to sustain growth for a long period, and good responsive tissue, i.e., petioles and somatic embryos, were found to be good explant source material.

Important advances have also been made in shoot regeneration from petioles of *in vitro* grown shoots of several cultivars. The regeneration capacity is still low; only one shoot can be recovered from one petiole; therefore the application in biotechnology is not immediate.

Much progress has been made in somatic embryogenesis using immature zygotic embryos and radicles of mature ones. Furthermore, the way of maintaining long-term high embryogenic potential of zygotic embryos through harvest and storage of the immature fruits has been determined. The correlation among embryogenesic potential and tissue type, period of seed harvesting and storage of fruit, and the effect of exogenous growth regulators has also been observed, but biochemical and physiological studies are still needed to compare juvenile and mature tissues. Particular attention should be paid to the exogenous auxin/cytokinin ratio which is strictly related to the stage of maturation and type of tissue. It has clearly been demonstrated how auxin can inhibit or stimulate embryogenesis according to the tissue used. However, the high heterozygosity of the olive embryos could limit the immediate agronomic use of the derived improved plants and therefore require further breeding work.

Great progress has been made with somatic embryogenesis from the mature tissues of "Canino" and "Moraiolo" cultivars. The protocol found for the cv. Canino which allows cyclic embryogenesis directly from the embryo epidermis, also allows for rapid and efficient mass propagation and easy transformation via *Agrobacterium* and regeneration of plantlets resistant to kanamycin.

The recent framework has already been laid down, but additional research is needed in order to find a protocol which achieves embryogenesis from many other cultivars, both from diploid and haploid tissues. Homozygous plants should be of great interest in speeding up genetic improvement by crossing. They cannot be obtained by traditional method due to prevalence of self-incompatibility of the cultivars and their offsprings. Waiting for a good regeneration protocol from haploid tissues, fertilization of flowers with irradiated pollen and subsequently the use of embryo rescue could be a good transitory alternative to somatic embryogenesis in obtaining haploid plantlets (James, 1987). As both fruit set and germination percentages are low, methods of embryo rescue will be of great advantage. This is because with several cultivars and probably in the interspecific crossing, the seed often die in the early stage of fruit development as well as probably in intraspecific crossing. Physical or chemical mutagenesis of micropropagated buds and subsequent isolation of mutated parts should not be ignored. Sufficient evidence is available to support the hypothesis that studies should be carried out to create a germplasm bank of shoot apices which could represent the ideal material for cryopreservation. In fact, there are a large number of important cultivars for use in breeding programs which otherwise would be lost due to the current practice of using only a few cultivars with intensive cultivation. Breeding with non-Mediterranean species should also be attempted in order to transfer some particular features.

Finally, obtaining transgenic plants with the *Bt* gene of *Bacillus thuringensis* to control *Dacus oleae* and *Prays oleae* should be attempted.

## 5. References

Altamura Betti, M.M., L. Altamura and G. Mazzolani, 1985. Elements for the revision of the genus *Olea* (Tourn.) L. IV. The taxa present in Oceania which can be ascribed to *Olea* and allied genera. Ann. Bot. 43: 45–52.

Bao, Z.-H., Y.-F. Ma, J.-F. Liu, K.-J. Wang, P.-F. Zhang, D.-X. Ni and W.-Q Yang, 1980. Induction of plantets from the hypocotyl of *Olea europaea* L. *in vitro*. Acta Bot. Sinica 2: 96–97.

Bartolini S., A. Minnocci and C. Vitagliano, 1991. Morphological studies on pollen in some clones of olive "Leccino" cultivar. Agr. Med. 122: 282–296.

Bellini, E. and E. Giordani, 1992. Miglioramento genetico dell'olivo: comportamento ereditario di alcuni caratteri. II. Contributo. In: Atti Qualità Olio Extravergine di Oliva, Firenze, 1–3 December, 1982, pp. 273–275.

Berenguer, A.G. 1978. Seleccion clonal en Olivo (*Olea europaea* L.). Olea 1: 7–15.

Bogani, P., R. Petruccelli, L. Polsinelli and G. Roselli, 1993. Identification of olive tree cultivars by using random amplified polimorphic DNA. In: The 2th International Symposium on Olive Growing, Jerusalem, September 6–10, 1993, p. 22.

Bourgin, J.P. and J.P. Nitsch, 1967. Obtention de Nicotiana haploides à partir d'étamines cultivées *in vitro*. Ann. Physiol. Veg. 9: 377–383.

Cañas, L.A. and A. Benbadis, 1988. Plant regeneration from cotyledon fragments of the olive tree (*Olea europaea* L.). Plant Sci. 54: 65–74.

Cañas, L.A., A.M. Wyssmann and M.C. Benbadis, 1987. Isolation, culture and division of olive (*Olea europaea* L) protoplasts. Plant Cell Rep. 5: 369–371.

Chevalier, A., 1948. L'origine de l'0livier cultivé et ses variations. Revue Botanique Appliquée 303–304: 1–25.

Ciferri, R., 1942. Recenti progressi degli Studi botanico-agrari sull'olivo. Convegno di Studi Olivicoli. Regia Accademia Georgofili, Firenze, pp. 49–95.

Durante, M., R. Petruccelli, G. Bartolini and R. Bernardi, 1992. Impiego delle proteine di riserva per la identificazione delle cultivar di olivo (*Olea europaea* L.). In: Atti Qualità Olio Extravergine di Oliva, Firenze, 1–3 December, 1992, pp. 57–60.

FAO (1991). Production Yearbook Series N. 76, Rome.

Fiorino, F and A.R. Leva, 1986. Investigation on the micropropagation of the olive (*Olea europaea* L.) Influence of some mineral elements on the proliferation and rooting of explants. Olea 17: 101–104.

Fodale, A.S., Mulè, R. and A. Tucci, 1993. Callus initiation from *in vitro* culture of olive ovules. In: The 2th International Symposium on Olive Growing, Jerusalem, September 6–10, 1993, p. 25.

Fontanazza, G., 1987. Presentiamo la cultivar "I-77". Terra e Vita (Speciale Olio e Olivo) 46: 10–11.

James, D.J., 1987. Cell and tissue culture technology for the genetic manipulation of temperature fruit trees. Biotechnology and Genetic Engineering Review Newcastle 5: 33–79.

Johnson, L.A.S., 1957. A review of the family *Oleaceae*. Contribution from the National Herbarium 2: 395–418.

Knoblauch, E., 1895. Oleaceae die naturlichen Pflanzenfamilien. In: Engler A., Prantl R. (eds.) 4: 1–16.

Lavee, S. and N. Avidan, 1982. Growth responses of tree callus to chlorogenic acid and related phenolic substances. In: A. Fujiwara (Ed.), Proceedings Plant Tissue Culture 1982, pp. 165–168. Maruzen, Tokyo.

Lavee, S. and G. Messer, 1969. The effect of growth-regulating substances and light on olive callus growth *in vitro*. J. Exp. Bot. 20: 604–614.

Lavee, S. and N. Adiri, 1974. The effects of abscisic acid and gibelleric acid on the development of apple and olive callus. In: Plant Growth Substances, Part VI. Tissue Culture Proceedings 18th International Conference, Hirokawa, Tokyo, pp. 1141–1148.

Lavee, S., A. Haskal and M. Wodner, 1986. Barnea: a new olive cultivar from first breeding generation. Olea 17: 95–99.

Leva, A.R., R. Petruccelli and A. Benelli, 1993. Plant regeneration and somatic embryogenesis in *Olea europaea* L. 6$^{th}$ European Congress on Biotechnology, Firenze, June 13–17, 1993, p. 364 (Abstr.).

Leva, A.R., R. Petruccelli, R. Goretti and M. Panicucci, 1992. Ruolo di alcuni microelementi e carboidrati nella proliferazione *in vitro* di cv. di olivo (*Olea europaea* L.) In: Atti Qualità Olio Extravergine di Oliva, Firenze, 1–3 December, 1992, p. 333.

Loukas, M. and C.B. Krimbas, 1983. History of olive cultivars based on their genetic distances. J. Hort. Sci. 58: 121–127.

Mazzolani, G. and M.M. Altamura Betti, 1976/77. Elementi per la revisione del genere *Olea* (Tourn) Linn. 1 - Nota introduttiva. Ann. di Bot. 35–36: 463–469.

Mazzolani, G. and M.M. Altamura Betti, 1979. Elementi per la revisione del genere Olea (Tourn). Linn. III - Olea Crysophylla Lam: forme affini. Ann. di Bot. 28: 15–31.

Mazzolani, G. and M.M. Altamura Betti, 1980. Elementi per la revisione del genere Olea (Tourn) Linn. IV. Ciclo di Olea capensis L. Ann. di Bot. 28: 149–172.

Mencuccini, M. and M. Lucchetti, 1992. Coltura di protoplasti isolati da differenti tessuti di cultivar di olivo (*Olea europaea* L.) In: Giornate Scientifiche S.O.I., Ravello, 8–10 April, 1992, pp. 146–147.

Mencuccini, M. and E. Rugini, 1993. *In vitro* shoot regeneration from olive (*Olea europaea* L.) cultivar tissues. Plant Cell Tiss. Org. Cult. 32: 283–288.

Mitrakos, K., A. Alexaki and P. Papadimitriou, 1992. Dependance of olive morphogenesis on callus origin and age. J. Plant Physiol. 139: 269–273.

Morettini, A., 1972. Olivicoltura: Ramo Editoriale degli Agricoltori, Roma.

Mulè, R., A.S. Fodale and A. Tucci, 1993. Attemps to generate haploid plants from *in vitro* cultures of *Olea europaea* L. anthers. In: The 2th International Symposium on Olive Growing, Jerusalem, September 6–10, 1993, p. 24.

Murashige, T. and F. Skoog, 1962. A revised medium for rapid growth and bioassay with tobacco tissue culture. Physiol. Plant 15: 473–492.

Orinos, T. and K. Mitrakos, 1991. Rhizogenesis and somatic embryogenesis in calli from wild olive [*Olea europaea* var. *sylvestris* (Miller) Lehr] mature zygotic embryos. Plant Cell Tiss. Org. Cult. 27: 183–187.

Pannelli, G., F. Famiani and E. Rugini, 1992a. Effects of the change on ploidy levels on anatomical, cytological, reproductive, growth performance and polyamine contents, in mutants of olive irradiated olive plants. Acta Hortic. 317: 209–218.

Pannelli, G., R. Sirianni and M. Servili, 1992b. Preliminari osservazioni su una popolazione di accessioni di olivo selezionata in Umbria. In: Atti Germoplasma Frutticolo Alghero, 21–25 September, 1992, pp. 273–276.

Parlati, M.V., E. Bellini and C. Menna, 1986. Selezione clonale dell'olivo nel Lazio. Primi risultati ottenuti nel Viterbese sulla cv. Canino Informatore Agrario 13: 25–29.

Perri, E., M.V. Parlati and E. Rugini, 1993. Isolation and culture of olive (*Olea europaea* L.) In: The 2th International Symposium on Olive Growing, Jerusalem, September 6–10, 1993, p. 27.

Perri, E., M.V. Parlati, A. Palapoli and A. Napolitano, 1992. Polimorfismo isoenzimatico in olivo. In: Giornate Scientifiche S.O.I., Ravello, 8–10 April, 1992, pp. 152–153.

Pontikis, C.A., M. Loukas and G. Kousounis, 1980. The use of biochemical markers to distinguish olive cultivars. J. Hort. Sci. 55: 333–343.

Prevost, G., G. Bartolini and C. Messeri, 1993. Cultivar Italiane di Olivo e Loro Sinonimi. Menegazzo, Lucca.

Roselli, G. and B. Donini, 1982. "Briscola" Nuova cultivar di olivo a sviluppo compatto. Rivista Ortofrutticola Italiana 66: 103–114.

Roselli, G., P. Rossi and G.G. Vendramin, 1992. Differenze isoenzimatiche in cultivar di olivo (Olea europea L.). Atti Congresso su Germoplasma Frutticolo, Alghero, 21–25 September, 1992, 113–118.

Roselli, G., 1977. Osservazioni sulla scultura della exina del polline di alcune specie da frutto, 1. Olivo. Riv. Ortoflorofrutt. It. 3: 157–163.
Roselli, G., 1990. Miglioramento genetico dell'olivo:la selezione del materiale di moltiplicazione e la tutela del germoplasma. Convegno "Nuove prospettive del vivaismo olivicolo". Floricoltura Pesciatina 10: 2–6.
Rugini, E. and S. Lavee, 1992. Olive Biotechnology. In: F.A. Hammerschlag and R. Litz (Eds.), Biotechnology of Perennial Fruit Crops, pp. 371–382. C.A.B. International Wallingford, Oxon U.K.
Rugini, E., A. Pellegrineschi, M. Mencuccini and D. Mariotti, 1991b. Increase of rooting ability in the woody species Kiwi (*Actinidia deliciosa* A. Chev.) by transformation with *Agrobacterium rhizogenes* T-DNA rol genes. Plant Cell Rep. 10: 291–295.
Rugini, E. and E. Fedeli, 1990. Olive (*Olea europaea* L.) as an oilseed crop. In: Y.P.S. Bajaj (Ed.), Biotechnology in Agriculture and Forestry, Vol 10, pp. 593–641. Springer-Verlag, Berlin.
Rugini, E. and G. Pannelli, 1992. Preliminary results on increasing fruit set in olive (*Olea europaea* L.) by chemical and mechanical treatments. Acta Hortic. 329: 209–210.
Rugini, E. and P. Tarini, 1986. Somatic embryogenesis in olive (*Olea europaea* L.). Conference Fruit Tree Biotechnology, October 14–15, 1986, Paris, p. 62.
Rugini, E., A. Jacoboni and M. Luppino, 1992. Role of basal shoot darkening and exogenous putrescine treatments on *in vitro* rooting and on endogenous polyamines changes in difficult-to-root woody species. Sci. Hort. 53: 63–72.
Rugini, E., 1981. Propagazione *in vitro* di una cultivar di olivo (*Olea europaea* L.). Valutazione di varie citochinine ed auxine. Atti Congresso Fitofarmaci in Agricoltura, Firenze, pp. 171–180.
Rugini, E., 1984. *In vitro* propagation of some olive (*Olea europaea* L.) cultivars with different root-ability, and medium development using analytical data from developing shoots and embryos. Sci. Hort. 24: 123–134.
Rugini, E., 1988. Somatic embryogenesis and plant regeneration in Olive (*Olea europaea* L.). Plant Cell Tiss. Org. Cult. 14: 207–214.
Rugini, E., 1992. Involvement of polyamines in auxin and *Agrobacterium rhizogenes*-induced rooting of fruit trees *in vitro*. J. Amer. Hort. Sci. 117: 532–536.
Rugini, E., G. Pannelli, M. Ceccarelli, M. Muganu and X. Quin, 1993. Selezione *in vitro* ed *in vivo* di piante poliploidi da mutanti mixoploidi di olivo cv Leccino e Frantoio. In: XXXVII Convegno Annuale della Società di Genetica Agraria, Orvieto, 11–14 October, 1993, p.166 (Abstr.).
Rugini, E., M. Luppino, M. De Agazio and S. Grego, 1991a. Endogenous polyamine and root morphogenesis variations under different treatments in cuttings and *in vitro* explants of olive. Acta Hort. 300: 225–232.
Rugini, E. and G. Caricato, 1994. Embriogenes: somatica organogenesi e produzione di piante do tessuti maturi di cultivar di olivo (*Olea europaea* L.) In: Atti Convegno su Tecniche, Norme e Qualita in olivicoltura, December 15–17, 1993 Potenza, p. 632–644.
Rugini, E. and G. Caricato, 1995. Somatic embryogenesis and plant recovery from mature tissues of olive cultivars (*Olea europaea* L.) "Canino" and "Moraiols", Plant Cell Reports (in press).
Rugini, E., 1986. Olive. In: Y.P.S. Bajaj (Ed.), Biotechnology in Agriculture and Forestry Trees, pp. 253–267. Springer-Verlag, Berlin.
Sax, K. and E.C. Abbe, 1932. Chromosome numbers and the anatomy of the secondary xylem in the Oleaceae. J. Arnold Arboretum 23: 37–48.
Scaramuzzi, F. and G. Roselli, 1986. Olive genetic improvement. Olea 17: 1–17.
Schenk, R.V. and A.C. Hildebrandt, 1972. Medium and techniques for induction and growth of monocotyledonous and dicotyledonous plant cell cultures. Can. J. Bot. 50: 199–204.
Taylor, H., 1945. Cyto-taxonomy and phylogeny of the *Oleaceae*. Brittonia 5: 337–367.
Wang, K.J., P.F. Zhang, D.X. Ni, X.Z. Zhu, W.Q. Yang and Z.H. Bao, 1981. Callus formation and organ regeneration in tissue culture of woody plants. Acta Bot. Sinica 23: 97–108.
White, P.R., 1963. The Cultivation of Animal and Plant Cells. Ronald Press, New York.

# 12. Somatic embryogenesis in grape

D.J. Gray

**Contents**



## 1. Introduction

Grape (Vitaceae, *Vitis* spp.) is a deciduous temperate fruit crop of ancient origin. *Vitis* contains two subgenera, *Euvitis* Planch., the bunch grape species that all contain 38 somatic chromosomes, and *Muscadinia* Planch., the muscadine grapes with 40 somatic chromosomes (Einset and Pratt, 1975; Winkler *et al.*, 1974). Species in each subgenus are interfertile but are only partially fertile between subgenera (Jelenkovic and Olmo, 1969). Estimates of the number of *Euvitis* species range from 28 to 43. These are separated into an American group of 18 to 28 species, an Asian group of 10 to 15 species and a European or central Asian group of one species (*Vitis vinifera* L.) (Einset and Pratt, 1975). *Muscadinia* contains three species; *Vitis rotundifolia* Michaux., *Vitis munsoniana* Simpson, both native to the southeastern United States and *Vitis popenoeii* Fennell, native to central America, which comprise the muscadine grape varieties. With the exception of muscadine grapes, which are morphologically and genetically distinct, most cultivated grapes are either pure strains or hybrids of *V. vinifera* and account for the vast majority of world production. Hybrids of *Vitis labrusca* L., an eastern North American native (Galet, 1988), and *V. vinifera*, some with additional native species parentage, are in widespread production in the northeastern United States (Cahoon, 1986). Complex hybrids between *V. vinifera* and various native American species form a series of disease resistant French-American hybrids (Einset and Pratt, 1975). Certain species and interspecific hybrids are utilized exclusively as rootstocks. A number of other *Vitis* species are used in breeding programs to adapt high-quality *vinifera* germplasm to suboptimal environmental regions (Alleweldt and Possingham, 1988).

*S. Jain, P. Gupta & R. Newton (eds.), Somatic Embryogenesis in Woody Plants, Vol. 2,* 191–217.

Grape production ($65 \times 10^6$ metric tons) exceeds that of any other temperate fruit crop and ranks after *Citrus* and banana among all fruit crops worldwide (FAO Production Yearbook, 1990). Grape surpasses all other fruit crops in value due to its multiple uses for fresh fruit, juice, jelly, raisins and wine. Wine production demonstrates the ultimate value of a fruit crop, with prices sometimes exceeding thousands of dollars per bottle.

### 1.1. *Conventional genetic improvement*

Although *V. vinifera* generally is considered to have the most desirable fruit quality, it is susceptible to many pests and diseases, including anthracnose, black rot, botrytis bunch rot, crown gall, downy mildew, eutypa dieback, various nematodes, phomopsis cane and leaf spot, phylloxera, Pierce's disease and powdery mildew. Hybridization with resistant species has been the only method available to produce resistant cultivars (Einset and Pratt, 1975; Galet and Morton, 1990). Other improvement objectives for grape include increasing fruit quality and yield.

However, improvement is slow and time-consuming because grape is heterozygous and exhibits pronounced inbreeding depression (Einset and Pratt, 1975). Therefore, an individual genotype cannot be reproduced by seed and all propagation of outstanding selections must be by vegetative means. The life cycle is relatively long, with a juvenility period ranging from one to six years, depending on environmental conditions and management practices (Einset and Pratt, 1975), such that fruit quality of progeny cannot be assessed for several years. Furthermore, resistance to many diseases cannot be confirmed until after several fruit-bearing years have passed. Another problem with the breeding approach, is that certain quality attributes of specific *vinifera* varieties, such as the color stability and flavor of "Cabernet Sauvignon", cannot easily be transferred to disease-resistant hybrids. Outstanding characteristics of varieties such as "Cabernet Sauvignon" were selected in ancient times, are genetically fixed, and are maintained by exclusive use of clonal propagation. Because of these obstacles to genetic improvement, alternative approaches to bypass sexual hybridization in order to add specific traits while maintaining the genetic integrity of such desirable varieties is especially attractive for grape.

### 1.2. *Genetic improvement through biotechnology*

For grape, *in vitro* genetic manipulation of somatic and/or zygotic cells holds great promise by either supplementing and enhancing existing breeding techniques or by circumventing conventional methods altogether. For example, somatic hybridization via protoplast fusion potentially might be utilized to circumvent the sterility inherent in crosses between *Euvitis* and *Muscadinia*, perhaps allowing seedlessness to be introduced into the former. While routine recovery of haploid and/or dihaploid plants via anther or ovule culture has not been accomplished in grape [haploid callus (Gresshoff and Doy, 1974) and plants (Zou and Li, 1981) were isolated reports], the potential of fertile dihaploids would be great, since all recessive genes would be expressed. Genetic mapping would be greatly simplified by use of such homozygous material. Genetic transformation offers significant opportunities for grape improvement by allowing specific traits to be introduced into otherwise traditional cultivars. Possible early applications of genetic transformation, include the addition of virus resistance by integration of viral coat protein genes (Nejidat *et al.*, 1990; Stark and Beachy, 1989), conferring resistance to insect pests by incorporation of genes that code for *Bacillus thuringensis* toxin (Barton *et al.*, 1987; Delannay *et al.*, 1989) and addition of genes to obtain herbicide tolerance (Oxtoby and Hughes, 1990; Shah *et al.*, 1986). Synthetic seed technology of grape, which would allow clonal germplasm of grape to be conserved in seed repositories, is relatively well-advanced as described below.

Implementation of these improvement strategies requires ability to manipulate and achieve regeneration from grape cells and tissues *in vitro*. In this chapter, the phenomenon of grape embryogenesis is reviewed in light of existing literature by exploring past and present approaches to culture initiation and maintenance, manipulation of embryo and plant development and specific uses for embryogenic culture systems.

## 2. Somatic embryogenesis

### 2.1. *Background and historical perspective*

Both embryogenic and, less commonly, organogenic regeneration pathways occur and previously were compared (Gray and Meredith, 1992). While only somatic embryogenesis is discussed here, it must be noted that difficulty in distinguishing between embryogenesis and organogenesis led to past confusion, since abnormal development during somatic embryogenesis can resemble shoot or root organogenesis. For example, examination of early literature concerning grape regeneration suggests that somatic embryogenesis was described as organogenesis in at least two instances. Shoot development from anther-derived callus of *Vitis thunbergii* Sieb. et Zucc. (Hirabayashi *et al.*, 1976) actually may have been an example of somatic embryogenesis. Although it is impossible to confirm, the regenerated structures resembled somatic embryos, based on an assessment of published photographs. Further, a paper by Favre (1977) is often mistakenly cited as an early report of shoot organogenesis, possibly due to the fact that the paper is in French, leading to misinterpretation by non-French speaking researchers. Translated to English, the critical text from this paper describes "bud formation that is first manifested by the appearance of small white organs. These develop into bipolar structures with an apical pole having more or less flattened pieces".

*Table 1.* Chronological listing of selected publications concerning grape somatic embryogenesis.

| Date | Species/Variety | Explant[1] | Medium[2] | Auxin[3] (μM) | Cytokinin[4] (μM) | Culture age[5] | Germination treatment | Reference |
|---|---|---|---|---|---|---|---|---|
| 1976 | *V. vinifera* "Cabernet Sauvignon" | O-Nu | N | 5–20 NOA | 5–10 BA | NG | 1 GA + 5 IP | Mullins and Srinivasan |
| 1977 | *Vitis* hybrid "Seyval Blanc" | S, L, F | MS | 4.5 2,4-D | 0.4 BA | 6 mo. | none | Krul and Worley |
| 1979 | *V. vinifera* × *V. rupestris* | A | N | 5 2,4-D | 1 BA | NG | 4C | Rajasekaran and Mullins |
| 1980 | *V. longii*, *V. rupestris*, *V. vinifera* "Grenache", *V. vinifera* × *V. rupestris* | A | N | 5 2,4-D | 1 BA | NG | 4C | Mullins and Rajasekaran |
| 1980 | *V. vinifera* "Cabernet Sauvignon", "Grenache", *V. vinifera* × *V. rupestris* | O-Nu | N | 5 NOA or 5 2,4-D then 10 NOA | 1 BA | NG | Plants not obtained | Srinivasan and Mullins |
| 1983 | *V. longii*, *V. rupestris*, *V. vinifera* "Grenache", "Sumoll" × "Cabernet Sauvignon", *V. vinifera* × *V. rupestris*, *Vitis* hybrid "Villard Noir", "Villard Blanc", M.G. 60–44 | A | N | 5 2,4-D | 1 BA | NG | NG | Rajasekaran and Mullins (a and b) |
| 1985 | *Vitis* hybrid "Seyval Blanc" | NG | NG | NG | NG | Indefinite | 1 BA | Krul |
| 1986 | *V. vinifera* "Cabernet Sauvignon" | A | MS | 4.5 2,4-D then 0.5 NAA | 1 BA | NG | 4C + cotyledon removal | Mauro *et al.* |
| 1987 | *V. longii* | A and Ov | MS | 5 2,4-D | 1 BA | 20 mo. | 1 BA or dehydration | Gray (b and d); Gray and Mortensen |
| 1988 | *V. rupestris*, *V. vinifera* | L | N | 4.5 NOA | 0–4.4 BA | NG | none or various growth regulators | Stamp and Meredith (a) |
| 1988 | *V. rupestris*, *V. vinifera* "Cabernet Sauvignon", "Cardinal", "Grenache", | A | N | 4.5 2,4-D or 4.5–7 NOA | 0.9–4.4 BA | NG | none or various growth regulators | Stamp and Meredith (a) |

| | | | | | | | | |
|---|---|---|---|---|---|---|---|---|
| | "Sauvignon Blanc", "Thompson Seedless", "White Riesling", *V. vinifera* × *V. rupestris* | | | | | | | |
| 1988 | *V. longii*, *V. vinifera* "Chardonnay", "French Colombard", "Grenache", "White Riesling" | Z | N | 4.5 2,4-D or 4.4–18 NOA | 0–4.4 BA | 12 mo. | none or various growth regulators | Stamp and Meredith (b) |
| 1989 | *V. longii*, *V. vinifera* "Thompson Seedless" | A, Ov, L | MS | 5 2,4-D | 1 BA | 48 mo. | 1 BA or dehydration | Gray |
| 1989 | *Vitis* hybrids | O | MS | 10 IAA[6] | 1 BA | 42 mo. | 1 BA or dehydration | Gray |
| 1989 | *V. vinifera* "Koshusanjaku" | L | N, MS | 5–10 2,4-D w./wo. transfer to 1 2,4-D | 5–10 BA or 5–10 TDZ or 5–10 KT | 24 mo. | none | Matsuta and Hirabayashi |
| 1992 | *V. rotundifolia* five varieties | O then Z | NN | 1 mg/l NOA | 0.2 mg/l BA | 12 mo. | MS + 1 μM BA | Gray |
| 1992 | *V. aestivalis*, *V. vinifera* "Koshusanjaku" | L | NN | 10 μM 2,4-D or 2,4,5-T | 10 μM KT | NG | NG | Matsuta |
| 1993 | *V. rotundifolia* 11 varieties | O then Z | NN | 1 mg/l NOA | 0.2 mg/l BA | NG | MS + 1 μM BA | Gray and Hanger |
| 1993 | *V. rupestris* | L | MS | 0.1 or 1 mg/l 2,4-D | 1 mg/l BA | 24 mo. | 4C + 1 mg/l BA + 1 mg/l IBA | Martinelli *et al.* |
| 1993 | *V. rotundifolia* 2 varieties | L | NN | 9 μM 2,4-D, then 10.7 μM NAA | 4.4 μM BA, then 0.9 μM BA | NG | 1 μM BA | Robacker |
| 1993 | *V. vinifera* "Riesling" | A, F | NN | 4.5 μM 2,4-D | 10 μM BA or 10 μM TDZ | NG | NG | Harst-Langenbucher and Alleweldt |

[1] A – anthers; F – floral tissue; L – leaves (includes use of lamina and/or petiole); NG – explant not given; O – fertilized ovule; O-Nu – unfertilized ovule cultured – nucellus confirmed to be source of embryogenic cells; Ov – unfertilized ovaries; S – shoot tip; Z – zygotic embryos.
[2] MS – Murashige and Skoog (1962); N – Nitsch and Nitsch (1969). Note that original medium formulae were often modified.
[3] 2,4-D – 2,4-dichlorophenoxyacetic acid; IAA – indole-3-acetic acid; NAA – α-naphthaleneacetic acid; NOA – naphthoxyacetic acid.
[4] BA – 6-benzyladenine; KT – N-(2-chloro-4-pyridyl)-N′-phenylurea; TDZ – thidiazuron – N-(1,2,3-thiadiazol-5-yl)-N′-phenylurea.
[5] Culture age – Time period over which embryogenic cultures could be maintained.
[6] Embryogenic culture initiation medium with IAA also contained 1 μM gibberellic acid.

The author states that these structures "resembled the plantlets found in seeds". It is clear from this description that somatic embryos were obtained (compare with description of somatic embryogenesis in the section below).

Grape somatic embryogenesis was first unequivocally reported by Mullins and Srinivasan (1976), who cultured unfertilized ovules of *V. vinifera* "Cabernet Sauvignon" in liquid medium containing napthoxyacetic acid (NOA) and 6-benzyladenine (BA). Proliferating nucellus-derived callus was extruded through the ovular micropyle and eventually gave rise to somatic embryos. Somatic embryos germinated into plants when transferred to agar-solidified medium that contained gibberellic acid ($GA_3$) and 2-isopentenyladenine (2-iP). Long term maintenance of cultures was not described. Krul and Worley (1977) documented somatic embryogenesis from calli derived from leaf, petiole and stem segments of the French-American hybrid "Seyval". This culture protocol differed from that of Mullins and Srinivasan (1976), since solidified medium was used that contained 2,4-dichlorophenoxyacetic acid (2,4-D) to induce callusing. Embryos developed after transfer of callus to medium containing α-naphthaleneacetic acid (NAA) and BA. Embryos germinated into plants after transfer to medium devoid of growth regulators. The original embryogenic line could be maintained indefinitely through secondary embryogenesis on growth regulator-free medium. A US patent describing this method of embryogenesis and plant production was eventually awarded (Krul, 1985). In the intervening years, research on grape somatic embryogenesis was reported by a number of researchers (Table 1).

### 2.2. *Culture initiation*

*Genotypic specificities.* Initiation of embryogenic cultures in grape is highly genotype-dependent, with all but a few successful reports emanating from only four species: *Vitis aestivalis* Michx., *V. longii* Prince, *V. rotundifolia*, *V. rupestris* Scheele or *V. vinifera* (Table 1). Most hybrids that produced embryogenic cultures are composed of at least one of these species. For example, *Vitis* hybrids "Seyval Blanc", "Villard Blanc" and "Villard Noir" have both *V. rupestris* and *V. vinifera* parentages. Additionally, genotypic specificities are noted within species. This is illustrated by *V. vinifera*, where very few cultivars were documented to produce somatic embryos. Genotypic specificity is currently an obstacle to implementing a broad range of biotechnological procedures, since regenerative cultures are not yet available for many important cultivars.

*Sex of donor vine.* Sex of the donor vine was suggested to be a contributing factor in the ability of particular explants to yield embryogenic cultures. Rajasekaran and Mullins (1983b) noted that the propensity of anthers from male vines to form somatic embryos was greater than that of hermaphroditic vines and that anthers from female vines did not form somatic embryos at all. Treatment of a male vine with cytokinin to feminize the flowers inhibited anthers while it stimulated the ovules' ability to produce somatic embryos (Rajasekaran and Mullins, 1983b; Srinivasan and Mullins, 1980). However, Gray and Mortensen (1987) found that both anthers and ovaries of untreated female *V. longii* "Microsperma" yielded embryogenic cultures. Therefore, the importance of sex expression on the ability of an explant to produce somatic embryos is somewhat unclear. However, growth responses of tissues from male and female vines clearly can be manipulated with growth regulators.

*Explant source.* Explant tissue type and condition are critical to successful culture initiation. Most successful attempts have relied on floral tissues, particularly anthers or unfertilized ovules (with the nucellus documented to be the source of embryogenic cells in the latter); however, leaves, shoot tips and zygotic embryos were also utilized (Table 1). Recently, preculture of fertilized ovules was employed to obtain immature zygotic embryos, which were used as explants to develop embryogenic cultures of *V. rotundifolia* (Gray, 1992). Subsequently, it was demonstrated that ovules cultured at maturities ranging from 20 to 40 days after pollination were equally effective in development of embryogenic cultures from 11 *V. rotundifolia* varieties (Gray and Hanger, 1993). Explant developmental stage as well as preconditioning may be important in certain instances. For example, anthers approximately 0.5 mm long were optimal for inducing embryogenic cultures of *V. vinifera* "Cabernet Sauvignon" (Mauro *et al.*, 1986). Anther translucence and color (translucent yellow-green) served as an indicator for determining the proper stage of both anther and unfertilized ovary explants (Goussard *et al.*, 1991; Newton and Goussard, 1990). Only leaves 1.5 to 5.0 mm in length produced embryogenic cultures for several varieties (Stamp and Meredith, 1988a). For anthers, chilling at 4°C for 72 h (Rajasekaran and Mullins, 1979) or 96 h (Harst-Langenbucher and Alleweldt, 1993) prior to culture resulted in an increase of callusing and embryogenesis when compared to no cold treatment. Perhaps other, as yet untested, methods of pretreating explants would allow extension of the embryogenic response to additional species and cultivars.

*Origin of embryogenic cells.* The specific origin of embryogenic cells in primary explants of grape is obscure. Rajasekaran and Mullins (1983a) suggested that anther somatic tissues and not gametic cells gave rise to embryogenic callus, since haploids and homozygous diploids were not obtained among regenerants. This observation was reinforced by Newton and Goussard (1990) who described embryogenic callus growing from the filament, interlocular parenchymatous tissue and/or anther wall epidermis in *V. rupestris*. Embryogenic callus appeared to arise from the nucellus in unfertilized ovules (Mullins and Srinivasan, 1976); whereas most polyembryos from fertilized ovules were of zygotic embryo origin (Durham *et al.*, 1989). In cultured leaves, embryogenic callus arose from mid-vein and adjacent lamina tissue

(Stamp and Meredith, 1988a). However, these studies did not pinpoint the actual tissue origin(s) of the embryogenic cells.

*Culture medium.* Basic embryogenic culture medium formulations have received relatively little attention. We are aware of only two, MS medium (Murashige and Skoog, 1962) and Nitsch's medium (Nitsch and Nitsch, 1969), or modifications thereof, that were successfully utilized (Table 1). However, details of medium formulation differ among various researchers. For example, Mullins and coworkers (Mullins and Srinivasan, 1976; Rajasekaran and Mullins, 1979; Mullins and Rajasekaran, 1980) utilized a liquid medium system, whereas others used agar-solidified medium. Stamp and Meredith (1988a) compared both liquid and solid methods for leaf culture and found only the latter to be suitable. Other media addenda are similar to those commonly used in cell culture of many other species, such as sucrose at 20 to 60 $g \cdot l^{-1}$. Study of the effects of 1 $mg \cdot l^{-1}$ adenine, 100 $mg \cdot l^{-1}$ glutamine and 10 $mg \cdot l^{-1}$ phenylalanine on embryogenesis from anthers showed that a combination of all three was better than any two or none at all (Mauro *et al.*, 1986).

Generally, the phenoxy-auxins 2,4-D, 2,4,5-trichlorophenoxyacetic acid (2,4,5-T) or NOA, combined with the cytokinin BA, are utilized to initiate embryogenic cultures (Table 1); however, indole-3-acetic acid (IAA) combined with $GA_3$ was used to produce long-lived embryogenic cultures from fertilized ovules (Gray, 1989). Presumably, in the latter instance, embryogenic cultures stemmed from immature zygotic embryos (i.e., from preexisting embryogenic cells) and an induction stage was unnecessary. Cytokinins thidiazuron (TDZ) and N-(2-chloro-4-pyridyl)-N′-phenylurea (KT-30) were shown to be superior to BA for leaf explants of *V. vinifera* "Koshusanjaku" (Matsuta and Hirabayashi, 1989) whereas more embryogenic cultures were obtained with BA than TDZ for anther explants of *V. vinifera* "Riesling", suggesting either an explant and/or genotypic difference in response to cytokinin.

*Culture conditions.* Reports in the literature of physical culture conditions vary considerably to the degree that specific requirements are uncertain. For example, Stamp and Meredith (1988a) cultured anthers in total darkness, whereas a 16/8 h light dark photoperiod was used for leaf explants. For anthers of *V. vinifera* "Riesling", an incubation temperature of 27°C resulted in increased embryogenesis when compared to 15°C or 32°C temperatures (Harst-Langenbucher and Alleweldt, 1993).

*Examples of culture protocols.* Two distinct culture protocols are given that have proven to be useful for multiple cultivars. *Protocol* 1. This culture protocol, described by Goussard *et al.* (1991), utilized unfertilized ovaries, obtained from inflorescences forced from dormant wood cuttings and was similar to that previously described by Mullins and Srinivasan (1976). Dormant canes 40 cm in length were treated with 2% Captan and conveniently stored for several months until needed at 2–3°C. The bases of canes sectioned into 12–15 cm lengths, each with 2 internodes and three buds, were placed into 80 ml water and kept at 28–30°C under constant illumination. After bud burst at 5–8 days, all vegetative organs were removed from shoots bearing inflorescences to encourage flower development, which were harvested when anthers were translucent yellow-green in color (maturation status of microspores was not considered). Twenty ovaries were cultured per 100 ml Nitsch and Nitsch liquid medium with 5 μM BA, 2.5 μM NOA and 2.5 μM 2,4-D. Incubation was at 80 rpm, 25°C in dark. After 30 days, tissues were transferred to the same medium, except that it was solidified, and incubated as before for 30 days. Embryogenesis occurred after transfer to the same medium, but lacking growth regulators. This procedure was used to routinely obtain somatic embryogenesis from seven *V. vinifera* varieties (P.G. Goussard, personal communication). *Protocol* 2. A multi-step solid medium culture system was used to induce embryogenesis from laminae and petioles of two *V. rotundifolia* varieties (Robacker, 1993), a species that previously was regarded as recalcitrant. Following culture on Nitsch's medium containing 9 μM 2,4-D and 4.4 μM BA, explants were transferred to medium with 10.7 μM NAA and 0.9 μM BA to encourage callusing. The cultures were then transferred to growth regulator-free medium. All cultures were maintained under a 16/8 h light/dark cycle at 27–30°C. Up to 19% of laminae and 91% of petioles eventually produced embryogenic cultures. Of the embryogenic cultures obtained, 10% of those from laminae and 50% of those from petioles were noted on the second medium; however, most embryogenic cultures did not emerge from the callus mass until after transfer to basal medium. Thus, this study suggested that an intermediate treatment to encourage cell division ultimately resulted in a higher frequency of embryogenesis.

*Culture morphology*. Embryogenic callus or embryos typically emerged some time after nonembryogenic callus had formed on initiation medium. Fig. 1A shows the emergence of embryogenic callus from a callused, necrotic leaf explant of *V. vinifera* "Thompson Seedless". Embryogenic callus is granular and white to yellow in color. Often a highly heterogenous callus, composed of white, yellow and/or green, soft, friable, dry to wet sectors develops from the explant (Gray and Mortensen, 1987). Cells of the friable callus were highly variable in shape and size. Often, this initial callus declines and becomes brown prior to emergence of embryogenic callus and somatic embryos (Gray, 1992). In rarer instances, embryogenic callus emerged directly from seemingly uncallused explants (D.J. Gray, personal observation). Embryogenic callus was typically white to pale yellowish and emerged from distinct sectors of the heterogenous callus. Newly-initiated embryogenic callus of grape was composed of clusters of small isodiametric cells, interspersed with larger vacuolate cells (Gray and Mortensen, 1987). The embryogenic cells were easily identified by their prominent nuclei and nucleoli as well as

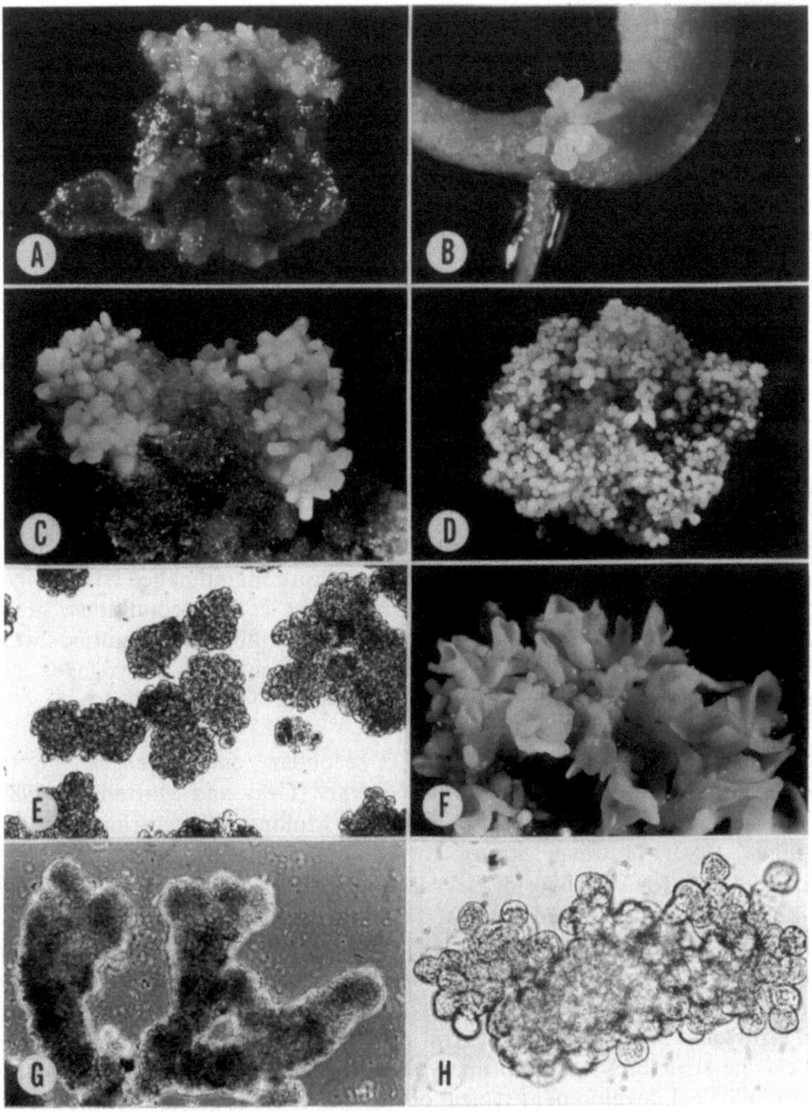


*Figure 1.* Somatic embryogenesis in grape. (A) White granular embryogenic callus emerging from cultured leaf of *Vitis vinifera* "Thompson Seedless". Note necrotic edge of lamina visible at bottom left (× 10). (B) Direct somatic embryogenesis from hypocotyl-radicle boundary of previously-formed embryo (× 14). From Gray and Mortensen (1987). (C) Somatic embryogenesis occurring from sectors of heterogenous callus maintained on medium with 2,4-D and BA (× 20). (D) Non-sectored embryogenesis from homogenous callus maintained on growth regulator-free medium (× 8). (E) Embryogenic cell clusters that compose homogenous embryogenic callus (× 85). (F) Embryo development on homogenous embryogenic callus (× 15). G) Organized embryogenic cell clusters typical of cultures maintained on medium with 2,4-D and BA (× 50). H. Uniform embryogenic cell culture typical of cultures maintained on growth regulator-free medium (× 160).

densely stained cell walls and cytoplasm (Krul and Worley, 1977). Embryogenic culture material typically contained distinct cell clusters, somatic embryos and highly abnormal somatic embryos that appeared as amorphous masses of white, compact, highly organized tissue embedded in a watery matrix (Gray and Mortensen, 1987; Gray, 1992). Microscopic examination revealed that the individual cells of embryogenic callus were highly uniform, being small, densely cytoplasmic with amyloplastids and lipoplastids (Gray, 1992).

### 2.3. *Culture maintenance*

Early reports by Mullins and coworkers (Table 1) suggested that somatic embryogenesis was a terminal event in their culture system and that regenerative cultures could not be maintained. However, Krul and Worley (1977) demonstrated that cultures could be maintained for at least six months and, subsequently, "indefinitely" (Krul, 1985). A *V. longii* cell line, homogenously composed of embryogenic-type cells, has been in continuous culture for eight years, although few somatic embryos now develop (D.J. Gray, unpublished data). Other documented time periods for culture maintenance have ranged from 12 to 48 months (Table 1). Thus, under certain conditions, grape embryogenic cultures are quite long-lived. Grape embryogenic cultures were maintained via two distinct methods: direct secondary embryogenesis or indirect embryogenesis via culture of embryogenic cells.

*Direct secondary embryogenesis.* Direct secondary embryogenesis was reported for several grape species and cultivars (Gray and Mortensen, 1987; Krul and Worley, 1977; Rajasekaran and Mullins, 1979; Srinivasan and Mullins, 1980). However, its use for culture maintenance was adequately explored only for *Vitis* hybrid "Seyval Blanc" (Krul, 1985; Krul and Worley, 1977) and *V. longii* "Microsperma" (Gray and Mortensen, 1987). With this regeneration pathway, new somatic embryos grew directly from previously-formed somatic embryos. Interestingly, the preponderance of new embryos emerge from at or near the hypocotyl-radicle boundary as illustrated in Fig. 1B (Gray and Mortensen, 1987; Krul, 1985). Secondary embryos first appear as clavate structures on the primary embryo surface (Fig. 2A) and undergo morphological development typical of zygotic embryos (Fig. 2B), eventually becoming fully mature in appearance (Fig. 2C). All stages of embryogeny are often present at a given time (Fig. 2D). A distinct suspensor is not obvious; however, secondary embryos are barely connected to primary embryos, suggesting a suspensor-like attachment. It is not known whether or not cells predisposed to develop into embryos exist at the hypocotyl-radicle boundary or if, perhaps, endogenous hormone levels that favor induction of embryogenic cells are optimized at this location.

For direct secondary embryogenesis, cultures were maintained by transferring either single embryos or clumps of embryos. After 4–8 weeks, secondary

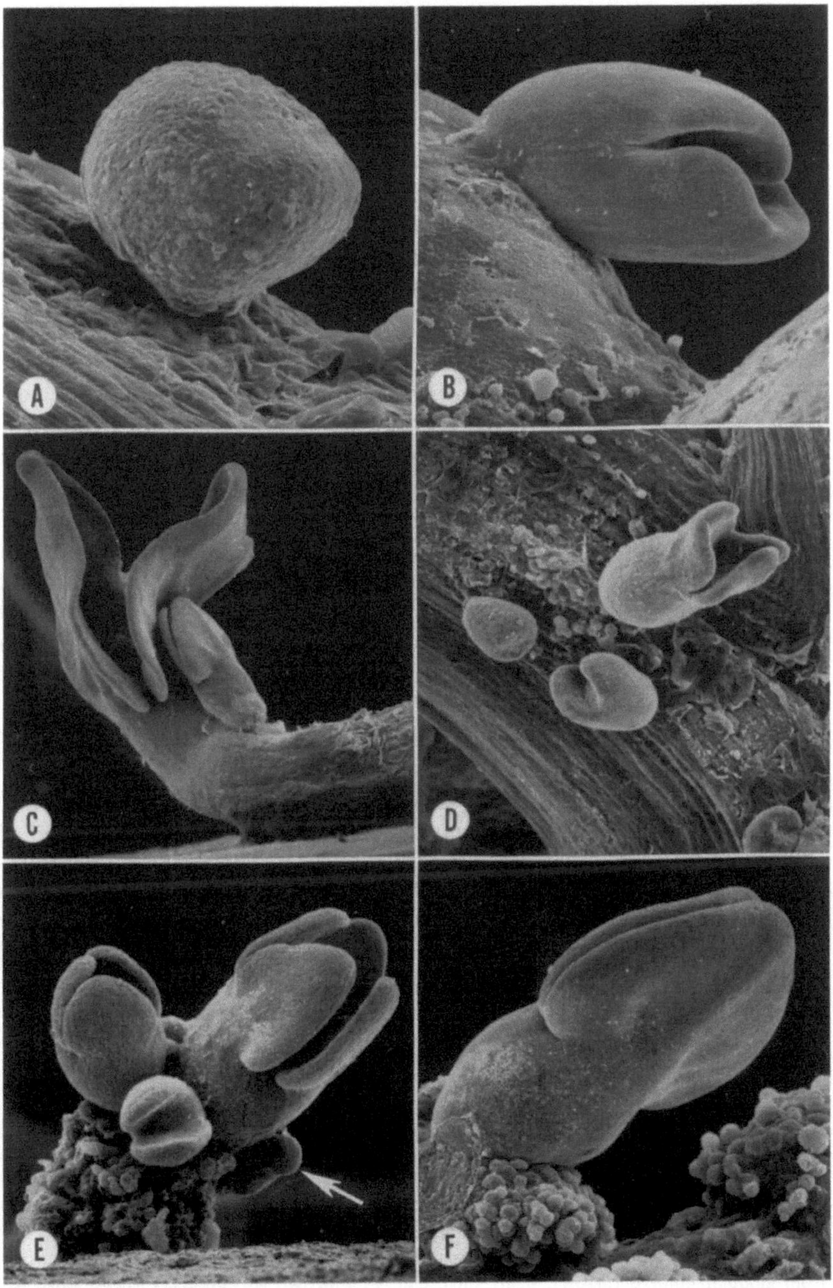
A
B
C
D
E
F

embryos were harvested and transferred to fresh medium. While both MS and Nitsch's media were used for maintenance, growth regulators were not required for continuous secondary embryogenesis to occur (Gray and Mortensen, 1987; Krul, 1985). Ability to undergo direct secondary embryogenesis in the absence of exogenously-supplied growth regulators appears to be possible in many, if not all, grape genotypes from which somatic embryogenesis has been induced (D.J. Gray, personal observation). However, propagation via direct secondary embryogenesis is relatively unwieldy and increase of embryogenic culture mass is slow, when compared to indirect embryogenesis from embryogenic cell culture. Furthermore, certain biotechnical applications, such as protoplast fusion and genetic transformation may be accomplished more easily with cell cultures.

*Indirect embryogenesis.* For indirect embryogenesis, cell cultures were maintained by either of two methods: 1) continual transfer on medium containing auxin with or without cytokinins, or 2) subculture on medium devoid of growth regulators.

Gray and Mortensen (1987) noted that callus maintained on medium with 2,4-D and BA continually sectored into embryogenic and non-embryogenic portions (Fig. 1C), which required careful separation during transfer; whereas Stamp and Meredith (1988a) and Matsuta and Hirabayashi (1989) did not report this phenomenon. The latter researchers reported that embryogenic callus was maintained by omitting vitamins, glycine and inositol from maintenance medium.

Embryogenic cell culture on growth regulator-free medium was first reported by Gray (1989) for *V. longii*, *V. vinifera* "Thompson Seedless" and three hybrid grape lines. The medium utilized was solidified MS with 6% sucrose and 0.2% activated charcoal. The latter ingredient was added to reduce culture browning. The same medium, except modified to contain $\frac{1}{4}$ of the normal MS nitrogen concentration, subsequently was utilized to maintain embryogenic cell cultures of several members of subgenera *Euvitis* and *Muscadinia*. Culture growth and maintenance was enhanced by utilizing high medium-to-headspace volume ratios in culture vessels, low light intensities and careful discrimination and selection of embryogenic cell type during transfers (Gray and Benton, 1991). Subsequently, embryogenic cultures of

*Figure 2.* Direct (A-D) and indirect (E-F) somatic embryogenesis in *Vitis longii*. (A) Globular embryo on somatic embryo hypocotyl (× 290). (B) Late torpedo-stage embryo on somatic embryo hypocotyl (× 140). (C) Mature embryo on germinated somatic embryo (× 20). (D) Globular, heart and late torpedo stage embryos growing from somatic embryo (× 40). (E) Four embryos at different stages and cotyledonary abnormalities growing from callus. Note clockwise from globular embryo (arrow): heart stage embryo with two cotyledons, torpedo stage with three cotyledons and mature embryo with four cotyledons (× 36). From Gray and Mortensen (1987). (F) Mature stage embryo growing from callus (× 50).

five *V. rotundifolia* varieties were developed utilizing this culture system (Gray, 1992). Selection of embryogenic cultures is accomplished by manually isolating the embryogenic cell masses and/or somatic embryos from initial embryogenic calluses. It was also possible to recover embryogenic cell culture lines from the watery residual material formed during direct somatic embryogenesis. This watery material was "scraped" from the agar surface and combined into small clumps, which would develop increasingly more cellular material and somatic embryos through subsequent transfers as described below. The embryogenic cultures of grape produced with this system were termed "perennial" due to their remarkable longevity of several years (Gray and Benton, 1991), in contrast to many other embryogenic culture systems (i.e., Gray *et al.*, 1993, for *Cucumis melo* L.) in which embryogenic potential declines rapidly over time.

Development of perennial-type embryogenic cell cultures that persist on growth regulator-free medium was dependent on stringent selection criteria utilized during transfer. In newly-initiated cultures, embryogenic cell masses were carefully separated from the adjacent tissue. All other material, including somatic embryos beyond the globular stage were excluded. Embryogenic cell material was combined into single clumps of approximately 3 mm in diameter. Selection of similar material and exclusion of somatic embryos was continued on successive transfers until a distinct cell line was obtained that was homogenously composed of proliferating cell clusters and young somatic embryos (Fig. 1D, E). Torpedo and cotyledonary-stage embryos were obtained by allowing globular embryos to remain in the culture mass through successive transfers (Fig. 1F); this resulted in masses of singulated embryos. As with direct secondary embryogenesis, somatic embryos that developed in this culture system passed through typical embryogenic stages (Figs. 2E, F). Utilizing these selection criteria, we developed perennial lines of over twenty species, hybrids and cultivars (D.J. Gray, unpublished data).

Comparison of cultures maintained on the aforementioned medium with or without 5 μM 2,4-D and 1 μM BA showed clear differences in morphology of embryogenic cell clusters (compare Figs. 1G and H). Embryogenic cell clusters from the heterogenous cultures that developed on medium with growth regulators were highly organized and embryo proliferation appeared to originate by a process of budding (Fig. 1G), whereas, cultures maintained on growth regulator-free medium were composed of small, relatively unorganized clusters of embryogenic cells (Fig. 1H) in which somatic embryos appeared to arise by organized divisions of single cells (Fig. 3A). Furthermore, only those cultures maintained without growth regulators persisted and could be regarded to be the "perennial" type.

It is unclear whether prolonged culture of embryogenic cells will result in mutation or other genetic abnormalities in grape as noted for some crops. The opportunity to maintain cultures with or without growth regulators may allow us to better define the effects of these substances on genetic stability. Embryo development from embryogenic cell cultures tended to decline

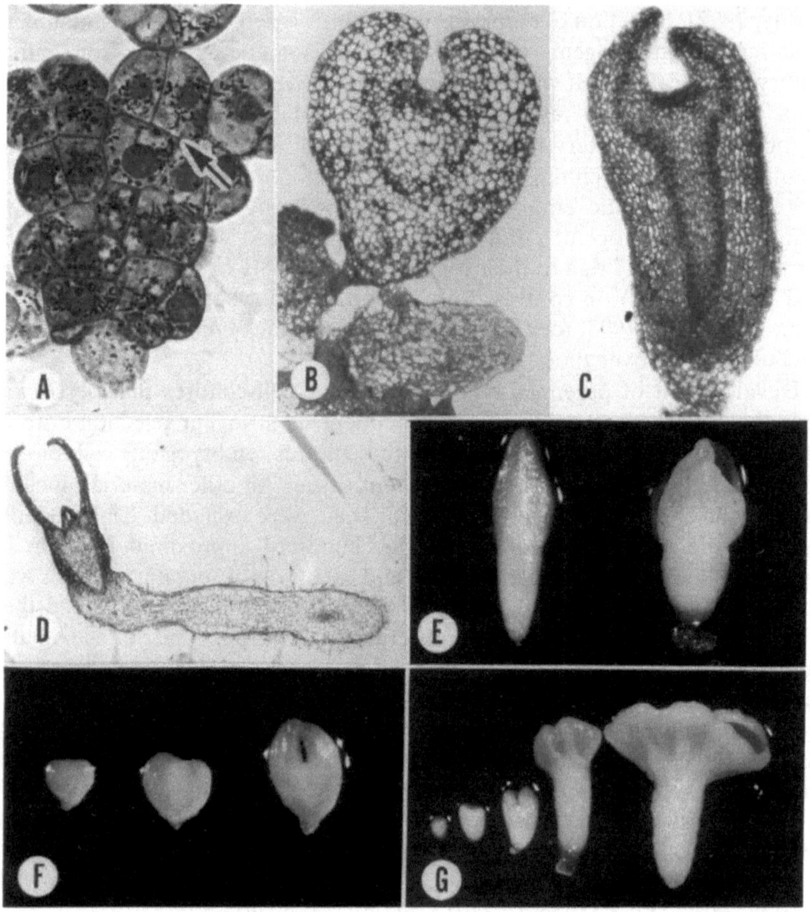


*Figure 3.* Somatic embryo development in *Vitis longii* "Microsperma", *Vitis rotundifolia* "Carlos" and *Vitis longii* "Thompson Seedless". (A) Three cell-stage somatic embryo (arrow) in embryogenic cell cluster of "Thompson Seedless" (× 1200). (B) Heart-stage somatic embryo of "Microsperma" (× 210). From Gray and Mortensen (1987). (C) Torpedo-stage somatic embryo of "Thompson Seedless" (× 70). (D) Germinated somatic embryo of "Thompson Seedless" (× 14). (E) Comparison of zygotic embryo (left) with somatic embryo (right). Note flattened appearance of zygotic embryo (× 15). (F) Morphologically correct somatic embryo development from cell cultures of "Thompson Seedless". Note suspensors at base of embryos (× 80). (G) Abnormal cotyledon development in somatic embryos from cell cultures of "Carlos". Note early development of fused cotyledons in second embryo from right. From Gray (1992).

gradually over prolonged time periods in perennial cultures; however, the typical embryogenic cell morphology did not change, suggesting possible emergence of mutant cell lines. We have never observed the development of non-embryogenic-type callus from this system. Embryo development often could be reinitiated from cell cultures that are at least nine years old by providing a pulse on any of the initiation media described above (D.J. Gray, unpublished data); this suggests that stable embryogenic cells may persist over very long time periods. Further support for genetic stability of somatic embryos comes from studies of somatic embryo-derived poinsettia (*Euphorbia pulcherrima* Willd ex Klotzsch) plants obtained from genetically unstable long-term suspension cultures, which were morphologically uniform and cytogenetically normal (Geier *et al.*, 1992). The issue of genetic stability is further explored below under "Growth in the Field".

### 2.4. *Embryo development*

Somatic embryos appeared to develop from single embryogenic cells via a suspensor (Figs. 3A, B, F) (Gray and Mortensen, 1987; Krul and Worley, 1977). Somatic embryos passed through recognizable developmental stages, possessed normal epidermal, cortical and vascular tissues and tended to be opaque white in color from the early torpedo stage to germination (Figs. 2A-F and 3A-G). Somatic embryos anatomically and morphologically faithful to all zygotic embryo stages were commonplace; however, a number of developmental abnormalities often occurred. The most common abnormality was the development of more than two, often fused, cotyledons (Figs. 2E and 3G) or somatic embryos that were larger or less flattened than zygotic embryos (Fig. 3E) (Gray and Mortensen, 1987). The development of fused cotyledons could be visualized in developing embryos at a very early stage (note second embryo from left in Fig. 3G). Differences in size and shape of somatic vs. zygotic embryos is likely due to absence of developing seed tissues, which exert a profound influence on zygotic embryo morphology. More subtle developmental abnormalities, such as no-to-poor development of the shoot apex (Faure, 1990) also occurred. These abnormalities were not likely to be related to genetic changes, since plants of normal appearance were recovered from such embryos.

Histochemical staining showed that starch [determined with the PARS reaction (amyloplasts stain pink) in tissue that was fixed, dehydrated and embedded in water-soluble plastic resin and iodine (amyloplasts stain black) in fresh specimens] and storage proteins [determined with napthol blue-black staining (protein deposits stain black) in embedded material] were present in somatic embryos, but did not exist in appreciable amounts (Fig. 4A) (Gray, 1988; Gray and Mortensen, 1987). However, lipids, which were lost during conventional processing of specimens for histochemical analysis into water-soluble plastic resin, were abundant in fresh material when stained with osmium tetroxide or sudan black-B (lipoplasts stain black) (Fig. 4B)

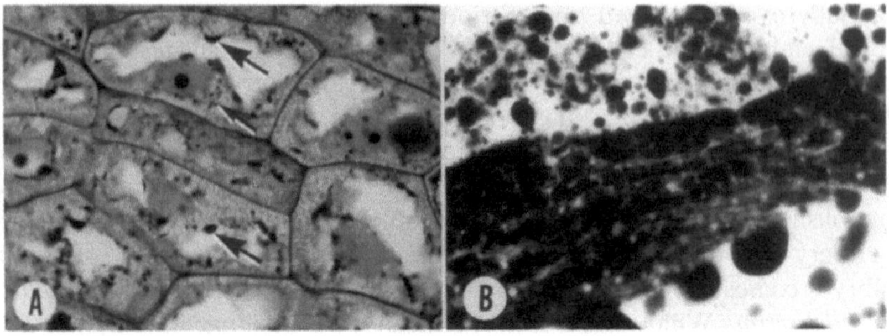


*Figure 4.* Histochemical localization of storage compounds in somatic embryos of *Vitis vinifera* "Thompson Seedless". (A) Storage proteins identified via staining with napthol blue-black (large arrows) and starch identified via the PARS reaction (small arrow) are not abundant in mature somatic embryos (× 1400). (B) Profuse lipids emerge from crushed cotyledon stained with sudan black-B (× 140).

(Gray, 1988). Grape is an oil seeded crop and it appears that the somatic embryos are faithful to zygotic embryos in this regard, although analytical comparisons of lipid type and amount have yet to be accomplished.

### 2.5. *Embryo germination and plant development*

At least some degree of plant recovery was possible from the first report of grape somatic embryogenesis (Mullins and Srinivasan, 1976). However, actual plant recovery rates were low and specific germination requirements varied widely among subsequent reports. Faure (1990) noted that, although 27% of *V. rupestris* somatic embryos had a morphology and anatomical structure identical to that of zygotic embryos, only 3% germination and development into plants was obtained. Examination of the literature shows that somatic embryos were often subjected to a pretreatment to obtain germination (Table 1). Pretreatments included cold stratification and addition of auxins, cytokinins, or gibberellic acid. In rarer instances, no treatment at all was required to induce germination and recover plants. However, somatic embryos produced on growth regulator-free medium always required some type of pretreatment to germinate (Gray, 1987c, 1989; Gray and Mortensen, 1987; Krul, 1985). To better understand germination requirements of grape somatic embryos, it is useful to consider germination of seeds. Grape seeds exhibit a type of dormancy that is alleviated by cold stratification and they germinate after a prolonged treatment at 4°C (Flemion, 1937). Grape somatic embryos also are dormant since germination is improved by cold treatments (Gray and Mortensen, 1987). Where no pretreatment was required for grape somatic embryo germination, culture systems that utilized auxins with or without cytokinins were used to produce somatic embryos (Table 1). Thus,

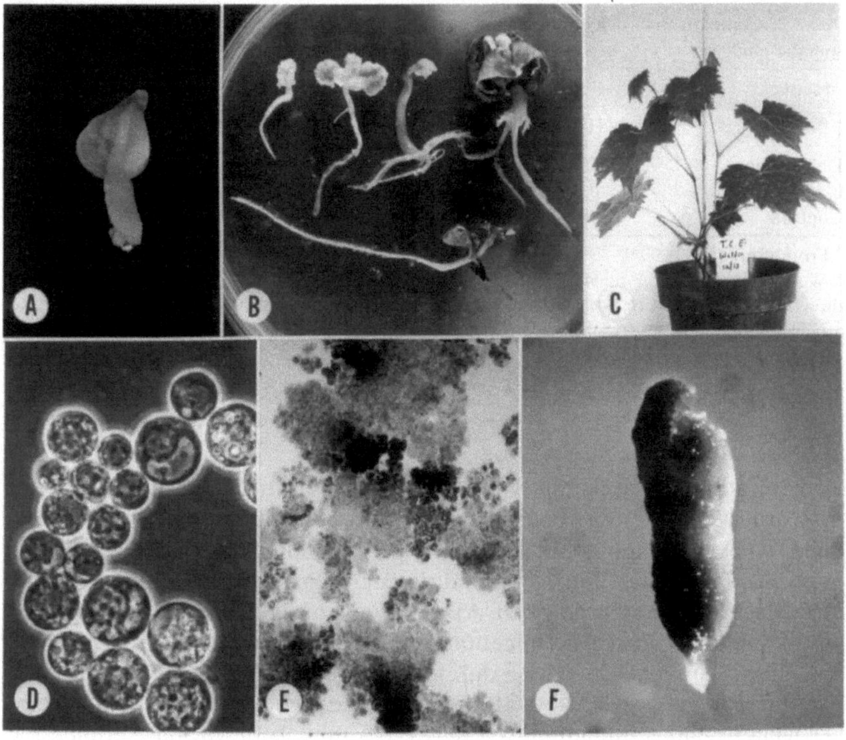


*Figure 5.* Plant recovery from somatic embryos and applications of somatic embryogenesis. (A) Dehydrated somatic embryo of *Vitis vinifera* "Thompson Seedless". Note shrivelled, discolored appearance (× 14). (B) Effect of growth regulators and dehydration on germination of "Thompson Seedless" somatic embryos. From left: no growth regulator (control), $GA_3$, ABA, BA. Bottom: dehydration (× 1). (C) Plant derived from somatic embryo of *Vitis rotundifolia* "Welder" (× 0.2). (D) Protoplasts derived from embryogenic cells of "Thompson Seedless" (× 1300). (E) Transient expression of GUS in embryogenic cell cultures of "Thompson Seedless" via particle bombardment. Dark areas denote sites of intense blue GUS reaction (× 40). (F) Transient expression of GUS in a "Thompson Seedless" somatic embryo. The entire side of the embryo facing particle discharge became intense blue in coloration (× 25).

these embryos were, in essence, subjected to a dormancy-breaking pretreatment.

The grape embryogenic culture system is the most studied example of somatic embryo dormancy. Although grape somatic embryos become very well developed and are morphologically similar to zygotic embryos, they germinate poorly. This demonstrates that the block to germination is at the physiological level and is not due to problems intrinsic to somatic embryo ontogeny. Studies of ABA concentration in grape embryogenic cultures showed a rapid increase during embryo development, which reached a peak

*Table 2.* Comparison of dehydration and benzyladenine (BA) for inducing germination in grape somatic embryos.[1]

| Treatment[2] | Percent germination response[3] | | | |
|---|---|---|---|---|
| | Hypocotyl | Root | Cotyledon | Shoot |
| Dehydration | 77 | 68 | 65 | 34 |
| 0.5 uM BA | 100 | 92 | 98 | 12 |
| Control | 76 | 88 | 36 | 0 |

[1] From Gray (1989).
[2] Well-developed embryos were either dehydrated for 21 days at 70% RH and 27°C, placed directly on medium with BA, or placed on basal medium (control).
[3] Germination response was based upon either enlargement and greening of hypocotyls and cotyledons or emergence of roots or shoots.

at maturation (Rajasekaran *et al.*, 1982). Cold stratification of somatic embryos resulted in a rapid decrease in ABA concentration. Exogenously supplied ABA inhibited somatic embryo germination (Gray, 1989). ABA is implicated as a controlling factor of dormancy in many types of seeds (Bewley and Black, 1985). Contrary to ABA, exogenously supplied $GA_3$ induced grape somatic embryo germination and the concentration of endogenous $GA_3$-like compounds increased during cold stratification (Pearce *et al.*, 1987; Takeno *et al.*, 1983). These studies illustrate endogenous control of embryogenesis and germination, whereby ABA inhibits precocious germination and thus promotes normal development while $GA_3$ promotes germination. Exogenously-supplied ABA was shown to increase plant recovery in one line (Goebel-Tourand *et al.*, 1993) but not in others (Gray, 1989). Another related study showed that embryogenic suspension cultures of a *V. vinifera* × *V. berlanderi* rootstock exhibited arrested development of embryos at the heart-stage unless the culture medium was refreshed daily (Coutos-Thevenot *et al.*, 1992a), in which the authors suggested to be due to secretion of a > 10 kDa macromolecule (Coutos-Thevenot *et al.*, 1992b). However, a direct relationship between the extracellular material and embryo development was not shown.

Grape somatic embryos were shown to be able to survive severe dehydration, similar to that of seeds (Gray, 1987b, 1990a; Gray and Purohit, 1991b). During dehydration, the embryos became reduced in size and more discolored and brittle in consistency (Fig. 5A) (Gray, 1987b). Their water content equilibrated to 13% when stored at 70% RH and they resumed a normal appearance after rehydration. Genotypic differences in response were noted; culture lines that produced relatively well developed somatic embryos were most responsive (Gray, 1989). After 21 days of dehydrated storage, 34% of embryos from one grape genotype produced plants following imbibition (Table 2). Dehydrated embryos germinated to produce plants immediately after imbibition, whereas no plants were recovered from nondehydrated controls. Thus, dehydration served as a dormancy-breaking pretreatment.

The germination pattern of previously-dehydrated grape somatic embryos became remarkably altered (Gray 1987d, 1989). Fig. 5B compares germination patterns from embryos placed on various media. The untreated control embryo (Fig. 5B, embryo on left) remained white and produced only a shoot. Treatment with $GA_3$ resulted in limited germination (second embryo from left), whereas ABA caused profuse adventitious rooting and an epicotyl hook reaction (third from left). BA caused dramatic enlargement and adventitious rooting (embryo on right) and was the most reliable treatment for producing shoots. However, only previously-dehydrated somatic embryos were capable of germinating and growing in a synchronous pattern of root to shoot emergence resembling that of a seedling (Fig. 5B, embryo on bottom). This was in contrast to germination of nondehydrated embryos, which germinated abnormally. Although plants were obtained, notably with BA pretreatment, root and shoot emergence were not synchronous. Regardless of germination treatment utilized, all of over 300 plants that we have recovered from somatic embryos appear to be morphologically normal (ex. Fig. 5C).

An alternative method of plant recovery from somatic embryos was described by Vilaplana and Mullins (1989), who showed that shoot organogenesis from hypocotyls and cotyledons was a possible regeneration route.

### 2.6. *Growth in the field*

While it has not been difficult to establish somatic embryo-derived grapevines in the field, relatively little data exists. Krul and Mowbray (1984) compared somatic embryo-derived vines of *V. vinifera* "Seyval Blanc" with normally propagated vines and found that those from somatic embryos had darker leaves, reddish vs. green canes and cylindrical rather than conical fruit clusters. Furthermore, cuttings from regenerated vines rooted and grew more rapidly. The vines yielded 5 tons of fruit per acre three years after planting (Krul and Mowbray, 1984) and eventually 7.3 tons per acre (Krul, 1984). Although the somatic embryo-derived vines differed from existing, conventionally-propagated plantings, they corresponded closely to the original description of "Seyval Blanc". This prompted the authors to suggest that the basis for the observed differences was due to elimination of latent viruses during *in vitro* culture. Thus, in this instance, it appeared that regenerated vines were genetically stable. Such observations were further clarified by Goussard and Wiid (1992), who produced embryogenic cultures from vines infected with grape fanleaf virus in which cells and somatic embryos were shown to be infected. Immunosorbent electron microscopy and ELISA demonstrated that cultures were freed of viral infection by heat treatment at 35°C. Plants regenerated from the heat treated somatic embryos had normal, uninfected leaf morphology (Goussard and Wiid, 1992) and are now growing vigorously, producing exceptional fruit, with no virus symptoms or signs of somaclonal variation (P.G. Goussard, personal communication). These observations of somatic embryo-derived grapevines and similar reports for other crops (i.e., Geier *et al.*, 1992), suggest that the process of somatic embryogenesis is remarkably resistant to genetic deviation, reinforcing the possibilities of controlled manipulation of the somatic genome.

### 2.7. *Applications of somatic embryogenesis*

Embryogenic culture technology is pivotal in applying many biotechnologies to grape improvement. A number of applications may rely on embryogenic cell cultures as described below.

*Somatic hybridization.* Success in somatic hybridization of grape will rest on our ability to regenerate plants from protoplasts. While there has been progress in isolation and culture methods, plant regeneration has yet to be obtained from grape protoplasts.

Leaf mesophyll tissue is the most commonly used source of protoplasts (Barbier and Bessis 1987, 1988, 1990; Lee *et al.*, 1989; Lee and Wetzstein, 1988; Nishimura *et al.*, 1984; Shimizu, 1985; Theodoropoulos and Roubelakis-Angelakis, 1990; Wright, 1985). Although isolation and culture protocols vary widely (Krul, 1988), most did not result in cell wall regeneration, cell division and sustained culture growth (Gray and Meredith, 1992). Therefore, it is difficult to determine the effect of specific protocols on the developmental events that lead to plant regeneration. The lack of progress in plant regeneration from grape protoplasts may be related to source tissue used. For example, success in plant regeneration from protoplasts of the Poaceae came only after development of specialized embryogenic suspension cultures that produced high yields of regenerative protoplasts (Gray, 1990b). A requirement for highly regenerative cell cultures in order to achieve plant regeneration from protoplasts is logical since, in practice, plant recovery rates from protoplasts are typically less than 10%. Currently, culture of *in vitro* leaves on regeneration medium designed to induce either embryogenesis or organogenesis yields very little plant regeneration when considering the large number of actual explant cells that are contained in a plated leaf. The chance of inducing a rare event such as plant regeneration from such leaves after protoplast isolation and culture seems remote indeed. With apparently adequate methodologies in place for isolation and culture, development of suitable culture lines is the next step toward achieving plant regeneration from protoplasts of grape. High yield of embryogenic protoplasts and regrowth back to cell cultures were obtained from the perennial-type cell cultures described above (Fig. 5D) (El-Gindy and Gray, unpublished data), however, plant regeneration has not yet been achieved. The current roadblock to plant regeneration must be removed before somatic hybridization, such as combining marginally sexually compatible *V. rotundifolia* and *V. vinifera*, can be accomplished.

*Genetic transformation.* Due the compelling applications for grapevine improvement, development of protocols to produce transgenic plants are underway worldwide. Transgenic cells and tissues were obtained by cocultivation with various *Agrobacterium* strains (Baribault *et al.*, 1989; Colby *et al.*, 1991; Gribaudo and Schubert, 1990; Guellec *et al.*, 1990); with the recovery of entire transgenic plants has been accomplished from organogenic cultures of *V. rupestris* (Mullins, 1990; Mullins *et al.*, 1990).

Baribault *et al.* (1990) produced shoots of *V. vinifera* "Cabernet Sauvignon" that contained both transformed and untransformed cells. Use of embryogenic cells or somatic embryos may be one approach to producing stably transformed *V. vinifera*. Embryogenic cells (Fig. 5E) and somatic embryos (Fig. 5F) of *V. vinifera* "Thompson Seedless" bombarded with DNA-coated tungsten particles were shown to transiently express the GUS gene (Gray and Songstad, unpublished data). Intact plants of transformed *V. vinifera* have been obtained by particle bombardment wounding, followed by *Agrobacterium* cocultivation (Scorza *et al.*, 1995). The "Thompson Seedless" cell line described herein was also transformed with this system. Transformed plants are GUS-positive and resistant to the antibiotic selection agent Kanamycin (Scorza *et al.*, in preparation). The coat protein-encoding gene for grape chrome mosaic nepovirus was inserted into tobacco (*Nicotiana tabacum* L.) and transgenic plants produced coat protein (Brault *et al.*, 1993). These plants were shown to be more resistant to inoculated infection than controls, providing evidence that a similar approach may be applicable in grape.

*Synthetic seeds.* Synthetic seeds are somatic embryos that are of commercial use in plant propagation (Gray, 1987a; Gray and Purohit, 1991a). While synthetic seeds of grape may not be useful for propagation, since adequate means are available and many vines are grafted, a significant application involves use of synthetic seeds for germplasm conservation (Gray and Compton, 1992). Because grape is self-incompatible, seeds cannot be used in germplasm conservation of distinct varieties or selections. Instead, germplasm conservation is accomplished by establishing plants in field germplasm repositories (orchards and vineyards), which are expensive and difficult to maintain over long time periods. For example, in the US, grape germplasm is preserved at the National Clonal Germplasm Repository (NCGR) at Davis, CA, and Geneva, NY. These locations house a combined total of 3900 grape accessions representing 41 and 34 species at each site, respectively. Two vines of each accession are maintained at a yearly cost of $ 25.00 per accession; however, the actual cost approaches $ 70.00 when administrative, research and evaluation cost are added (Gray and Compton, 1992). The problem of maintaining entire plants to accomplish clonal germplasm storage is so great as to have stifled or completely inhibited the collection of germplasm for many crops. These problems have prompted exploration of *in vitro* methods for conservation of clonal germplasm. Use of dehydrated quiescent somatic embryos, as described above, which could then be conserved in seed repositories, represents a cost effective solution. Conservation via somatic embryogenesis would maintain germplasm purity and allow the amount of space required to house valuable collections to be reduced from several acres to just a few square feet of storage space. Such a conservation system would be preferable, since it would utilize existing seed storage facilities and techniques.

## 3. Conclusion

Grape is the worlds most valuable fruit crop due to its multiple uses and, as such, is the subject of considerable research emphasis. Somatic embryogenesis was first reported in 1976 for *Vitis vinifera* "Cabernet Sauvignon". Since that time, somatic embryogenesis has been obtained for an increasing number of additional species, hybrids and/or cultivars. Somatic embryo-derived grapevines appear to be normal in every respect. However, problems with initiating embryogenic cultures due to genotypic specificities persist and plant recovery rates from grape somatic embryos remain relatively low. Further technical development is compelling, since the embryogenic culture system is pivotal to applying a number of biotechnical procedures, including somatic hybridization, genetic transformation and synthetic seed technology, to grapevine improvement.

## 4. Acknowledgements

This is an updated and expanded version of pertinent portions of a previously published chapter (Gray and Meredith, 1992). Support was provided by the Program in Science and Technology Cooperation, Office of the Science Advisor, United States A.I.D. (grant # DHR-5600-G-00-0057-00). Florida Agricultural Experiment Station No. R-03386.

## 5. References

Alleweldt, G. and J.V. Possingham, 1988. Progress in grapevine breeding. Theo. App. Genet. 75: 669–673.

Barbier, M. and R. Bessis, 1987. Isolation and culture of grape leaf protoplasts (*Vitis vinifera* L., var. Chardonnay). Bulletin de l' Office International de la vigne et du Vin: 679–680, 765–775 (In French with English summary).

Barbier, M. and R. Bessis, 1988. Effets de differents facteurs contributant a l'amelioration de l'isolement de protoplasts a' partir de feuilles de vigne (*Vitis vinifera* L.). Bulletin de la Societe Botanique de France 135: 251–261 (In French with English summary).

Barbier, M. and R. Bessis, 1990. Isolation and culture of grapevine cv. Chardonnay leaf protoplasts. Euphytica 47: 39–44.

Baribault, T.J., K.G.M. Skene and S.N. Steele, 1989. Genetic transformation of grapevine cells. Plant Cell Rep. 8: 137–140.

Baribault, T.J., K.G.M. Skene, P.A. Cain and S.N. Steele, 1990. Transgenic grapevines: regeneration of shoots expressing $\beta$-glucuronidase. J. Exp. Bot. 41: 1045–1049.

Barton, K.A., H.R. Whiteley and N.S. Yang, 1987. *Bacillus thuringensis* delta-endotoxin expressed in transgenic *Nicotiana tabacum* provides resistance to lepidopteran insects. Plant Physiol. 85: 1103–1109.

Bewley, J.D. and M. Black, 1985. Seeds: Physiology of Development and Germination. Plenum, New York.

Brault, V., T. Candresse, O. le Gall, R.P. Delbos, M. Lanneau and J. Dunez, 1993. Genetically engineered resistance to grapevine chrome mosaic nepovirus. Plant Mol. Biol. 21: 89–97.

Cahoon, C.A., 1986. The concord grapes. Fruit Varieties J. 40: 106–107.

Colby, S.M., A.M. Juncosa and C.P. Meredith, 1991. Cellular differences in *Agrobacterium* susceptibility and regenerative capacity restrict the development of transgenic grapevines. J. Amer. Soc. Hort. Sci. 116: 356–361.

Coutos-Thevenot, P., I. Goebel-Tourand, M.-C. Mauro, J.-P. Jouanneau, M. Boulay, A. Deloire and J. Guern, 1992a. Somatic embryogenesis from grapevine cells. I – Improvement of embryo development by changes in culture conditions. Plant Cell Tiss. Org. Cult. 29: 125–133.

Coutos-Thevenot, P., O. Maes, T. Jouenne, M.-C. Mauro, M. Boulay, A. Deloire and J. Guern, 1992b. Extracellular protein patterns of grapevine cell suspensions in embryogenic and non-embryogenic situations. Plant Sci. 86: 137–145.

Delannay, X., B.J. LaVallee, R.K. Proksch, R.L. Fuchs, S.R. Sims, J.T. Greenplate, P.G. Marrone, R.B. Dodson, J.J. Augustine, J.G. Layton and D.A. Fischhoff, 1989. Field performance of transgenic tomato plants expressing the *Bacillus thuringensis* var. *kurstaki* insect control protein. BioTechnol. 7: 1265–1269.

Durham, R.E., G.A. Moore, D.J. Gray and J.A. Mortensen, 1989. The use of leaf GPI and IDH isozymes to examine the origin of polyembryony in cultured ovules of seedless grape. Plant Cell Rep. 7: 669–672.

Einset, J. and C. Pratt, 1975. Grapes. In: J. Janick and J.N. Moore (Eds.), Advances in Fruit Breeding, pp. 130–153. Purdue University Press, West Lafayette.

Faure, O., 1990. Embryons somatiques de *Vitis rupestris* et embryons zygotiques de *Vitis* sp.: Morphologie, histologie, histocimie et developpement. Can. J. Bot. 68: 2305–2315.

Favre, J.-M., 1977. First results on *in vitro* production of bud-neoformations in grapevine. Ann. l'Amelior. des Plantes 27: 151–169 (In French with English summary).

Flemion, F, 1937. After-ripening at 5°C favours germination of grape seeds. Contributions of the Boyce Thompson Institute 9: 7–15.

Food and Agricultural Organization of the United Nations, 1990. Production Yearbook, Rome.

Galet, P., 1988. Cépages et Vignobles de France. Vol. I. Les Vignes Américaines. Imprimerie Charles Déhan, Montpellier.

Galet, P. and L.T. Morton, 1990. Introduction: The family Vitaceae and *Vitis* speciation. In: R.C. Pearson and A.C. Goheen (Eds.), Compendium of Grape Diseases, pp. 2–3. APS Press, St. Paul.

Geier, T., A. Beck and W. Preil, 1992. High uniformity of plants from cytogenetically variable embryogenic cultures of poinsettia (*Euphorbia pulcherrima* Willd ex Klotzch). Plant Cell Rep. 11: 150–154.

Goebel-Tourand, I., M.-C. Mauro, L. Sossountoz, E. Miginiac and A. Deloire, 1993. Arrest of somatic embryo development in grapevine: histological characterization and the effect of ABA, BAP and zeatin in stimulating plantlet development. Plant Cell Tiss. Org. Cult. 33: 91–103.

Goussard, P.G. and J. Wiid, 1992. The elimination of fanleaf virus from grapevines using *in vitro* somatic embryogenesis combined with heat therapy. S. Afr. J. Enol. Vitic. 13: 81–83.

Goussard, P.G., J. Wiid and G.G.F. Kasdor, 1991. The effectiveness of *in vitro* somatic embryogenesis in eliminating fanleaf virus and leafroll associated viruses from grapevines. S. Afr. J. Enol. Vitic. 12: 77–83.

Gray, D.J., 1987a. Introduction to the symposium. Proceedings, Symposium on Synthetic Seed

Technology for the Mass Cloning of Crop Plants: Problems and Perspectives. HortSci. 22: 796–797.
Gray, D.J., 1987b. Quiescence in monocotyledonous and dicotyledonous somatic embryos induced by dehydration. Proceedings, Symposium on Synthetic Seed Technology for the Mass Cloning of Crop Plants: Problems and Perspectives. HortSci. 22: 810–814.
Gray, D.J., 1987c. Concluding remarks. Proceedings, Symposium on Synthetic Seed Technology for the Mass Cloning of Crop Plants: Problems and Perspectives. HortSci. 22: 814.
Gray, D.J., 1987d. Effects of dehydration and other environmental factors on dormancy in grape somatic embryos. HortSci. 22: 1118 (Abstr.).
Gray, D.J., 1988. Ontogeny of grape somatic embryos. HortSci. 23: 807 (Abstr.).
Gray, D.J., 1989. Effects of dehydration and exogenous growth regulators on dormancy, quiescence and germination of grape somatic embryos. In Vitro Cell. Dev. Biol. 25: 1173–1178.
Gray, D.J., 1990a. Synthetic seed for clonal production of crop plants. In: R.B. Taylorson (Ed.), Recent Advances in the Development and Germination of Seeds, pp. 29–45. Plenum, New York.
Gray, D.J., 1990b. Somatic cell culture and embryogenesis in the Poaceae. In: M.J. Kasperbauer (Ed.), Biotechnology in Tall Fescue Improvement, pp. 25–57. CRC Press, Boca Raton.
Gray, D.J., 1992. Somatic embryogenesis and plant regeneration from immature zygotic embryos of muscadine grape (*Vitis rotundifolia*) cultivars. Amer. J. Bot. 79: 542–546.
Gray, D.J. and C.M. Benton, 1991. Perennial embryogenic cultures of grape. HortSci. 26: 772.
Gray, D.J. and M.E. Compton, 1992. Grape somatic embryo dormancy and quiescence: potential of dehydrated synthetic seeds for germplasm conservation. In: K. Redenbaugh (Ed.), Synseeds: Applications of Synthetic Seeds to Crop Improvement, pp. 367–379. CRC Press, Boca Raton.
Gray, D.J. and L.A. Hanger, 1993. Effect of ovule maturity on recovery of zygotic embryos and embryogenic cultures from muscadine grape. HortSci. 28: 227.
Gray, D.J., D.W. McColley and M.E. Compton, 1993. High-frequency somatic embryogenesis from quiescent seed cotyledons of *Cucumis melo* cultivars. J. Amer. Soc. Hort. Sci. 118: 425–432.
Gray, D.J. and C.P. Meredith, 1992. Grape. In: F.A. Hammerschlag and R.E. Litz (Eds.), Biotechnology in Perennial Fruit Crops, pp. 229–262. C.A.B. International, Wallingford.
Gray, D.J. and J.A. Mortensen, 1987. Initiation and maintenance of long term somatic embryogenesis from anthers and ovaries of *Vitis longii* "Microsperma". Plant Cell Tiss. Org. Cult. 9: 73–80.
Gray, D.J. and A. Purohit, 1991a. Embryogenesis and development of synthetic seed technology. Crit. Rev. Plant Sci. 10: 33–61.
Gray, D.J. and A. Purohit, 1991b. Quiescence and dormancy in somatic embryos. In: Y.P.S. Bajaj (Ed.), Biotechnology in Agriculture and Forestry, Vol. 17, pp. 382–394. Springer-Verlag, Berlin.
Gresshoff, P.M. and C.H. Doy, 1974. Derivation of a haploid cell line from *Vitis vinifera* and the importance of the stage of meiotic development of anthers for haploid culture of this and other genera. Z. Planzenphysiol. 73: 132–141.
Gribaudo, I. and A. Schubert, 1990. Grapevine root transformation with *Agrobacterium rhizogenes*. Proceedings, Fifth International Symposium on Grape Breeding. Vitis Special Issue, 412–418.
Guellec, V., C. David, M. Branchard and J. Tempé, 1990. *Agrobacterium rhizogenes* mediated transformation of grapevine (*Vitis vinifera* L.). Plant Cell Tiss. Org. Cult. 20: 211–215.
Harst-Langenbucher, M. and G. Alleweldt, 1993. Einfluß verschiedener Vorbehandlungen auf die Induktion somatischer Embryogenese an Antheren der Rebsorte Riesling. Vitis 32: 1–7.
Hirabayashi, T., I. Kozaki and T. Akihama, 1976. *In vitro* differentiation of shoots from anther callus in *Vitis*. HortSci. 11: 511–512.
Jelenkovic, G. and H.P. Olmo, 1969. Cytogenetics of Vitis. V. Allotetraploids of *V. vinifera* L. × *V. rotundifolia* Michx. Vitis 8: 265–279.

Krul, W.R., 1984. Recent advances in cloning and genetic engineering of the grapevine. The Maryland Grapevine 4: 10–13.

Krul, W.R., 1985. *In vitro* propagation of grape. United States Patent No. 4,532,733.

Krul, W.R., 1988. Recent advances in protoplast culture of horticultural crops: small fruits. Sci. Hort. 37: 231–245.

Krul, W.R. and G.H. Mowbray, 1984. Grapes. In: W.R. Sharp, D.A. Evans, P.V. Ammirato, and Y. Yamada (Eds.), Handbook of Plant Cell Culture, Vol 2, pp. 396–434. Macmillan, New York.

Krul, W.R. and J.F. Worley, 1977. Formation of adventitious embryos in callus cultures of "Seyval", a French hybrid grape. J. Amer. Soc. Hort. Sci. 102: 360–363.

Lee, N. and H.Y. Wetzstein, 1988. Protoplast isolation and callus production from leaves of tissue-cultured *Vitis* spp. Plant Cell Rep. 7: 531–534.

Lee, N., H.Y. Wetzstein and C.H. Bornman, 1989. Cortical microtubule organization in *Vitis* protoplasts as affected by concentration of enzyme isolation medium and duration of incubation. Physiol. Plant 77: 27–32.

Martinelli, L., P. Bragagna, V. Poletti and A. Scienza, 1993. Somatic embryogenesis from leaf- and petiole-derived callus of *Vitis rupestris*. Plant Cell Rep. 12: 207–210.

Matsuta, N., 1992. Effect of auxin on somatic embryogenesis from leaf callus in grape (*Vitis* spp.). Jpn. J. Breed. 42: 879–883.

Matsuta, N. and T. Hirabayashi, 1989. Embryogenic cell lines from somatic embryos of grape (*Vitis vinifera* L.). Plant Cell Rep. 7: 684–687.

Mauro, M.Cl., C. Nef and J. Fallot, 1986. Stimulation of somatic embryogenesis and plant regeneration from anther culture of *Vitis vinifera* cv. Cabernet-Sauvignon. Plant Cell Rep. 5: 377–380.

Mullins, M.G., 1990. Applications of tissue culture to the genetic improvement of grapevines. Proceedings, Fifth International Symposium on Grape Breeding. Vitis Special Issue: 399–407.

Mullins, M.G. and K. Rajasekaran, 1980. Plantlets from cultured anthers of vitis species and hybrids. In: Proceedings of the Third International Symposium on Grape Breeding, Davis, pp. 111–119.

Mullins, M.G. and C. Srinivasan, 1976. Somatic embryos and plantlets from an ancient clone of grapevine (cv. Cabernet-Sauvignon) by apomixis *in vitro*. J. Exp. Bot. 27: 1022–1030.

Mullins, M.G., F.C.A. Tang and D. Facciotti, 1990. *Agrobacterium*-mediated genetic transformation of grapevines: transgenic plants of *Vitis rupestris* Scheele and buds of *Vitis vinifera* L. BioTechnol. 8: 1041–1045.

Murashige, T. and F. Skoog, 1962. A revised medium for rapid growth and bioassays with tobacco tissue cultures. Physiol. Plant. 15: 473–497.

Nejidat, A., W.G. Clark and R.N. Beachy, 1990. Engineered resistance against plant virus diseases. Physiol. Plant. 80: 662–668.

Newton, D.J. and P.G. Goussard, 1990. The ontogeny of somatic embryos from *in vitro* cultured grapevine anthers. S. Afr. J. Enol. Vitic. 11: 70–81.

Nishimura, M., I. Hara-Nishimura and S.P. Robinson, 1984. Isolation of metabolically competent protoplasts from grapevine leaves. Plant Sci. Lett. 37: 171–175.

Nitsch, J.P. and C. Nitsch, 1969. Haploid plants from pollen grains. Science 163: 85–87.

Oxtoby, E. and M.A. Hughes, 1990. Engineering herbicide tolerance into crops. Trends in Biotechnol. 8: 61–65.

Pearce, D., R.P. Pharis, K. Rajasekaran and M.G. Mullins, 1987. Effects of chilling and ABA on [$^3$H]Gibberellin $A_4$ metabolism in somatic embryos of grape (*Vitis vinifera* L. × *V. rupestris* Scheele). Plant Physiol. 80: 381–385.

Rajasekaran, K. and M.G. Mullins, 1979. Embryos and plantlets from cultured anthers of hybrid grapevines. J. Exp. Bot. 30: 399–407.

Rajasekaran, K. and M.G. Mullins, 1983a. The origin of embryos and plantlets from cultured anthers of hybrid grapevines. Amer. J. Enol. Vitic. 34: 108–113.

Rajasekaran, K. and M.G. Mullins, 1983b. Influence of genotype and sex-expression on formation of plantlets by cultured anthers of grapevines. Agronomie 3: 233–238.

Rajasekaran, K., J. Vine and M.G. Mullins, 1982. Dormancy in somatic embryos and seeds of *Vitis*: changes in endogenous abscisic acid during embryogeny and germination. Planta 154: 139–144.

Robacker, C., 1993. Somatic embryogenesis and plant regeneration from muscadine leaf explants. HortSci. 28: 53–55.

Scorza, R., J.M. Cordts, D.J. Gray, D.W. Ramming and R.L. Emershad, 1995. Transformation of "Thompson Seedless" grape. (in preparation).

Scorza, R., J.M. Cordts, D.W. Ramming and R.L. Emershad, 1995. Transformation of grape (*Vitis vinifera* L.) Zygotic-derived somatic embryos and regeneration of transgenic plants. Plant Cell Rep. (in press).

Shah, D.M., R.B. Horsch, H.J. Klee, G.M. Kishore, J.A. Winter, N.E. Tumer, C.M. Hironaka, P.R. Sanders, C.S. Gasser, S. Aykent, N.R. Siegel, S.G. Rogers and R.T. Fraley, 1986. Engineering herbicide tolerance in transgenic plants. Science 233: 478–482.

Shimizu, J.-I., 1985. Cell regeneration and division of grape mesophyll protoplasts. J Plant Physiol. 119: 419–424.

Srinivasan, C. and M.G. Mullins, 1980. High-frequency somatic embryo production from unfertilized ovules of grapes. Sci. Hort. 13: 245–252.

Stamp, J.A. and C.P. Meredith, 1988a. Somatic embryogenesis from leaves and anthers of grapevine. Sci. Hort. 35: 235–250.

Stamp, J.A. and C.P. Meredith, 1988b. Proliferative embryogenesis from zygotic embryos of grapevine. J. Amer. Soc. Hort. Sci. 113: 941–945.

Stark, D.M. and R.N. Beachy, 1989. Protection against potyvirus infection in transgenic plants – evidence for broad spectrum resistance. BioTechnol. 7: 1257–1262.

Takeno, K., M. Koshioka, R.P. Pharis, K. Rajasekaran and M.G. Mullins, 1983. Endogenous gibberellin-like substances in somatic embryos of grape (*Vitis vinifera* × *Vitis rupestris*) in relation to embryogenesis and the chilling requirement for subsequent development of mature embryos. Plant Physiol. 73: 803–808.

Theodoropoulos, P.A. and K.A. Roubelakis-Angelakis, 1990. Progress in protoplast isolation and culture from virus-free axenic shoot cultures of *Vitis vinifera* L. Plant Cell Tiss. Org. Cult. 20: 15–23.

Vilaplana, M. and M.G. Mullins, 1989. Regeneration of grapevines (*Vitis* spp.) *in vitro*: Formation of adventitious buds on hypocotyls and cotyledons of somatic embryos. J. Plant Physiol. 134: 413–419.

Winkler, A.J., J.A. Cook, W.M. Kliewer and L.A. Lider, 1974. General Viticulture, Revised Edition. University of California Press, Berkeley.

Wright, D.C., 1985. Factors affecting isolation of protoplasts from leaves of grape (*Vitis vinifera*). Plant Cell Tiss. Org. Cult. 4: 95–100.

Zou, C.J. and P.F. Li, 1981. Induction of pollen plants of grape (*Vitis vinifera* L.). Acta Bot. Sinica 23: 79–81 (In Chinese with English description of figures and table).

# 13. Somatic embryogenesis in *Salix*

Lotta Grönroos

**Contents**



## 1. Introduction

*Salix* is a diverse genus of several hundred species, some being tall trees, others creeping brushwood while most are medium-sized bushes. They are popularly known as willows and in some cases as sallows and osiers. Almost all are dioecious, with the numerous, insect-pollinated flowers in cylindrical inflorescences, catkins. Each flower is situated behind a bract, or flower scale, on the catkin, and is composed of one or more nectaries and one pistil (in female plants) or a few anthers (in male plants). The seeds, several in each capsule, have conspicuous seed hairs and are wind dispersed. The seed is made up of a well developed embryo surrounded by a thin seed coat. It usually loses its viability unless germination takes place within a few days, since it has nearly no food reserves. The seedlings are delicate and cannot cope with competition from other plants. The leaves are simple and arranged spirally, with stipules that sometimes fall off at an early stage. Buds are protected by a single bud-scale.

Willow species are present in almost every part of the world, although mostly they are indigenous in temperate to arctic regions of the northern hemisphere (Newsholme, 1992). Willows belong to the family Salicaceae. Besides the small genera *Chosenia* and *Toisusu*, which may actually be subgenera within *Salix*, the only other genus in this family is *Populus* (aspens and poplars). *Salix* is considered taxonomically a complicated genus due to the variability observed within species. Thus, it is difficult to estimate the exact number of *Salix* species. Estimates vary from 300 to 500 or more.

*S. Jain, P. Gupta & R. Newton (eds.), Somatic Embryogenesis in Woody Plants, Vol. 2,* 219–234.

Hybridization between willow species is common. A taxonomic guide can be found in the book on willows by Newsholme (1992).

An interesting summary of the history of willow culture, the uses of willows, their cultivation, biology and taxonomy has been given by Larsson and Bremer (1991) in a taxonomic study of willows in the *Salix viminalis* group. Willows have been cultivated for more than 2000 years. The Romans used the willows (probably *Salix viminalis*, *S. fragilis* and *S. purpurea*) mainly for packaging. Since trade was wide-spread at the time, there was a great demand for willow basketry. Willow cultivation waned with the fall of the Roman empire but again became popular in the 19th century when industrialization in Europe put an increased demand on packaging materials. However, willows were soon replaced by cheaper, imported rattan from Southeast Asia. Nevertheless, willows continued to be cultivated in smaller stands for other purposes. Traditional uses of willows have been for wickerwork, packaging, basketry, furniture, decorations, fencing, windbreaks, boats, cricket bats, charcoal, fodder, tannin and salicylic acid production (Larsson and Bremer, 1991; Newsholme, 1992; Ledin and Alriksson, 1992). Nowadays, willows are also finding uses as noise reducing walls, lake shore protection, wickerwork for riverbanks, wildlife refuges, for beekeeping, for sludge cleaning and municipal waste-water recycling and for energy production (Ledin and Alriksson, 1992). Several *Salix* species play an important role in energy forestry, due to fast growth and wood production. The interest in energy forestry is increasing in several countries and willow plantations are becoming economically important again. I believe we are on the verge of a breakthrough for this renewable energy source.

Willows for energy production are cultivated as short rotation forests. Cuttings, taken from young shoots of fast-growing clones during winter, are planted in spring. The cuttings produce several shoots which grow to a height of five to seven meters in a few years. After three to five years, the shoots can be harvested and processed for combustion. New shoots grow from the stumps so that the forest can be harvested several times, at three- to five-year intervals, before re-planting is necessary.

Much attention has recently been paid to genetic improvement of willows for energy plantations. Previously, suitable plant material was merely selected from wild populations and from existing genetic collections (Ager *et al.*, 1986). Breeding programs for willows are now well under way (Gullberg, 1993). Tissue culture techniques can offer new dimensions to genetic improvement. Techniques with potential applications include (1) zygotic embryo culture to circumvent incompatibility between maternal tissues and embryos (embryo rescue), (2) pollen culture to produce haploids and homozygotes, (3) protoplast fusion to create hybrids between sexually incompatible species, (4) *in vitro* selection and regeneration of plants resistant to specific pathogen toxins for example, and (5) genetic transformation to introduce specific genes into the plant genome. If these techniques are to be of any practical importance applied to *Salix*, it is of utmost importance that the plants can satisfactorily be multiplied in large numbers using *in vitro* propagation.

In spite of their increasing economic importance, willows are subjected to research in only a few tissue culture laboratories. This paucity may reflect the reluctancy of willows to differentiate anything but roots. In *Populus*, the other energy forestry genus from the Salicaceae family, shoots could be obtained from callus cultures as early as 1964 (Mathes, 1964). Numerous articles have since been published on plant regeneration (review: Ahuja, 1987) and somatic embryogenesis in *Populus* (see Chapter 6 in this volume) and genetically transformed poplar plants have even been produced (Filatti *et al.*, 1987). *Salix* species were propagated *in vitro* only via meristem or shoot cultures (e.g., Bhojwani, 1980; Dhir *et al.*, 1984; Bergman *et al.*, 1985; Grönroos *et al.*, 1990; Tormala and Saarikko, 1985). Recently, however, it has been shown that somatic embryogenesis is possible even in this previously recalcitrant genus. Moreover, the *Salix* somatic embryogenesis system can surely become an excellent model system for genetic studies because willows: 1) are easy to propagate vegetatively via cuttings with no obvious ageing problems, 2) have a short generation time, about 2 years, 3) have a small genome, and 4) are already the subject of breeding programs (Gullberg, 1993).

*Salix exigua* Nutt. was the first willow to regenerate plants via *de novo* shoot production from callus (Stoehr *et al.*, 1989). Callus was initiated from leaf tissues on culture medium supplemented with benzyladenine (BA) and 2,4-dichlorophenoxyacetic acid (2,4-D). Numerous shoot primordia were produced in two *Salix exigua* clones after transfer of callus to Woody Plant Medium (WPM; Lloyd and McCown, 1981) with 0.44 μM BA. The shoots elongated on WPM with 0.22 μM BA and were subsequently excised from the callus and rooted on half strength MS medium (Murashige and Skoog, 1962) without auxin. The regenerated plantlets were transferred to non-sterile conditions with ambient humidity and their survival rate was 80%.

*Salix daphnoides* Vill. and a *Salix caprea* L. hybrid have shown potential for direct *de novo* shoot formation (Grönroos, 1989). Whole leaves were excised from shoots cultured *in vitro*, making sure that no axillary bud was attached. The shoot cultures had been initiated from vegetative buds and subsequently cultured on WPM without hormones as described by Bergman *et al.* (1985). The excised leaves were cultured upright with the petiole submerged in solidified WPM with 1 μM BA. Adventitious shoots developed within 4 weeks from a compact, nodular tissue at the petiole or leaf base. Although the regeneration frequency was low (4–5%) and the experiment was small (20 leaves of *Salix daphnoides* and 100 leaves of the *S. caprea* hybrid), these results demonstrated that adventitious buds could be produced.

*Salix amygdaloides* stem explants have been shown to produce a yellow callus with potential for organo- and embryogenesis (Liu, 1992).

For commercial production of trees *in vitro*, propagation via somatic embryogenesis is potentially more useful than shoot culture or adventitious shoot production, since large numbers of somatic embryos can be handled (Gupta *et al.*, 1993). Other advantages are that root and shoot meristems are present in the same unit, embryos can be grown freely floating in liquid medium (facilitating scale-up and transfers), cryopreservation is possible and it may in the future be possible to make artificial seeds. However, much research remains to be done before somatic embryogenesis can be used for practical purposes in *Salix*.

*Salix viminalis* L. is the only willow shown to produce somatic embryos. In the following text, I will summarize the published papers on somatic embryogenesis in this species (Grönroos, 1989; Grönroos *et al.*, 1989a,b) and also add some previously unpublished results.

## 2. Culture initiation

Embryogenic callus was not produced directly from willow explants. Other types of callus (Type 1 and Type 2, see the Sections Establishment of Callus Types and Description of Callus Types below) were produced first. These could be recognized and separated from each other ca 6 weeks after culture initiation. One of the callus types (Type 2) later gave rise to embryogenic callus.

### 2.1. *Callus initiation* (*Grönroos, 1989*; *Grönroos* et al., *1989a*)

Floral buds of ten female *Salix viminalis* clones were collected from field-grown plants during the non-growing season (November – March). After surface sterilization of the buds, pistils were isolated and cultured in darkness at 25°C on MS medium containing 5 μM 2,4-D + 0.05 μM BA. Pistils of all clones produced at least some subculturable callus within 4 weeks, although clonal differences in callus initiation were found: The number of pistils with callus varied from 13 to 98% in different clones. The original position of the pistil in the catkin did not affect callus production. However, pistils from buds shorter than 5 mm usually yielded callus less frequently than pistils from larger buds. In an experiment with one clone, pistils were cultured on media with 2.5, 5 or 10 μM 2,4-D with or without 0.05 μM BA. In all combinations, callus was initiated on at least 88% of the pistils. The sucrose concentration was not critical for callus initiation, if kept in the range 1–8%. The sucrose concentration normally used was 3%.

Callus could also be initiated on other explants, including female and male catkin segments, zygotic embryos at developmental stages more advanced than the heart-shaped stage (i.e., torpedo-shaped to mature), seedling tissues, vegetative buds and leaf pieces from shoot cultures. These tissues were isolated from several *Salix* species (*S. viminalis*, *S. caprea*, *S. dasyclados*, *S. schwerinii* and *S. smithiana*) and cultured in the same manner as pistils

(Grönroos, 1989). *Salix caprea* explants did not produce callus. Leaf pieces from shoot cultures of *Salix schwerinii* produced callus on media with 0.5 – 25 μM 2,4-D with or without 0.05 μM BA. On media with low 2,4-D concentration (0.25 μM 2,4-D with 0, 0.05 or 0.5 μM BA) and on media with 0.5 μM BA + 0.25–25 μM 2,4-D, only roots or root-like nodules were produced from the leaf pieces.

## 2.2. *Establishment of callus types* (*Grönroos, 1989*; *Grönroos* et al., *1989a*)

The explants and callus tissues produced were subcultured to fresh medium every $2\frac{1}{2}$–4 weeks. Whenever enough callus was produced, it was separated from the explant and the different callus types were separated from each other and cultured independently. Two main callus types (Types 1 and 2), both non-embryogenic, could be recognised (see Section Description of Callus Types below) and separated from each other 6 weeks after culture initiation. Type 2 callus was a forerunner to embryogenic callus (see Sections Establishment of Embryogenic Callus and Description of Callus Types below). Type 2 callus was not produced as frequently as Type 1 callus. Only 6 out of 10 clones yielded Type 2 callus from pistils. Neither the isolation date (1st November – 19th March), nor the original position of the pistil in the catkin, nor the bud size had any considerable effect on Type 2 callus yield from pistils. Catkin segments containing pistils also yielded Type 2 callus but it was considered preferable to isolate pistils from the catkins in order to avoid, as far as possible: seasonal variations, effects of bud size, positional effects, any unfavourable influence from the catkin tissues, possible restrictions in nutrient and growth regulator supply to pistils and difficulties in observing the origins of callus. In addition, the isolation of several pistils from each catkin segment was possible and their culture required less space. Zygotic embryos produced only Type 1 callus, wheras the other explants produced either Type 1, Type 2 or both callus types. The presence of BA (0.05 μM) together with 2,4-D in the medium stimulated the production of Type 2 callus from pistils and leaf pieces. Sucrose concentrations between 1–6% were optimal for Type 2 callus formation from pistils.

## 2.3. *Establishment of embryogenic callus* (*Grönroos, 1989*; *Grönroos* et al., *1989a*)

More than six months after callus initiation and after numerous subcultures, Type 2 callus from pistils of one *Salix viminalis* clone yielded a third callus type, Type 3, which was embryogenic (see Section Description of Callus Types below). In many plant species, the choice of explant is critical when callus is initiated. Zygotic embryos and floral tissues have proven to be excellent sources of embryogenic callus in several angiosperms (Ammirato, 1983). Embryogenic callus from *Salix* was obtained only from pistils.

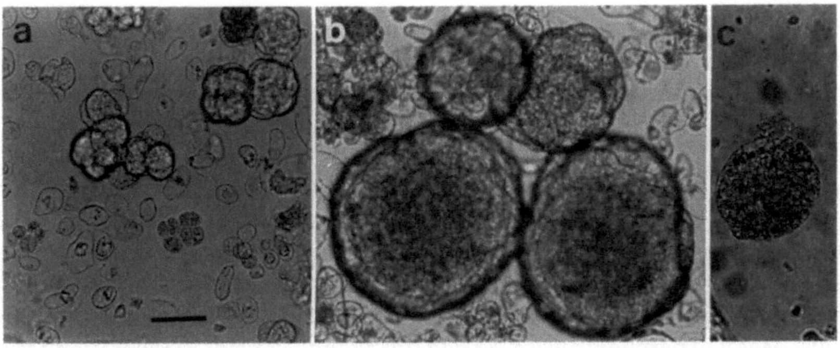


*Figure 1.* Embryogenic callus and zygotic embryo of *Salix viminalis* suspended in a few drops of water on a microscope slide. (a-b) Cell aggregates in embryogenic callus. (c) Globular zygotic embryo. Scale bar 0.05 mm for a-c.

The fact that only one out of ten clones produced embryogenic callus verifies the importance of genotypic differences when establishing tissue cultures and regenerating plants. This has been emphasized by several authors (e.g., Oelck and Schieder, 1983; Tomes, 1985; Mathias and Simpson, 1986). However, it is still possible that the other *Salix viminalis* clones might produce embryogenic callus under other culture conditions or at much lower frequencies. In one experiment, 149 pistils were isolated, 99% of which produced callus. Type 2 callus was produced by 57% of the pistils but only one pistil explant yielded Type 2 callus that subsequently produced Type 3 callus.

The conversion of Type 2 callus to the embryogenic state has not been studied. Formation of embryogenic callus can sometimes benefit from high (6–12%) sucrose concentrations. For example, in *Zea mays*, high sucrose concentrations increased the frequency of callus formation and the quality of the embryogenic callus (Lu *et al.*, 1983). However, the embryogenic callus production in *Salix* could not be stimulated by raising the sucrose concentration. Furthermore, the transition time was unpredictable and difficult to manipulate, occurring more than half a year after callus initiation. Nevertheless, transition from Type 2 to embryogenic callus was obtained three times, from pistil explants isolated in three consecutive years.

### 2.4. *Description of callus types* (*Grönroos, 1989*; *Grönroos* et al., *1989a*)

Type 1 callus was soft with a smooth, wet-looking surface and was composed mostly of long, tubular, highly vacuolated cells. The cytoplasm was often packed with rough endoplasmic reticulum but other organelles were relatively few in number. No organized structures could be differentiated from this type of callus.

Type 2 callus was composed of white or pale yellow, fast-growing, friable, non-compact callus with an uneven surface as well as compact, yellow nodules of different sizes. The proportions of friable callus and nodules varied. The nodules were composed of tightly associated cells and had either a distinct, smooth surface or else they were covered with larger, loose cells. Although their anatomy was not studied in great detail, they were apparently similar to the poplar nodules described by McCown *et al.* (1988). The friable callus consisted of large, rounded cells, usually highly vacuolated. Type 2 callus was fast-growing and rhizogenic. Roots could be seen within 3 weeks after transfer to hormone-free medium. The roots developed from nodules in the callus. Many experiments were carried out to induce plant regeneration from Type 2 callus (Grönroos, 1989; Grönroos *et al.*, 1989a; Vahala *et al.*, 1989) but no buds, shoots or embryo-like structures were obtained.

Type 3 callus was a fast-growing, friable, bright yellow, embryogenic callus made up of small, compact cell clumps. The cell clumps were sometimes separate from each other but most often they were grouped in small or large aggregates. Cell aggregates (Fig. 1a-b) resembling zygotic embryos at various stages of early development up to the globular stage (Fig. 1c) were observed in embryogenic callus of *Salix viminalis*. The cells at the periphery of the aggregates were small, densely cytoplasmic and had large nuclei with prominent nucleoli, whereas the more central cells were larger and more vacuolated. Embryogenic callus from willow shared characteristic anatomical and ultrastructural features with embryogenic callus of several other angiosperms. The ultrastructure of embryogenic callus cells of willow was also similar to that of cells from apical shoot meristems of willow (Berggren, 1985), the main difference being the absence of starch-filled plastids in meristems.

The identification and separation of different callus types from each other is extremely important and has greatly contributed to the maintenance of cultures with embryogenic potential, for example in the case of cereals (Tomes, 1985). The appearance of embryogenic callus varies from species to species. For example, in a *Populus* hybrid, embryogenic callus was tightly compact and clear (Michler and Bauer, 1991), not at all friable as in *Salix*. In *Populus maximowiczii*, embryogenic callus from anthers was soft and translucent with a mucilaginous, slimy or gel-like consistency and consisted of highly vacuolated microspore cells (Stoehr and Zsuffa, 1990).

### 3. Culture maintenance

Embryogenic callus was maintained on the same medium and under the same culture conditions as during its initiation, i.e., MS medium with 5 μM 2,4-D + 0.05 μM BA, in darkness at 25°C (Grönroos *et al.*, 1989a). The embryogenic callus had a tendency to become necrotic or to produce Type 1 callus. The conversion of embryogenic cultures to non-embryogenic cultures with elongated, highly vacuolated cells has also been reported in *Populus*

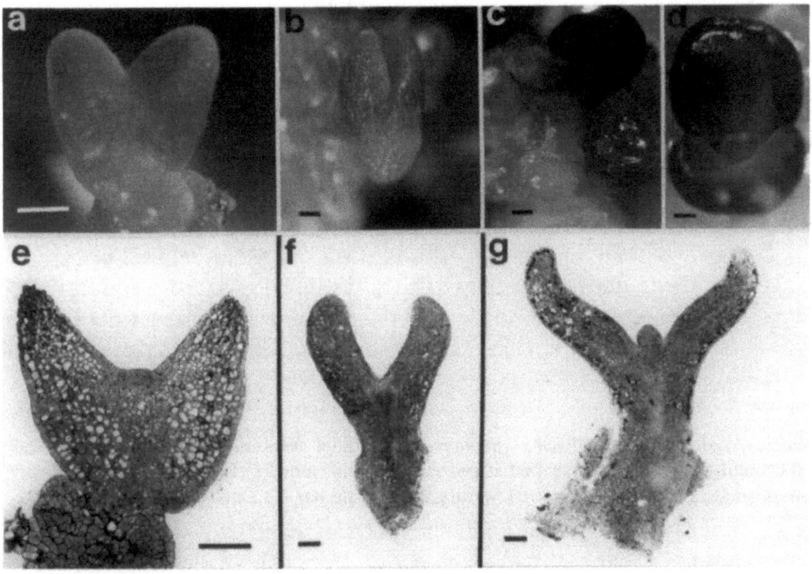


*Figure 2.* Somatic embryos of *Salix viminalis*. (a) Heart shaped stage. (b) Cotyledonary stage. (c-d) A mature somatic embryo with shoot apex. (e-g) Longitudinal sections. (e) Heart shaped stage. (f) Cotyledonary stage. (g) Mature stage. Scale bars 0.1 mm for a-g.

*ciliata* (Cheema, 1989). By selecting the proper callus type during each subculture and by making sure that the callus had good contact with the medium, embryogenic *Salix viminalis* callus was maintained for more than four years. The necessary contact with the medium was achieved by spreading the callus in rows on the medium surface. Addition of 7.6 μM abscisic acid (ABA) to the callus medium prevented callus browning to some extent and seemed to stimulate callus growth.

## 4. Embryo development

Somatic embryos were observed after transfer of embryogenic (Type 3) callus to 2,4-D-free medium. They developed at the edges of the callus, in close contact with the medium surface. For optimal embryo development, it was important to separate the callus into small pieces and to spread them out on the medium surface. Every 2 or 3 weeks, the callus and developing embryos were subcultured and again the callus was separated into smaller pieces to minimize competition between the embryos (Grönroos *et al.*, 1989a).

After 1–2 weeks on hormone-free medium in dim light (14–20 $\mu Em^{-2}$ $s^{-1}$, 16 h day photoperiod), the embryos were observed under the dissection microscope as compact, smooth-surfaced, white to yellow globules or cup-

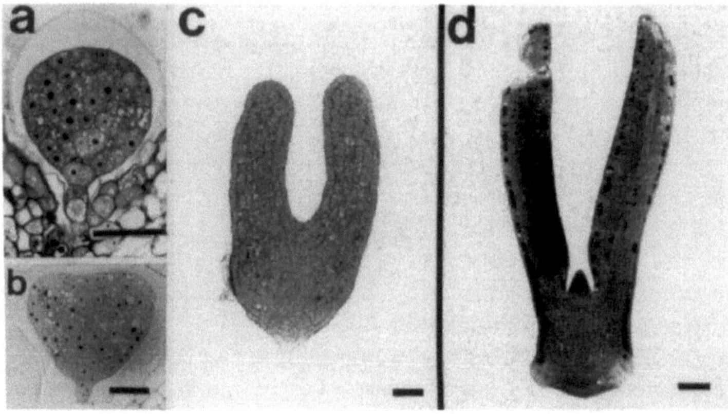


*Figure 3.* Zygotic embryos of *Salix viminalis*, longitudinal sections. (a) Globular stage, scale bar 0.05 mm. (b) Early heart shaped stage, scale bar 0.05 mm. (c) Late heart shaped to cotyledonary stage, scale bar 0.1 mm. (d) Mature stage, scale bar 0.1 mm.

shaped structures. They soon aquired green glossy cotyledons with a tint of red towards the white radicle, thus closely resembling zygotic embryos (Grönroos *et al.*, 1989a). Somatic embryos developed up to a mature stage within 6 weeks (Fig. 2a-d). Light microscope sections, prepared principally according to Fowke (1984), revealed that the development of somatic embryos (Fig. 2e-g) resembled the development of zygotic embryos (Fig. 3) with the formation of radicle, cotyledons, procambium and shoot apex. Although many embryos were observed to be bipolar, in most cases the radicle was embedded in callus or concealed by long root hairs, or else the embryos were organised into clusters. Light microscope studies, however, showed that the somatic embryos were indeed bipolar, with a procambium connecting the root apex and the cotyledons (Grönroos *et al.*, 1989a). There was no vascular connection between the embryo and the mother tissue at any stage during the development of somatic embryos. The most obvious difference between zygotic and somatic embryos was that, instead of a suspensor, the somatic embryos had a region of meristematic cells at their base and a meristematic region sometimes extended along the surface of the embryo (Grönroos *et al.*, 1989a). Another difference was that somatic embryos were sometimes slightly larger than zygotic embryos, especially at the globular and heart-shaped stages. Such size differences have also been described in *Populus* (Michler and Bauer, 1991).

In spite of the similarities between zygotic and somatic embryos, many deviations from the normal morphology of zygotic embryos were also observed, for example fused cotyledons, the possession of more than two cotyledons, poor development of cotyledons and secondary embryo proliferation (Grönroos *et al.*, 1989a). The formation of atypical embryos occurs in most species in which somatic embryogenesis has been reported, as reviewed by Ammirato (1987). The process of repetitive embryogenesis in *Salix viminalis* resulted in embryogenic lines that could be maintained on hormone-free medium for more than 7 months (Grönroos *et al.*, 1989a). Somatic embryos that failed to grow into plantlets often generated secondary embryos. Instead of participating in the coordinated growth of one embryo, individual cells may themselves become embryogenic growth centers (Ammirato, 1987). The meristematic regions at the base of the *Salix viminalis* embryos were probably embryogenic growth centers that caused secondary embryo production. Another fate that often overtook somatic embryos of *Salix viminalis* was overgrowth of callus. Roots without connections to embryos as well as root hairs also developed from the callus. The roots often emerged from callus regions that consisted of red cells (Grönroos *et al.*, 1989a).

In darkness, the somatic embryos remained white to yellow but were otherwise similar to the embryos in dim light. In strong light (170–190 $\mu Em^{-2} s^{-1}$, 16 h day photoperiod), stunted, red embryos with a rough surface developed. More embryos and fewer roots were formed on MS medium than on WPM (Grönroos *et al.*, 1989a).

In order to optimize embryo development, reduce the number of abnormal embryos as well as to induce embryo germination, different culture media were evaluated (Table 1). In several species, it has been demonstrated that somatic embryos respond differently to exogenously applied hormones, depending on their developmental stage. Therefore, different media were studied at different time periods during *Salix viminalis* somatic embryo development, i.e., at different developmental stages. In different species, embryo development and germination can be promoted by cytokinins or gibberellic acid ($GA_3$), and ABA has been demonstrated to selectively inhibit many abnormalities or to be essential or beneficial for somatic embryo development (Vasil and Vasil, 1981; Lu and Vasil, 1981; Ammirato, 1983; von Arnold and Hakman, 1988). The ammonium concentration for optimal embryo development has been shown to vary from species to species (Stuart *et al.*, 1985) and abnormal embryo development can sometimes be inhibited by raising the sucrose level (e.g., Green, 1982). All these parameters were tested on *Salix viminalis* embryogenesis (Table 1).

Embryogenic callus (Type 3) growing on MS medium with 5 μM 2,4-D + 0.05 μM BA + 7.6 μM ABA was transferred to the different MS regeneration media (Table 1). An average of 40 mg callus was transferred to each petridish (diameter 5 cm), spreading it in rows on the medium surface. A rough estimation of the developing embryos after 2 weeks was done by counting the number of embryo aggregates. The embryos were usually clustered in aggregates due to secondary embryogenesis. The general appearance of the embryos and callus tissues was also recorded.

Normal strength (1 × ) MS medium was best for embryo development at all developmental stages (Table 1, experiment 2). Ammonium concentrations

*Table 1.* Screening experiments for optimization of somatic embryo development and germination in *Salix viminalis*. Embryogenic callus was transferred from MS medium with 5 μM 2,4-D + 0.05 μM BA + 7.6 μM ABA to modified or supplemented MS media. Callus was transferred to 4 petridishes per regeneration medium, except in experiment 1, in which a total of 103 petridishes were used on 9 separate occasions, in experiment 4.1, in which 26 petridishes of medium with 10% sucrose were used on 6 separate occasions, in experiment 8.1, in which 8 petridishes of medium with 7.6 μM ABA were used on 2 separate occasions and in experiment 7.1, in which 3 petridishes per treatment were used. Unless otherwise stated, the embryogenic tissue was subcultured to the same regeneration medium for several subculture periods. Figures in parentheses () indicate the number of embryo aggregates per petridish visible from above after 2 weeks on regeneration medium, respectively. Standard deviation was calculated if the treatment was repeated three times or more.

| Experiment no. | Treatment |
|---|---|
| 1 | Hormone-free medium[1] (48.6 ± 35.4) |
| 2.1 | Hormone-free medium at 1/8×, 1/4×, 1/2×, 1× and 2× concentrations of all components except sucrose, which was kept at 3% (0, 0, 2.2, 26.0, 0) |
| 2.2 | As in 2.1 after preculture for 2, 3 and 6 w. on normal strength (1×) medium |
| 3.1 | Hormone-free medium with the following ammonium concentrations: 0, 1, 2, 3, 4, 5, 10, 20.6 and 30 mM. The $NH_4^+$ concentration was varied by changing the $NH_4NO_3$ concentration and the $NO_3^-$ level was kept constant by adjusting with $KNO_3$ (0, 0, 0.8, 1.8, 6.0, 13.0, 27.5, 18.5, 26.8) |
| 3.2 | As in 3.1 after preculture for 2 and 4 w. on 20.6 mM $NH_4^+$ |
| 4.1 | Hormone-free medium with 0, 1, 2, 3[1], 6[1], 10 or 20% sucrose (0, 11.0, 34.4, 62.2, 65.4, 77.9 ± 48.7, 23.9) |
| 4.2 | As in 4.1 after preculture for 2 w. on 3% sucrose |
| 5.1 | 0.1, 1, 5 or 50 μM BA (42.2, 53.1, 26.8, 0) |
| 5.2 | 5 μM BA after preculture for 5 w. on hormone-free medium |
| 5.3 | Hormone-free medium after preculture for 2 w. on 0.1, 1 and 5 μM BA. |
| 6.1 | 0.5, 5 or 50 μM zeatin (79.8, 53.7, 0.5) |
| 6.2 | 5 μM zeatin after preculture for 3 w. on hormone-free medium |
| 6.3 | Hormone-free medium after preculture for 3 w. on 0.5 and 5 μM zeatin |
| 7.1 | 0.3, 3 or 15 μM $GA_3$ as well as 0.3, 3 and 15 μM $GA_{4/7}$ (95.0, 31.0, 81.7; 66.7, 78.3, 94.7) |
| 7.2 | As in 7.1 after preculture on hormone-free medium for 3 w. |
| 7.3 | Hormone-free medium after preculture for 3 w. on 0.3 or 15 μM $GA_3$ |
| 8.1 | 0.076, 0.76, 7.6 and 38 μM ABA (60.5, 73.5, 30.2, –) |
| 8.2.1 | As in 8.1 after preculture for 3 w. on hormone-free medium |
| 8.2.2 | Hormone-free medium for 3 w., then 2 or 7 w. on 7.6 μM ABA, then hormone-free medium again[1] |
| 8.3 | 7.6 μM ABA after preculture for 4, 5, 6, 8, 10 and 12 w. on hormone-free medium |
| 8.4 | Hormone-free medium after preculture for 2, 3, 4, 5[1], 6[1] and 8 w. on 7.6 μM ABA |
| 8.5.1 | 7.6 μM ABA after preculture for 2–5 w. on 0.1, 1 or 5 μM BA, 0.5 or 5 μM zeatin, 0.3 or 15 μM $GA_3$ or hormone-free medium with 10% sucrose |
| 8.5.2 | As in 8.5.1. After 3 w. on ABA, the embryos were transferred to hormone-free medium[1] |
| 8.6.1 | 0.1, 1, 5, 25 or 50 μM BA or 0.3, 3[1] or 15 μM $GA_3$ or 0.3[1], 3 or 15 μM $GA_{4/7}$ after preculture for 5 w. on 7.6 μM ABA |
| 8.6.2 | 0.1, 1[1], 5[1], 25 or 50[1] μM BA after preculture for 5 w. on 7.6 μM ABA. After 2 w. on BA, the embryos were transferred to hormone-free medium |

[1] indicates examples of treatments that resulted in embryo germination and plant regeneration.

of 10–30 mM supported the embryo development adequately at all developmental stages (Table 1, experiment 3). The lower the ammonium concentration, the greater the number of roots that developed. More and greener embryos were produced with 6 and 10% sucrose than with 3% sucrose (Table 1, experiment 4.1). A student t-Test indicated that the number of embryo aggregates was significantly higher with 10% sucrose than with 3% sucrose ($p = 0.001$). However, if callus was initially cultured on 3% sucrose for 2 weeks, and callus with developing embryos was then transferred to 6 or 10% sucrose (Table 1, experiment 4.2), fewer embryos developed. Sucrose concentrations below 3% were suboptimal and on 20% sucrose, only a few small embryos developed (Table 1, experiment 4.1).

Cytokinins (1 and 5 μM BA and 5 μM zeatin) caused the development of red, stunted embryos which did not develop normally after transfer to a hormone-free medium; 0.1 μM BA and 0.5 μM zeatin did not seem to affect embryo development. The presence of 50 μM BA or 50 μM zeatin in the medium inhibited embryo development almost completely (Table 1, experiments 5 and 6).

$GA_3$ and $GA_{4/7}$ treatment (Table 1, experiment 7) resulted in root and callus growth as well as in abnormal elongation growth of embryo hypocotyls and cotyledons. These embryos could not develop normally after transfer to hormone-free medium.

Abscisic acid at the concentration of 7.6 μM reduced undesired features such as abnormal development, callus growth, secondary embryo production and root hair development, but did not inhibit them completely. Embryos on 7.6 μM ABA were usually greener than on hormone-free medium and they had a smooth surface. These effects appeared within 2 weeks on medium containing ABA. The number of embryos produced was lower than on hormone-free medium (Table 1, experiments 1 and 8.1). The effects of 0.76 μM ABA on embryo development were similar to those of 7.6 μM ABA but 0.76 μM ABA did not reduce secondary embryogenesis as efficiently. Abscisic acid at 0.076 μM did not have any visible effect on embryo development and 38 μM ABA reduced or completely inhibited embryo formation (Table 1, experiment 8.1). The favourable ABA effects were observed when embryos (including secondary embryos) were transferred to medium containing 7.6 μM ABA after 0–12 weeks on hormone-free medium or after 2–5 weeks on medium containing BA, zeatin, $GA_3$ or 10% sucrose, but only on young embryos (mostly secondary embryos) that had not started to show undesirable features (Table 1, experiments 8.2.1, 8.3 and 8.5.1). The ABA effects were reversible even after 8 weeks of culture on ABA: the undesired features were seen on young embryos after transfer to hormone-free medium (Table 1, experiments 8.2.2, 8.4, 8.5.2 and 8.6).

It can be concluded that normal strength (1 × ) MS medium with normal ammonium concentration (20.6 mM) was satisfactory for *Salix viminalis* embryo development. The best culture protocol, so far, for embryo development from embryogenic callus is as follows: transfer from MS medium with 5 μM

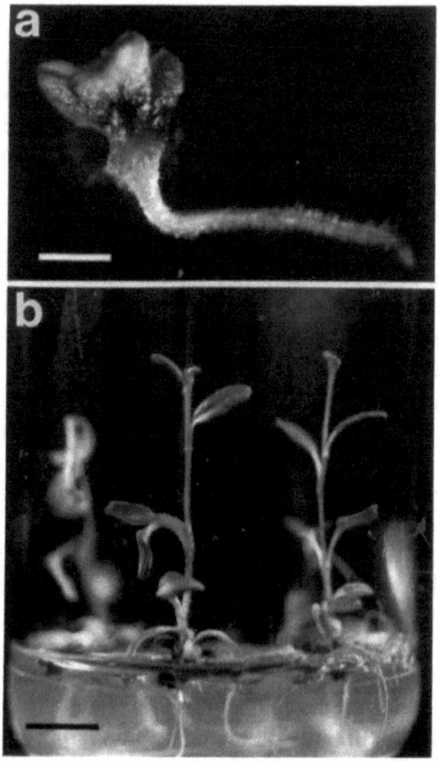


*Figure 4.* (a) Germinating somatic embryo of *Salix viminalis*, scale bar 1 mm. (b) Plants from somatic embryos of *Salix viminalis*, scale bar 10 mm.

2,4-D + 0.05 μM BA + 7.6 μM ABA to MS medium with 10% sucrose for 2 weeks, then MS medium with 7.6 μM ABA until the embryos have matured.

## 5. Embryo germination

Somatic embryos germinate (Fig. 4a) and develop into plants (Fig. 4b) with a variety of treatments (Table 1, experiments 1, 4.1, 8.2.2, 8.4, 8.5.2 and 8.6). More research is required to determine the optimal treatment for plant regeneration. Hormones were not essential for plantlet production, although ABA treatment increased the number of embryos that had the capacity to develop into plantlets. The embryos did not germinate on medium with ABA; they had to be transferred to ABA-free medium. It was important to isolate embryos from each other and from callus tissue as soon as possible.

Not all of the embryogenic callus lines produced plants. It seemed that callus lines could lose this capacity. Instead of producing shoots, apparently mature embryos would then grow to an abnormal size, become red, obtain a rough surface, become overgrown with callus or produce secondary embryos. Nevertheless, more than 120 plants of *Salix viminalis* were obtained. These plants were transferred to 200 ml glass jars, containing ca 40 ml MS medium, where they reached a height of about 8 cm. Transfer to *ex vitro* conditions was not attempted.

## 6. Conclusions

Much research remains to be done before somatic embryogenesis can be applied commercially or used as a routine tool for genetic improvement of willows. Somatic embryogenesis has so far been achieved in only one *Salix viminalis* clone. More attention needs to be paid to the other clones and species. The embryogenic callus could rapidly be propagated so that a very large number of somatic embryos could be obtained within a short period of time. Efforts should be made to stimulate the formation of embryogenic callus earlier and to study the regulation of transition from non-embryogenic (Type 2) to embryogenic (Type 3) callus. Plants regenerated via somatic embryos must be field tested to evaluate the occurence of physiological and/or genetic changes. The stage has also been set for basic physiological and molecular genetic studies on somatic embryogenesis in *Salix*.

## 7. Acknowledgements

Research was supported by the National Energy Administration in Sweden. I am grateful to Ms. Annette Axén for skillful technical assistance with preparation and sectioning of embryos for light microscopy. I thank Prof. Sara von Arnold and Dr. Roland Grönroos for invaluable comments on the manuscript and for making computer facilities available to me.

## 8. References

Ager, A., A.C. Rönnberg-Wästljung, J. Thorsén and G. Sirén, 1986. Genetic improvement of willows for energy forestry in Sweden. Swed. Univ. Agr. Sci., Dept. of Ecology and Environmental Research. The Energy Forestry Project, Report No. 43.

Ahuja, M.R., 1987. *In vitro* propagation of poplar and aspen. In: J.M. Bonga and D.J. Durzan (Eds.), Cell and Tissue Culture in Forestry, Vol. 3, pp. 207–223. Martinus Nijhoff, Dordrecht.

Ammirato, P.V., 1983. Embryogenesis. In: D.A. Evans, W.R. Sharp, P.V. Ammirato and Y. Yamada (Ed.), Handbook of Plant Cell Culture, Vol. 1, pp. 82–123. MacMillan Publishing Co., New York/London.

Ammirato, P.V., 1987. Organizational events during somatic embryogenesis. In: C.E. Green,

D.A. Somers, W.P. Hacket and D.D. Biesboer (Eds.), Plant Tissue and Cell Culture, pp. 57–81. Alan R. Liss, New York.

Arnold, S. von and I. Hakman, 1988. Regulation of somatic embryo development in *Picea abies* by abscisic acid (ABA). J. Plant Physiol. 132: 164–169.

Berggren, B., 1985. Ultrastructure of the histological zones in growing vegetative buds of Salix spp. Nord. J. Bot. 4: 771–789.

Bergman, L., S. von Arnold and T. Eriksson, 1985. Effects of $N^6$-benzyladenine on shoots of five willow clones (*Salix* spp.) cultured *in vitro*. Plant Cell Tiss. Org. Cult. 4: 135–144.

Bhojwani, S.S., 1980. Micropropagation method for a hybrid willow (*Salix matsudana* × *alba* NZ-1002). N.Z. J. Bot. 18: 209–214.

Cheema, G.S., 1989. Somatic embryogenesis and plant regeneration from cell suspension and tissue cultures of mature himalayan poplar (*Populus ciliata*). Plant Cell Rep. 8: 124–127.

Dhir, K.K., R. Angrish and M. Bajaj, 1984. Micropropagation of *Salix babylonica* through *in vitro* shoot proliferation. Proc. Indian Acad. Sci. (Plant Sci.) 93: 655–660.

Filatti, J.J., J. Sellmer, B. McCown, B. Haissig and L. Comai, 1987. *Agrobacterium* mediated transformation and regeneration of *Populus*. Mol. Gen. Genet. 206: 192–199.

Fowke, L.C., 1984. Preparation of cultured cells and tissues for transmission electron microscopy. In: I.K. Vasil (Ed.), Cell Culture and Somatic Cell Genetics of Plants, Vol. 1, pp. 728–737. Academic Press, Orlando/San Diego/New York/London/Toronto/Montreal/Sydney/Tokyo.

Green, C.E., 1982. Somatic embryogenesis and plant regeneration from the friable callus of *Zea mays*. In: A. Fujiwara (Ed.), Proc. 5th Intl. Congr. Plant Tissue and Cell Culture, pp. 107–108. Maruzen Co., Ltd., Tokyo.

Grönroos, L., 1989. Shoot regeneration and somatic embryogenesis in *Salix* cultured *in vitro*. Acta Univ. Upsaliensis, Comprehensive Summaries of Uppsala Dissertations from the Faculty of Science 190, 48 pp.

Grönroos, L., S. von Arnold and T. Eriksson, 1989a. Callus production and somatic embryogenesis from floral explants of basket willow (*Salix viminalis* L.). J. Plant Physiol. 134: 558–566.

Grönroos, L., S. von Arnold and T. Eriksson, 1989b. Somatic embryos from callus of *Salix viminalis* L. In: E. Dreyer, G. Aussenac, M. Bonnet-Masimbert, P. Dizengremel, J.M. Favre, J.P. Garrec, F. Le Tacon and F. Martin (Eds.), Forest Tree Physiology. Proceedings of the International Symposium 25–30 September, 1988, Nancy. Elsevier/INRA, Paris. Ann. Sci. For. 46, Suppl.: 108–109.

Grönroos, L., C. Hardner, U. Gullberg and S. von Arnold, 1990. Field performance of *Salix* clones propagated via shoot cultures *in vitro*. Scand. J. For. Res. 5: 487–492.

Gullberg, U., 1993. Towards making willows pilot species for coppicing production. The Forestry Chronicle (in press).

Gupta, P.K., G. Pullman, R. Timmis, M. Kreitinger, W.C. Carlson, J. Grob and E. Welty, 1993. Forestry in the 21st century. The biotechnology of somatic embryogenesis. Bio/technol. 11: 454–459.

Larsson, G. and B. Bremer, 1991. Korgviden – nyttoväxter förr och nu. [The *Salix viminalis* group – useful plants then and now.] Svensk Bot. Tidskr. (Lund) 85: 185–200.

Ledin, S. and A. Alriksson, 1992. Handbook on How to Grow Short Rotation Forests. 1992. Section of Short Rotation Forestry, Swedish University of Agricultural Sciences.

Liu, F., 1992. Variation in organogenic capacity within and among three willow species under conditions of *in vitro* tissue culture. Thesis submitted in Conformity with the Requirements for the Degree of Master of Science in Forestry at the Graduate Department of Forestry in the University of Toronto, Toronto, Canada.

Lloyd, G. and B. McCown, 1981. Commercially-feasible micropropagation of mountain laurel, *Kalmia latifolia*, by use of shoot-tip culture. Comb. Proc. Intern. Plant Prop. Soc. 30: 421–427.

Lu, C. and I.K. Vasil, 1981. Somatic embryogenesis and plant regeneration from leaf tissues of *Panicum maximum* Jacq. Theor. Appl. Genet. 59: 275–280.

Lu, C., V. Vasil and I.K. Vasil, 1983. Improved efficiency of somatic embryogenesis and plant regeneration in tissue cultures of maize (*Zea mays* L.). Theor. Appl. Genet. 66: 285–289.
Mathes, M.C., 1964. The *in vitro* formation of plantlets from isolated aspen tissue. Phyton 21: 137–141.
Mathias, R.J. and E.S. Simpson, 1986. The interaction of genotype and culture medium on the tissue culture responses of wheat (*Triticum aestivum* L. em. thell) callus. Plant Cell Tiss. Org. Cult. 7: 31–37.
McCown, B.H., E.L. Zeldin, H.A. Pinkalla and R.R. Dedolph, 1988. Nodule culture: a developmental pathway with high potential for regeneration, automated micropropagation, and plant metabolite production from woody plants. In: J.W. Hanover and D.E. Keathley (Eds.), Genetic Manipulation of Woody Plants. Basic Life Sciences, Vol. 44, pp. 149–166. Plenum Press, New York.
Michler, C.H. and E.O. Bauer, 1991. High frequency somatic embryogenesis from leaf tissue of *Populus* spp. Plant Sci. 77: 111–118.
Murashige, T. and F. Skoog, 1962. A revised medium for rapid growth and bio assays with tobacco tissue cultures. Physiol. Plant. 15: 473–497.
Newsholme, C., 1992. Willows: the genus *Salix*. B.T. Batsford Ltd., London.
Oelck, M.M. and O. Schieder, 1983. Genotypic differences in some legume species affecting the redifferentiation ability from callus to plants. Z. Pflanzenzüchtg. 91: 312–321.
Stoehr, M.U., M. Cai and L. Zsuffa, 1989. *In vitro* plant regeneration via callus culture of mature *Salix exigua*. Can. J. For. Res. 19: 1634–1638.
Stoehr, M.U. and L. Zsuffa, 1990. Induction of haploids in *Populus maximowiczii* via embryogenic callus. Plant Cell Tiss. Org. Cult. 23: 49–58.
Stuart, D.A., J. Nelsen, C.M. McCall, S.G. Strickland and K.A. Walker, 1985. Physiology of the development of somatic embryos in cell cultures of alfalfa and celery. In: M. Zaitlin, P. Day and A. Hollaender (Eds.), Biotechnology in Plant Science, pp. 35–47. Academic Press, Orlando.
Tomes, D.T., 1985. Cell culture, somatic embryogenesis and plant regeneration in maize, rice, sorghum and millets. In: S.W.J. Bright and M.G.K. Jones (Eds.), Cereal Tissue and Cell Culture, pp. 175–203. Martinus Nijhoff/Dr. W. Junk, Dordrecht/Boston/Lancaster.
Tormala, T. and E. Saarikko, 1985. Micropropagation of willows (Salix spp.). In: W. Palz, J. Coombs and D.O. Hall (Eds.), Energy from Biomass, 3rd E.C. Conference, pp. 427–429. Elsevier Applied Science, London/New York.
Vahala, T., P. Stabel and T. Eriksson, 1989. Genetic transformation of willows (*Salix* spp.) by *Agrobacterium tumefaciens*. Plant Cell Rep. 8: 55–58.
Vasil, V. and I.K. Vasil, 1981. Somatic embryogenesis and plant regeneration from suspension cultures of pearl millet (*Pennisetum americanum*). Ann. Bot. 47: 669–678.

# 14. Somatic embryogenesis in *Camellia* spp.

Ana M. Vieitez

**Contents**



## 1. Introduction

### 1.1. *Distribution and history*

*Camellia* (Theales, Theaceae) is a genus of tropical or subtropical trees and shrubs native to Eastern Asia. Its antiquity has been confirmed by fossils belonging to the Cretaceous period of the Secondary Era. The genus name was created by Linnaeus in the honour of the Jesuit missionary G.J. Kamel, who first recommended to grow the "Japanese rose" in Europe. In his *Systema Naturae* of 1735, Linnaeus gave the name *Camellia tsubaki* to the plant indegenous to Japan as the *tsubaki* and it only acquired its present systematic name, *Camellia japonica*, in his *Species Plantarum*, in which he also described the tea plant, *C. sinensis*, under the name *Thea sinensis*. The systematics of the genus have since varied as the number of known members has increased; in particular, Chang and Bartolomew (1984) have carried out a major revision, grouping some 200 recognized species in four subgenera and twenty sections.

All these 200 species appear to have originated in the area bounded by longitudes 85° and 150° E and latitudes 37° N and 10° S, but their distribution within this area is not uniform. Apparently, 90% originated in South and

*S. Jain, P. Gupta & R. Newton (eds.), Somatic Embryogenesis in Woody Plants, Vol. 2*, 235–276.

Southeast of China, especially in the provinces of Yunnan, Guangxi and Guangdong on either side of the Tropic of Cancer (Samartín and Samartín, 1988). *Camellia japonica*, however, originated in Japan, as did three other species, whereas *C. lancelolata* is the only species present in the Philippines, Borneo, Sumatra and Java.

In the East, the chief species of *Camellia* were spread from country to country by Buddhist monks; thus *C. sinensis* was introduced into Japan in the 12th century. Their introduction into Europe began some 300 years ago, when the first attempts were made to cultivate tea plants brought by sea from the East. Though the tea plant never succeeded as a commercial crop in Europe, the ornamental camellias that were brought with it flourished all over starting from England to France, Belgium and Italy, and from Portugal to Galicia (N.W. Spain) in their acid soil and temperate humid climate (Samartín and Samartín, 1988). England was also the immediate source of camellia introduction in Australia during the mid 19th century, before this time they were unknown in this continent, in spite of its proximity to the natural area of distribution of the genus. They had already been introduced into the United States at the beginning of the 18th century. Tea plants were introduced into East Africa early in this century.

### 1.2. *Importance*

The economic importance of the genus *Camellia* is largely due to *C. sinensis* cvs. "Sinensis" and "Assamica", whose young leaves are used to prepare tea. In S.E. China, the custom of tea drinking goes back to at least 1000 B.C., and its spread to other countries led to the development of major trade routes. Tea plants currently occupy some 1.7 million ha of land, and annual world tea production is about 2.2 million tons (Kato, 1989a), 70% of which is black tea and 30% green. The main black tea producing countries are India, Sri Lanka and Indonesia, and the primary green tea producing countries are China, Japan and parts of the former USSR (Shizuoka Tea Experiment Station, 1992).

Other economically important members of this genus include *C. oleifera*, *C. semiserrata* and *C. chekiangolomy* (all of the Section Oleifera), their seed oil is widely used in Chinese cookery. An increasing interest in ornamental camellias has led to the continuous efforts by researchers, growers and breeders to produce new varieties and hybrids with improved floral and growth characteristics. The ornamental species include mainly *C. japonica* (the species with most varieties) and also *C. reticulata*, *C. sasanqua* and *C. saluenensis*.

### 1.3. *Biotechnology in* Camellia: *somatic embryogenesis*

Since the propagation of some *Camellia* species by conventional methods is inefficient, other means have been sought. Bennett (1977, 1978) was among the first to suggest to use *in vitro* tissue culture for rapid clonal propagation of *Camellia* species of economic importance. Use of tissue culture technology offers a potential adjunct for *Camellia* propagation and breeding. Recent advances in biotechnology have increased the prospects for improving tea yield and quality, through the development of essential lines as micropropagation, protoplast fusion and genetic engineering (Jain and Newton, 1990). Though pursuit of the latter two approaches remains for the near future. Since 1980's, *in vitro* propagation of different species of *Camellia* has been accomplished (Kato, 1989a; Vieitez *et al.*, 1992).

Three micropropagation methods have been used for *in vitro* regeneration of *Camellia*: i) the development of axillary buds from shoot-tips and nodal explants; ii) the induction of adventitious buds; and iii) somatic embryogenesis. Somatic embryogenesis is considered to be, in general, the most efficient means of clonal multiplication of plants (Vasil, 1985). The applications of somatic embryos in clonal propagation, artificial seed production (utilizing bioreactors), and cryopreservation for long term storage leading to the establishment of genebanks, would be beneficial for both tea plants (Jain and Newton, 1990) and ornamental camellias. In particular, the induction of somatic embryos on zygotic embryos produced by interspecific cross-pollination (which are heterogenic) allows the propagation of hybrids whose seeds either fail to develop or have poor viability. Examples of these applications are provided by the success of cotyledon culture of various hybrids of *C. chrysantha* (Nadamitsu *et al.*, 1986; Yamaguchi *et al.*, 1987). In both reports, somatic embryos were induced on few immature seeds, produced before abortion. Rapid disease-free clonal multiplication of hybrid seeds or selected clones by somatic embryogenesis can thus accelerate breeding programmes for both ornamental *Camellia* species and *C. sinensis* and other related species (Kato, 1989a; Jain and Newton, 1990). Furthermore, adventive embryogenesis can also afford androgenic embryos and haploid plants (Chen and Liao, 1990) or, through secondary somatic embryos and polyploid formation (Kato, 1989b). Both approaches are significantly important for the improvement of *Camellia* species.

In order to achieve all of these objectives, one needs to undertake basic research on the induction and development of somatic embryos, and a reliable technology to scale-up plant multiplication from somatic embryos.

## 2. Culture initiation

Within the genus, somatic embryogenesis has mainly been investigated using economically important species of *C. sinensis*, *C. japonica* and *C. reticulata*, though preliminary reports on other species or hybrids have already been published (Table 1). In most studies, cotyledons have been used as the source of initial explants. The large cotyledons of *Camellia* seeds appear to have a great capacity for morphogenesis of both adventitious buds (see Tian-Ling,

*Table 1.* Summary of *in vitro* somatic embryogenesis studies in *Camellia*.

| Species | Explant source | Induction medium[a] | Germination medium[a] | Response | Reference |
|---|---|---|---|---|---|
| *C. brevistela* | Mature cotyledons | MS + BA (3) | – | Somatic embryos | Nakamura, 1988 |
| *C. chrysantha* | Cotyledons | MS + BA (1) + NAA (0.2–0.5) | – | Somatic embryos, plantlet regeneration | Zhuang and Liang, 1985b |
| *C. japonica* | Mature cotyledons | Knop + BA (0.5–5) | – | Indirect embryogenesis, plantlet regeneration | Bennett and Scheibert, 1982 |
| | Mature cotyledons | MS + BA (0–5) + IBA (0–2) | MS + $GA_3$ (1) | Direct embryogenesis, plantlet regeneration | Kato, 1986a |
| | Mature cotyledons | MS + BA (3) | – | Somatic embryos | Nakamura, 1988 |
| | Primary somatic embryos | MS + $GA_3$ (1) + colchicine (0.1%) | MS + $GA_3$ (1) | Secondary polyploid embryos, polyploid plants | Kato, 1989b |
| | Mature and immature cotyledons, embryonic axes | MS + BA (0–2) + IBA (0–2) | MS + $GA_3$ (3–5) + IAA (1–2) | Direct embryogenesis, plantlet regeneration | Vieitez and Barciela, 1990 |
| | *In vitro* plantlets (roots) | MS + Z(1) or MS + BA (0–2) + IBA (0–2) | MS + $GA_3$ (5) + IAA (1–2) | Direct embryogenesis, plantlet regeneration | Vieitez *et al.*, 1991 |
| | Cotyledons | MS + BA (1) | – | Origin and development of somatic embryos | Barciela and Vieitez, 1993 |
| | Leaves from shoot cultures | MS + BA (8) + IBA (0.5) or WPM + BA (8) + IBA (0.5) | MS + $GA_3$ (3) + IAA (1) | Callus, somatic embryos and adventitious buds, plantlet regeneration | San-José and Vieitez, 1993 |
| *C. oleifera* | Cotyledons | – | – | Embryoids | Yan *et al.*, 1984 |
| *C. reticulata* | Mature cotyledons | MS + BA (1) + NAA (0.2) | $B_5$ or liquid MS + BA (0.1–0.2) + IAA (0.1–0.5) | Direct embryogenesis, plantlet regeneration | Zuang and Liang, 1985a |
| | Mature and immature cotyledons, embryonic axes | MS + IBA (0.5–1) MS + BA (1) + IBA (0.5) | MS + $GA_3$ (3–5) + IAA (1–2) | Direct embryogenesis, indirect embryogenesis, plantlet regeneration | Plata and Vieitez, 1990; Plata, 1993 |
| | Primary somatic embryos | MS + BA (0.5) + IBA (0.1) | – | Origin and development of secondary embryos | Plata *et al.*, 1991 |

| | | | | | |
|---|---|---|---|---|---|
| *C. sasanqua* | Mature cotyledons | MS + BA (0–10) + IBA (0–2) | – | Direct embryogenesis | Kato, 1986b |
| | Cotyledons | MS + BA (1) + NAA (0.2–0.5) | MS + BA (2) + IAA (0.5) + ABA (0.2) + Glutamine (500) or MS + $GA_3$ (1) | Direct embryogenesis, plantlet regeneration | Zhuang *et al.*, 1988 |
| | Mature cotyledons | MS + BA (3) | – | Somatic embryos | Nakamura, 1988 |
| *C. sinensis* | Mature cotyledons | MS + K (10) + IAA (1) | – | Indirect embryogenesis? Plantlet regeneration | Wu, 1976; Wu *et al.*, 1981 |
| | Mature cotyledons | – | – | Embryoids | Yan and Ping, 1983 |
| | Mature cotyledons | MS + BA (0–10) + IBA (0–2) | MS + BA (10) + IBA (0.5) | Direct embryogenesis, plantlets regeneration | Kato, 1982, 1986a |
| | Immature cotyledons | MS (basal medium) MS + BA + IBA | MS + BA + NAA | Direct and indirect embryogenesis, plantlet regeneration | Abraham and Raman, 1986 |
| | Mature and immature cotyledons | MS + BA (1–5) | – | Direct embryogenesis | Nakamura, 1988 |
| | Mature zygotic embryos | MS + 2,4–D (3) + K (0.2) MS + BA (3–6) + K (0.5) | – | Callus, indirect embryogenesis, multiple shoots | Das and Barman, 1988, 1992 |
| | Shoots from embryoids | Liquid MS 1/2 + IBA (0.5) or pulse IBA (50) 2 h | – | Rooting, plantlet regeneration | Jain *et al.*, 1990, 1991 |
| | Shoots from somatic embryos | MS 1/4 + IBA (0.5–3) | – | Rooting, plantlet regeneration | Nakamura, 1991 |
| | Mature and immature cotyledons | MS + K (0.05) + 2,4D (0.5) | MS 1/2 + K (0.05) + AC (0.2%) + Glucose (1.5%) | Direct embryogenesis, plantlet regeneration | Bano *et al.*, 1991 |
| | Immature cotyledons | MS + BA (10) + IBA (0.5) + Adenine Sulfate (80) | $B_5$ + BA (3) + IAA (2) | Direct embryogenesis, plantlet regeneration | Jha *et al.*, 1992 |

*Table 1.* Continued.

| Species | Explant source | Induction medium[a] | Germination medium[a] | Response | Reference |
|---|---|---|---|---|---|
| *C. japonica* × *C. chrysantha* | Immature zygotic embryos | MS + K (0.1–0.5) + NAA (0.5–1) + YE (1g/l) | A + 2iP (0.2–0.5) + $GA_3$ (5) + PVP (5g/l) | Somatic embryos, plantlet regeneration | Yamaguchi *et al.*, 1987 |
| *C. japonica* × *C. granthamiana* | Mature cotyledons | MS + BA (3) | – | Somatic embryos | Nakamura, 1988 |
| *C. vietnamensis* × *C. chrysantha* | Mature cotyledons | MS + BA (3) + NAA (1) | MS + $GA_3$ (1) + CM (10%) | Somatic embryos, adventitious buds, plantlet regeneration | Nadamitsu *et al.*, 1986 |

[a] Quantities in brackets are expressed in mg/l unless otherwise stated.

Mineral media: A – Anderson (1984); $B_5$ – (Gamborg *et al.*, 1968); Knop – (Tabachnik and Kester, 1977); MS – Murashige and Skoog (1962); WPM – Woody Plant Medium (Lloyd and McCown, 1980).

Supplements: IAA – indole-3–acetic acid; IBA – 3–indolebutyric acid; NAA – naphtaleneacetic acid; 2,4–D – 2,4–Dichlorophenoxyacetic acid; BA – Benzyladenine; 2iP – iso-pentenylaminopurine; K – Kinetin; Z – Zeatin; ABA – Abscisic acid; $GA_3$ – Gibberellic acid; CM – Coconut milk; PVP – Polyvinylpyrrolidone; YE – Yeast Extract.

1982, for a study of *C. oleifera*) and somatic embryos (Table 1), either directly on cotyledon surface or indirectly via callus.

### 2.1. *C. sinensis*

Among *Camellia* species, *C. sinensis* has been extensively studied for the induction of somatic embryogenesis. The first successful report on the regeneration of tea plantlets from the cotyledon-derived calli was published by Wu and coworkers (Wu, 1976; Wu *et al.*, 1981). Furthermore, they did not clarify whether the observed process was via somatic embryogenesis or due to the differentiation of adventitious buds that subsequently formed roots. Although 2–8 mg/l 2,4-dichlorophenoxyacetic acid (2,4-D) was necessary for callus induction, organogenesis seldom occurred in the presence of this auxin. Indole-3-acetic acid (IAA) (1 mg/l) and kinetin (10 mg/l) played an important role in the formation of roots and shoots. However, for some varieties, it became essential to increase the concentration up to 4 and 20 mg/l, respectively.

Kato (1982) and Yan and Ping (1983) reported induction of somatic embryos on mature cotyledon sections cultured on MS medium (Murashige and Skoog, 1962) containing 6-benzyladenine (BA) and 3-indolebutyric acid (IBA). In a later report, Kato (1986a) clearly described the induction of somatic embryogenesis. Mature seeds of *C. sinensis* cv. "Yabukita" were sterilized and embryos excised. The cotyledons were cut in 7–10 pieces and these cotyledon slice explants were inoculated on MS medium supplemented with 0–10 mg/l BA, 0–2 mg/l IBA and 0–2 mg/l naphthaleneacetic acid (NAA). After two months, adventitious embryos were formed directly on swollen, callus-free surfaces of 17% of the explants, regardless of presence or absence of exogenous growth regulators in the medium. In the same report (Kato, 1986a), similar observations were noticed for *C. japonica*, although the response rate was 50%.

Direct differentiation of embryos on cotyledon pieces from immature seeds, cultured on MS medium lacking growth regulators, has been achieved for two South Indian tea clones (Abraham and Raman, 1986). In this study, indirect embryo generation was also reported; inoculation of cotyledon explants on a medium containing 2,4-D, casein hydrolisate and coconut milk, resulted in the production of an anthocyanin-rich callus that later on differentiated into embryos upon transfer on a basal medium supplemented with IBA and BA for 10–12 weeks. Mature tea embryos and cotyledon explants also produced a white compact callus when cultured in MS nutrient medium amended with 3 mg/l 2,4-D and 0.2 mg/l kinetin (Das and Barman, 1988, 1992). Calli were subcultured in a medium devoid of auxin and supplemented with 3–6 mg/l BA and 0.5 mg/l kinetin. In this second medium, spongy tissue was produced after eight weeks, some of which differentiated into somatic embryos. Shoots were differentiated from these embryos after 12 weeks.

In cotyledon cultures of cv. "Yabukita" (Fig. 1a), the same cultivar used

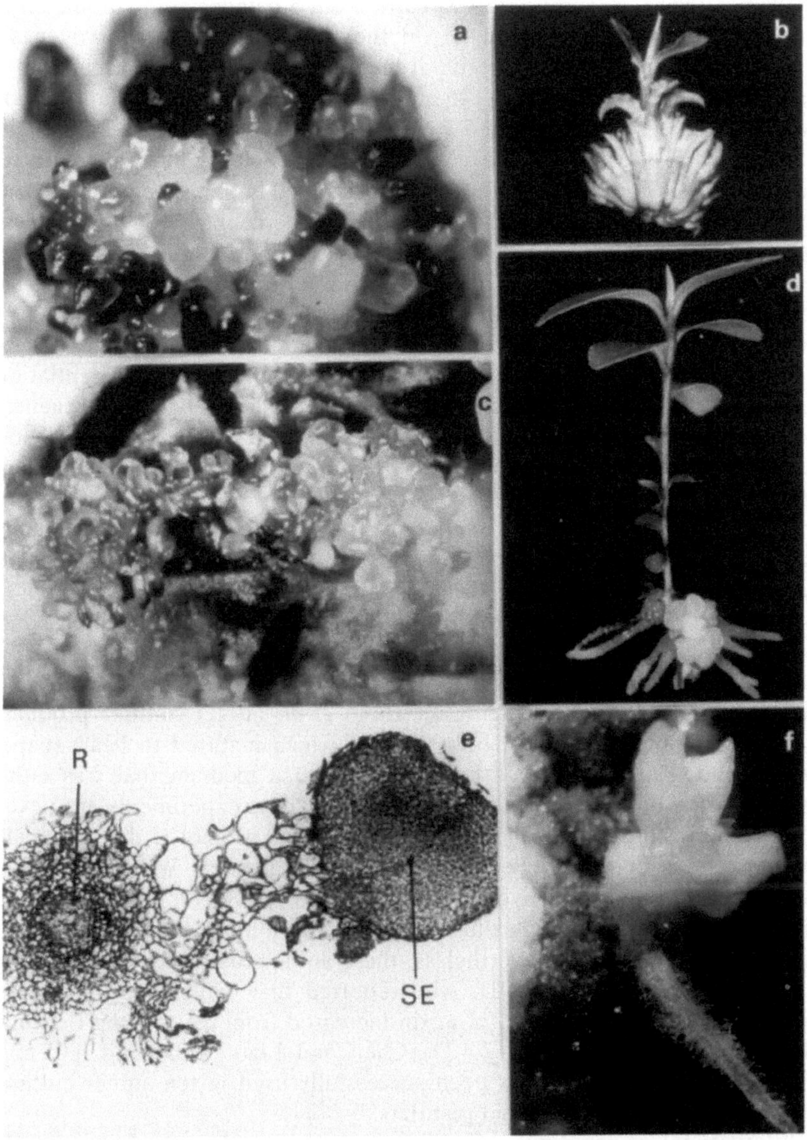


*Figure 1.* Initiation of somatic embryos in *Camellia.* (a) Embryo formation from cotyledon culture of *C. sinensis* in medium containing 3 mg/l BA. (b-c) Torpedo and cotyledonary stage *C. japonica* embryos originated directly on the hypocotyl of a embryonic axis explant (b) and on a cotyledon explant (c). X9. (d) Induction of somatic embryogenesis on the roots of *C. japonica* plantlets raised *in vitro*. (e) Transverse section of an adventitious root (R) bearing a globular somatic embryo (SE). Note the parenchymatic linkage between root and somatic embryo. X92. (f) Somatic embryos differentiated from callus from a leaf explant of *C. japonica* cultured in MS medium containing 8 mg/l BA + 0.5 mg/l IBA; one of them shows elongated radicle. X9. (a; Nakamura, 1988; d; Vieitez *et al.*, 1991).

by Kato (1986a), Nakamura (1988) carried out a detailed study on the effect of BA concentration (0–20 mg/l) on the rate of somatic embryogenesis, which ranged from 16% in the basal medium without growth regulators to greater than 30% when BA (1–5 mg/l) was added to the medium [almost twice the success rate achieved by Kato (1986a)]. Nakamura (1988) also observed that the size of the cotyledon slice explants markedly affected embryogenesis and that larger explants responded better. The optimum timing of cotyledon culture, which resulted in high percentage of embryogenesis (45–50%), was observed to be from late September to mid October when matured seeds (12–13 mm in diameter) were used in Shizuoka, Japan. The rate of embryogenesis fell to less than 15% when immature seeds (August) or stored mature seeds (5°C for 2–3 months) were used. Remarkable varietal differences were observed in the rates of somatic embryo differentiation, ranging from 0 to 50% in 13 different cultivars studied. High embryogenesis percentages were obtained with "Yabukita" and "Kurasawa" whereas, those of "Houryoku", "Da-Yeh Oolong", "Manipuri 9", "Chi-Shim Dah-Pan" and "Satumabeni" were less than 6%.

Unlike Nakamura (1988), Bano *et al.* (1991) achieved a higher somatic embryogenesis rate (70%) with immature cotyledons (3–8 mm size) as compared to 29% by culturing mature cotyledons (13 mm size) on MS medium containing 0.5 mg/l 2,4-D and 0.05 mg/l kinetin. The apparent discrepancy between these findings and Nakamura's may have been due to either different growth regulator combinations employed and/or intervarietal differences in embryogenic capacity. Furthermore, Bano *et al.* (1991) cultures produced masses of globular embryos, only 30% of which matured to heart-shaped and torpedo-shaped embryos upon transfer to a medium that was either hormone-free or contained 0.05 mg/l kinetin. These experiments are nevertheless interesting due to the fact that 2,4-D has promoted direct somatic embryogenesis on the cotyledons of a *Camellia* species. Wu *et al.* (1981), Abraham and Raman (1986) and Das and Barman (1988) reported that media containing 2,4-D promoted callus formation and inhibited the expression of somatic embryogenesis. Nevertheless, these authors used higher amounts of 2,4-D than Bano *et al.* (1991), who reported that explant response rates declined as the concentration of auxin increased from 0.5 to 4 mg/l. It may also be noted that 0.5 mg/l 2,4-D (Chen and Liao, 1983) and 0.05 mg/l (Shimokado *et al.*, 1986) has been successfully used in tea anther cultures for the regeneration of haploid plantlets.

Intervarietal or genotypic differences may also explain the failure of Jha *et al.* (1992) to achieve somatic embryo formation on cotyledonary explants of an elite Darjeeling tea clone (T-78) cultured on MS medium, either without growth regulators or containing 1–10 mg/l NAA, IBA or 2,4-D. Some 20% of explants produced embryos after 14–16 weeks of culture in a medium amended with 10 mg/l BA, but 1–5 mg/l BA (the optimal concentration for Nakamura, 1988) was ineffective to produce somatic embryos. Another possible explanation, however, is the physiological state of the plant material. Jha *et al.* (1992) used cotyledons from 6–8 week-old seedlings, whereas those used in the studies mentioned above had been obtained from ungerminated seeds. The mean number of somatic embryos per explant rose from 16.5 to 33.8 when the MS medium was supplemented with 80 mg/l adenine and 0.5 mg/l IBA in addition to 10 mg/l BA.

## 2.2. *C. japonica*

### 2.2.1. *Somatic embryogenesis from zygotic embryo explants*

Somatic embryogenesis in *C. japonica* has been achieved both indirectly via callus (Bennett and Scheibert, 1982) and directly on the cotyledon surface which is similar to *C. sinensis* (Kato, 1986a; Nakamura, 1988; Vieitez and Barciela, 1990). The basal medium used generally has been MS, similar to cultures of *C. sinensis*, though Bennett and Scheibert (1982) used Knop's mineral salts (Tabachnik and Kester, 1977). The two species differ, however, *C. japonica* appears to have a greater competence for embryogenesis than *C. sinensis*. In Kato's (1986a) experiments, some 50% of the explants produced somatic embryos with a mean of 31 embryos per embryogenic explant regardless of the combination of growth regulators (BA, NAA and IBA) in the basal medium; whereas Nakamura (1988) achieved response rates of 48–58% with 3 mg/l BA.

A rather comprehensive optimization of somatic embryogenesis of *C. japonica* was carried out by Vieitez and Barciela (1990), who cultured cotyledons and embryonic axes from both mature and immature zygotic embryos on MS medium alone or with various combinations of BA (1–4 mg/l) and IBA (0.5–2 mg/l) (Fig. 1b,c). Immature seeds collected during July were divided into two size classes (2–5 and 6–8 mm). The September and October embryos (12–14 mm long) were each divided into the following five explants: two distal semicotyledons, two proximal semicotyledons, and the embryonic axis, which was attached to the remainders of cotyledon tissue by the cotyledonary petioles. Embryogenesis rates showed no clear dependence on the concentration of the growth regulators, though high concentrations such as the 4 mg/l BA + 2 mg/l IBA combination used by Kato (1986a) were inhibitory. Successful embryogenesis was clearly dependent on the physiological maturity of the zygotic embryos and the type of initial explant used (embryonic axis or cotyledon segment). In seeds collected during July, the success rate was 25% for the 2–5 mm explants and 40% for the 6–8 mm explants. The number of somatic embryos per embryogenic explant ranged from 2 to 5. The embryogenic potential of July seeds is of particular interest in view of viability that interspecific crosses liable to abort. The success rate of embryonic axis explants from seeds collected in September was 94%, significantly higher than that of cotyledon explants; among the seeds collected during October, the embryogenesis rate was 20% for embryonic axes and not significantly different for cotyledon explants (Table 2). The superiority of embryonic axes over cotyledon explants for September seeds was indepen-

*Table 2.* Percentage of somatic embryogenesis in embryonic axis and cotyledon explants from *Camellia japonica* seeds collected in September and October. Results after two months culture in different combination of BA and IBA. Figures in brackets represent the number of somatic embryos per embryogenic explant (Vieitez and Barciela, 1990).

| Medium (mg/l) | September | | | | October | | | |
|---|---|---|---|---|---|---|---|---|
| | Embryonic axes | Proximal cotyledon | Distal cotyledon | Mean | Embryonic axes | Proximal cotyledon | Distal cotyledon | Mean |
| Basal medium | 100 (73.0) | 14 (1.0) | 14 (1.0) | 42.6[a] | 33 (9.0) | 30 (14.5) | 30 (18.8) | 31 |
| BA 1 | 100 (18.0) | 75 (13.1) | 75 (8.0) | 83.3 | 0 (0) | 46 (10.5) | 45 (5.2) | 30 |
| BA 2 | 100 (19.6) | 67 (14.2) | 25 (28.0) | 64 | 25 (4.5) | 50 (4.5) | 50 (6.8) | 42 |
| IBA 1 | 100 (64.6) | 71 (7.5) | 86 (8.0) | 85.6 | 29 (5.8) | 22 (10.5) | 0 (0) | 17[b] |
| IBA 2 | 100 (60.0) | 40 (20.4) | 40 (5.7) | 60 | 20 (3.5) | 30 (13.1) | 40 (4.5) | 30 |
| BA 0.5 + IBA 0.5 | 100 (40.3) | 43 (17.6) | 29 (5.7) | 57[a] | 33 (8.0) | 38 (7.3) | 13 (12.5) | 28[b] |
| BA 0.5 + IBA 1 | 100 (31.5) | 45 (9.0) | 36 (13.6) | 60 | 40 (5.8) | 56 (12.5) | 78 (11.2) | 58 |
| BA 1 + IBA 0.5 | 100 (56.7) | 45 (4.0) | 27 (6.2) | 57[a] | 0 (0) | 0 (0) | 40 (2.2) | 13[b] |
| BA 1 + IBA 1 | 57 (33.2) | 0 (0) | 21 (7.0) | 26[a] | 17 (1.0) | 13 (1.0) | 25 (1.0) | 18[b] |
| BA 4 + IBA 2 | 80 (46.8) | 43 (5.7) | 0 (0) | 41[a] | 0 (0) | 14 (3.5) | 29 (5.8) | 14[b] |
| Mean | 94 | 44.3 | 35.3 | 57.7 | 20 | 30 | 35 | 28 |

[a] Significant lower (P = 0.05) than with 1 mg/l IBA, by chi-square test.
[b] Significant lower (P = 0.05) than with 0.5 mg/l BA + 1 mg $l^{-1}$ IBA, by chi-square test.

dent of the growth regulators used, although the mean number of somatic embryos produced per embryonic axes was significantly inhibited by BA in the absence of IBA. Growth regulators in seeds collected in October had more influence on embryogenesis rates than did the type of explant used. Growth regulators did not influence the embryogenic frequency, but had a significant effect on the number of differentiated embryos per embryogenic explant. These results were subsequently confirmed by Jha *et al.* (1992) for a Darjeeling tea clone.

Most of the somatic embryos differentiated on embryonic axes developed on hypocotyl region, though some also emerged on the cotyledonary petioles (Vieitez and Barciela, 1990). The decline in the embryogenic capacity of embryonic axes observed between September and October could be attributed to the loss of competence associated with the maturation of the zygotic embryo. September was found to be the optimum time of the year for collecting *C. japonica* seeds for embryogenesis, whereas Nakamura (1988), who used only cotyledon explants, had found October to be the best month for *C. sinensis* seed collection. The inclusion of embryonic axis as an initial explant must be accounted for in defining this difference. Because of the high embryogenesis rate among embryonic axes during September, our procedure was more efficient than any of the cotyledon-based procedures developed for *C. sinensis* and *C. japonica* (Kato, 1986a; Nakamura, 1988; Bano *et al.*, 1991; Jha *et al.*, 1992).

Embryogenic capacity is conserved by *C. japonica* seedling cotyledons, but is largely lost by the axis. We studied (unpublished results) somatic embryogenesis on cotyledons, cotyledonary nodes, hypocotyls, basal root segments and 1st and 2nd shoot internodes from *in vitro* grown 1-month-old *C. japonica* seedlings. The explants were cultured for 8 weeks in MS mineral medium with or without 1 mg/l BA and 0.1 mg/l IBA. The highest embryogenesis rates (about 50%) were recorded for cotyledons; hypocotyl segments and cotyledonary nodes had 11% and 10% rates respectively, whereas root and internode segments failed to produce somatic embryos. The presence of growth regulators in the medium had little effect on embryogenesis rates when compared with differences among explant types. Somatic embryos formed directly on hypocotyl and cotyledonary node explants, but on cotyledon explants they differentiated from protuberances that partially covered the explants.

#### 2.2.2. *Somatic embryogenesis from adventitious roots*

Most of the embryogenic systems described in the *Camellia* genus have utilized mature or immature seeds as an initial material (Table 1), but expansion to a wider range of older explants would be desirable. We have also investigated somatic embryogenesis from explants derived from plantlets regenerated from *in vitro* shoot multiplication cultures of three *C. japonica* clones, two of them of juvenile origin and one derived from an adult specimen of *C. japonica* "Alba Plena" (Vieitez *et al.*, 1991). For the regeneration of whole plantlets, 2–3 cm shoots were excised from multiplication cultures for rooting by dipping their basal ends in 1 g/l IBA for 15 min and subsequently cultured in a basal medium either MS or Woody Plant Medium (WPM; Lloyd and McCown, 1980) with half-strength macronutrients and no growth regulators (Samartín *et al.*, 1986; Vieitez *et al.*, 1989a,b). After 1 month in a rooting medium, rooted shoots were transferred to embryogenic media consisting of basal medium alone or with various combinations of zeatin (1 mg/l), BA (1–2 mg/l) and IBA (0.1–1.0 mg/l). Eight weeks in these media did not induce embryogenesis in the plantlets of mature "Alba Plena" origin, but did induce the direct formation of yellowish, white somatic embryos on the roots of plantlets of the two clones of juvenile origin (Fig. 1d). The overall embryogenesis rate was 26% for one of these clones and 10% for the other clone. As in the case of explants derived from zygotic embryos, the growth regulators used apparently did not influence embryogenesis, which also occurred in their absence. Embryos differentiated either singly or in groups of maximum up to seven. Their subsequent development was not synchronized, but most progressed to the cotyledonary stage after 8 weeks in embryogenic culture medium. Somatic embryos appeared to differentiate from epidermic cells and to be connected to the root via a few parenchymatic cells making them easy to remove. Vascular connections between root and embryo were not observed (Fig. 1e).

### 2.2.3. *Somatic embryogenesis from leaf explants*

The embryogenic competence of *C. japonica* leaves has also been investigated. The leaves of shoot cultures are an abundant source of tissue and also have been used for genetic transformation of several plant species (Klee *et al.*, 1987). We have found that leaf cultures of *C. japonica*, like those of woody species of genera such as *Populus* (Park and Son, 1988) or *Rubus* (Fiola *et al.*, 1990), can develop both somatic embryos and adventitious buds (San-José and Vieitez, 1993). In these experiments, 3–4 fully expanded apical leaves from stock axillary shoot cultures, originally derived from a juvenile *C. japonica* plant, were utilized. These leaves were cut perpendicularly to the midvein and both apical and basal halves were cultured with the abaxial side down on MS or WPM medium, both supplemented with BA (4.4–35.5 μM), IBA (0.5–9.8 μM) or NAA (2.6–5.3 μM). Light inhibited both callus formation and morphogenesis, but somatic embryogenesis and/or adventitious bud formation occurred on explants grown in the dark for at least 14 weeks in the presence of BA and an auxin (IBA or NAA). The highest differentiation rate of 33% was achieved with leaf sections cultured in the dark for 20 weeks on MS medium containing 35.5 μM BA + 2.5 μM IBA (Fig. 1f). Two of the seven treatments that induced somatic embryogenesis simultaneously induced adventitious bud formation, although the differentiation rate in these cases was only 7%. Bud and embryo formation occurred predominantly on the distal region of the basal sections and the proximal region of apical sections on callus developed from the cut surface of the explant. The fact that differentiation was indirect, via callus, may explain why it required the presence of auxin and a high concentration of cytokinin (Sharp *et al.*, 1980), whereas the direct somatic embryogenesis discussed previously was only slightly influenced by growth regulators and occurred on the basal medium. These experiments concluded with the successful multiplication of the somatic embryos and the development of adventitious buds, and the regeneration of plantlets from the multiplicated cultures (San-José and Vieitez, 1993). Hence, this system could allow the application of *in vitro* selection, mutant isolation and genetic transformation techniques for *Camellia*.

### 2.3. *C. reticulata*

Zhuang and Liang (1985a) reported the formation of somatic embryos directly from the surface of *C. reticulata* cotyledon segments. After 35–45 days of culture in the MS medium containing 1 mg/l BA and 0.2 mg/l NAA, the average embryogenesis rate was 43% in all the four cultivars studied.

Plata and Vieitez (1990) achieved similar direct embryogenesis results with cotyledon sections and embryonic axes from *C. reticulata* "Mouchang" seeds, collected during September (Fig. 2a–c), a month before complete maturity. The embryogenesis rate was 31% on MS medium without growth regulators, and reached a maximum of 51% with 0.5 mg/l IBA. In general, there was a little difference between the rates obtained with the sixteen combinations of BA, IBA and NAA tried, although the success rate was increased by 0.5–1.0 mg/l IBA and decreased by 0.2–1.0 mg/l NAA in the culture medium. In the cv. "Early Crinsom", Zhuang and Liang (1985a) reported that the embryogenic medium was amended with 1 mg/l BA and 0.2 mg/l NAA, but they did not consider the effects of IBA on embryogenesis.

Plata (1993) found that 0.5–1.0 mg/l 2,4-D, with or without 0.5 mg/l BA, almost totally inhibited embryogenesis on cotyledon slices though the cut surfaces of the cotyledons developed a compact white callus, the subsequent subculture of the 2,4-D treated explants for 6 weeks in basal medium with or without 0.5 mg/l BA failed to induce the differentiation of somatic embryos. This behaviour by *C. reticulata* contrasted with that of *C. sinensis* cotyledon cultures, which likewise produced callus, but not somatic embryos in the presence of 2,4-D, and instead differentiated embryos from the callus after transfer to a medium devoid of 2,4-D (Abraham and Raman, 1986; Das and Barman, 1988).

The embryogenic competence of *C. reticulata* seeds, like that of *C. japonica* seeds (Vieitez and Barciela, 1990), depends on the time of year at which the seeds are collected. An experiment was conducted to test the significance of timing in relation to the type of explant cultured in a basal medium with or without 0.5 mg/l IBA (Plata, 1993). Embryogenesis rates were zero for July seeds and maximum for September seeds. For August and September seeds, IBA significantly increased both the embryogenesis percentage and the

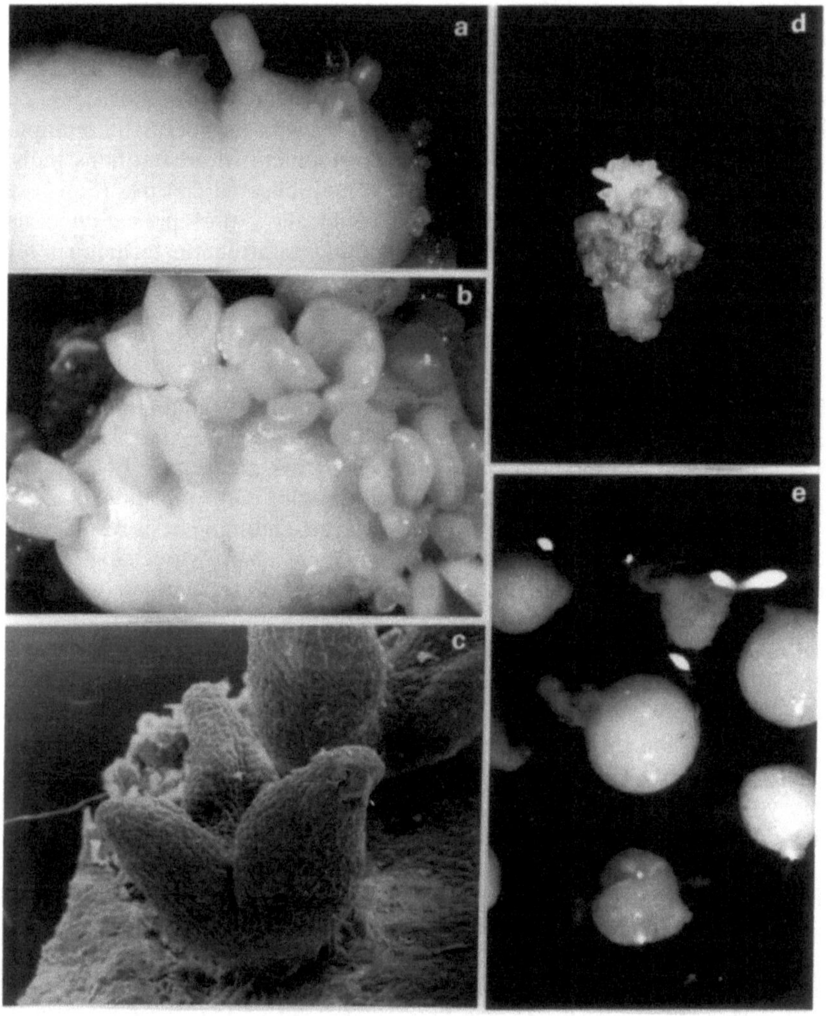


*Figure 2.* Initiation of somatic embryogenesis in *C. reticulata*. (a,b) Different stages of somatic embryos differentiated on a swollen tissue formed on the cotyledon explant after 4 (a) and 8 (b) weeks of culture. X10. (c) Scanning electron micrograph showing an embryo with three cotyledons. X100. (d) Indirect embryogenesis on cotyledon callus subcultured in 0.5 mg/l BA + 1 mg/l IBA. (e) Globular and heart-stage somatic embryos differentiated in embryogenic suspension cultures with 0.5 mg/l BA + 1 mg/l IBA. X10. (c; Plata and Vieitez, 1990, reproduced by permission of the Editor J. Hort. Sci.).

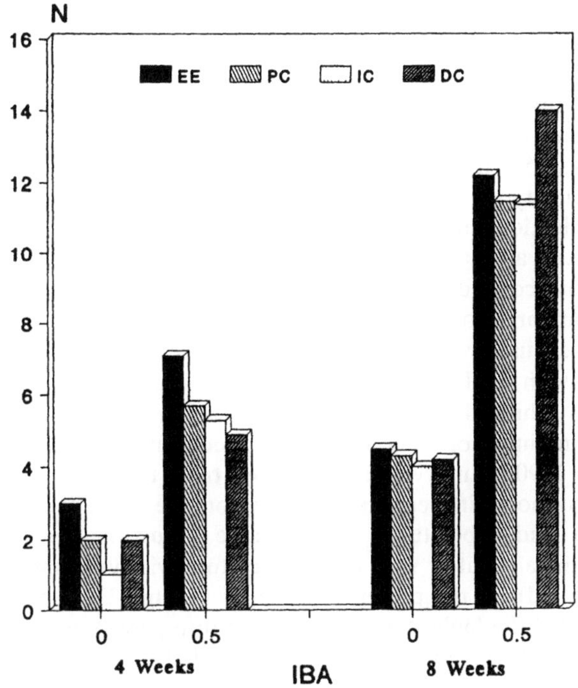


*Figure 3.* Mean numbers of somatic embryos per explant (N) recorded after 4 and 8 weeks' culture of various kinds of explant from *C. reticulata* seeds in basal medium with or without 0.5 mg/l IBA; seeds were collected in September. The explant types were embryonic axis (EE) and proximal (PC), intermediate (IC) and distal (DC) cotyledon segments (Courtesy E. Plata).

*Table 3.* Indirect somatic embryogenesis and callus growth after 3 successive subcultures using callus sections (125–150 mg) derived from initial cotyledon explants of *C. reticulata*. The frequency of embryogenesis (%) and the mean number of somatic embryos (N) was recorded after six weeks' culture using 8 callus explants per treatment and 3 replicates (Plata and Vieitez, 1990, reproduced by permission of the Editor, J. Hort. Sci.).

| Treatment (mg/l) | | | Embryogenesis | | Callus growth[a] |
|---|---|---|---|---|---|
| BA | IBA | NAA | % | N | |
| – | – | – | 18 | 11.8 | + |
| 1 | – | – | 24 | 4 | + |
| 0.5 | 1 | – | 41 | 19 | + + + |
| 1 | 0.5 | – | 63 | 21.3 | + + |
| – | 1 | – | 36 | 18 | + + + |
| – | – | 1 | 23 | 2.9 | + |
| 1 | – | 0.2 | 26 | 1.9 | + |

[a] + scant, + + moderate, + + + high.

number of embryos per explant, without regard to the types of explant used (embryonic axis and proximal, intermediate and distal cotyledon segments). In fact, embryo production in September on IBA medium was triple that obtained on auxin-free medium (Fig. 3). The most responsive explant was embryonic axes in IBA medium after 4 weeks and distal cotyledon sections after 8 weeks, but these differences in embryogenic capacity between the various explant types were not significant in comparison with the effects of IBA and collection time. The difference in embryogenic capacity between the abaxial and adaxial cotyledon surfaces was more significant important; 87% of reactive cotyledon sections developed somatic embryos on the abaxial surface as compared to only 27% developing embryos on the adaxial surface.

Although callus developed on cotyledon slices was not embryogenic in experiments with 2,4-D medium (Plata, 1993), indirect somatic embryogenesis has been achieved via callus developed on the cut surfaces of cotyledon explants maintained for 3 months in the media as listed in Table 3 (Plata and Vieitez, 1990). Embryos differentiated from the callus were similar in appearance to those differentiated directly from the cotyledon surface. Callus was isolated and subcultured in the same induction medium; embryos emerged from callus after four weeks in culture without transfer to another medium (Fig. 2d). In the presence of 0.5 or 1.0 mg/l IBA, callus grew well and also yielded the highest embryogenic culture percentage and number of embryos. Like direct embryogenesis on cotyledon explants, indirect embryogenesis was inhibited by NAA.

The development of embryogenic callus allows the establishment of suspension cultures which can be served as a source of embryogenic cells or protoplasts for genetic engineering. Plata (1993) investigated the embryogenic capacity of suspension cultures initiated from the fragments of embryogenic callus obtained from *C. reticulata* cotyledon explants cultured in half-strength MS medium containing 0.5 mg/l BA and 1 mg/l IBA. After 4 weeks, cell aggregates 1 mm in diameter and globular, heart- and torpedo-shaped somatic embryos were formed (Fig. 2e). Embryos in the suspension cultures had an absolute requirement for IBA for differentiation and development.

The morphogenetic capacity of leaves from axillary shoot cultures of two juvenile clones of the cv. "Mouchang" was also investigated, but none of the treatments used with *C. japonica* leaves induced somatic embryogenesis in the *C. reticulata* material. Only one of the two clones showed adventitious bud differentiation (San-José and Vieitez, 1993). These findings confirm that *C. reticulata* is less embryogenic than *C. japonica* and is dependent on the type of explant used (leaves or embryonic axes).

### 2.4. *Other species*

The following experiments on cotyledon culture for somatic embryogenesis have also been carried out with other species: *C. brevistela* (Nakamura,

1988), *C. chrysantha* (Zhuang and Liang, 1985b), *C. oleifera* (Yan *et al.*, 1984) and *C. sasanqua* (Nakamura, 1988; Zhuang *et al.*, 1988), and with new hybrids whose development is hindered by poor fertility or embryo viability (Nadamitsu *et al.*, 1986; Yamaguchi *et al.*, 1987). In all these reports, somatic embryogenesis was achieved using MS medium containing a cytokinin (usually BA), with or without auxin (usually NAA) (Table 1). However, none of these studies involved comprehensive experiments to determine optimum conditions for embryogenesis.

### 2.5. *Androgenic embryos*

Haploid tea plants derived from anther cultures were obtained by Chen and Liao (1983). Furthermore, they mentioned the differentiation of buds on a shiny callus proliferated from microspores and the shoots were then rooted. However, the differentiation of true pollen embryos and regeneration of haploid plants was described by Raina and Iyer (1974, 1983) and Shimokado *et al.* (1986), who established the importance of haploid plants for the production of homozygous plants in tea breeding.

Bearing in mind that this chapter deals with somatic embryogenesis, we do not intend to further discuss haploid embryo regeneration, since they are androgenic embryos, formed by the male gametophyte (microspores or pollen grains) and, as such, do not correspond to the category of somatic embryos formed by the sporophytic cells (Bhojwani and Razdan, 1983). For more details of tea haploid embryo differentiation, see Kato (1989a) and Chen and Liao (1990).

## 3. Embryo development

In most of the embryogenic systems reported in *Camellia*, both directly or indirectly induced, the development and maturation of embryos occurred in the induction medium. Unlike those of many other species (Tulecke, 1987; Thorpe, 1988), the development of *Camellia* somatic embryos did not require two phase culture procedures; they passed through typical stages of development culminating in the cotyledonary stage on the induction medium, without any further subculture (Nakamura, 1988; Vieitez and Barciela, 1990; Plata and Vieitez, 1990). An exception to this generalization occurred when Bano *et al.* (1991) induced the formation of pro-embryonic cell masses in a medium containing 0.5 mg/l 2,4-D plus 0.05 mg/l kinetin. Thirty percent of embryos matured heart or torpedo stage on a 2,4-D free medium.

The time taken by somatic embryos of *C. sinensis* to develop on explants has been 8–10 weeks (Abraham and Raman, 1986; Kato, 1986a) or 14–16 weeks (Jha *et al.*, 1992). Somatic embryos of *C. japonica* (Vieitez and Barciela, 1990) and *C. reticulata* (Plata and Vieitez, 1990) began to emerge from cotyledon cultures after 4–5 weeks (Fig. 2a) and much earlier from embryonic axis explants. Indirect somatic embryogenesis is much slower; embryogenic callus takes 12–14 weeks to develop on *C. reticulata* cotyledons (Plata and Vieitez, 1990) and 14–16 weeks to develop on *C. japonica* leaf explants (San-José and Vieitez, 1993).

### 3.1. *Origin of the embryos*

The histological and anatomical aspects of somatic embryogenesis in *Camellia* have been mentioned by Kato (1986a), Plata and Vieitez (1990), Vieitez *et al.* (1991), and Bano *et al.* (1991) without giving details on the cellular events and anatomical changes that occurred during embryogenesis. Barciela and Vieitez (1993) made a detailed study on the origin and anatomical development of somatic embryos differentiated on *C. japonica* cotyledon sections. This study used computer-aided image analysis for cytological quantification, and the measurements of cell starch and protein contents by staining cell areas with periodic acid Schiff (PAS) stain and by mercuric bromophenol blue, respectively.

Barciela and Vieitez (1993) observed morphologically that small protuberances or nodules began to appear on the abaxial epidermis of the cotyledons after seven days in culture. The nodules continued to develop (Fig. 4a), and by day 30 they were 4–6 mm in diameter became moderately prominent, and bore embryos in several different stages of development (Fig. 2a). After 2 months of culture, embryos were 6–8 mm long and could be isolated either for germination or secondary embryogenesis. Only the abaxial surface of the cotyledon was morphogenetically competent; somatic embryos only formed on this side regardless of whether the explant had been placed abaxial or adaxial side down on the culture medium. To determine whether the embryogenic nodules could be maintained indefinitely in cultures, they were isolated from the initial cotyledons, removed from *in vitro* produced somatic embryos and then cultured for 6 months with monthly transfer to a fresh medium. The parenchymatic tissue of the nodule failed to proliferate or grow and turned progressively necrotic.

At day 0, histological sections of cotyledon explants (Fig. 4b) showed a single-layer epidermis and the abaxial subepidermis. The cotyledonary parenchyma, which was formed of cells that were longer than the subepidermal cells, had a nucleus-to-cell area ratio (N/C) of only 1.6% (Table 4). Vascular bundles traversed the parenchyma. There were clear differences between abaxial and adaxial subepidermal cells, the latter being longer and arranged more regularly, as in the palisade layer of the true leaf. Furthermore, storage proteins were detected in the subepidermal layer on the adaxial side but not on the abaxial side. These differences between the two subepidermal layers suggest that the abaxial subepidermis had not differentiated to the same degree as the adaxial subepidermis. Therefore, the morphogenetic competence of cotyledon explants may thus depend on the relatively unspecialized nature of abaxial tissues.

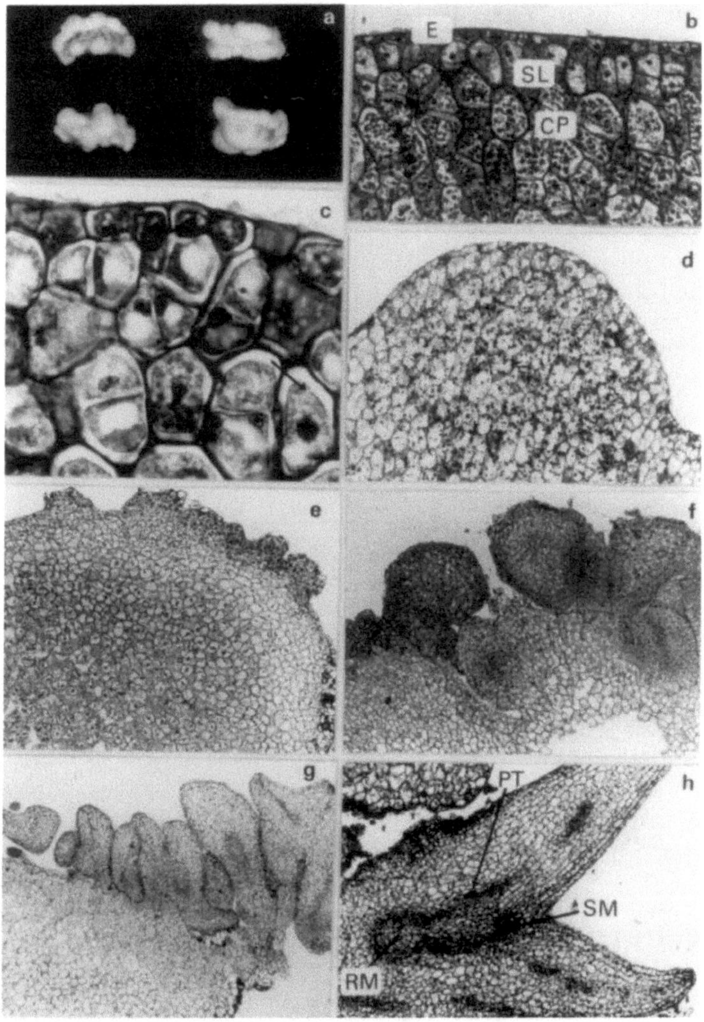


*Figure 4.* Somatic embryo development in *C. japonica*. (a) Formation of protuberances or nodules on cotyledon explants after 18 days culture. (b) Transverse section of a cotyledon prior to culture showing cotyledonary parenchyma (CP), subepidermal (SL) and epidermal (E) layers in abaxial surface. X295. (c) A 3 days cultured cotyledon showing dedifferentiated cells and mitotic activity in subepidermal layers (arrow). Note that cell divisions occur in all spatial planes. X590. (d) Formation of a protuberance in the cotyledon abaxial surface after 7 days of culture, showing accumulation of starch grains in the nodule parenchyma. X118. (e) Differentiation of preembryogenic cells in superficial cell layers of the nodule. X92. (f) Multicellular proembryonal complexes. X92. (g) Heart-shaped and torpedo-shaped somatic embryos differentiated by day 24. X48. (h) Longitudinal section of a cotyledonary somatic embryo at day 30, showing shoot (SM) and root (RM) meristems and procambial strands (PT). X92. (b,f; Barciela and Vieitez, 1993).

*Table 4.* Cell area (C), nuclear area (N) and nucleus-to-cell area ratio (N/C), expressed as a percentage, in cell types located in initial cotyledon explants or differentiated during the embryogenic process of *C. japonica*. C and N values are means ± SE (Barciela and Vieitez, 1993).

| Parameters | Initial cotyledon | | | Embryogenic process | | | |
|---|---|---|---|---|---|---|---|
| | Epidermal cells | Subepidermal cells | Parenchymatous cells | Nodule parenchyma | Deep nodule layers | Preembryogenic cells | Embryogenic cells |
| C ($\mu m^2$) | 166.3 ± 44.1 | 510.1 ± 120.2 | 1380.7 ± 386.5 | 398.5 ± 111.7 | 4170.1 ± 1651.1 | 215.2 ± 62.9 | 126.6 ± 31.6 |
| N ($\mu m^2$) | 25.1 ± 4.1 | 22.8 ± 4.3 | 22.6 ± 5.6 | 49.6 ± 10.6 | 32.2 ± 9.7 | 46.5 ± 9.5 | 53.9 ± 10.5 |
| N/C (%) | 15.1 | 4.5 | 1.6 | 12.5 | 0.7 | 21.6 | 42.6 |

After 1–2 days of culture, cell dedifferentiation began in the abaxial epidermis and subepidermis. A number of cells appeared which were more isodiametric and had a relatively less condensed nuclei and more visible nucleoli than the cells of the initial cotyledon. By day 3, cell dedifferentiation was more evident, and the onset of cell division was observed in cells of the abaxial region (Fig. 4c). There was a clear difference in cellular activity between the abaxial and adaxial regions during the first few days of culture; on the abaxial side, the mitotic activity was intense with different division planes, whereas on the adaxial side the mitosis rate was much lower and only periclinal divisions occurred. After 7–9 days, the proliferation in the abaxial region had given rise to protuberances, or lenticular nodules, consisting of thin-walled vacuolated parenchyma cells (Fig. 4d) that were about the same size as initial subepidermal cells but with nuclei more than twice as large (N/C = 12.5%; Table 4). During this time the explants underwent a rapid decline in protein body content, which was high in the original cotyledon but had almost disappeared from the surface regions by day 7. On the contrary, starch content, which is a minor component of the initial cotyledon, increased especially in the abaxial layers. In particular, the PAS-positive proportion of nodular regions during days 7–9 was about four times that of equivalent regions of initial cotyledons (Fig. 4d).

After 12–15 days, successive cell divisions at the surface of the nodular protuberances resulted in the formation of embryogenic precursor cells (Fig. 4e), which were smaller, less vacuolated, had densely stained cytoplasm and a N/C ratio of 21.6%. Continued division of these preembryogenic cells led to further dedifferentiation into true embryogenic cells by day 18, which appeared as groups of isodiametric, microvacuolated, starch-containing cells with protein-rich cytoplasm and large nuclei (N/C = 42.6%). The final value of the N/C ratio, which increased progressively throughout these early stages of embryogenesis, may reflect the degree of cell dedifferentiation to reach the embryogenic level (Margara, 1982). The relative nuclear size also increased during dedifferentiation leading to adventitious bud formation (Flinn *et al.*, 1989).

By day 21, few-celled proembryos compact multicellular proembryonal complexes (Fig. 4f) and globular somatic embryos, could be distinguished. Most somatic embryos have apparent multicellular origin arising from the proembryonal cell complex, but the presence of few-celled proembryos within a thick common wall suggests that the origin of some embryos may be unicellular. Starch deposits declined in the nodule parenchyma underlying embryogenic areas, reaching a minimum upon the development of the somatic embryos. Protein bodies were not observed in nodule tissue, but embryogenic cells and early embryos stained densely with mercuric bromophenol blue for total proteins.

In other species, the formation of embryogenic areas is preceded by heavy deposition of starch in the surrounding cells (El Maâtaoui *et al.*, 1990; Raj Bhansali, 1990). The decline in starch content as embryogenesis progresses suggests that starch metabolism fulfils the intense energy requirements of dividing cells (Stamp, 1987). The decrease of protein reserves during the first days of culture may also reflect a metabolic shift associated with the intense cell activity in the abaxial tissues of the cotyledon. Similarly, a rapid decrease in the storage protein content coincides with the onset of caulogenic determination in conifer cotyledon cultures (Patel and Thorpe, 1984; Flinn *et al.*, 1989).

Between days 24 and 27, heart-shaped, torpedo-shaped, and early cotyledonary embryos were present (Fig 4g). By day 30, many embryos had reached the full cotyledonary stage, with shoot and root meristems and differentiation of procambial tissue (Fig. 4h). However, embryos developed asynchronously at all stages of development were observed in successive days. Vascular connections between embryos and subjacent tissues were not observed.

The above histological analysis suggests that the nodules associated with the occurence of somatic embryogenesis can be considered as a small localized callus tissue, which is necessary for the redetermination of embryogenic cells (Sharp *et al.*, 1980). Histological observations of embryogenesis in *C. sinensis* (Fig. 5a) (Kato, 1986a; Nakamura, 1988) and *C. reticulata* (Plata and Vieitez, 1990) suggested that somatic embryos develop directly from cotyledons without any apparent callus phase. However, the differentiation of the embryos was nevertheless related to the developing swollen parts (Kato, 1986a), swollen whitish areas (Plata and Vieitez, 1990) or compact bulging tissue of cotyledon explants (Bano *et al.*, 1991). Such swellings might be equivalent to the nodules observed in *C. japonica* (Vieitez and Barciela, 1990; Barciela and Vieitez, 1993). Be that as it may, embryogenesis should be regarded as indirect only if a true callus phenotype is formed (Wann, 1988). Though morphologically visible, the nodules that develop on *C. japonica* cotyledons cannot be ascribed as a true callus which is able to be subcultured. A limited proliferation before the differentiation of somatic embryos in *Theobroma cacao* was considered to be an intermediate process between direct and indirect embryogenesis (Duhem *et al.*, 1989).

### 3.2. *Embryo morphology*

The morphology of the *Camellia* somatic embryos is influenced by the concentration of cytokinin in the medium. In case of *C. japonica* (Vieitez and Barciela, 1990; Vieitez *et al.*, 1991), most embryos could be classified into the following two clearly distinct types: i) "seed-like" embryos, which were yellowish-white with large cotyledons similar to mature zygotic *Camellia* embryos (Fig. 5b); ii) "bud-like" embryos, which were green with cotyledons resembling true leaves. They generally developed in media with a relatively high BA concentration (Fig. 5c). Ammirato (1985) stated that reasonably high levels of cytokinins partially or totally inhibit the development of somatic embryo cotyledons, and that the shoot apex grows out to form the first

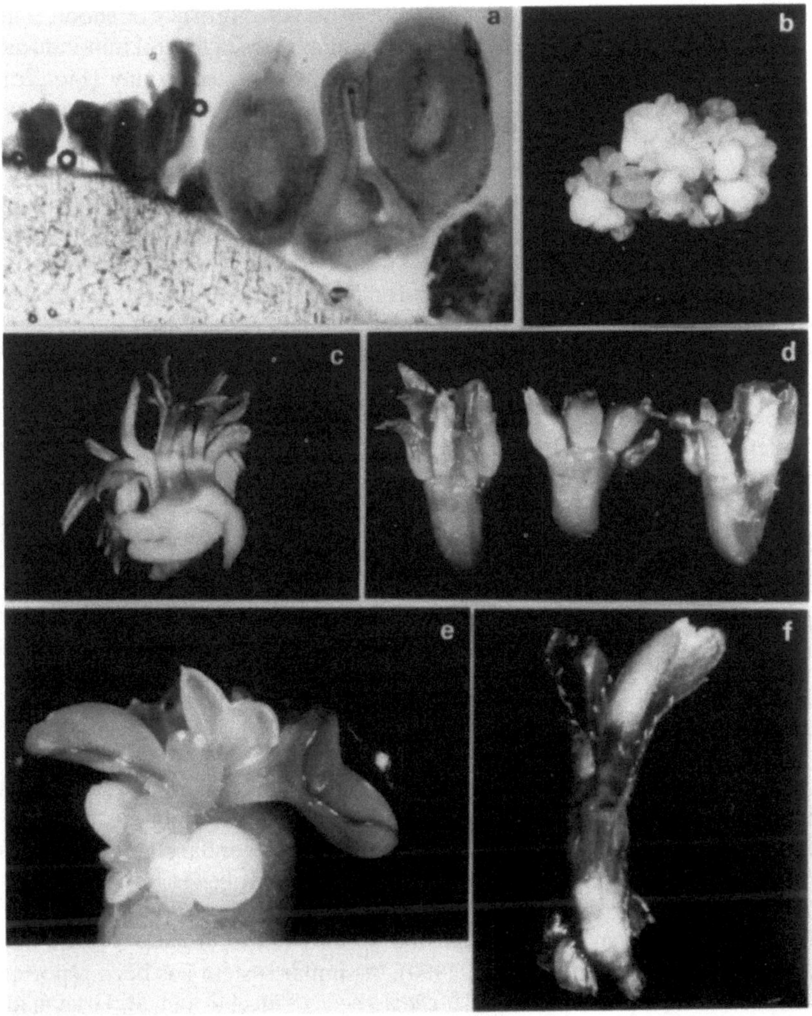


*Figure 5.* (a) Somatic embryos at different stages differentiated in *C. sinensis* cotyledon. Note the similarity to somatic embryos development in *C. japonica* (Fig. 4g). (b-d) Morphological types of *C. japonica* somatic embryos. Clusters of seed-like (b) and bud-like (c) somatic embryos. (d) Fasciated embryos showing fusion at the hypocotyl/root level. X9. (e-f) Secondary embryogenesis. (e) Clusters of secondary embryos on the hypocotyl surface of a primary somatic embryo. X9. (f) Typical well-developed bud-like embryo with secondary embryos located in a "crown like" position in the hypocotyl region. X9..(a; courtesy Y. Nakamura).

mature leaves, so that the somatic embryo looks more like a shoot. The observed "bud-like" embryos may be an example of such cytokinin-induced premature shoot emergence. Anomalies such as polycotyledony (Fig. 2c), cotyledon hypertrophy or fasciation (Fig. 5d) were observed to various extents among both seed-like and bud-like embryos, but both kinds were genuinely bipolar having both shoot and root meristems.

Jha *et al.* (1992) identified three major distinct types of *C. sinensis* somatic embryos: i) white "seed-like" embryos, similar to the seed-like *C. japonica* embryos (Vieitez and Barciela, 1990). Two cotyledons were usually produced and sometimes the cotyledons were partially or completely fused. The seed-like embryos generally failed to produce whole plants; ii) Green "cup-shaped" embryos, which rarely exhibited bipolar development; and iii) "Globular" embryos, which followed a normal development pattern showing heart, torpedo and cotyledonary-shaped stages and formed normal rooted plantlets. However, the formation of these different types of embryos were not related to the growth regulator in the media (Jha *et al.*, 1992).

## 4. Culture maintenance

There are two different kinds of developmental patterns in the maintenance of embryogenesis in *Camellia*. One is the so-called repetitive somatic embryogenesis. In the other system, there is a clear callus phase before embryo formation, and the multiplication of somatic embryos is based on continuous growth of the embryogenic callus.

When somatic embryo differentiation occurs directly from the explant without an intervening embryogenic callus, then the embryogenic competence can be maintained through repetitive or secondary embryogenesis (Wann, 1988). The phenomenon of repetitive embryogenesis by direct production of new embryos on the surface of previously differentiated embryos (Fig. 5e) is quite common in *Camellia* species (Kato, 1986a; Vieitez and Barciela, 1990; Plata and Vieitez, 1990). A similar system has been reported in other woody species such as *Juglans regia* (Tulecke and McGranahan, 1985), *Robinia pseudoacacia* (Merkle and Wiecko, 1989) and *Eucalyptus citriodora* (Muralidharan *et al.*, 1989). However, detailed reports on *Camellia* secondary embryogenesis and the consequent long-term maintenance of the embryogenic capacity are limited.

To induce the differentiation of secondary embryos of *C. sinensis*, somatic embryos can be subcultured on the same fresh medium used for the induction of embryogenesis on cotyledon cultures (Abraham and Raman, 1986). Secondary embryos form on the hypocotyl of primary embryos subcultured in a medium containing 1 mg/l gibberellic acid ($GA_3$), and furthermore, the embryogenic competence can be maintained through successive generations of embryos (Kato, 1986a). However, the best results on secondary embryogenesis in a Darjeeling *C. sinensis* clone were obtained by culturing

*Table 5.* Secondary embryogenesis of bud-like and seed-like somatic embryos of *C. japonica* cultured on various media for 8 weeks (Vieitez *et al.* 1991).

| Medium (mg/l) | Bud-like type | | | | Seed-like type | | | |
|---|---|---|---|---|---|---|---|---|
| | Forming embryos (%) | $N^{z,y}$ | Embryos formed$^{x}$ (%) | | Forming embryos (%) | $N^{z,y}$ | Embryos formed$^{x}$ (%) | |
| | | | Bud-like | Seed-like | | | Bud-like | Seed-like |
| Basal medium | 83 | 45.5 a | 8.9 | 91.1 | 85 | 26.6 ac | 4.9 | 95.2 |
| BA (1) | 71 | 19.0 cd | 46.3 | 53.7 | 60 | 16.3 b | 9.9 | 90.1 |
| BA (2) | 75 | 28.2 bcd | 74.4 | 25.6 | 53 | 10.3 bd | 15.1 | 84.9 |
| BA (1) + IBA (0.1) | 62 | 27.1 bcd | 61.4 | 38.6 | 67 | 30.9 a | 23.4 | 76.6 |
| BA (1) + IBA (1) | 84 | 37.5 ab | 58.2 | 41.8 | 68 | 29.7 ac | 24.8 | 75.2 |
| BA (2) + IBA (0.1) | 43 | 15.7 c | 29.8 | 70.2 | 45 | 12.2 bd | 15.4 | 84.6 |
| BA (2) + IBA (1) | 43 | 34.9 abc | 46.4 | 53.6 | 28 | 5.9 d | 11.5 | 88.5 |
| IBA (0.1) | 83 | 51.1 a | 4.5 | 95.5 | 82 | 32.9 a | 3.0 | 97.0 |
| IBA (1) | 72 | 38.7 ab | 2.7 | 97.3 | 75 | 20.1 bc | 2.7 | 97.3 |

[z] Mean number of bud-like and seed-like secondary embryos per responding explant.
[y] Mean separation whithin columns by Duncan's multiple range test, $P = 0.05$.
[x] Percentage of embryos formed according to morphological type.

primary embryos in a medium containing BA (3 mg/l) + IAA (0.1 mg/l) + $GA_3$ (5 mg/l) (Jha *et al.*, 1992). On this medium, embryogenic competence was maintained for over 30 months, although the productivity of the secondary embryos was not estimated. Bano *et al.* (1991) maintained embryogenic *C. sinensis* cultures by a variant of the recursive secondary embryogenesis method in which clumps of globular embryo masses were subcultured (on the induction medium). Embryogenic capacity was maintained for 12 months, with an average of 30–35 somatic embryos per gram of a 60-day-old culture.

Somatic embryos of *C. japonica* exhibit great capacity for secondary embryogenesis regardless of whether they are derived from cotyledon or embryonic axis cultures (Kato, 1986a; Vieitez and Barciela, 1990), from roots (Vieitez *et al.*, 1991) or from leaves of plantlets raised *in vitro* (San-José and Vieitez, 1993). When somatic embryos, produced by any of these explants, are excised and transferred to MS medium with or without growth regulators, direct secondary embryogenesis occurs within 3–4 weeks. The secondary embryos are mainly located in a "crown-like" position on the hypocotyl region of the primary embryo (Fig. 5f), although some also appear on the cotyledons. The embryogenic capacity of isolated secondary embryos is highest if they are subcultured when they are 7–9 mm long, and it declines when embryos are 3–4 mm long. Clusters of these small immature embryos can successfully be subcultured. The two morphological types of embryos (seed-like and bud-like) are distinct among primary and secondary embryos. Repetitive embryogenesis was evaluated with different growth regulators in the media for both embryo types (Table 5) (Vieitez *et al.*, 1991). Bud-like embryos produced significantly more secondary embryos than did seed-like embryos, and they had a greater germination capacity. The medium significantly influenced both the secondary embryogenesis percentage and the number of secondary embryos per responsive embryo, independent of the embryo type. The best results were achieved with a basal medium or a basal medium supplemented with 0.1 mg/l IBA. For maintenance of the multiplication system, we adopted a medium containing 1 mg/l BA + 0.1 mg/l IBA, which favoured development of the more productive and more readily germinated bud-like embryos. Medium containing 2 mg/l BA afforded more embryos, but it also favoured fasciation. The embryogenic capacity has been maintained for more than 3 years by subculturing secondary embryos at 7–8 week interval.

Isolated somatic embryos of *C. reticulata* were also able to generate adventive embryos, both on their cotyledons and on their hypocotyl region, especially the latter. Like *C. japonica*, *C. reticulata* has been maintained *in vitro* by subculture of secondary embryos, although productivity is lower than with the former species. A high incidence of secondary embryogenesis (65%) and a mean number of about 10 secondary embryos formed per explant on a medium containing 0.5 mg/l BA and 0.5 mg/l IAA which significantly doubled the rate obtained in the basal medium. The best results, with multi-

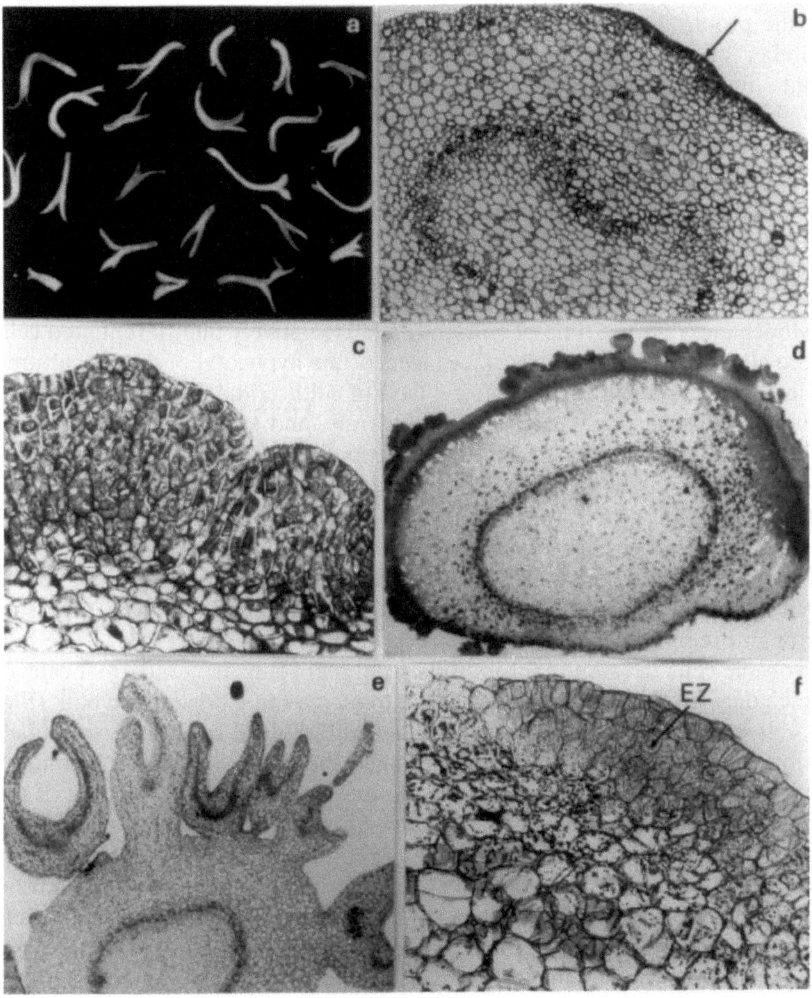


*Figure 6.* Development of secondary embryos in the hypocotyl of primary somatic embryos of *C. reticulata*. (a) Isolated somatic embryos at the time of culture (primary embryos). (b) Transverse section of the hypocotyl at the time of excision. Note the relatively undifferentiated zone (arrow) located at the periphery of the explant. X92. (c) Embryogenic tissue in superficial cell layers of the hypocotyl after 10 days of culture. Note the embryogenic cells with a high nucleus-to-cell ratio. X233. (d) Transverse section of the hypocotyl at day 14 showing a superficial embryogenic area from which secondary embryos at globular and heart stages developed. X18. (e) Section showing torpedo and early cotyledonary-shaped embryos at day 22. Note the continuity between the epidermis of both primary and secondary embryos. X36. (f) Cells of a embryogenic zone (EZ) and underlying cortex at day 8 showing a heavy deposition of starch. X296. (a,c; Plata *et al.*, 1991).

plication coefficients of 6.1–6.4 (the multiplication coefficient is the product of the proportion of explants with secondary embryogenesis and the mean number of secondary embryos per explant), are obtained using medium containing 0.1–0.5 mg/l BA and 0.5 mg/l IAA (Plata and Vieitez, 1990). The combination of 2 mg/l BA + 1 mg/l IBA provided more embryos per explant (13.2), but the low secondary embryogenesis rate (26%) meant a multiplication coefficient of only 3.4 and many embryos were fused together like secondary embryos of *C. japonica* when produced on media containing 2 mg/l BA.

Plata *et al.* (1991) studied the cellular origin of secondary embryos and the developmental events during secondary embryogenesis of *C. reticulata* cv. "Mouchang". At the time of culture (day 0), relatively undifferentiated cells were already present in the surface layers of the hypocotyl of the subcultured primary embryo (Fig. 6a,b). After 2 days of culture, the number of undifferentiated cells had increased in the epidermis and subjacent subepidermis, which soon exhibited a clearly dedifferentiating region of preembryogenic cells. After 8–10 days of culture, repeated division and dedifferentiation of preembryogenic cells had given rise to true embryogenic cells having the similar characteristic features defined by Williams and Maheswaran (1986). A computer-assisted image analysis revealed a gradient in the nucleus-to-cell area ratio from differentiated cells of the subepidermal layer (11%), passing through preembryogenic cells (23%) to embryogenic cells (31%). The embryogenic cells developed into a superficial meristem, the epidermal cells of which only divided anticlinally to adjust to increasing meristematic bulk (Fig. 6c). It was clear that the epidermis did not rupture to let the secondary embryos pass through, but instead became the epidermis of the secondary embryos which developed from indentations and subsequent budding of the embryogenic tissue. After about 12 days of culture, further growth had led to the differentiation of the globular stage embryos. Embryogenesis was thus direct, and the secondary embryos apparently had a multicellular origin from embryogenic areas originating in both the epidermal and subepidermal layers of the hypocotyl region (Fig. 6d). Progression to the heart and torpedo stages (days 14–16) involved the division, enlargement and differentiation of the cells of the new secondary embryos. The cotyledonary stage embryos that were identifiable after 19–21 days possessed meristematic shoot and root poles connected by procambial strands, and none had vascular connections with parent tissue (Fig. 6e).

In the above process, the formation of embryogenic areas was preceded by accumulation of starch in the surrounding hypocotyl cortical cells (Fig. 6f). The embryogenic cells and developing embryos also contained starch deposits, although in the form of smaller grains. However, starch content declined as the secondary embryos developed, suggesting that starch is mobilized and used as an energy source in the developing embryos. These changes in starch content were very similar to those observed during the development of somatic embryos on initial explants (see Section Embryo Development).

*Table 6.* Effect of cold storage on the performance of secondary somatic embryos of *C. japonica* after transfer to standard conditions and in successive eight-week subcultures. Survival % = Frequency of surviving secondary embryo clusters. N = Mean number of non-necrotic secondary embryos per cluster.

| Time in cold (months) | Transfer to standard conditions | | Subculture number (N) | | | | |
|---|---|---|---|---|---|---|---|
| | Survival (%) | N | 1 | 2 | 3 | 4 | 5 |
| 3 | 91.4 | 51.6 | 13.7 | 7.0 | 7.1 | 4.5 | 4.2 |
| 6 | 80.9 | 21.2 | 9.3 | 14.8 | 7.8 | 5.8 | 4.8 |
| 9 | 40.0 | 10.7 | 3.1 | 6.3 | 6.5 | 6.6 | – |
| 12 | 50.0 | 4.7 | 4.2 | 1.3 | – | – | – |

It is worth emphasizing the difference between the direct embryogenic process described above and the process that occurred with initial cotyledon explants (see Section Embryo Development). In the latter, the embryogenic area arose at the surface of the parenchymatous tissue which had previously proliferated to form a nodule or protuberance on the cotyledon surface. However, in the secondary embryogenesis process, the morphogenetic competence of primary embryos may depend on the presence of relatively undifferentiated cells in their hypocotyl epidermis at the time of explantation. According to Williams and Maheswaran (1986) the hypocotyl epidermis is embryogenically determined, or sufficiently close to this state for minimal redetermination.

Although most of embryogenic multiplication systems of *Camellia* species involves repetitive embryogenesis, the maintenance of embryogenic callus has also been tried. Abraham and Raman (1986) and Das and Barman (1988) did not clearly state the maintenance of the embryogenic cotyledon callus from which they obtained somatic embryos of *C. sinensis*. In the indirect embryogenic system of *C. reticulata*, the embryogenic competence is maintained through callus subculture in MS medium containing 1 mg/l BA plus 0.5 mg/l IBA (Table 3). This treatment yielded the greatest number of embryos per subcultured callus, more than the number of embryos produced by secondary embryogenesis in this species. The ability of embryogenic callus to produce somatic embryos did not decline after ten subcultures at 6-week intervals (Plata and Vieitez, 1990).

### 4.1. *Cold storage*

Cold storage may be a desirable procedure for the maintenance of embryogenic cultures. We studied (unpublished results) the effects of cold storage (3–4°C) for 3, 6, 9 and 12 months on *C. japonica* secondary embryo cultures. This experiment has shown that increasing cold storage time decreases the frequency of surviving embryo clusters and the mean number of surviving embryos per cluster, recorded 2 weeks after transfer to standard culture conditions. The most noteworthy point was the reduction of repetitive embryogenesis capacity in successive generations of embryos (Table 6). Although cold storage has been successful in the maintenance of cultures of other woody species, cold-stored *C. japonica* embryos cannot be efficiently multiplied.

## 5. Embryo germination

So far, we have discussed the initiation and maintenance of embryogenic multiplication systems. However, the efficiency of such systems for plantlet production depends on the efficiency with which the embryos germinate. The germination process can be defined as the emergence of the radicle accompained by the greening and elongation of the cotyledons. Although most reports on somatic embryogenesis in *Camellia* have mentioned conversion of somatic embryos into plantlets, few details have usually been given as to germination conditions and efficiency.

It has already been mentioned that seed-like somatic embryos of *C. japonica* exhibit less germination capacity than bud-like embryos (Vieitez and Barciela, 1990). Embryo conversion was also dependent on the distinct morphological types differentiated in *C. sinensis* cultures (Jha *et al.*, 1992); globular embryos developed to normal rooted plantlets when transferred to the germination medium, but none of the combinations of growth regulators used succeeded in inducing the germination of seed-like or cup-shaped embryos. Root elongation was sometimes achieved but without development of shoots.

A striking variety of culture media have been used to germinate *Camellia* embryos (Table 1), perhaps as a consequence of little research work has been on this stage of the embryogenic process. Unlike the somatic embryos of other woody plants (Rugini, 1988; Stamp and Meredith, 1988), *Camellia* species have low germination rates on culture media without growth regulators (Plata and Vieitez, 1990; Vieitez *et al.*, 1991; Jha *et al.*, 1992). For several species and hybrids, germination is stimulated by $GA_3$ (Table 1), but this growth regulator is unsuitable for *C. sinensis* (Kato, 1986a; Jha *et al.*, 1992). The medium containing 10 mg/l BA + 0.5 mg/l IBA proved to be useful for inducing the concomitant development of shoots and roots in *C. sinensis* embryos (Kato, 1986a); however the author did not state the germination rate actually achieved. The best germination rates for *C. sinensis* somatic embryos were reported by Bano *et al.* (1991). According to them, normal embryo development and germination occurred in 40% of the heart and torpedo-shaped embryos after transfer to the half-strength MS medium supplemented with 0.05 mg/l kinetin, 0.2% activated charcoal and 1.5% glucose. Jha *et al.* (1992) obtained 20% germination on B5 medium (Gam-

*Figure 7.* Germination and plant conversion. (a) Secondary embryogenesis on the hypocotyl zone of a germinating *C. japonica* somatic embryo. (b) Plantlets derived from *C. reticulata* somatic embryos after 6 weeks in germinating medium containing 4 mg/l $GA_3$ + 2 mg/l IAA. (c) Germination of two *C. reticulata* embryos in 5 mg/l $GA_3$ + 0.5 mg/l IAA + 0.5 mg/l BA medium showing development of axillary shoots. (d) Somatic embryo derived plants of *C. japonica* three years after being transferred to soil.

*Table 7.* Effect of $GA_3$ and IAA on germination of somatic *C. japonica* embryos producing whole plantlets, rootless shoots, and shootless roots. Only shoots and roots longer than 5 mm were recorded after 5 weeks of culture[z] (Vieitez *et al.*, 1991).

| Medium (mg/l) | Plantlets[y] (%) | LS[x] | LR[w] | Rootless shoots[v] (%) | LS[x] | Shootless roots[u] (%) | LR[w] |
|---|---|---|---|---|---|---|---|
| $GA_3$ (1) | 6.5 c | 10.0 d | 11.1 | 29.0 ab | 13.3 | 6.5 | 9.0 |
| $GA_3$ (2.5) | 22.5 b | 13.8 cd | 11.9 | 35.5 ab | 19.1 | 6.5 | 5.0 |
| $GA_3$ (5) | 26.5 b | 30.9 a | 14.6 | 41.2 a | 19.2 | 14.7 | 15.6 |
| $GA_3$ (1) + IAA (1) | 28.6 b | 12.2 cd | 10.3 | 28.6 ab | 14.1 | 17.9 | 14.0 |
| $GA_3$ (1) + IAA (2) | 19.2 b | 16.2 cd | 9.3 | 38.5 ab | 13.1 | 7.7 | 9.5 |
| $GA_3$ (2.5) + IAA (1) | 40.6 ab | 21.4 b | 12.6 | 34.4 ab | 20.6 | 6.3 | 9.0 |
| $GA_3$ (2.5) + IAA (2) | 48.3 a | 17.2 bc | 15.3 | 20.7 ab | 21.7 | 6.9 | 10.5 |
| $GA_3$ (5) + IAA (1) | 62.1 a | 21.3 b | 12.7 | 20.1 b | 21.3 | 6.7 | 21.0 |
| $GA_3$ (5) + IAA (2) | 60.7 a | 17.8 bc | 14.2 | 17.9 b | 29.4 | 7.1 | 13.0 |

[z] Mean separation within columns by Duncan's multiple range test, $P = 0.05$: lack of letters indicates factor nonsignificant.
[y] Percentage of plantlets developed.
[x] Mean length (mm) of shoots.
[w] Mean length (mm) of roots.
[v] Percentage of embryos producing only shoots.
[u] Percentage of embryos producing only roots.

borg *et al.*, 1968) containing high concentrations of cytokinin and auxin (3 mg/l BA + 2 mg/l IAA).

Adventitious embryos of *C. japonica* produced shoots and roots on the medium supplemented with 1 mg/l $GA_3$ (Kato, 1986a). Vieitez *et al.* (1991) carried out experiments to identify the best gibberellin/auxin combination for the germination of 5–9 mm, bud-like somatic embryos of *C. japonica*. In our experiments, germination was deemed to have occurred only if both the radicle and the shoot developed. Whole plantlets as well as partially regenerated plantlets lacking either root or shoot were produced after five weeks on the germination medium. The germination rate on a medium lacking growth regulators was poor, 9%, but generally was improved by adding $GA_3$ (1–5 mg/l), with or without IAA (1–2 mg/l) in the culture medium (Table 7). The best results were obtained with 5 mg/l $GA_3$ + 1 mg/l IAA; after 5 weeks on this medium, 62% of embryos had germinated, which compares well with the germination rates reported for other woody species (Wann, 1988). During the first two weeks after the transfer of the somatic embryos to a germination medium, cotyledons grew and thickened and a single tap root began to emerge; the shoot developed later. On a medium containing 1 mg/l $GA_3$, the first leaves that developed on the shoot tended to thicken and turn into cotyledon-like structures whereas media with 5 mg/l $GA_3$ induced thinner shoots with normal leaves. Similar structures, intermediate between cotyledons and normal leaves, have developed during the germination of red oak (*Quercus rubra*) embryos. They appeared to be transitional between embryonic and germinative phases (Gingas and Lineberger, 1989) indicating that a certain kind of immaturity in germinating embryos. In keeping with this lack of maturity, which is often associated with the failure of the embryo to develop as a normal plantlet (Williams and Maheswaran, 1986), secondary embryogenesis sometimes occurred on the hypocotyl of germinating *C. japonica* embryos (Fig. 7a). It was more frequent when low $GA_3$ concentration was used (Vieitez *et al.*, 1991).

According to our germination experiments with *C. reticulata* (Plata and Vieitez, 1990; Plata, 1993), 11% germination rate was attained on a medium lacking growth regulators but this frequency increased significantly (up to 30%) by supplementing the medium with $GA_3$ with or without IAA (Fig. 7b). The best rate was achieved on a medium containing 5 mg/l $GA_3$ + 2 mg/l IAA. After 5 weeks of culture, this medium induced the germination in 51% of the embryos, a figure similar to that obtained with *C. japonica*. Addition of 0.1–0.5 mg/l BA to a medium containing $GA_3$ favoured the development of axillary shoots although the root remained blocked (Fig. 7c). As with *C. japonica*, the incidence of secondary embryogenesis was reduced in germination media containing relatively high amounts of $GA_3$.

The improvement of germination rates by $GA_3$ that was observed in our experiments with *C. japonica* and *C. reticulata* has also been reported for other woody species, including *Aesculus hippocastanum* (Radojević, 1988), *Citrus unshiu* (Nito and Iwamasa, 1990) and *Juglans nigra* × *J. regia* (Deng and Cornu, 1992). The involvement of $GA_3$ in germination control is well documented for zygotic embryos (LePage-Degivry *et al.*, 1990), and seems to extend to somatic embryos too. It has been suggested that germination requires both a decrease in endogenous abscisic acid levels and an increased sensitivity to $GA_3$ (LePage-Degivry *et al.*, 1990; Deng and Cornu, 1992).

Plata (1993) investigated the effect of the physical state of the medium on the germination of somatic embryos of *C. reticulata*, comparing solid versus liquid media with paper bridge. This difference in culture conditions did not affect plantlet conversion frequency, but did affect plantlet morphology. By the end of the experiment, shoots were an average 7 mm longer on the solid medium than on the liquid medium. The liquid medium favoured the development of longer, thinner and straighter roots. The formation of lateral roots was not common on either medium.

The influence of light on the germination of somatic embryos of *C. japonica* was investigated by Vieitez *et al.* (1991). Root growth was promoted, and secondary embryogenesis rates were lower, when the first 7 or 15 days on germination medium were spent in the dark. However, plantlet conversion rates were significantly decreased because this treatment also inhibited shoot development. A similar effect of darkness was observed in the germination of *C. reticulata* embryos (Plata, 1993).

### 5.1. *Effect of cold storage*

As pointed out in the previous section (Culture Maintenance), cold storage reduced the capacity for secondary embryogenesis, this process being inversely proportional to the maturation state of the embryo. Since maturity seems to be an important factor in germination, we determined the favourable response of cold storage on the germination response of somatic embryos. Six-week-old *C. japonica* and *C. reticulata* embryo cultures obtained by recursive secondary embryogenesis were stored at 3–4° C for 1 or 2 months, after that secondary embryos were isolated from the primary embryos and transferred to a germination medium. After 2 month cold treatment, the germination capacity improved significantly in both the species. The germination rate of *C. japonica* embryos jumped from 54% to 86%. Germination rate was also faster after cold storage. After 5 weeks on a germination medium, cold-stored embryos had developed shoots twice as long as those on controls. Finally, the incidence of secondary embryogenesis during germination decreased after cold storage. The effect of this treatment may be due to the breaking of embryo dormancy as suggested by Tulecke and McGranahan (1985) in *Juglans nigra* × *J. regia*.

### 5.2. *Adventitious rooting of somatic embryo derived shoots*

As mentioned above, germination medium containing cytokinin tended to inhibit root development, but promotes the development of axillary shoots. In the absence of an efficient germination system for a particular *Camellia* species or variety, an alternative possibility is to regenerate plantlets from shoots isolated from the partially germinated embryos on a cytokinin containing medium. Among *Camellia* species, this alternative has chiefly been employed for the regeneration of *C. sinensis* cultivars in which the early development of embryo axillary buds resulted in multiple shoot structures on somatic embryos in the presence of BA (Nakamura, 1988; Das and Barman, 1988). Addition of 1 mg/l IBA induced rooting in 30%–40% of such shoots (Jha *et al.*, 1992). Rooting of regenerated shoots derived from *C. sinensis* embryos was recalcitrant on a solid MS medium, although 45% shoots formed roots on paper bridge wetted with half-strength MS liquid medium having 0.5 mg/l IBA (Jain *et al.*, 1990). Nakamura (1991) used the same paper bridge system, with a medium containing 1–3 mg/l IBA, to root shoots obtained from somatic embryos of *C. sinensis* cv. "Yabukita", achieving up to 100% rooting frequency for shoots 20 mm long.

The rooting method can have a great influence on the survival of rooted plantlets after transfer to soil. Jain *et al.* (1991) found that shake cultures of *C. sinensis* shoots in the liquid rooting medium was efficient in promoting root initiation (50%) and growth, but none of the plantlets survived in ambient conditions. On the other hand, treatment of shoots for 2 h with 50 mg/l IBA followed by transplant to a soil:peat mixture for 2 months, not only increased the rooting rate to 97% but also achieved a survival rate of 85% in the nursery.

## 6. Growth in the field. Somaclonal variation

Though several reports of somatic embryogenesis in *Camellia* species state that somatic embryo derived plantlets were transplanted to soil (Kato, 1989a; Bano *et al.*, 1991; Jha *et al.*, 1992), they give few details on survival and growth in the field.

In our experiments (Vieitez *et al.*, 1991), 6–7 week-old *C. japonica* plantlets recovered from germinated embryos (emblings) were transferred to pots containing a mixture of peat:quartz sand (1:1) and placed under a plastic tunnel with a mist and fog system in the greenhouse. Hoagland's mineral solution (Hoagland and Arnon, 1950) was applied once a week. Eight weeks after the transplantation, the plants were removed from the tunnel and grown in the greenhouse in pots containing a mixture of peat:quartz sand:soil (1:1:1). Survival rates ranged from 40% to 78% depending on the clone. Acclimatized plants appeared morphologically similar to seedlings; three years later they had not yet flowered (Fig. 7d).

Kato (1989a) transferred *C. sinensis* plantlets derived from cotyledons to a mixture of vermiculite and soil. After 4 weeks under plastic covers their growth was healthy and similar to that of seedlings under natural conditions. In this report, information is not given on the acclimatization success. Jha *et al.* (1992) hardened tea plantlets for 8 weeks in quarter-strength MS salts before transplantation to pots containing a mixture of peat and soil (1:1) achieving an acclimatization rate of 70%, very similar to the rate we achieved with *C. japonica*.

Although the information on growth of emblings in the field is limited, no morphological differences or abnormalities were observed (Zhuang and Liang, 1985a; Zhuang *et al.*, 1988; Bano *et al.*, 1991). Jha *et al.* (1992) also reported that their embryo derived plants were true-to-type to mother plants, with a chromosome number of $2n = 30$, suggesting a stable nature. The conformity of these type of plants is important whenever somatic embryogenesis is to be considered as a true-to-type method of micropropagation of selected genotypes.

In contrast with the above findings, the most detailed report on the performance of the regenerated plants in the field (Wu *et al.*, 1981) also mentions the morphological variation. In this case, regeneration of tea plantlets was indirect, via cotyledon callus, and it is not even clear whether the plantlets were regenerated from true somatic embryos or from adventitious buds. The plantlets grew vigorously in soil, but were considerably different morphologically, especially leaves and flowers from the naturally-grown plants of the same age. Plantlets derived from calli were more vigorous, with a greater number of flowers per bush and larger seeds. Furthermore, the floral type was exceptionally favourable for early control of crossing. In view of these findings, these plantlets were used as females in crosses with a Chin-Hsin Oolong variety that produces high quality tea and has poor growth rate. Altogether, 171 new F1 hybrids selected for better yield and vigour were transplanted directly from nursery beds to the field.

Various studies indicated that callus culture produces abundant cytological anomalies affecting both structural and numerical chromosome constitution, which can be correlated with many phenotypic abnormalities (Groose and Bingham, 1984; Evans and Sharp, 1986). A wide range of chromosomal variations were observed in the callus cells derived from tea leaf and cotyledon explants cultured in MS nutrient medium supplemented with 6 mg/l 2,4-D (Das, 1992). More than 60% of the cells were diploid in the callus after 8 weeks of inoculation, and the rest were triploid, tetraploid and aneuploid. Among the polyploids, cells with more than 60 chromosomes were also observed. Das (1992) emphasized the possibility of producing polyploids through tea callus culture which regenerated *in vitro* shoots through embryogenesis. However, the percentage recovery of polyploids was very low.

Secondary embryogenesis has also been used for the production of polyploid *Camellia*. The systematic formation and propagation of polyploids through secondary embryo culture were reported in *C. japonica* (Kato, 1989b). When somatic embryos were treated with 0.1% colchicine for one week, secondary embryos formed on the hypocotyl of 32% of primary embryos, after culture for 60 days on MS medium containing 1 mg/l $GA_3$. Secondary embryos proliferated in a similar manner during subsequent subcultures. Diploid, tetraploid and, occasionally, triploid or hexaploid chromosome numbers were observed in cells of the secondary embryos produced. In tetraploid embryos, the chromosome number was stable through successive generations of embryos for a 3-year period. These polyploid embryos developed into plantlets with autopolyploid morphological characteristics.

All these observations show that, in *Camellia*, variation from the mother plant can be generated during the formation of somatic embryos, especially if they are produced indirectly via a callus phase.

## 7. Concluding remarks

So far, previous studies have shown that *Camellia* species appear to have a considerable inherent capacity for direct somatic embryogenesis, at least in their juvenile phase. This capacity is especially strikingly shown by the production of somatic embryos in the basal medium without growth regulators. This independence of exogenous auxins for somatic embryogenesis is, relatively rare in woody species. However, growth regulators, particularly BA and IBA, do modulate the embryogenic process in *Camellia*, increasing the induction percentage and/or the number of embryos in several cases. In most of the reports, somatic embryo induction was achieved using the MS mineral medium either nonsupplemented with growth regulators or supplemented with high cytokinin and low auxin, or a cytokinin alone. At this stage, it is relevant to point out that for the induction of somatic embryogenesis in most of the other plants, 2,4-D is very essential. This is, however, partially true with *Camellia* tissue cultures, where NAA or IBA have been found to be more effective (Table 1). Success also depends on both the physiological maturity of the initial explant and the type of explant used as well as the *Camellia* species or variety.

Indirect somatic embryogenesis via callus developed on initial explants can also be induced in *Camellia* cultures. However, unlike direct embryogenesis, indirect embryogenesis appears to require the inclusion of auxin or auxin plus cytokinin in the medium.

The ease with which *Camellia* form somatic embryos directly on explants may reflect that these species, when physiologically juvenile, possess an inherent capacity to produce "pre-embryogenic determined cells" from which somatic embryos develop (Sharp *et al.*, 1980). Indirect embryogenesis, which requires auxin and/or cytokinin (Abraham and Raman, 1986; Das and Barman, 1988; Plata and Vieitez, 1990; San-José and Vieitez, 1993) may involve the redetermination of callus cells as "induced embryogenic determined cells" (Sharp *et al.*, 1980).

In general, initial cultures for somatic embryo induction have been grown under a 16 h day, 8 h night regime at temperatures of 25–28°C. Continuous dark (Nakamura, 1988) and daily 8 h of diffuse light (Bano *et al.*, 1991) have also been used for tea cotyledon cultures, although in neither of these studies, the results were of the chosen light conditions compared with those of a 16 photoperiod. Culture in the dark, for at least 14 weeks, was essential for the induction of somatic embryogenesis on *C. japonica* leaves; light inhibited both callus formation and morphogenesis.

The long-term maintenance of the embryogenic capacity, especially by repetitive embryogenesis, makes the continuous supply of somatic embryos with an adequate multiplication rate possible. Moreover, it should be noted that most embryogenic systems for *Camellia* species involve direct embryogenesis, which theoretically affords the guarantee of genetic stability which is important in clonal propagation.

The regeneration capacity of juvenile tissues in *Camellia* species is of considerable interest. Since the embryo is heterogenic in hybrid seeds, the somatic embryogenesis by cotyledon and embryo culture may constitute a viable means of rapid clonal propagation of crossing materials which have poor viability, or selected *Camellia* clones (Kato, 1989a). However, to optimize the scale-up plant micopropagation from the somatic embryos, the *in vitro* methods should be improved in four aspects: i) It is necessary to study the development of the distinct morphological types of somatic embryos, investigating the endogenous control of their formation in order to obtain high-frequency production of morphologically normal somatic embryos with the capacity to germinate; ii) Efforts need to be directed on the factors influencing the maturation and germination phases of the embryos, especially in *C. sinensis*. Cold storage, desiccation, solidifying agents, physical and chemical components of the media must be tested to promote embryo germination and to improve the efficiency of conversion into plantlets; iii) The suitable conditions for hardening the emblings after transfer to soil have to be determined. In order to establish whether they are true-to-type, especially if they are produced by an indirect callus phase, field testing should be accomplished; iv) It would be desirable to extend the research to mature material to achieve a large-scale propagation of selected adult *Camellia* clones.

Indirect somatic embryogenesis systems, and the development of suspension cultures of embryogenic cells, are other areas which may be of great importance in *Camellia* genetic engineering, and they need to be studied more extensively.

## 8. Acknowledgements

We are grateful to Drs. Y. Nakamura, Japan, and S.C. Das, India, for sending their reprints. Further thanks are given to Dr. Y. Nakamura for the sending of additional prints and slides.

The financial support from CICYT, Spain (BIO89–0529), and EEC (ECLAIR, Contract AGRE 0067) is gratefully acknowledged.

## 9. References

Abraham, G.C. and K. Raman, 1986. Somatic embryogenesis in tissue culture of immature cotyledons of tea (*Camellia* spp.). In: D.A. Somers, B.G. Gengenbach, D.D. Biesboer, W.P. Hackett and C.E. Green (Eds.), Abstr. 6th Int. Congr. of Plant Tissue and Cell Culture, pp. 294. Univ. Minnesota, Minneapolis.

Ammirato, P.V., 1985. Patterns of development in culture. In: R.R. Henke, K.W. Hughes, M.P. Constantin and A. Hollaender (Eds.), Tissue Culture in Forestry and Agriculture, pp. 9–29. Plenum Press Corporation, New York.

Anderson, W.C., 1984. A revised tissue culture medium for shoot multiplication of *Rhododendron*. J. Amer. Soc. Hort. Sci. 109: 343–347.

Bano, Z., S. Rajarathnam and B.D. Mohanty, 1991. Somatic embryogenesis in cotyledon cultures of tea (*Thea sinensis* L.). J. Hort. Sci. 66: 465–470.

Barciela, J. and A.M. Vieitez, 1993. Anatomical sequence and morphometric analysis during somatic embryogenesis on cultured cotyledon explants of *Camellia japonica* L. Ann. Bot. 71: 395–404.

Bennett, W.Y., 1977. Tissue culture for Camellias? Am. Camellia Yearb.: 188–190.

Bennett, W.Y., 1978. Tissue culture for Camellias II. Am. Camellia Yearb.: 102–104.

Bennett, W.Y. and P. Scheibert, 1982. *In vitro* generation of callus and plantlets from cotyledons of *Camellia japonica*. Camellia J. 37: 12–15.

Bhojwani, S.S. and M.K. Razdan, 1983. Plant Tissue Culture: Theory and Practice. Elsevier Science Publishers B.V., Amsterdam.

Chang, H.T. and B. Bartolomew, 1984. Camellias. B.T. Badsfort Ltd., London.

Chen, Z. and H. Liao, 1983. A success in bringing out tea plants from the anthers. China Tea 5: 6–7.

Chen, Z. and H. Liao, 1990. Tea: Anther culture. In: Z. Chen, D.A. Evans, W.R. Sharp, P.V. Ammirato and M.R. Sondhal (Eds.), Handbook of Plant Cell Culture, Vol. 6, Perennial Crops, pp. 475–479. McGraw-Hill Publishing Company, New York.

Das, S.C., 1992. Non-conventional techniques of regenerating polyploids in tea. Proc. 31st Tocklai Conf., TRA, Jorhat, pp. 26–30.

Das, S.C. and T.S. Barman, 1988. Current state and future potential of tissue culture in tea. Proc. 30th Tocklai Conf., TRA, Jorhat, pp. 90–94.

Das, S.C. and T.S. Barman, 1992. Tea shoot regeneration from embryo callus. In: N.S. Subba Rao, C. Balagopalan and S.V. Ramakrishna (Eds.), New Trends in Biotechnology, pp. 81–86. Oxford and IBH Publishing Co. Pvt. Ltd., New Delhi/Bombay.

Deng, M-D. and D. Cornu, 1992. Maturation and germination of walnut somatic embryos. Plant Cell Tiss. Org. Cult. 28: 195–202.

Duhem, K., N. Le Mercier and P. Boxus, 1989. Donnés nouvelles sur l'induction et le développement d'embryons somatiques chez *Theobroma cacao* L. Café Cacao Thé 33: 9–14.

El Maâtaoui, M., H. Espagnac and N. Michaux-Ferrière, 1990. Histology of callogenesis and somatic embryogenesis induced in stem fragments of cork oak (*Quercus suber*) cultured *in vitro*. Ann. Bot. 66: 183–190.

Evans, D.A. and W.R. Sharp, 1986. Somaclonal and gametoclonal variation. In: D.A. Evans, W.R. Sharp and P.V. Ammirato (Eds.), Handbook of Plant Cell Culture, Vol. 4: Techniques and Applications, pp. 97–132. Macmillan Publishing Company, New York.

Fiola, J.A., M.A. Hassan, H.J. Swartz, R.H. Bors and R. McNicols, 1990. Effect of thidiazuron, light fluence rates and kanamycin on *in vitro* shoot organogenesis from excised *Rubus* cotyledons and leaves. Plant Cell Tiss. Org. Cult. 20: 223–228.

Flinn, B.S., D.T. Webb and W. Newcomb, 1989. Morphometric analysis of reserve substances and ultrastructural changes during caulogenic determination and loss of competence of Eastern White pine (*Pinus strobus*) cotyledons *in vitro*. Can. J. Bot. 67: 779–789.

Gamborg, O.L., R.A. Miller and K. Ojima, 1968. Nutrient requirements of suspension cultures of soybean root cells. Exp. Cell. Res. 50: 151–158.

Gingas, V.M. and R.D. Lineberger, 1989. Asexual embryogenesis and plant regeneration in *Quercus*. Plant Cell Tiss. Org. Cult. 17: 191–203.

Groose, R.W. and E.T.Bingham, 1984.Variation in plants regenerated from tissue culture of tetraploid alfalfa heterozygous for several traits. Crop Sci. 24: 655–658.

Hoagland, D.R. and D.I. Arnon, 1950. The water culture method for growing plants without soil. Cal. Agri. Exp. Stn. Circ. Berkeley 347: 1–32.

Jain, S.M. and R.J. Newton, 1990. Prospects of biotechnology for tea improvement. Proc. Indian Natn. Sci. Acad. B56: 441–448.

Jain, S.M., S.C. Das and T.S. Barman, 1990. Enhancement of root induction from regenerated shoots of tea (*Camellia sinensis* L.). In: Abstr. VIIth Int. Congr. Plant Tissue and Cell Culture, Amsterdam, June 24–29, p. 107.

Jain, S.M., S.C. Das and T.S. Barman, 1991. Induction of roots from regenerated shoots of tea (*Camellia sinensis* L.). Acta Hort. 289: 339–340.

Jha, T.B., S. Jha and S.K. Sen, 1992. Somatic embryogenesis from immature cotyledons of an elite Darjeeling tea clone. Plant Sci. 84: 209–213.

Kato, M., 1982. Results of organ culture on *Camellia japonica* and *C. sinensis*. Jpn. J. Breed. 32 (Suppl. 2): 267–277.

Kato, M., 1986a. Micropropagation through cotyledon culture in *Camellia japonica* L. and *C. sinensis* L. Jpn. J. Breed. 36: 31–38.

Kato, M., 1986b. Micropropagation through cotyledon culture in *Camellia sasanqua*. Jpn. J. Breed. 36 (Suppl. 1): 82–83.

Kato, M., 1989a. *Camellia sinensis* L. (Tea): *In vitro* regeneration. In: Y.P.S. Bajaj (Ed.), Biotechnology in Agriculture and Forestry, Vol. 7: Medicinal and Aromatic Plants II, pp. 82–89. Springer-Verlag, Berlin/Heidelberg.

Kato, M., 1989b. Polyploids of camellia through culture of somatic embryos. HortSci. 24: 1023–1025.

Klee, H., R. Horsch and S. Rogers, 1987. *Agrobacterium*- mediated plant transformation and its further applications to plant biology. Ann. Rev. Plant Physiol. 38: 467–486.

Le Page-Degivry, M.T., P. Barthe and G. Garello, 1990. Involvement of endogenous abscisic acid in onset and release of *Helianthus annuus* embryo dormancy. Plant Physiol. 92: 1164–1168.

Linnaeus, C., 1735. Sistema Naturae. Leiden.

Lloyd, G. and B.H. McCown, 1980. Commercially-feasible micropropagation of mountain laurel, *Kalmia latifolia*, by use of shoot tip culture. Proc. Int. Plant Prop. Soc. 30: 421–427.

Margara, J., 1982. Bases de la multiplication vegetative. Les méristèmes et l'organogenèse. Institut National de la Recherche Agronomique, Paris.

Merkle, S.A. and A.T. Wiecko, 1989. Regeneration of *Robinia pseudoacacia* via somatic embryogenesis. Can. J. For. Res. 19: 285–288.

Muralidharan, E.M., P.K. Gupta and A.F. Mascarenhas, 1989. Plantlet production through high frequency somatic embryogenesis in long term cultures of *Eucalyptus citriodora*. Plant Cell Rep. 8: 41–43.

Murashige, T. and F. Skoog, 1962. A revised medium for rapid growth and bioassays with tobacco tissue cultures. Physiol. Plant. 15: 473–497.

Nadamitsu, S., Y. Andoh, K. Kondo and M. Segawa, 1986. Interspecific hybrids between *Camellia vietnamensis* and *C. chrysantha* by cotyledon culture. Jpn. J. Breed. 36: 309–313.

Nakamura, Y., 1988. Efficient differentiation of adventitious embryos from cotyledon culture of *Camellia sinensis* and other *Camellia* species. Tea Res. J. 67: 1–12.

Nakamura, Y.,1991. *In vitro* propagation techniques of tea plants. Jpn. Agric. Res. Q. 25: 185–194.

Nito, N. and M. Iwamasa, 1990. *In vitro* plantlet formation from juice vesicles of satsuma (*Citrus unshiu* Marc.). Plant Cell Tiss. Org. Cult. 20: 137–140.

Park, Y.G. and S.H. Son, 1988. *In vitro* organogenesis and somatic embryogenesis from punctured leaf of *Populus nigra* × *P. maximowiczii*. Plant Cell Tiss. Org. Cult. 15: 95–105.

Patel, K.R. and T.A. Thorpe, 1984. Histochemical examination of shoot initiation in cultured cotyledon explants of radiata pine. Bot. Gaz. 145: 312–322.

Plata, E., 1993. Morfogénesis *in vitro* de *Camellia reticulata*: Procesos de embriogénesis somática y regeneración de plantas. Doctoral Thesis, University of Santiago de Compostela, Santiago de Compostela.

Plata, E. and A.M. Vieitez, 1990. *In vitro* regeneration of *Camellia reticulata* by somatic embryogenesis. J. Hort. Sci. 65: 707–714.

Plata, E., A. Ballester and A.M. Vieitez, 1991. An anatomical study of secondary embryogenesis in *Camellia reticulata*. In Vitro Cell. Dev. Biol. 27P: 183–189.

Radojević, L., 1988. Plant regeneration of *Aesculus hippocastanum* L. (horse-chestnut) through somatic embryogenesis. J. Plant Physiol. 132: 322–326.

Raina, S.K. and R.D. Iyer, 1974. A study of response to anther culture. Indian J. Genet. 34A: 283.

Raina, S.K. and R.D. Iyer, 1983. J. Plantation Crops, C.S. Venkata Ram Commen 11 (Suppl.): 63.

Raj Bhansali, R., 1990. Somatic embryogenesis and regeneration of plantlets in pomegranate. Ann. Bot. 66: 249–253.

Rugini, E., 1988. Somatic embryogenesis and plant regeneration in olive (*Olea europea* L.). Plant Cell Tiss. Org. Cult. 14: 207–214.

Samartín, A., A.M. Vieitez and E. Vieitez, 1986. Rooting of tissue cultured camellias. J. Hort. Sci. 61: 113–120.

Samartín, M.C. and A.P. Samartín, 1988. La Camelia. Un Regalo Oriental para Occidente. Editorial Everest, S.A., Leon.

San-José, M.C. and A.M. Vieitez, 1993. Regeneration of *Camellia* plantlets from leaf explant cultures by embryogenesis and caulogenesis. Sci. Hort. 54: 303–315.

Sharp, W.R., M.R. Sondhal, R.S. Caldas and S.B. Maraffa, 1980. The physiology of *in vitro* asexual embryogenesis. Hort. Rev. 2: 268–310.

Shimokado, T., T. Murata and Y. Miyaji, 1986. Formation of embryoid by anther culture of tea. Jpn. J. Breed. 36 (Suppl. 2): 282–283.

Shizuoka Tea Experiment Station, 1992. Japanese Green Tea in Shizuoka. Kurofune Printing Co., Ltd., Shizuoka.

Stamp, J.A., 1987. Somatic embryogenesis in cassava: The anatomy and morphology of the regeneration process. Ann. Bot. 59: 451–459.

Stamp, J.A. and C.P. Meredith, 1988. Proliferative somatic embryogenesis from zygotic embryos of grapevine. J. Amer. Soc. Hort. Sci. 113: 941–945.

Tabachnik, L. and D.E. Kester, 1977. Shoot cultures for almond and almond-peach hybrid clones *in vitro*. HortSci. 12: 545–547.

Thorpe, T.A., 1988. *In vitro* somatic embryogenesis. ISI Atlas of Sci.: Anim. Plant Sci. 1: 81–88.

Tian-Ling, L., 1982. Regeneration of plantlets in cultures of immature cotyledons and young embryos of *Camellia oleafera* Abel. Acta Biol. Exp. Sin. 15: 393–403.

Tulecke, W., 1987. Somatic embryogenesis in woody perennials. In: J.M. Bonga and D.J. Durzan (Eds), Cell and Tissue Culture in Forestry. Vol. 2, pp. 61–69. Martinus Nijhoff Publishers, Dordrecht.

Tulecke, W. and G. McGranahan, 1985. Somatic embryogenesis and plant regeneration from cotyledons of walnut, *Juglans regia* L. Plant Sci. 40: 57–63.

Vasil, I.K., 1985. Somatic embryogenesis and its consequence in Gramineae. In: R.H. Randol, K.W. Hughes, M.L. Constantin and H. Alexander (Eds.), Tissue Culture in Forestry and Agriculture, Vol. 32, pp. 41–47. Plenum Press, New York.

Vieitez, A.M. and J. Barciela, 1990. Somatic embryogenesis and plant regeneration from embryonic tissues of *Camellia japonica* L. Plant Cell Tiss. Org. Cult. 21: 267–274.

Vieitez, A.M., J. Barciela and A. Ballester, 1989a. Propagation of *Camellia japonica* cv. "Alba Plena" by tissue culture. J. Hort. Sci. 64: 177–182.

Vieitez, A.M., M.C. San-José and A. Ballester, 1989b. Progress towards clonal propagation of *Camellia japonica* cv. "Alba Plena" by tissue culture techniques. J. Hort. Sci. 64: 605–610.

Vieitez, A.M., M.C. San-José, F.J. Vieitez and A. Ballester, 1991. Somatic embryogenesis from roots of *Camellia japonica* plantlets cultured *in vitro*. J. Amer. Soc. Hort. Sci. 116: 753–757.

Vieitez, A.M., M.L. Vieitez, A. Ballester and E. Vieitez, 1992. Micropropagation of *Camellia* spp. In: Y.P.S. Bajaj (Ed.), Biotechnology in Agriculture and Forestry, Vol. 19: High-Tech and Micropropagation III, pp. 361–387. Springer-Verlag, Berlin/Heidelberg.

Wann, S.R., 1988. Somatic embryogenesis in woody species. Hort. Rev. 10: 153–181.

Williams, E.G. and G. Maheswaran, 1986. Somatic embryogenesis: factors influencing coordinated behaviour of cells as an embryogenic group. Ann. Bot. 57: 443–462.

Wu, C.T., 1976. Studies on the tissue culture of tea plants. J. Agric. Assoc. China 90 (LII): 30–42.

Wu, C.T., T.K. Huang, G.R. Chen and S.Y. Chen, 1981. A review on the tissue culture of tea plants and on the utilization of callus-derived plantlets. In: A.N. Rao (Ed.), Tissue Culture of Economically Important Plants, pp. 104–106. Proc. COSTED Symp., Singapore.

Yamaguchi, S., T. Kunitake and S. Hisatomi, 1987. Interspecific hybrid between *Camellia japonica* cv. Chochidori and *C. chrysantha* produced by embryo culture. Jpn. J. Breed. 37: 203–206.

Yan, M.Q. and C. Ping, 1983. Studies on development of embryoids from the culture cotyledons of *Thea sinensis* L. Sci. Silv. Sin. 19: 25–29.

Yan, M.Q., C. Ping, M. Wei and Y.H. Wang, 1984. Tissue culture and transplanting of *Camellia oleifera*. Sci. Silvae Sin. 20: 341–350.

Zhuang, C. and H. Liang, 1985a. *In vitro* embryoid formation of *Camellia reticulata* L. Acta Biol. Exp. Sin. 18: 275–281.

Zhuang, C. and H. Liang, 1985b. Somatic embryogenesis and plantlets formation in cotyledon culture of *Camellia chrysantha*. Acta Bot. Yunn. 7: 446–450.

Zhuang, C., J. Duan and J. Zhou, 1988. Somatic embryogenesis and plantlets regeneration of *Camellia sasanqua*. Acta Bot. Yunn. 10: 241–244.

# 15. Somatic embryogenesis in *Rosa* spp.

A.V. Roberts, K. Yokoya, S. Walker and J. Mottley

**Contents**



## 1. Introduction

### 1.1. *The genus* Rosa

The genus *Rosa* includes over 100 species, distributed throughout the temperate and subtropical regions of the Northern Hemisphere. According to Rehder (1960) the genus consists of 4 subgenera; 3 are monotypic and the fourth, *Eurosa*, includes all the remaining species grouped into 10 sections. The genus exhibits a typical polyploid series with a basic chromosome number of 7. Euploids range from $2n = 2x = 14$, to $2n = 8x = 56$ (Darlington and Wylie, 1955) and aneuploids are rare (Rowley, 1960).

### 1.2. *Roses in cultivation*

Roses are grown throughout the world as ornamental plants. World-wide, 200 million bushes and 2 billion cut-flowers are sold per annum. Multipetalled varieties with a short flowering season have been cultivated in Europe and the Middle East since the days of the Roman Empire and are still grown in parks and rose collections. "Perpetual-flowering" roses, which have been cultivated from even earlier times in China, were introduced into Europe in about 1800 AD. Hurst (1941) and Wylie (1954, 1955a,b) have reconstructed the developments from early crosses between these two groups of roses to the origin of the modern Hybrid Tea (large-flowered) roses which are used for greenhouse production of cut flowers, and Polyantha (small cluster-flowered) and Floribunda (large cluster-flowered) roses which are the predominant garden roses. Roses are also used by the perfume industry. The annual production of the beautifully scented attar of rose, which is distilled from petals of *Rosa* × *damascena* and *R. centifolia*, is valued at US $ 12 million.

*S. Jain, P. Gupta & R. Newton (eds.), Somatic Embryogenesis in Woody Plants, Vol. 2,* 277–289.

Hybridization has been extensively and intensively practised for centuries and it seems unlikely that easily accessible sources of diversity remain unexplored. Attention is turning, therefore, to protoplast fusion and genetic transformation as methods of introducing new genes. Growers are particularly interested in resistance to disease. Blackspot disease, caused by the fungus *Diplocarpon rosae*, is a particular problem in the garden (Horst, 1983). Powdery mildew and grey mold caused by the fungi *Sphaerotheca pannosa* and *Botrytis cinerea*, respectively, are damaging diseases in the greenhouse (Horst, 1983). Improvement in the vigour of rootstock varieties (Van der Mark *et al.*, 1990) is also an objective of potential importance. Some sources of resistance to disease and improved vigour exist within the genus *Rosa*. Other sources within the family Rosaceae may be accessed by protoplast fusion and genetic transformation may be used to introduce genes from still more diverse taxonomic groups.

### 1.3. *Propagation of roses*

Most cultivated roses are propagated by the *in vivo* method of grafting (greenhouse plants) or budding (garden plants) onto rootstocks such as *R. multiflora* and *R. canina* "Inermis" which have been selected for their vigour (De Vries, 1993). Roses are also amenable to micropropagation and the extensive literature on this subject has been reviewed by Skirvin *et al.* (1990) and Short and Roberts (1991). Multiplication based on shoot proliferation can be induced on media such as MS salts and vitamins (Murashige and Skoog, 1962) supplemented with 30–40 g/l sucrose, cytokinin and auxin and, according to some authors, $GA_3$ (gibberellic acid). Rooting is improved by the use of media containing lower concentrations of ammonium ions than the MS formulation. This requirement can be met by either using the MS formulation at half or quarter strength, or reducing the proportion of ammonium nitrate in the salt mixture (Hyndman *et al.*, 1982). Another requirement for rooting is the use of a lower concentration of cytokinin than that in the multiplication medium. A good, representative protocol for the micropropagation of roses is that used by Douglas *et al.* (1989), which included MS salts and vitamins, 1 mg/l BA (6-benzyladenine), 0.1 mg/l NAA (1-naphthaleneacetic acid) and 0.1 mg/l $GA_3$ in the multiplication medium and 0.25 strength MS salts and vitamins, and 0.1 mg/l IAA (3-indoleacetic acid) in the rooting medium. Both media contained 30 g/l sucrose.

Many scion varieties do not sustain vigorous growth on their own roots and are unsuited to micropropagation as currently practised. It is possible, however, that they could be grafted onto rootstocks *in vitro* if this proves to be feasible on a commercial scale.

### 1.4. *Adventitious regeneration*

The earliest record of adventitious regeneration was that of Hill (1967) who did not indicate whether embryos or shoots were induced. Other authors have specified that they obtained either shoots (Arene *et al.*, 1993; Burger *et al.*, 1990; Ishioka and Tanimoto, 1990; Lu, 1993; Lloyd *et al.*, 1988; Roberts *et al.*, 1990; Rout *et al.*, 1992; Valles and Boxus, 1987) or embryos (De Wit *et al.*, 1990; Kunitake *et al.*, 1993, Matthews *et al.*, 1991; Noriega and Söndahl, 1991; Rout *et al.*, 1991), although, from some explants, Arene *et al.* (1993) obtained shoots and embryos on the same medium. Callus induced from *in vitro* roots of *R. persica* × *xanthina* in medium supplemented with 3 mg/l 2,4-D can be directed either towards the formation of adventitious shoots or embryos. Shoots are formed when callus is transferred to a medium supplemented with 0.3 mg/l $GA_3$ (gibberellic acid) and embryos are formed on a medium supplemented with 1 mg/l 2,4-D (2,4-dichlorophenoxyacetic acid), 1 mg/l ABA (abscisic acid) and 0.3 mg/l $GA_3$ (Roberts *et al.*, 1990). In this example, the pathway leading to development of shoots is preferable because it proceeds more rapidly than development of embryos (2 weeks vs. 10 weeks). In our experience, protocols based on media without 2,4-D do not maintain long-term regenerative potential and there are no published claims to the contrary. Protocols leading only to short–term regeneration are unlikely to be useable for the regeneration of plants from protoplast-derived cultures. Further investigation is needed into the common elements in the organogenic and embryogenic pathways. This may lead to the development of flexible protocols that combine speed of adventitious plant production with long–term retention of regenerative potential.

Plant regeneration after both protoplast fusion and genetic transformation of protoplasts with *Agrobacterium tumefaciens* have been achieved in the hybrid *R. persica* × *xanthina* (Matthews, 1993) using the protocol of Matthews *et al.* (1991). Plants transformed by *A. tumefaciens*, and *A. rhizogenes* have been regenerated from callus of cv Royalty (Firoozabady *et al.*, 1991) using the protocol of Noriega and Söndahl (1991). Roots of *R. canina* which were transformed with *A. rhizogenes* (van der Mark *et al.*, 1990) have now been induced to regenerate shoots (D.P. De Vries, personal communication). Clearly, the development of dependable methods of regeneration is needed to underpin further work on the biotechnology of roses.

## 2. Initiation of embryogenic callus

Embryogenic cultures have been initiated from *in vivo* leaves, internodes, filaments of stamens and zygotic embryos, and from *in vitro* leaflets, internodes and roots (Table 1). A variety of salts, vitamins, amino acids and growth regulators were used, and it is not clear to what extent the chosen media reflect specific requirements of different source tissues and different

*Figure 1.* Somatic embryogenesis in *Rosa hybrida* "Frensham" induced from root-derived callus using procedures of Matthews *et al.* (1991). (A) Friable callus on medium containing 3 mg/l 2,4-D, (B) Cotyledonary stage embryo producing secondary embryos on medium containing 3 mg/l 2,4-D; cotyledonary zone, hypocotyl bearing secondary embryos and radicle indicated by single, double and triple arrows, respectively, (C) Embryo development on medium without growth regulators; embryos with one cotyledon and two cotyledons indicated by single and double arrows, respectively; a precociously germinated embryo with three leaves is seen on right.

genotypes. It does appear, however, that not all responses observed by different authors on their chosen initiation media were equivalent. For example, De Wit *et al.* (1990) obtained somatic embryos from callus derived from leaflets on the same medium that was used for callus initiation, whereas Noriega and Söndahl (1991) subcultured callus derived from stamen filaments through a series of different media before obtaining normal embryos (Table 1). The calli that are initially induced differ in colour according to the tissue of origin. Green calli are obtained from leaves, yellow from stamens and white from internodes, roots and embryos. Although the texture of the initial callus varies from soft to hard, several authors describe a transition to a pale, friable and sometimes mucilaginous callus before somatic embryogenesis is initiated (Fig. 1A). One factor which might be integrated into existing protocols with advantage is the discriminating use of sugars. Kunitake *et al.* (1993) obtained rapid callus growth using galactose but found that different sugars, e.g., fructose, were needed to induce embryo development. Arene *et al.* (1993) based the culture medium for regeneration from leaves on that used in the present authors' laboratory (Lloyd *et al.*, 1988) but cut the leaf lamina into 3–4 mm strips. The callus they obtained was more regenerative, which suggests that endogenous responses to wounding may be important.

## 3. Culture maintenance

Long-term maintenance of callus in an embryogenic condition has been reported only on media containing 2,4-D (Matthews *et al.*, 1991; Noriega and Söndahl, 1991; Roberts *et al.*, 1990) (Table 1). In the authors' laboratory, root-derived calli have been maintained without loss of embryogenic potential on media containing 3 mg/l 2,4-D for 5 years in *R. persica* × *xanthina* and, so far for 0.5–2 years, in 12 cultivars representing a wide range of genotypes. Possibly, embryo formation on media containing 2,4-D is less dependent on the influence of endogenous hormones in the tissue of origin. Such influence is seen in the ability of zygotic embryos of *R. rugosa* to generate callus and then somatic embryos on media without exogenous growth regulators (Kunitake *et al.*, 1993) although this embryogenic capacity did not persist after 6 months.

The data of Arene *et al.* (1993), which are discussed in Section 5, indicates that genetic variation is generated in the callus phase. However, long-term culture may be an unavoidable requirement if the objective of an investigation is somatic hybridization or some other manipulation involving protoplast culture and subsequent regeneration of plants.

*Table 1.* Sources of embryogenic rose callus and stage of development induced on specified culture media. (Where authors tested more than one medium, only the best is shown.)

| Authors<br>Species/cultivar | Source of embryogenic callus.<br>Stages of development | Culture media used to induce stage specified in previous column<br>Salts | Vitamins and amino acids (mg/l)<br>m-inositol | nicotinic acid | pyridoxin -HCl | thiamine -HCl | glycine | others | Growth regulators (mg/l) | Carbohydrates, gels and other ingredients (g/l) |
|---|---|---|---|---|---|---|---|---|---|---|
| Arene *et al.* (in press)<br>cv Meirutral | ***In vitro* leaves & roots:**<br>Callus initiation embryo development* & germination | MS | 100 | 0.5 | 0.5 | 0.1 | 2.0 | - | NAA (0.2), BA (3) | sucrose (40), agar (7) |
| | ***In vitro* internodes:**<br>Callus initiation | MS | 100 | 0.5 | 0.5 | 0.1 | 2.0 | - | NAA(0.0005), BA (1.0) | sucrose (40), agar (7) |
| | Embryo development* & germination | MS | 100 | 0.5 | 0.5 | 0.1 | 2.0 | - | NAA (0.2), BA (3) | sucrose (40), agar (7) |
| | **Zygotic embryos:**<br>Callus initiation, embryo development* & germination | MS | 100 | 0.5 | 0.5 | 0.1 | 2.0 | - | 2,4-D (1), BA (0.5) | sucrose (40), agar (7) |
| De Wit *et al.* (1990)<br>cv Domingo &<br>cv Vickey Brown | ***In vitro* leaflets:**<br>Callus initiation & embryo development | ½MS | 50 | 0.25 | 0.25 | 0.05 | 1.0 | - | NAA (0.01), K (0.1) | sucrose (20), agar (7) |
| | Germination | MS | 100 | 1.0 | 1.0 | 0.1 | 2.0 | - | NOA (0.1), K (0.1) | sucrose (20), agar (7) |
| Kunitake *et al.* (1993) | **Zygotic embryos:** | | | | | | | | | |
| *R. rugosa* | Callus initiation | MS | 100 | 1.0 | 1.0 | 0.1 | 2.0 | - | - | galactose (36), gellan gum (2) |

*Table 1.* Continued.

| Authors<br>Species/cultivar | Source of embryogenic callus.<br>Stages of development | | Salts | m-inositol | nicotinic acid | pyridoxin -HCl | thiamine -HCl | glycine | others | Growth regulators (mg/l) | Carbohydrates, gels and other ingredients (g/l) |
|---|---|---|---|---|---|---|---|---|---|---|---|
| | | | Culture media used to induce stage specified in previous column | | | | | | | | |
| | | | | Vitamins and amino acids (mg/l) | | | | | | | |
| | Embryo development | | MS | 100 | 1.0 | 1.0 | 0.1 | 2.0 | - | K (1) | fructose (18), gellan gum (2) |
| | Germination | | MS | 100 | 1.0 | 1.0 | 0.1 | 2.0 | - | - | sorbitol (18), gellan gum (2) |
| Matthews *et al* (1991) | ***In vitro* roots:** | | | | | | | | | | |
| *R. persica x xanthina* | Callus initiation & maintenance | | SH | 1000 | 5.0 | 0.5 | 5.0 | - | - | 2,4D (3) | sucrose (30), agar (7) |
| | Embryo development & germination | | SH | 1000 | 5.0 | 0.5 | 5.0 | - | - | - | sucrose (30), agar (7) |
| Noriega & Söndahl (1991) | ***In vivo* stamens:** | | | | | | | | | | |
| | Callus initiation | a. | B5+ $NH_4SO_4$ (329mg/l) | 100 | 1.5 | 1.5 | 5.0 | 2.0 | - | NAA (0.25), Z (1.5), $GA_3$ (1) | sucrose (30), Gelrite (2.4) |
| | | b. | MS | 100 | 1.5 | 1.5 | 5.0 | 2.0 | - | NAA (0.25), Z (1.5), $GA_3$ (1) | sucrose (30), Gelrite (2.4) |
| | Maintenance | | N6 | 100 | 1.5 | 1.5 | 5.0 | 2.0 | - | 2,4-D (5), Z (1), $GA_3$ (1) | sucrose (20), Gelrite (2.4) |
| | Embryo development | a. | MS | 20 | 5.0 | 1.5 | 1.5 | 2.0 | - | $GA_3$ (1), ABA (2) | sucrose (20), Gelrite (2.4) |
| | | b. | MS | 100 | 5.0 | 1.5 | 1.5 | 2.0 | KM8P | $GA_3$ (1), ABA (0.2) | sucrose (20), Gelrite (2.4) coconut water (100) |

*Table 1.* Continued.

| Authors<br>Species/cultivar | Source of embryogenic callus.<br>Stages of development | | Salts | Culture media used to induce stage specified in previous column<br>Vitamins and amino acids (mg/l)<br>m-inositol | nicotinic acid | pyridoxin -HCl | thiamine -HCl | glycine | others | Growth regulators (mg/l) | Carbohydrates, gels and other ingredients (g/l) |
|---|---|---|---|---|---|---|---|---|---|---|---|
| | Germination | a. | ½MS | - | - | - | 1.0 | - | - | | sucrose (20), Gelrite (2.4) |
| | | b. | ¾MS | 100 | 1.5 | 1.5 | 0.5 | 2.0 | - | IAA (0.3), BA(3) | sucrose (30), agar (6) |
| Roberts *et al* (1990) | ***In vitro* roots:** | | | | | | | | | | |
| *R. persica x xanthina* | Callus initiation & maintenance | | SH | 1000 | 5.0 | 0.5 | 5.0 | - | - | 2,4-D (3) | sucrose (30), agar (8) |
| | Embryo development | | SH | 1000 | 5.0 | 0.5 | 5.0 | - | - | 2,4-D (1), $GA_3$(0.3), ABA (1) | sucrose (30), agar (8) |
| | Germination | | SH | 1000 | 5.0 | 0.5 | 5.0 | - | - | IBA (0.1), phloroglucinol (16) | sucrose (30), agar (8) |
| Rout *et al* (1991) | ***In vivo* leaves:** | | | | | | | | | | |
| 'Landora' | Callus initiation | | ½MS | 50 | 0.25 | 0.25 | 0.05 | 1.0- | - | 2,4-D (0.5), NAA (1.0) BA (0.5) | sucrose (30), agar (8) |
| | Embryo development | a. | ½MS | 50 | 0.25 | 0.25 | 0.05 | 1.0 | proline (600) | NAA (0.01), BA(0.5), $GA_3$ (0.1) | sucrose (30), agar (8) |
| | | b. | ½MS | 50 | 0.25 | 0.25 | 0.05 | 1.0 | - | NAA (0.01), BA(0.5), $GA_3$ (0.1) | sucrose (30), agar (8) |
| | Germination | | ½MS | 50 | 0.25 | 0.25 | 0.05 | 1.0 | - | BA(0.5),adenine sulphate (4.5), $GA_3$ (0.1) | sucrose (30), agar (8) |

*Table 1.* Continued.

| Authors<br>Species/cultivar | Source of embryogenic callus.<br>Stages of development | Salts | m-inositol | nicotinic acid | pyridoxin -HCl | thiamine -HCl | glycine | others | Growth regulators (mg/l) | Carbohydrates, gels and other ingredients (g/l) |
|---|---|---|---|---|---|---|---|---|---|---|
| | | | Vitamins and amino acids (mg/l) | | | | | | | |
| | ***In vivo* stems:**<br>Callus initiation | ½MS | 50 | 0.25 | 0.25 | 0.05 | 1.0 | - | 2,4-D (2.0), NAA (1.0), BA (0.5) | sucrose (30), agar (8) |
| | Embryo development | | | | as for leaves | | | | | |
| | Germination | | | | as for leaves | | | | | |

(Column headers m-inositol through Growth regulators and Carbohydrates are under the spanning head "Culture media used to induce stage specified in previous column".)

* adventitious embryos and shoots both observed on the same medium.

Abbreviations

Ca pant-calcium pantothenate
L-glut -L glutamine
**Vitamins:**
KM8P-formulation of Kao, KN & Michayluk, MR (1975). Planta 126 : 105-110.

**Salts:**
B5-formulation of Gamborg, OL, Miller, RA & Ojuna, K. (1968). Exp. Cell Res. 50 : 151-158.
MS-formulation of Murashige, T & Skoog, F (1962). Physiol. Plant. 15 : 473-497.
N6-formulation of Chu, C.C. *et al.* (1975). Scientia Sinica 18 : 659- 668.
SH-formulation of Schenk, RU & Hildebrandt, AC (1972). Can. J. Bot. 50 : 199-204.

**Growth regulators:**

2,4-D - 2,4-dichlorophenoxyacetic acid, IAA - 3-indoleacetic acid, IBA - indole -3-butyric acid, NAA - 1-naphthaleneacetic acid, NOA - 2-naphthoxyacetic acid, BA - 6-benzyladenine, Z - zeatin, $GA_3$ - gibberellic acid, ABA - abscisic acid.

## 4. Embryo development

Protocols for the maintenance of callus on media containing 2,4-D (Matthews *et al.*, 1991; Noriega and Söndahl, 1991; Roberts *et al.*, 1990) (Table 1) involve a reduction in the concentration or elimination of 2,4-D in the embryo development medium. Otherwise, initial stages of embryogenesis revert to recallusing or secondary embryos may form on the cotyledons and hypocotyl (Fig. 1B). This lowering of the auxin level to induce embryo development has often been used in other genera (Stafford and Warren, 1991). Both Noriega and Söndahl (1991) and Roberts *et al.* (1990) included ABA and $GA_3$ in the embryo development medium although at different concentrations (Table 1). Rout *et al.* (1991) reported that proline, optimally at a concentration of 600 mg/l, stimulated embryo initiation but that its exclusion from the embryo development medium reduced the frequency of developmental abnormalities.

During embryo development, white globular stage embryos (0.5–1 mm diameter) elongate (1–3 mm) while the cotyledons expand and open. Developmental abnormalities are often encountered, including the formation of a single cotyledon, more than 2 cotyledons and fused cotyledons (Fig. 1C). Mature somatic embryos are considerably smaller than zygotic embryos which are typically about 10 mm in length after emergence from the achene. As in zygotic embryos, greening of the cotyledons precedes germination.

## 5. Embryo germination

Media which have been used for embryo germination reflect a need for an adjustment to maintain co-ordinated development of shoot and root. The exposure of embryos to previous cultural regimes influences the adjustment required and it is, therefore, difficult to identify common trends in the protocols of different authors (Table 1). Strategies of some authors (Table 1) included changes in the type of auxin, cytokinin or sugar used. Thus De Wit *et al.* (1990) switched from 0.01 mg/l NAA in the development medium to 0.1 mg/l NOA (2-naphthoxyacetic acid) in the germination medium, Rout *et al.* (1991) included 4.5 mg/l adenine sulphate in the germination medium and Kunitake *et al.* (1993) switched from fructose (18 g/l) in the embryo development medium to sorbitol (18 g/l) in the germination medium (Table 1). In our laboratory, replacement of 30 g/l sucrose by 30 g/l maltose led to fewer developmental abnormalities and better germination.

Germination frequencies are often low but reported estimates range from 12% (Noriega and Söndahl, 1991; Rout *et al.*, 1991) to 60% (De Wit *et al.*, 1990). Noriega and Söndahl (1991) pointed out that successful germination could only be accomplished if embryos possessed normal cotyledons and a distinct hypocotyl. It is possible that differing estimates of germination rates reflect differing approaches to the selection of embryos for transfer to germination medium.

One procedure which accords with common practice in the germination of zygotic embryos and which could be complementary to the various protocols described in Table 1, is that of chilling somatic embryos before germination. Rout *et al.* (1991) found that treatment at 8°C for 4 days raised germination rates from 0% to 12% and chilling at 4°C for 2 weeks more than doubled germination rates in our laboratory.

## 6. Growth in the greenhouse and field

Several authors have reported transfer of *in vitro* regenerated roses to soil in the greenhouse or field but there have been no indications of physiologically abnormal development that would, for example, distinguish adventitiously regenerated plants from micropropagated plants.

The only detailed and systematic study of genetic variation amongst adventitiously regenerated roses has been that of Arene *et al.* (1993) on plants derived from cv Meirutral. In this study, no genetic variants were found in a sample of 130 plants regenerated by direct organogenesis. However, in a sample of 120 plants regenerated by adventitious organogenesis and embryogenesis from callus of leaves, roots and internodes, 21.7% differed in morphology from the original clone. Also, in a sample of 180 plants regenerated from callus induced from zygotic embryos, 70% differed in morphology from cv. Meirutral. As the zygotic embryos were obtained by self-fertilization of cv. Meirutral, some variation would be expected as a result of genetic segregation at meiosis, but the authors believed that a part of the variation could be attributed to somaclonal variation. Differences in the number, shape and colour of petals, and in growth habit were described but the data presented did not distinguish between plants regenerated from callus of vegetative and zygotic origin. Variants recorded in the first flush of flowers were propagated by cuttings and the expression of the variant characters was found to be stable in two succeeding flushes. This evidence strongly suggests that the expression of changed characters was genetically based and that variants of vegetative origin were thus somaclonal variants. These results indicate that direct organogenesis may give a sufficiently low frequency of "off types" to be considered as a method of micropropagation and that regeneration from callus could be used to generate somaclonal variants to improve or diversify existing cultivars.

## 7. Conclusions

Recent publications from several laboratories indicate that somatic embryogenesis can be induced in a range of rose genotypes. Regeneration can also be obtained by adventitious shoot production and it is not easy to predict which of these two approaches will give the best general method. Users will be attracted to protocols that can be applied to a wide range of genotypes and which minimise the time between initiation of cultures and production of regenerated plants. Other considerations, which may vary in importance according to the proposed application, include minimization or maximization of somaclonal variation and the maintenance of regenerative potential. Commercial interest in regenerating somaclonal variants, somatic hybrids and genetic transformants are likely to stimulate continuing interest in the refinement of protocols.

## 8. References


Arene, L., C. Pellegrino and S. Gudin, 1993. A comparison of the somaclonal variation level of *Rosa hybrida* L. cv. Meirutral plants regenerated from callus or direct induction from different vegetative and embryonic tissues. Euphytica 71: 83–90.

Burger, D.W., L. Lui, K.W. Zary and C.I. Lee, 1990. Organogenesis and plant regeneration from immature embryos of *Rosa hybrida* L. Plant Cell Tiss. Org. Cult. 21: 147–152.

Darlington, C.D. and A.P. Wylie, 1955. Chromosome Atlas of Flowering Plants. Allen and Unwin, London.

De Vries, D.P., 1993. The vigour of glasshouse roses. Doctoral Thesis, Agricultural University of Wageningen, Wageningen.

De Wit, J.C., H.F. Esendam, J.J. Honkanen and U. Tuominen, 1990. Somatic embryogenesis and regeneration of flowering plants in rose. Plant Cell Rep. 9: 456–458.

Douglas, G.C., C.B. Rutledge, A.D. Casey and D.H.S. Richardson, 1989. Micropropagation of floribunda, ground cover and miniature roses. Plant Cell Tiss. Org. Cult. 19: 55–64.

Firoozabady, E., C. Noriega, M.R. Söndahl and K.E.P. Robinson, 1991. Genetic transformation of rose (*Rosa hybrida* cv. Royalty) via *Agrobacterium tumefaciens*. In Vitro 27: 154A.

Hill, G.P., 1967. Morphogenesis of shoot primordia in cultured stem tissue of a garden rose. Nature 216: 596–597.

Horst, R.K., 1983. Compendium of Rose Diseases. Phytopathology Society, St. Paul, MN.

Hurst, C.C., 1941. Notes on the origin and evolution of our garden roses. J.R. Hort. Soc. (London) 66: 73–82, 242–250, 282–289.

Hyndman, S.E., P.M. Hasegawa and R.A. Bressan, 1982. Stimulation of root initiation from cultured rose shoots through the use of reduced of mineral salts. HortSci. 17: 82–83.

Ishioka, N. and S. Tanimoto, 1990. Plant regeneration from Bulgarian rose callus. Plant Cell Tiss. Org. Cult. 22: 197–199.

Kunitake, H., H. Imamizo and M. Mii, 1993. Somatic embryogenesis and plant regeneration from immature seed derived calli of rugosa rose (*Rosa rugosa* Thunb.). Plant Sci. 90: 187–194.

Lloyd, D., A.V. Roberts and K.C. Short, 1988. The induction *in vitro* of adventitious shoots in *Rosa*. Euphytica 37: 31–36.

Lu, C.Y., 1993. The use of thidiazuron in tissue culture. In Vitro Cell. Dev. Biol. 29P: 92–96.

Matthews, D., 1993. The use of protoplasts for the regeneration and genetic manipulation of roses. Ph.D. Thesis, University of East London, London.

Matthews, D., J. Mottley, I. Horan and A.V. Roberts, 1991. A protoplast to plant system in roses. Plant Cell Tiss. Org. Cult. 24: 173–180.

Murashige, T. and F. Skoog, 1962. A revised medium for rapid growth and bio-assays with tobacco tissue cultures. Physiol. Plant. 15: 473–497.

Noriega, C. and M.R. Söndahl, 1991. Somatic embryogenesis in Hybrid Tea roses. Bio/Technol. 9: 991–993.

Rehder, A., 1960. Manual of Cultivated Trees and Shrubs Hardy in North America, 2nd Ed. Springer/Berlin/Heidelberg/New York.

Roberts, A.V., I. Horan, D. Matthews and J. Mottley, 1990. Protoplast technology and somatic embryogenesis in *Rosa*. In: J. de Jong (Ed.), Integration of *In Vitro* Techniques in Ornamental Plant Breeding, pp. 100–115. CPO Centre for Plant Breeding Research, Wageningen.

Rout, G.R., B.K. Debata and P. Das, 1991. Somatic embryogenesis in callus cultures of *Rosa hybrida* L. cv. Landora. Plant Cell Tiss. Org. Cult. 27: 65–69.

Rout, G.R., B.K. Debata and P. Das, 1992. *In vitro* regeneration of shoots from callus cultures of *Rosa hybrida* L. cv. Landora. Indian J. Exp. Bot. 30: 15–18.

Rowley, G., 1960. Aneuploidy in the genus *Rosa*. J. Genetics 59: 258–68.

Short, K.C. and A.V. Roberts, 1991. *Rosa* spp. (Roses): *In vitro* culture, micropropagation and the production of secondary products. In: Y.P.S. Bajaj (Ed.), Biotechnology in Agriculture and Forestry, Vol. 15, pp. 376–397. Springer-Verlag, Berlin/Heidelberg.

Skirvin, R.M., M.C. Chu and H.J. Young, 1990. Rose. In: P.V. Ammirato, D.R. Evans, W.R. Sharp and Y.P.S. Bajaj (Eds.), Handbook of Plant Cell Culture Vol. 5, pp. 716–743. McGraw-Hill Publ. Co., New York.

Stafford, A. and G. Warren, 1991. Plant Cell and Tissue Culture. Open University Press, Milton Keynes.

Valles, M. and Ph. Boxus, 1987. Regeneration from *Rosa* callus. Acta Hort. 212: 691–696.

Van der Mark, F., J.P. Pijnacker-Hordijk, G.A.I. Varega, D.P. de Vries and J.J.M. Dons, 1990. *In vivo* transformation of clonal *Rosa canina* rootstocks with *Agrobacterium rhizogenes*. J. Genet. and Breeding 44: 263–268.

Wylie, A.P., 1954. The history of garden roses, Part 1. J.R. Hortic. Soc. (London) 79: 555–571.

Wylie, A.P., 1955a. The history of garden roses, Part 2. J.R. Hortic. Soc. (London) 80: 1024.

Wylie, A.P., 1955b. The history of garden roses, Part 3. J.R. Hortic. Soc. (London) 80: 77–87.

# 16. Somatic embryogenesis in cacao (*Theobroma cacao* L.)

Antonio Figueira and Jules Janick

**Contents**



## 1. Introduction

*Theobroma cacao* L. (cacao) is a major tropical crop, grown for its oil-rich seed, the unique source of cocoa solids and cocoa butter. Cacao is cultivated mainly by small-growers, and represents a significant export commodity for some developing countries. Major centers of cacao production are in West Africa (Ivory Coast, Ghana, Nigeria, Cameroon), Brazil, and more recently, Malaysia and Indonesia.

Cacao, a small understory tree, originated in the neotropical humid forests of Amazonia, and was domesticated in pre-Colombian times by Meso-American natives (Bergmann, 1969). Cacao exhibits a dimorphic type of growth. The orthotropic apical meristem ceases its growth when the plant reaches 1 to 2 m, leading to the development of lateral plagiotropic buds. The species is cauliflorous with inflorescences formed in old leaf scars on stems after hardening. After 2 to 3 years, the plant starts bearing fruits (pods) containing 30 to 50 seeds embedded in a sweet-acid mucilaginous pulp. This pulp has an important role in cacao quality, because precursors of the chocolate flavor develop during pulp fermentation. Cacao seeds are cold and desiccation sensitive (recalcitrant), and loose viability shortly after harvest. Because of seed recalcitrance, cacao germplasm needs to be preserved in clonal repositories, which utilize large areas and resources, and are highly susceptible to environmental risks.

Despite its economic importance, genetic improvement has not had a major impact on the cacao industry (Kennedy *et al.*, 1987). Cultivated cacao populations in the main producing areas have a narrow genetic base, and recent genetic improvement programs have been mainly based on progenies

*S. Jain, P. Gupta & R. Newton (eds.), Somatic Embryogenesis in Woody Plants, Vol. 2,* 291–310.

derived from crosses between local selections and a few genotypes collected near the crop center of diversity in the Amazon river headwaters. Disparities between predicted performance of these cacao "hybrids" (actually crosses of selected heterozygous genotypes) and actual yields in commercial plantations have inhibited the acceptance of this segregating material by growers, and have underscored the lack of basic genetic knowledge in cacao (Hunter, 1990). Additionally, a number of devastating cacao diseases (e.g., witches' broom caused by *Crinipellis perniciosa* Singer and monilia pod rot caused by *Monilia roreri* Ciferri in South America; swollen-shoot virus in West Africa; and vascular-streak dieback caused by *Oncobasidium theobromae* (Talbot and Kenea) in Southeast Asia) cannot be effectively controlled except for genetic resistance. Thus, new approaches are needed for genetic improvement of cacao.

The selection of asexually-propagated, superior genotypes appears to be the most efficient method for cacao genetic improvement. Grafting and budding have been used for over 90 years in cacao (Wood and Lass, 1987), and large scale commercial production of rooted cutting was developed in Trinidad in early the 1930s (Cope, 1976). More recently, budded plants have been used in Malaysia and Indonesia (Toxopeus, 1987). However, the commercial use of asexually propagated plants have been hampered by dimorphic growth, lack of efficient propagation technology, and high plant cost. Plagiotropic shoots are more widely available than orthotropic, and plants derived from plagiotropic budwood or cuttings, develop a low, spreading canopy making field work and harvesting more difficult. Production and the extensive pruning required for maintenance of rooted cuttings are costly, and higher technological input is required. Limited knowledge of possible rootstock-scion interactions, together with the possibility of chupon growth arising from the rootstock producing an undesirable tree, have limited the use of budded plants.

The development of a micropropagation system, generating a rapid increase in orthotropic propagules, offers the possibility of reducing cost per propagated plant, while overcoming the restriction imposed by plagiotropic growth habit. Low input agricultural systems employed by small cacao growers, should benefit enormously from the availability of disease-resistant, high-yielding orthotropic clones if they can be furnished at competitive prices. Further, *in vitro* techniques might allow germplasm preservation and distribution.

There are basically three morphogenic processes involved in rapid multiplication *in vitro* (micropropagation): organogenesis, shoot proliferation via rapid axillary branching, and somatic embryogenesis. Cacao, however, has been recalcitrant to tissue culture techniques.

### 1.1. *Organogenesis*

Early attempts to obtain organogenesis (morphogenesis) from *in vitro* culture of cacao had failed, and only callus growth was observed from various explant types: Archibald (1954) pioneered obtaining callus from bark cambium; callus and cell suspension cultures were obtained from stem and petioles (Townsley, 1974), and from *in vitro*-germinated seedling root, stem, hypocotyl, and cotyledon (Hall and Collin, 1975). Searles *et al.* (1976) used immature leaf and pod sections to produce callus. Calli were later obtained from pericarp, ovule, embryo (Pence *et al.*, 1979), and anthers (Esan, 1977). A large number of media salt formulations, supplemented with various plant growth regulators, concentrations, and mode of applications, as well as cacao plant extracts, and complex organic addenda (e.g. coconut water, alkaloids) have been tested in culture without any positive effects on organogenesis, with the exception of root initiation (Dublin, 1984).

### 1.2. *Shoot proliferation*

Restricted growth had been also a common observation for shoot tip and axillary bud cultures *in vitro*. There is a lack of consistency between inductive hormonal treatments with growth and development responses (Orchard *et al.*, 1979; Passey and Jones, 1983; Janick and Whipkey, 1985; Blake and Maxwell, 1984; Legrand *et al.*, 1984; Dufour and Dublin, 1985; Legrand and Mississo, 1986; Bertrand, 1987). Explant factors (size, donor tissue, physiological stage of donor shoot, bud placement, orientation in culture), media formulation (basal salts, carbon source concentration, type, vitamins, physical state), and physical conditions (light, gas exchange, temperature) have been intensively studied without substantial progress. Excessive callus formation is a common observation. In general only sporadic growth of the first flush of shoots had been obtained (Orchard *et al.*, 1979; Passey and Jones, 1983; Blake and Maxwell, 1984; Dufour and Dublin, 1985; Bertrand, 1987). Flynn *et al.* (1990) reported induction of budbreak and elongation of axillary buds from mature plants with continuous growth for six months. Promotive factors were described to be light quality, culture vessel humidity, frequent explant transfer, explant orientation, and decreased explant stress. Cotyledonary nodes can be considered an exception to the recalcitrance of cacao, because axillary shoots can proliferate and grow while attached to the original node (Janick and Whipkey, 1985; Kong and Rao, 1982; Legrand *et al.*, 1984; Legrand and Mississo, 1986). Subcultured axillary shoots, however, failed to grow when detached from the cotyledonary node (Janick and Whipkey, 1985).

Results in our laboratory at Purdue University indicated that detached axillary shoots, single-node cuttings from mature plants, as well as microcuttings from elongated shoots cultured *in vitro*, elongate and produce leaves under elevated $CO_2$, with further subculturing possible for the first time (Figueira *et al.*, 1991, 1994). Increasing $CO_2$ concentration from ambient (470 ppm) to 24,000 ppm (culture vessel levels) significantly improved total shoot elongation, number of leaves, leaf area per explant, and shoot dry and fresh weight (Figueira and Janick, 1994). Although light was necessary for the $CO_2$ response, the effect of various photon fluxes was not significant for the measured growth parameters. Although these advances indicate that microcuttings could be a promising way to clonally propagate cacao these techniques have not been tested on a commercial scale.

### 1.3. *Somatic embryogenesis*

Somatic embryogenesis, the process of embryo initiation and development from cells that are not the direct product of gametic fusion is a natural phenomenon in many species *in vivo*, that is generally restricted to intraovular tissues. The induction of somatic embryogenesis has become an important tool of biotechnology since its discovery by Steward *et al.* (1958) and by Reinert (1958). There are many benefits anticipated from somatic embryos based on their high potential for proliferacy, their bipolar structure with both root and shoot meristems, and their ability to be singulated for synthetic seed (Janick, 1993). Other uses include metabolite production, disease elimination, germplasm preservation, and the production of homozygous lines via androgenesis (somatic embryos from microspores). The fact that somatic embryos can regenerate from single cells provide opportunities for crop improvement exploiting biotechnological techniques such as cell selection, somatic hybridization, and transformation. The limiting factor for the exploitation of somatic embryos still remains to be the lack of reliable protocols for induction, development, and conversion (germination), especially in recalcitrant species.

## 2. Initiation and culture of cacao somatic embryos

Somatic embryogenesis has been reported from a number of tissues of cacao including zygotic embryos (Esan, 1977; Pence *et al.*, 1979, 1980), nucellar tissue (Söndahl *et al.*, 1989; Chatelet *et al.*, 1992; Figueira and Janick, 1993); young bud petals (Söndahl *et al.*, 1989; Lopez-Baez *et al.*, 1993); and leaves (Litz, 1986). Somatic embryos derived from immature zygotic embryos have limited value for propagation because they are untested genotypes. Thus, only somatic embryos derived from sporophytic tissue (such as nucellus, leaf, or petal) have value for germplasm conservation or clonal propagation. Somatic embryos derived from zygotic tissues however, can be used for metabolite production.

### 2.1. *Via zygotic embryos*

Somatic embryogenesis in cacao was first reported by E.B. Esan at the Vth International Cocoa Research Conference at Ibadan, Nigeria in September 1975, and the proceedings were published in 1977 (Esan, 1977). The phenomenon was reported independently by Pence *et al.* in 1979, and subsequently repeated by a great number of workers (Kong and Rao, 1982; Tahardi, 1984;

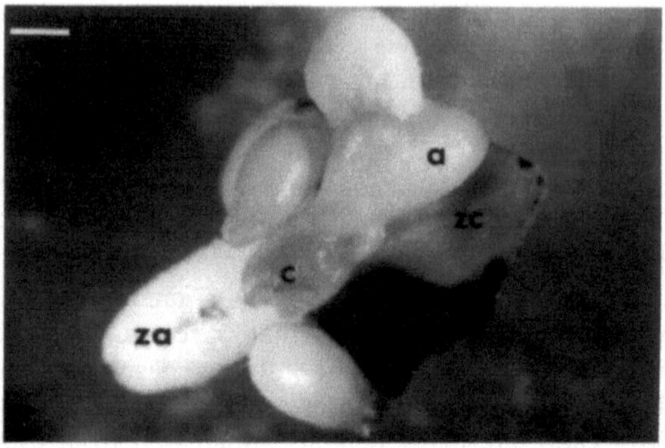


*Figure 1.* Somatic embryos initiated via the "non-budding" process by differential growth of cotyledonary tissue. Developing somatic embryo with a defined axis (a) and cotyledons (c) clearly distinguishable from the axis (za) and cotyledons (zc) of the zygotic embryo (bar = 1 mm), (Pence *et al.*, 1980).

Adu-Ampomah *et al.*, 1988; Santos and Machado, 1989; Duhem *et al.*, 1989). The protocols for the induction of somatic embryogenesis involve the following steps:

i. Extraction of immature embryos close to the 100-day stage after pollination (white to white/pink spade-shaped embryos). Maximum embryogenesis occurs in translucent white embryos about 6.5 to 7.0 mm (Pence *et al.*, 1979, 1980).
ii. Zygotic embryos initiate somatic embryos at low frequency on basal medium with casein hydrolysate, but the combination of auxin ($\alpha$-naphthalene-acetic acid, NAA; 2,4-dichlorophenoxyacetic acid, 2,4-D; or indole-3-acetic acid, IAA) and coconut water (10%) act synergistically to stimulate somatic embryogenesis. Alternatively, somatic embryogenesis has been initiated on basal media devoid of complex organic addenda (coconut water and casein hydrolysate) supplemented with 4 $\mu$m zeatin (Santos and Machado, 1989), or 4.44 $\mu$m 6-benzylaminopurine (BA), 0.03 $\mu$m gibberellic acid ($GA_3$), plus 0.57 $\mu$m IAA (Duhem *et al.*, 1989).
iii. Maintenance of embryos is carried out in auxin-free medium.

Somatic embryos arise from zygotic embryos through four pathways: (1) A "non-budding" process (Fig. 1) whereby structures resembling embryos or parts of embryos differentiate from internal meristematic cotyledonary tissue (Pence *et al.*, 1980); (2) a budding process (Fig. 2) in which somatic embryos pass through the normal stages of embryogenesis; globular structures appear on a suspensor-like stalk that arises from epidermal tissue on cotyledonary veins or the embryonic axes (Pence *et al.*, 1980); (3) a process referred to as high frequency somatic embryogenesis involving proembryonic cell masses

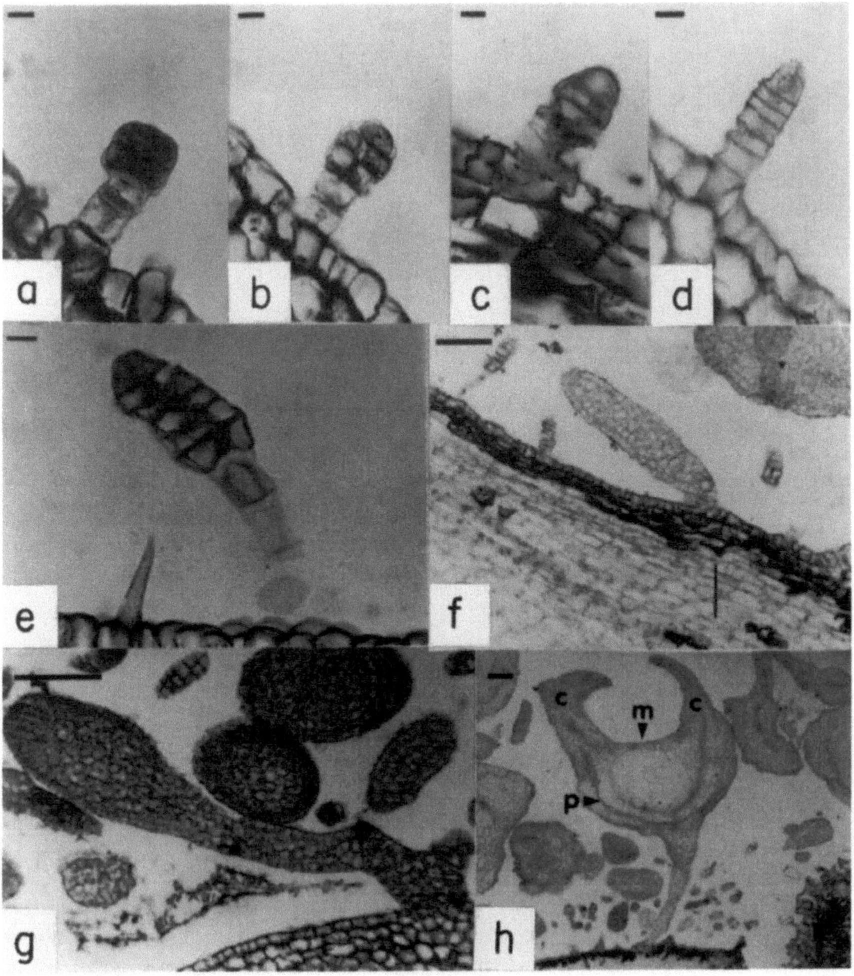


*Figure 2.* Photomicrographs illustrating the initiation of "budding" type somatic embryos. (a, b) Hair-like structure with 4- to 10-celled globular body on a 3- to 4-celled stalk (bar = 0.01 mm). (c, d) Hair-like structures formed by transverse divisions only (bar = 0.01 mm). (e-g) Developing somatic embryos still attached to the stalk with the upper portion becoming multicellular (e; bar - 0.01 mm; f and g; bar - 0.1 mm); and (h) developing a characteristic embryo morphology. Cotyledons (c) and apical meristem (m) are clearly discernable along with the procambial strands (p) (bar = 0.1 mm), (Pence *et al.*, 1980).

*Figure 3.* Proembryogenic callus mass.

*Figure 4.* Embryogenic callus showing various stages of somatic embryo development.

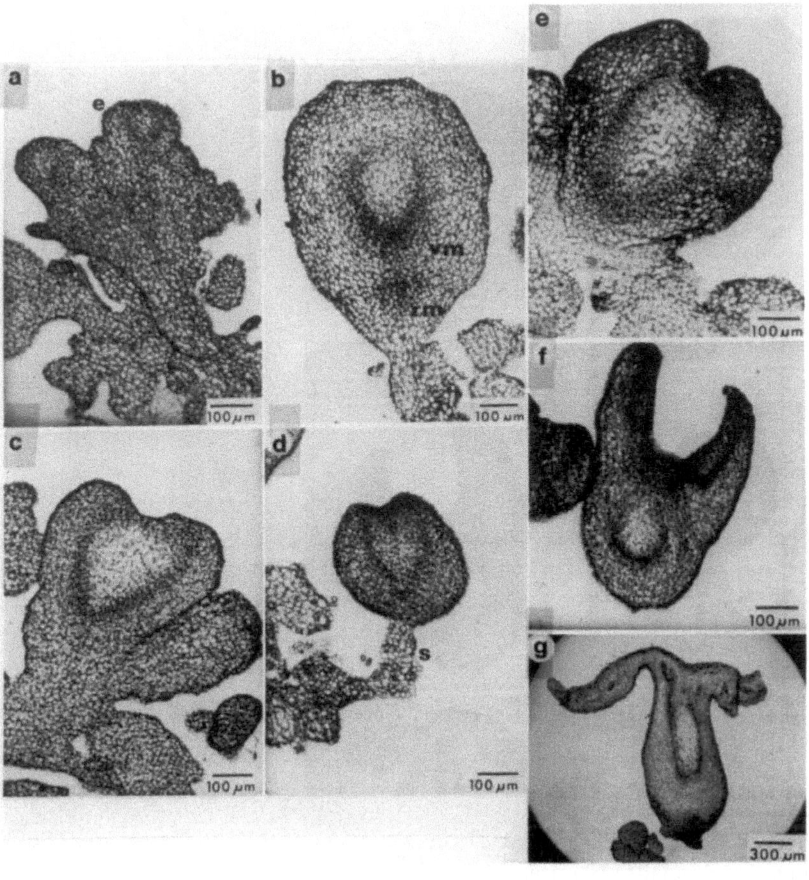


*Figure 5.* Development of somatic embryos of cacao via callus. (a) An older nodule showing an early stage of embryo (e) formation; (b) globular embryo showing the formation of root meristem (rm) and vascular meristem (vm); (c) heart-shaped embryo loosely connected to the callus mass; (d) heart-shaped embryo with a suspensor-like connection to the callus mass; (e) heart-shaped embryo loosely connected to the callus mass; (f) torpedo stage; (g) walking-stick embryo stage showing cotyledonary development (Kononowicz and Janick, 1984).

(Fig. 3) (Sharp *et al.*, 1980); and (4) a process where embryos develop from embryogenic callus (Figs. 4 and 5) (Kononowicz *et al.*, 1984).

## 2.2. *Via nucellar tissue*

The first attempt to culture nucellar tissue was also made by Esan, 1977. In his paper, he shows a figure (Fig. II) that suggests that abnormal somatic embryogenesis has occurred, but it is not clear if the somatic embryo is

*Figure 6.* Cotyledonary embryos (ce) and embryogenic callus (ec) arising from nucellar tissue (n) (Figueira and Janick, 1993).

derived from nucellar tissue or from a portion of the zygotic embryo. Subsequently, Söndahl (1991) and Söndahl *et al.* (1989, 1993) confirmed somatic embryogenesis from nucellar tissue. His protocol involved a multistep procedure involving basal media supplemented with auxin (2,4-D), cytokinin (6-($\gamma,\gamma$-dimethylallylamino)purine, 2iP), gibberellins, abscisic acid (ABA), and organics (coconut water, malt extract). Chatelet *et al.* (1992) reported embryogenic cell initiation (proembryos) derived from nucellar and inner integuments as early as 21 days after culture on an induction media; mature somatic embryos, however, were not obtained. Figueira (1992) and Figueira and Janick (1993) reported the initiation and development of somatic embryos from nucellar tissue (Fig. 6) using a procedure listed below:

Ovules (ca. 12 mm) are extracted from immature fruits (80 to 120 mm), about 60 days after pollination. At this stage zygotic embryos are globular and barely visible macroscopically. The basal end of the ovule containing the zygote is excised and liquid endosperm is drained away. The remaining ovular material is dissected into about 4 to 5 pieces, and explanted onto a somatic embryo induction media consisting of half-strength liquid Murashige and Skoog (MS) salts, vitamins (0.3 μM thiamine-HCl, 2.4 μM pyridoxine-

HCl, 0.6 mM *i*-inositol, 4.1 μM nicotinic acid, 26.6 μM glycine), and supplemented with 117 mM sucrose, 10% coconut water, 4 μM 2,4-D, 0.5 μM 2iP, 0.5 g · $l^{-1}$ malt extract, 0.5 g · $l^{-1}$ casein hydrolysate, and 0.2% polyvinylpyrrolidone mol wt 10,000 (PVP), with pH adjusted to 5.3 before autoclaving. The cultures are placed on a gyratory shaker in the dark for 60 days with one transfer to the same media after 30 days. The pieces of callus are transferred to a semi-solid developmental medium, consisting of half-strength MS media with the same vitamins as above, but supplemented with 117 mM sucrose, 2.5 μM 2iP, 10% coconut water, 0.1 g · $l^{-1}$ malt extract, 0.2% PVP 10,000, with pH adjusted to 5.3 before autoclaving for 60 days under light. Callus is recultured on semi-solid maintenance medium with half-strength MS supplemented with the same vitamins as above, 87.6 mM sucrose, 1 g · $l^{-1}$ casein hydrolysate, and 8 g · $l^{-1}$ agar, with pH adjusted to 5.3 prior to autoclaving.

Even with the presence of anti-oxidants extensive, browning occurs on ovular explants after 5 to 10 days in liquid induction media. Despite this browning, a yellowish callus with nodular appearance can arise during the first 60 days in culture from both the cut surface and intact areas of the nucellus. After transfer to semi-solid media all tissues turn completely brown and appear necrotic. However, after 150 days in culture, green outgrowths form and somatic embryos begin to appear in low frequency. Embryogenic budding centers consisting of proliferating globular embryos are produced and can be continually cultured on the maintenance medum. Although somatic embryos have been obtained from genotypes of various origins at low frequency, results indicate that nucellar embryony may not be easily obtained in all genotypes.

### 2.3. *Via leaves and petals*

The only report of induction of somatic embryogenesis from leaves is based on a preliminary report of Litz (1986), who reported low frequency somatic embryogenesis from young fully-expanded leaves, using a semi-solid half-strength MS medium supplemented with 316 mM sucrose, 0.2% activated charcoal, and high levels of 2,4-D (22.5 to 90 μM) and BA (444 to 666 μM). Somatic embryos did not develop beyond the late heart stage.

The induction of somatic embryogenesis from petals has been reported by Söndahl (1991) and Söndahl *et al.* (1989, 1993). Young flower bud petals (4 to 8 mM long) were first pre-treated at 10°C for 24 h before inoculation on a liquid induction medium containing cytokinin and 2,4-D, and then cultured in the dark. Direct embryogenesis was observed on the basal portion of young petal explants after 3 weeks. The response was genotype dependent.

## 3. Culture maintenance

There are three ways to maintain somatic embryogenic cultures: via budding centers, via proembryonic cell masses, and via embryogenic callus.

Somatic embryos of some clones continue to proliferate in the absence of growth regulators by budding (direct epidermally produced embryos via somatic embryos), which form continuously in certain propagating cultures. These have been referred to as *budding centers*. Of 50 clones originally established, two clones continued to proliferate for over five years through hypocotylary budding.

In many cases budding centers revert to a proembryonic cell mass, consisting of globular embryos from which rapid repetitive budding occurs. This proembryonic callus mass may loose its capacity for development and must be rejuvenated. Rejuvenation is promoted by coconut water and replacement of sucrose by glucose or fructose as the carbon source.

The other system for maintenance of somatic embryos is through embryogenic callus. Embryogenic callus is pale yellow and friable; an associated white callus which turns brown is nonembryogenic (Kononowicz *et al.*, 1984; Kononowicz and Janick, 1984c). The embryogenic callus may be induced to produce embryos by various treatments. The response of embryogenic calli to gibberellic acid is complex and clone specific (Kononowicz and Janick, 1984c). For example, $GA_3$ stimulated embryogenesis in one clone but not another. Under certain circumstances embryogenic callus completely reverts to embryos (Fig. 4, right), but the defining stimulus of this "high frequency" embryogenesis is unresolved. Embryogenesis from callus is stimulated when sucrose is replaced by glucose or fructose (Elhag *et al.*, 1987, 1988). Embryogenic callus retains embryogenic competence in liquid culture. Embryogenic liquid cultures have been established from nucellus and petals, and a frequency of 8.8% somatic embryos per colony was recorded 4 weeks after plating (Söndahl *et al.*, 1993).

## 4. Somatic embryo uses

Cacao somatic embryogenesis has been studied with two objectives. One is to develop somatic embryos metabolically into analogs of mature seed for the production of cocoa butter and cocoa solids (Janick, 1986, 1991). The other objective is to convert somatic embryos into plants for propagation. Because somatic embryos have the capacity to germinate precociously, it is unnecessary to complete embryo development for propagation purposes.

### 4.1. In vitro *production of cocao butter*

Embryos cultured *in vitro* can be shifted to the accumulation phase by the addition of high concentrations of sucrose to the basal medium (Pence *et al.*,

*Table 1.* Embryo growth and lipid development of somatic embryos of cacao cultured in low and high sucrose compared with zygotic embryos *in vivo* (Janick *et al.*, 1982; Kononowicz and Janick, 1984).

| Variable | Somatic embryo | | Zygotic embryo *in vivo* |
|---|---|---|---|
| | 3% Sucrose | 3–27% Sucrose[z] | |
| embryo growth | | | |
| Fresh wt (FW) (g) | 3.8 | 1.0 | 1.5 |
| Dry wt (DW) (g) | 0.3 | 0.4 | 1.0 |
| DW/FW (%) | 8.0 | 39.0 | 66.7 |
| Lipid production | | | |
| Lipid (L)/DW (%) | 6.9 | 25.9 | 48.0 |
| Fatty acid (FA)/DW (%) | 1.3 | 25.3 | 45.0 |
| FA/L | 18.8 | 97.7 | 93.8 |
| FA composition (mol%) | | | |
| Palmitic (P) | 30.6 | 26.4 | 28.8 |
| Stearic (S) | 8.7 | 28.8 | 33.5 |
| Oleic (O) | 12.3 | 34.9 | 34.0 |
| P + S + O | 51.6 | 89.1 | 96.3 |
| Linoleic | 40.4 | 9.4 | 3.2 |
| Linolenic | 7.5 | 0.4 | 0.3 |

*Table 2.* Triglycerides of asexual embryos of cacao cultured in low and high sucrose protocols compared to commercial cocoa butter (Kononowicz and Janick, 1984).

| Triglyceride[z] | Content (mol%) | | Cocoa butter |
|---|---|---|---|
| | 3% Sucrose | 3–27% Sucrose | |
| Monounsaturated | | | |
| POS | 11.4 | 32.1 | 36.8 |
| POP | 13.5 | 27.2 | 18.8 |
| SOS | 28.0 | 15.6 | 22.6 |
| Total | 52.9 | 74.9 | 78.2 |
| Diunsaturated | | | |
| SOO | 0.0 | 5.5 | 8.9 |
| POO | 0.0 | 8.4 | 8.9 |
| Total | 0.0 | 13.9 | 17.8 |
| Polyunsaturated | | | |
| PLP | 47.1 | 7.9 | 2.2 |
| PLO | 0.0 | 3.3 | 1.8 |
| Total | 47.1 | 11.2 | 4.0 |

[z] P – Palmitic acid; S – Stearic acid; O – Oleic acid.

1981a,b). The most efficient protocol (Pence *et al.*, 1981b) has been to place spade-shaped embryos (equivalent to 100- to 120-day-old zygotic embryos) in liquid basal medium containing 3% sucrose with culture tubes placed in a Rollodrum apparatus rotating at 5 rpm for 10 days and then shift to basal medium with elevated sucrose, i.e., two days each with 9, 15, 21% sucrose and 30 days with 27% sucrose for a total of 46 days in liquid medium. Growth of embryos in liquid is maximal with 3% sucrose. The high sucrose protocol after day 10 induces the suppression of growth and the simultaneous accumulation of anthocyanins (Wright *et al.*, 1982), lipids (Pence *et al.*, 1981b), alkaloids (Paiva and Janick, 1983), and proteins, mimicking normal development *in vivo*. Zygotic and somatic embryos respond similarly to this protocol (Kononowicz and Janick, 1984a). Fatty acid and triglyceride accumulation of somatic and zygotic embryos *in vitro* and zygotic embryos *in vivo* follow a similar pattern of development. This can be demonstrated by comparing embryos cultured *in vivo* and *in vitro* in respect to the mole percentage of each fatty acid (Table 1) or triglyceride (Table 2), or by the rate of accumulation of individual fatty acids as total fatty acids increase. Based on the accumulation pattern, the quality of lipids from somatic embryos should closely resemble commercial cocoa butter when fatty acid content on a gram dry weight basis approaches 50%, as found in mature zygotic embryos.

An analysis of lipid quality of embryos with a fatty acid content of about 30% by differential scanning calorimetry indicated that cocoa butter-like material was produced by *in vitro* culture of embryos, but some low melting components remained, indicating that lipid quality was not identical to cocoa butter. Fatty acid quality equivalent to cocoa butter cannot be expected until fatty acid in somatic embryos reaches at least 50% on a dry weight basis (Kononowicz and Janick, 1984a). The sucrose effect on the accumulation phase of embryo development is a general "turn on" characterized by simultaneous increases in anthocyanin, alkaloid (caffeine + theobromine), and lipid content on a dry weight basis. This developmental signal is independent of embryo size. However, stage is critical. For example, embryos grown beyond 10 days in 3% sucrose lose their responsiveness to the high sucrose signal.

Although sucrose, glucose, or fructose all support proliferation and growth of somatic embryos (Janick *et al.*, 1984), only sucrose stimulates the accumulation phase of embryo development. Not only does glucose not support lipid accumulation, but small amounts of glucose in combination with sucrose inhibit accumulation (Kononowicz and Janick, 1984b).

The protocol for development of somatic embryos is characterized by large variability from experiment to experiment in embryo dry weight as well as fatty acid production on a dry weight basis. Much of this variation appears to be related to embryo state, but even under the best conditions embryos, after 46 days in the high sucrose protocol, do not reach maturity characterized by a fatty acid content of 50% on a dry weight basis. The best individual somatic embryos reached 39% fatty acid, but because of embryo variability the highest mean in any single experiment was 25% fatty acids. This lack of complete development is not due to insufficient time in 27% sucrose. The development of somatic embryos in high sucrose reaches a peak at 30 days in a high sucrose liquid medium. When embryos are left for more than 30 days, fatty acid content on a dry weight basis does not increase appreciably or declines. Medium replenishment does not improve fatty acid accumulation.

Various factors affecting fatty acid content *in vitro* were tested on the assumption that some factor in the protocol apparently arrests or prevents embryo development. Nutrient modification included total MS salts, inorganic and organic nitrogen, potassium, and phosphorus as well as growth-regulating substances including coconut water, BA, NAA, GA, and ABA. Environmental factors included light and temperature. None of the nutrient modifications including organic and inorganic salts or growth substances increased fatty acid accumulation over the basal medium plus sucrose from 3 to 27% in the standard protocol (Wright *et al.*, 1984a). Light or darkness was not critical. A series of experiments with temperatures ranging from 10 to 35°C indicated that fatty acid quantity and quality was optimum at 26°C (Wright *et al.*, 1984b).

The fact that some embryos (zygotic or somatic) reach high levels of fatty acid content concomitant with high fatty acid quality indicates that the state of the embryo (embryo receptivity) is a critical factor. Clearly, stage or state of the embryo before initiation of the high sucrose protocol is important. However, the absence of complete development in any embryo indicates that the protocol does not allow complete development. The factor(s) that limit the system are unknown.

## 4.2. *Micropropagation via somatic embryogenesis*

Micropropagation via somatic embryogenesis would be feasible with somatic embryos derived from sporophytic tissues. The successful conversion of cacao somatic embryos to independent plants, however, has not been readily achieved. Immature cacao somatic embryos rarely germinate. Soaking spade-shaped somatic embryos in distilled water or renewing a liquid medium induced radical development in up to 40% of embryos suggesting that removal of one or more inhibitors promoted precocious germination (Wang and Janick, 1984). The leachate from somatic embryos inhibited seed germination of lettuce. The inhibitor was shown not to be ABA but resembled *p*-coumaric acid (Wang and Janick, 1985). Pence (1991a), however, detected high levels of ABA during development of zygotic embryos of cacao *in vivo*, peaking at early maturation phase. Somatic embryo conversion to seedlings was enhanced by cotyledonary excision (Novak *et al.*, 1986; Adu-Ampomah *et al.*, 1988; Duhem *et al.*, 1989), use of liquid medium (Adu-Ampomah *et al.*, 1988), and the inclusion of high levels of activated charcoal (Duhem *et al.*, 1989), protocol modifications all compatible with the hypothesis of the presence of an inhibitor. Micrografting immature somatic embryos on *in*

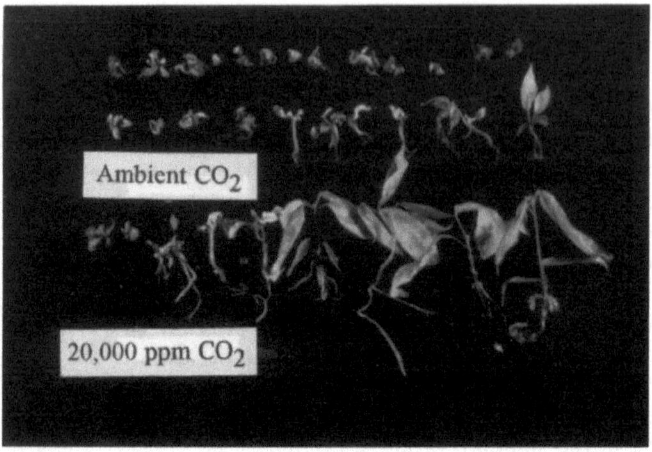


*Figure 7.* Somatic embryos of zygotic origin showing increasing leaf and root development in response to 20,000 ppm $CO_2$ as compared to ambient $CO_2$ (Figueira and Janick, 1993)

*vitro* germinated seedlings also increased conversion (Villalobos and Aguilar, 1990).

Figueira and Janick (1993) have shown that somatic embryo development responds to high $CO_2$ as do shoots (Fig. 7). High rates of conversion to plants were obtained based on the following protocol:

i. Globular to torpedo stage somatic embryos (2 to 5 mm) are transferred to a maturation half-strength liquid MS medium supplemented with 234 mM sorbitol and 29 mM sucrose for 15 days in a gyratory shaker (100 rpm).
ii. Somatic embryos are transferred to an embryo conversion medium consisting of semi-solid Woody Plant medium (Lloyd and McCown, 1980), supplemented with 88 mM fructose, placed in chambers receiving 20,000 ppm $CO_2$ (Figueira *et al.*, 1991; Figueira and Janick, 1993).
iii. Plantlets can be transferred to soil after 60 days.

Evaluation of cacao somatic embryo growth under field conditions has not been conducted, but normal growth and development have been observed under greenhouse conditions (Fig. 8). Presently, field trials with plants derived from nucellar and petal somatic embryos are under establishment in Costa Rica and Ecuador (M.R. Söndahl, Fitolink Co., personal communication).

*Figure 8.* Potted seedling derived from nucellar embryogenesis.

## 5. Cryopreservation

Cryopreservation, or storage under liquid nitrogen conditions, is the method of choice for long-term germplasm conservation (Withers, 1984). Freezing hydrated tissues, such as recalcitrant seeds, is usually lethal, and cryoprotectant solutions together with slow freezing procedures should be employed (Pence, 1991b). The main limitation to the use of desiccated embryos consist in survival after desiccation. Attempts to cryopreserve cacao aggregate cell suspensions, calli, shoot tips, mature zygotic embryos (Withers, 1984; Withers *et al.*, 1986), and dissected mature embryo axes (Withers, 1984; Withers *et al.*, 1986; Pence, 1991b) have failed. Fine cell cultures were successfully preserved, but because of the non-regenerative capacity, this result was of little impact (Withers, 1984). Withers (1984) and Withers *et al.* (1986) suggested cryopreservation of dormant buds of cacao recovered *in vitro* or by grafting, but no progress has been published.

Recently, Pence (1991b) evaluated slow hydrated freezing after cryoprotectant treatment (0.5 M sucrose and 10% DMSO), and desiccated fast freezing for the maintenance of embryogenic potential of immature zygotic embryos of cacao. Calli and somatic embryos were produced from hydrated, frozen and, at a lower rate, from desiccated, fast frozen immature zygotic embryos. The survival of immature embryos probably relates to more uniform cryoprotection or desiccation in small embryos (Pence, 1991b). Genetic stability was not evaluated.

The similarity between immature zygotic and somatic embryos of cacao, suggests that identical procedures for cryopreservation developed for immature zygotic embryos should be applicable for nucellar and petal somatic embryos. Successful cryopreservation of nucellar somatic embryos combined with a successful regeneration system should allow the establishment of a long-term preservation of clonal material.

## 6. Conclusion

Somatic embryogenesis in cacao has been induced from immature zygotic embryos and nucellar tissue and, with limited success, from other sporophytic tissues including young flowers and leaves. Somatic embryos, once induced, can proliferate indefinitely. Under appropriate protocols somatic embryos develop, either as physiologically mature cotyledons or toward acclimatized autotrophic "seedlings." Potential uses of somatic embryos include rapid clonal propagation, cryopreservation for germplasm preservation, and *in vitro* production of cotyledonary lipids (cocoa butter) and solids. The potential usefulness of somatic embryogenesis for micropropagation or germplasm preservation will depend on the development of efficient techniques to induce somatic embryos in sporophytic tissues and to encourage efficient development into self-sustaining plants. The exploitation of somatic embryos for metabolite production requires developing protocols to induce complete maturity of cotyledons under *in vitro* conditions. None of these potentioal benefits will be possible without extensive research on the physiology of somatic embryo morphogenesis.

## 7. References

Adu-Ampomah, Y., F.J. Novak, R. Afza, M. van Duren and M. Perea-Dallos, 1988. Initiation and growth of somatic embryos of cocoa (*Theobroma cacao* L.). Café Cacao Thé 32: 187–200.

Archibald, J.F., 1954. Culture *in vitro* of cambial tissue of cacao. Nature 173: 351–352.

Bergmann, J.F., 1969. The distribution of cacao cultivation in pre-columbian America. Ann. Assoc. Amer. Geographer 59: 85–96.

Bertrand, B., 1987. Quelques aspects du microbouturage du cacaoyer (*Theobroma cacao* L.). Café Cacao Thé 31: 175–182.

Blake, J. and P. Maxwell, 1984. Tissue culture propagation of cacao by the use of axillary buds. Proc. Int. Conf. Cocoa Coconuts, Kuala Lumpur, Malaysia.

Chatelet, P., N. Michaux-Ferriere and P. Dublin, 1992. Potentialites embryogenes du nucelle et du tegument interne de graines immatures de cacaoyer (*Theobroma cacao* L.). C.R. Acad. Sci. Paris Serie III. 315: 55–62.

Cope, F.W., 1976. Cacao. In: N.W. Simmonds (Ed.), Evolution of Crop Plants, pp. 285–289. Longman, London.

Dublin, P., 1984. Cacao. In: P.V. Ammirato, D.A. Evans, W.R. Sharp and Y. Yamada (Eds.), Handbook Plant Cell Culture, Vol. 3, pp. 541–563. Macmillan, New York.

Dufour, M. and P. Dublin, 1985. Quelques donnees sur la multiplication vegetative *in vitro* des cacaoyers cultives (*Theobroma cacao* L.). Café Cacao Thé 29: 235–244.

Duhem, K., N. Lemercier and P. Boxus, 1989. Données nouvelles sur l'induction et le developpement d'embryons somatiques chez *Theobroma cacao*. Café Cacao Thé 33: 9–14.

Elhag, H.M., A. Whipkey and J. Janick, 1987. Induction of somatic embryogenesis from callus in *Theobroma cacao* in response to carbon source and concentration. Rev. Theobroma 17: 153–162.

Elhag, H.M., A. Whipkey and J. Janick, 1988. Factors affecting asexual embryogenesis via callus in *Theobroma cacao* L. Arab. Gulf. J. Scient. Res. Agr. Biol. Sci. B6(1): 31–43.

Esan, E.B., 1977. Tissue culture studies on cacao (*Theobroma cacao* L.). A supplementation of current research. In: Proc. V Int. Cocoa Res. Conf., pp. 116–125. Cacao Res. Inst. of Nigeria, Ibadan.

Figueira, A., 1992. Biotechnological studies in *Theobroma cacao* L.: Micropropagation, characterization of polysaccharides, and estimation of genetic Diversity using molecular markers. Ph.D. Thesis, Purdue Univ., West Lafeyette, Indiana.

Figueira, A. and J. Janick, 1993. Development of nucellar somatic embryos of *Theobroma cacao* L. Acta Horticulturae 336: 231–236.

Figueira, A. and J. Janick, 1994. Optimization of $CO_2$ and light levels for *in vitro* culture of *Theobroma cacao* L. J. Amer. Soc. Hort. Sci. 119: 865–871.

Figueira, A., A. Whipkey and J. Janick, 1991. Increased $CO_2$ and light promote *in vitro* shoot growth and development of *Theobroma cacao* L. J. Amer. Soc. Hort. Sci. 116: 585–589.

Figueira, A., A. Whipkey and J. Janick, 1994. Elevated $CO_2$ facilitates micropropagation of *Theobroma cacao* L. Proc. 1991 Int. Cocoa Conf., Challenges in the 90s. Kuala Lumpur, Malaysia, 25–28 September 1991, pp. 360–371.

Flynn, W.P., L.J. Glicenstein and P.J. Fritz, 1990. *Theobroma cacao* L.: An axillary bud *in vitro* propagation procedure. Plant Cell Org. Tiss. Cult. 20: 111–117.

Hall, T.R.H. and H.A. Collin, 1975. Initiation and growth of tissue cultures of *Theobroma cacao* L. Ann. Bot. 39: 555–570.

Hunter, J.R., 1990. The status of cacao (*Theobroma cacao* L., Sterculiaceae) in the Western hemisphere. Econ. Bot. 44: 425–439.

Janick, J., 1986. Embryogenics: the technology of obtaining useful products from the culture of asexual embryos. In: O.J. Cracomo, W.R. Sharp, D.A. Evans, J.E. Bravo, F.C.A. Tavares and E.F. Paddock (Eds.), Biotechnology of Plants and Microorganisms, pp. 97–117. Ohio State Univ. Press, Columbus.

Janick, J., 1991. Production of seed lipids via culture of somatic embryos. In: J. Rattray (Ed.), Biotechnology of Plant Fats and Oils, pp. 76–104. Amer. Oil Chem. Soc., Champaign, IL.

Janick, J., 1993. Agricultural uses of somatic embryos. Acta Horticulturae 336: 207–215.

Janick, J., A.K. Kononowicz and S. Kitto, 1984. Effect of carbon source on proliferation and development of asexual embryos of *Theobroma cacao* L. HortScience 19: 859.

Janick, J. and A. Whipkey, 1985. Axillary proliferation of shoots from cotyledonary nodal tissue of cacao. Rev. Theobroma 15: 125–131.

Janick, J., D.C. Wright and P.M. Hasegawa, 1982. *In vitro* production of cacao seed lipids. J. Amer. Soc. Hort. Sci. 107: 919–922.

Kennedy, A.J., G. Lockwood, G. Mossu, N.W. Simmonds and G.Y. Tan, 1987. Cocoa breeding: past, present, and future. Cocoa Grower's Bull. 38: 5–22. Cadbury Schweppes, Birmingham, UK.

Kong, L.S. and A.N. Rao, 1982. Induction of callus and organogenesis in cocoa tissues. In: Proc. COSTED Symp. on Tissue Culture of Economically Important Plants, Singapore, pp. 107–112.

Kononowicz, A.K. and J. Janick, 1984a. *In vitro* development of zygotic embryos of *Theobroma cacao* L. J. Amer. Soc. Hort. Sci. 109: 266–269.

Kononowicz, A.K. and J. Janick, 1984b. The influence of carbon source on the growth and development of asexual embryos of *Theobroma cacao* L. Physiol. Plant. 61: 155–162.

Kononowicz, H. and J. Janick, 1984c. Response of embryogenics callus of *Theobroma cacao* L. to gibberellic acid and inhibitors of gibberellic acid synthesis. Z. Planzenphysiol. 113: 359–366.

Kononowicz, H., A.K. Kononowicz and J. Janick, 1984. Asexual embryogenesis via callus of *Theobroma cacao* L. Z. Planzenphysiol. 113: 347–358.

Legrand, B., C. Cilas and E. Mississo, 1984. Comportement des tissus de *Theobroma cacao* L. var. *Amelonado* cultives *in vitro*. Café Cacao Thé 28: 245–250.

Legrand, B. and E. Mississo, 1986. Influence de la taille des explantats et des regulateurs de croissance sur le developpement des tissus de *Theobroma cacao* L., var. *Amelonado* cultives *in vitro*. Café Cacao Thé 30: 239–246.

Litz, R.E., 1986. Tissue culture studies with *Theobroma cacao*. In: P.S. Dimick (Ed.), Proc. Cacao Biotechnol. Symp, pp. 111–120. Dept. Food Science, College of Agriculture, Pennsylvania State Univ., College Park.

Lloyd, G. and B. McCown, 1980. Commercially-feasible micropropagation of mountain laurel, *Kalmia latiflora*, by use of shoot-tip culture. Proc. Int. Plant Prop. Soc. 30: 421–427.

Lopez-Baez, O., H. Bollon, A. Eskes and V. Petiard, 1993. Embryogenése somatique de cocaoyer *Theobroma cacao* L. à partir de pièces florales. C.R. Acad. Sci. Paris Serie III. 316: 549–585.

Novak, F.J., B. Donini and G. Owusu, 1986. Somatic embryogenesis and *in vitro* plant development of cocoa (*Theobroma cacao*). In: Proc. Int. FAO/IAEA Symp. Nuclear Techniques and *In vitro* Culture for Plant Improvement, pp. 443–449. IEAE, Vienna.

Orchard, J.E., H.A. Collinand K. Hardwick, 1979. Culture of shoot apices of *Theobroma cacao*. Physiol. Plant. 47: 207–210.

Paiva, M. and J. Janick, 1983. *In vivo* and *in vitro* production of alkaloids in *Theobroma cacao* L. Acta Hort. 131: 265–273.

Passey, A.J. and O.P. Jones, 1983. Shoot proliferation and rooting *in vitro* of *Theobroma cacao* L. type *amelonado*. J. Hort. Sci. 58: 589–592.

Pence, V.C., P.M. Hasegawa and J. Janick, 1979. Initiation and development of asexual embryos *Theobroma cacao* L. J. Amer. Soc. Hort. Sci. 104: 145–148.

Pence, V.C., P.M. Hasegawa and J. Janick, 1980. Initiation and development of asexual embryos *Theobroma cacao* L. *in vitro*. Z. Pflanzenphysiol. 98: 1–14.

Pence, V.C., P.M. Hasegawa and J. Janick, 1981a. *In vitro* cotyledonary development and anthocyanin synthesis in zygotic and asexual embryos of *Theobroma cacao*. J.Amer. Soc. Hort. Sci. 106: 381–385.

Pence, V.C., P.M. Hasegawa and J. Janick, 1981b. Sucrose-mediated regulation of fatty acid composition in asexual embryos of *Theobroma cacao*. Physiol. Plant. 53: 378–384.

Pence, V.C., 1991a. Abscisic acid in developing zygotic embryos of *Theobroma cacao*. Plant Physiol 95: 1291–1293.

Pence, V.C., 1991b. Cryopreservation of immature embryos of *Theobroma cacao*. Plant Cell Rep. 10: 144–147.

Reinert, J., 1958. Untersuchungen über die Morphogenese an Gewebekulturen. Ber. Dtsch. Bot. Ges. 71: 15.

Santos, A.V.P. and R.D. Machado, 1989. A scanning electron microscope study of *Theobroma cacao* somatic embryogenesis. Ann. Bot. 64: 293–296.

Searles, B.R., P.T. Alvim and W.R. Sharp, 1976. Hormonal control of cellular proliferation in cultured callus derived from *Theobroma cacao* L. Rev. Theobroma 6(3): 77–81.

Sharp, W.R., M.R. Söndahl, L.S. Caldas and S.B. Maraffa, 1980. The physiology of *in vitro* asexual embryogenesis. Hort. Rev. 2: 268–310.

Söndahl, M., T.B. Sereduk, Z. Chen and C.M. Bellato, 1989. Somatic embryogenesis and plant regeneration of cacao. Patent 88/3078, Jan. 25 1989, Republic of South Africa.

Söndahl, M., 1991. Tissue culture of cacao, coffee and oil palm. In: Abstr. 4th Conf. Int. Plant Biotechnology Network (IPBNet), Biotechnology for Tropical Crop Improvement in Latin America, Jan. 14–18, 1991, San Jose, Costa Rica, p. 98.

Söndahl, M., S. Liu and A. Bragin, 1993. Cacao somatic embryogenesis. Acta Horticulturae 336: 245–248.

Steward, V.C., M.O. Mapes and K. Mears, 1958. Growth and organized development of cultured cells. II. Organization in cultures grown from freely suspended cells. Amer. J. Bot. 45: 653–704.

Tahardi, S., 1984. Tissue cultures of *Theobroma cacao*. Menara Perkebunan 52: 174–178.

Toxopeus, H., 1987. Botany, types and populations. In: G.A.R. Wood and R.A. Lass (Eds.), Cocoa. Longman, London.

Townsley, P.M., 1974. Chocolate aroma from plant cells. J. Inst. Can. Sci. Technol. Aliment. 7: 76–78.

Villalobos, V. and M.E. Aguilar, 1990. Plant regeneration of cacao (*Theobroma cacao* L.) by micrografting somatic embryos. In: Abstr. VIIth Int. Congr. Plant Tissue and Cell Culture. Int. Assoc. for Plant Tissue Culture, Amsterdam, p. 140.

Wang, Y.C. and J. Janick, 1984. Inducing precocious germination in asexual embryos of *Theobroma cacao* L. HortScience 19: 839–841.

Wang, Y.C. and J. Janick, 1985. Characterizing the germination inhibitor from asexual embryo leachate of *Theobroma cacao*. J. Amer. Soc. Hort. Sci. 110: 113–117.

Withers, L.A., 1984. A report on the *in vitro* culture of cocoa: a synthesis of current activities and opinions. Report of 2nd Meeting Cocoa Working Group, IBPGR, Rome, Italy.

Withers, L., J.A. Yidana and M.D. Atkinson, 1986. Cocoa germplasm—some novel approaches to its conservation. In: P.S. Dimick (Ed.), Proc. Symp. Cacao Biotechnology, pp. 97–110. Pennsylvania State Univ., College Park.

Wood, G.A.R. and R.A. Lass, 1987. Cocoa. Longman, London.

Wright, D.C., A.K. Kononowicz and J. Janick, 1984a. Factors affecting *in vitro* fatty acid content and composition in asexual embryos of *Theobroma cacao*. J. Amer. Soc. Hort. Sci. 109: 77–81.

Wright, D.C., P.M. Hasegawa and J. Janick, 1984b. Temperature effects on *in vitro* lipid accumulation in asexula embryos of *Theobroma cacao* L. Lipids 18:863–867.

Wright, D.C., W.D. Park, N.R. Leopold, P.M. Hasegawa and J. Janick, 1982. Accumulation of lipids, proteins, alkaloids, and anthocyanins during embryo development *in vivo* of *Theobroma cacao* L. J. Amer. Oil Chem. Soc. 59: 475–479.

# 17. Somatic embryogenesis in white ash (*Fraxinus americana* L.)

John E. Preece and Sharon Bates

**Contents**



## 1. Introduction

White ash (*Fraxinus americana* L.), a member of the Oleaceae, is an important hardwood tree species and the most common native ash found in North America. The fast growing dioecious trees can attain heights of more than 30 meters and trunk diameters of up to 1.3 meters. Younger trees have a pyramidal or ovoid shape and mature trees have rounded tops. In landscaping, white ash is well suited for large yards and open areas such as parks and along roadsides. The wood is excellent for tool handles and sporting equipment because it is hard, strong, stiff, shock resistant, and lightweight.

Variation among white ash trees within a stand has been reported to be low, making selection difficult (Clausen *et al.*, 1981). However, stand to stand variability was significant. Some of this variation among stands has been attributed to various ploidy levels: diploid, tetraploid, pentaploid, and hexaploid (Clausen *et al.*, 1981). The polyploid trees were found in the southern portions of the range and the diploid trees were present throughout the range. The authors concluded that the tetraploid trees were the best when plantings were restricted to the southern half of the natural range for white ash.

White ash has a long juvenility period of 20 years before it can produce seeds (Young and Young, 1992). Seed production has been hindered further because female trees flower heavily only every two or three years (Elias, 1980) resulting in three to five year intervals between seed crops (Young and Young, 1992).

Clonal propagation and use of biotechnology therefore have tremendous potential for white ash. The ability to clonally propagate trees that have

*S. Jain, P. Gupta & R. Newton (eds.), Somatic Embryogenesis in Woody Plants, Vol. 2*, 311–325.

been selected, bred, or genetically engineered could significantly advance the improvement of white ash. To use biotechnology, however, it is important that reliable adventitious regeneration systems also be developed.

The earliest attempt at establishing ash *in vitro* was by Gautheret (1934). He excised pieces of vascular cambium from several trees, including *Fraxinus excelsior* L. (European ash). Explants were placed on a medium containing Knop's solution; however, neither callus nor shoots were obtained.

By 1966 Wolter and Skoog developed a specific medium to support continuous callus growth of *Fraxinus pennsylvanica* Marsh. (green ash). This medium was an agar-solidified modified Reinert and White inorganic solution with sucrose, myo-inositol, pyridoxine, 2,4-dichlorophenoxyacetic acid (2,4-D), and kinetin. Naphthaleneacetic acid (NAA) could be substituted for 2,4-D and the addition of gibberellic acid ($GA_3$) improved yields while thiamine had no effect. Callus grew continuously, but not as fast as with herbaceous species. Neither somatic embryogenesis nor organogenesis were reported.

Micropropagation of white ash via proliferation of axillary shoots was first reported by Preece *et al.* (1987b). They stimulated branching from seedling shoot tips in stationary liquid Woody Plant Medium (Lloyd and McCown, 1980) supplemented with 22 or 44 μM benzyladenine (BA). Microshoots were rooted *in vitro* in agar-solidified medium or in vermiculite. Plants were acclimatized to the greenhouse and field. By 1989, Navarrete *et al.* (1989) had improved the axillary shoot multiplication rate from juvenile explants about three-fold. They used MS medium (Murashige and Skoog, 1962) with 1 μM BA, 3 μM thidiazuron (TDZ), and 1 μM indolebutyric acid (IBA). Microshoots were rooted *in vitro* then established in soil in a greenhouse.

In 1987 it was reported that white ash could be regenerated adventitiously via somatic embryogenesis (Preece *et al.*, 1987a,b). In both reports cut mature seeds were used as explants, and up to 22.5% of the cut seeds gave rise to somatic embryos.

## 2. Primary somatic embryogenesis

### 2.1. *Explants*

Cut immature and mature seeds have been used to obtain somatic embryogenesis from white ash (Bates *et al.*, 1992; Preece *et al.*, 1987a,b, 1989). It is more convenient to use mature seeds, because if stored properly they are available at all times; whereas immature seeds are present for only a short time during the years when white ash has a seed crop.

*Disinfestation*. To prepare seeds for disinfestation and placement *in vitro*, we removed the pericarps and cut off the apical 1 to 2 mm from each seed (end with the cotyledon tips) to "mark" that end for future reference. Seeds that were not adequately filled were discarded because they have proven to

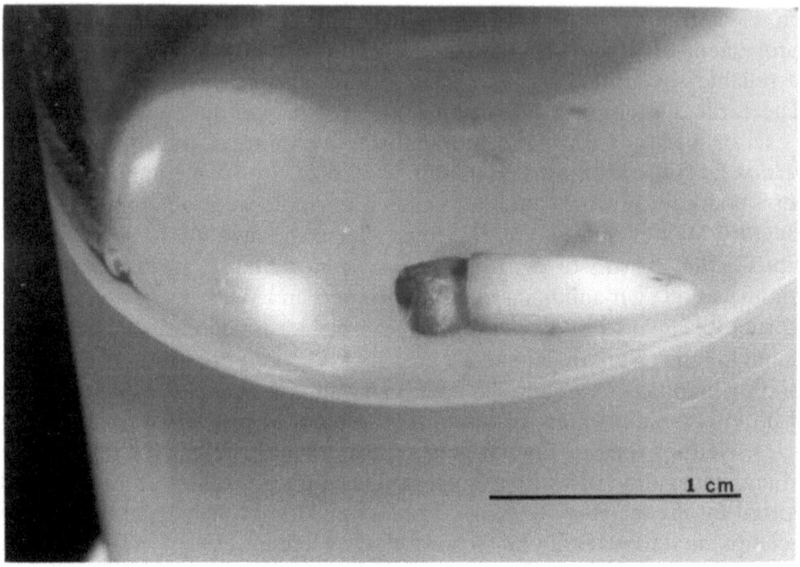


*Figure 1.* Germination of cut white ash seed. Note that the cotyledons, not radicle emerge first. Cotyledons emerge from the cut testa.

be non-viable. Seeds were surface disinfested for 30 min in 1.05% NaClO with 0.1% Tween 20. This was followed by three 5-minute rinses in sterile deionized water. Following the third rinse, the apical 1/3 (approximately 2 to 3 mm) opposite the radicle (the previously "marked" end) was excised and discarded, and the cut seed was then placed *in vitro*. We have found that it was not necessary to initially cut a portion of the seed prior to disinfestation if the terminal 1 mm from each end of the seed was excised before placing *in vitro* (Preece *et al.*, 1993). Typically, only the end where the cotyledon tips are located is cut.

*Germination*. Germination of intact, dormant seeds was poor, even with the addition of $GA_3$ to the *in vitro* medium (Preece *et al.*, 1989). *In vitro* ash seed germination could be synchronized and most seeds germinated if the seeds were cut after surface disinfestation (Preece *et al.*, 1989, 1993).

The pattern of germination differed from what is typical for most seeds because of the effect of cutting (Preece *et al.*, 1989; Fig. 1). The cotyledons emerged first and contacted the medium containing the plant growth regulators. Additionally, the cotyledons remained attached to the meristematic portions of the embryo (radicle and shoot apical meristem). It appears that this was critical for adventitious regeneration of ash. Excised cotyledons from seedlings that were germinated *in vitro* failed to regenerate. This was attributed to critical factors being produced in the developing seedling that were translocated to the cotyledons or inhibitory factors moving out of the cotyledons.

### 2.2. *Media*

*Primary media.* Published reports on somatic embryogenesis from white ash are summarized in Table 1. The best somatic embryogenesis was obtained when cut white ash seeds were cultured on DKW (Driver and Kuniyuki, 1984) primary medium containing 5 μM 2,4-D and 5 μM BA for four or six weeks (Preece *et al.*, 1987a,b, 1989). It made little difference whether the explants remained on the primary medium for four or six weeks (Preece *et al.*, 1987b). In fact, the first time that somatic embryogenesis was observed on white ash was on cotyledons of a seedling that had been on the primary medium for approximately 3 months (unpublished observation). However, incubate explants are now incubated on primary medium routinely for only four weeks because of convenience.

Huetteman and Preece (1993) have documented the potency of thidiazuron for woody plant tissue culture. When cut white ash seeds were incubated for four weeks on primary medium consisting of MS medium with 10 μM 2,4-D and 0.1 or 1.0 μM TDZ, somatic embryos formed on up to 10% of the explants after transfer to secondary medium (Bates *et al.*, 1992; Table 2). Therefore, TDZ can substitute for the cytokinin BA and somatic embryos will form.

*Organic nitrogen.* Casein hydrolysate is a source of reduced nitrogen and has been incorporated into different media that have stimulated embryogenesis in several plant species. White ash does not appear to respond to casein hydrolysate in a similar manner (tabular data not shown). No somatic embryogenesis was observed when media contained casein hydrolysate, but somatic embryos were obtained when media lacked casein hydrolysate and were supplemented with 5 or 10 μM 2,4-D. Casein hydrolysate was not necessary for somatic embryo development.

White ash was unlike many other plant species because an additional source of reduced nitrogen did not appear to be beneficial for somatic embryogenesis. Walnut and pecan species have been shown to produce somatic embryos when reduced forms of nitrogen, such as casein hydrolysate or glutamine, were incorporated into the medium (Merkle *et al.*, 1987; Neuman *et al.*, 1993; Tulecke and McGranahan, 1985). The lack of this requirement for white ash simplified the medium, and may have a bearing on the fact that white ash regeneration occurred on relatively intact seedlings growing *in vitro*.

*Secondary media.* The greatest amount of embryogenesis occurred when explants were transferred to a secondary medium consisting of MS (Murashige and Skoog, 1962) with 0.5 μM BA and 0.5 μM NAA or DKW (Driver

*Table 1.* Chronological summary of successful treatments that achieved somatic embryogenesis in white ash.[1]

| Nutrient salts | Plant growth regulators (μM) | | Percent genesis | Embryo germination | Greenhouse establishment | Reference |
|---|---|---|---|---|---|---|
| | Primary medium | Secondary medium | | | | |
| MS or DKW | 5 BA + 5 2,4-D | none or 0.5 BA + 0.5 NAA | 17.5–22.5 | yes | not reported | Preece *et al.*, 1987a |
| MS or DKW | 0 or 5 BA + 0.5 or 5 2,4-D | not reported | 2.6–16.7 | yes | not reported | Preece *et al.*, 1987b |
| MS or DKW | 0 or 5 BA + 0.5 or 5 2,4-D | 0 or 0.5 BA + 0 or 0.5 NAA | 0.0–20.0 | yes | not successful | Preece *et al.*, 1989 |
| MS[2] | 5 BA + 5 2,4-D | 0.5 BA + 0.5 NAA | from only one nodule | not reported | not reported | Preece and Bates, 1991 |
| MS | 0.1 or 1 TDZ + 10 2,4-D | 0.5 BA + 0.5 NAA | up to 10 | yes | yes, when epicotyls were rooted | Bates *et al.*, 1992 |
| MS or DKW | not reported | none, or 1 BA + 1 NAA and/or 1 TDZ | secondary genesis = up to 76.5% | yes | yes, when epicotyls were rooted | Bates *et al.*, 1993 |

[1] In all studies cut mature seeds were used as the explant.
[2] Somatic embryos formed on nodules.

*Table 2.* The influence of seed source, TDZ and 2,4–D on somatic embryogenesis after 12 weeks.[1]

| Seed source[2] | 2,4-D (μM) | TDZ (μM) | Percent embryogenesis | Total no. embryos |
|---|---|---|---|---|
| 10704–20 | 10.0 | 0.1 | 3.8 | 1 |
| | | 1.0 | 4.5 | 1 |
| 10780–2 | 0.1 | 10.0 | 2.6 | 4 |
| | 10.0 | 0.1 | 2.5 | 4 |
| | | 1.0 | 10.0 | 12 |
| 10810–1 | 10.0 | 0.1 | 8.8 | 4 |
| | | 10.0 | 2.6 | 3 |

[1] Each percentage based on 21–40 seedlings. These results represent the results from a 3×3 factorial combination of 0.1, 1.0, or 10.0 μM 2,4–D and 0.1, 1.0, or 10.0 μM TDZ. Treatment combinations not presented did not stimulate somatic embryogenesis. The large number of 0 values precluded statistical analysis of data.

[2] Three local trees assigned with North Central Forest Experiment Station (NCFES) accession numbers (the number before the hyphen indicates the stand and the number following the hyphen indicates the individual tree within the stand).

and Kuniyuki, 1984) with no plant growth regulators (Preece *et al.*, 1989). This indicates that nutrient salts can substitute for plant growth regulators for development of white ash somatic embryos.

### 2.3. *Origin of primary somatic embryos*

Somatic embryogenesis in *F. americana* has been shown to be both direct from the explant and indirect via callus (Preece *et al.*, 1989). Embryos appeared to go through globular and heart stages of development prior to elongation of cotyledons. Embryogenic capacity of explants persisted for several months and many somatic embryos formed from an explant (Preece *et al.*, 1989; Fig. 2).

### 2.4. *White ash genotype*

There were differences in the quality and quantity of adventitious regeneration among the seed sources (Table 2). The least organogenic seed source, 10780–2, was the most embryogenic and produced the most somatic embryos while 10704–20, which was the most organogenic, was the least embryogenic and produced the fewest embryos (for data on organogenesis from these sources, see Bates *et al.*, 1992). Seedlings from 10810–1 were intermediate for both somatic embryogenesis and shoot organogenesis.

There was a strong genotypic effect on somatic embryogenesis of white ash. This was similar to that reported for shoot organogenesis from the same genotypes (Bates *et al.*, 1992). Therefore, depending upon which white ash

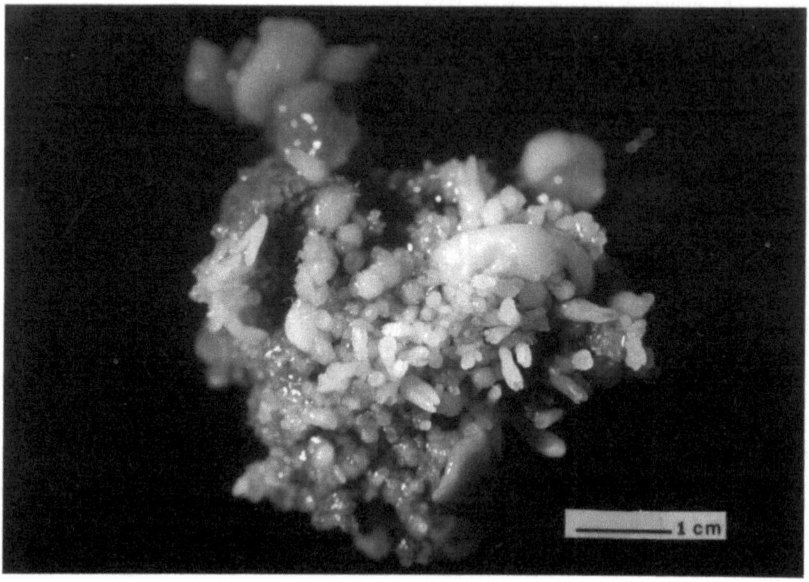


*Figure 2.* White ash somatic embryos in various stages of development growing from callus after several months *in vitro*.

seed source was used, to supply seeds, the percentage of explants producing somatic embryos differed.

### 2.5. *Incubation conditions*

Usually, white ash cultures are incubated under 16 h photoperiods provided by 40 W cool white fluorescent lamps at a photosynthetic photon flux of 20 $\mu$mol $s^{-1}$ $m^{-2}$ for liquid cultures and 35 to 47 $\mu$mol $s^{-1}$ $m^{-2}$ for solidified cultures. The effects of light on somatic embryogenesis with white ash have not been adequately tested. Observationally, somatic embryos have formed, developed, and germinated in darkness and in the light; however, growth has appeared more vigorous under a 16 h photoperiod.

Temperature also has not been tested systematically. It has been observed that *in vitro* white ash seeds germinate and regenerate very slowly when the culture room is $< 23$°C. White ash seeds respond faster and regeneration appears to be greater if day temperatures are maintained between 27 and 35°C.

*Figure 3.* White ash somatic embryos germinating *in vitro*.

## 2.6. *Transfers*

Typically, explants are transferred to fresh medium every four weeks. However, it is observed that white ash somatic embryos grow better and germinate when left on a media for a prolonged period of time without transfer. The culture of somatic embryos pictured in Fig. 2 was on an explant that had not been transferred to fresh medium in about three months. Likewise the somatic embryos that were germinating in Fig. 3 had been planted so that there were only a few spread over the surface of the medium and which had not been transferred for at least three months.

When white ash callus tissue was transferred to fresh medium monthly, as is the practice in our laboratory, somatic embryos such as those shown in Fig. 2 are not observed. When culture transfer is infrequent, they respond better with embryogenesis. Attempts have been made to duplicate this by increasing the sucrose concentration in the medium. It is postulated that over time a medium tended to dry, the constituents became more concentrated and this decreases the water potential. Increasing sucrose lowered the water potential and has been used commonly with somatic embryogenesis studies. However, increasing the sucrose content of the medium did not promote somatic embryo maturation in these experiments.

## 3. Secondary (repetitive) somatic embryogenesis

Secondary somatic embryos formed from other somatic embryos have been placed onto agar-solidified or liquid media (Bates *et al.*, 1993). Only very few of the secondary somatic embryos germinated into whole plants.

*Liquid suspension cultures.* Somatic embryos, and cell suspensions were placed into 25 or 50 ml liquid medium supplemented with 1 μM NAA and 1 μM BA. Transfers were made to the same medium on a weekly or biweekly basis. Callus placed in 50 ml of liquid medium remained at the bottom of the culture vessel, moved very little, and did not break up as well as callus placed in 25 ml of medium. There were no readily observable differences in the amount of cell growth comparing weekly and biweekly transfers regardless of the volume of medium in the culture vessel. From these observations, subsequent transfers and subcultures were made on a biweekly basis in vessels containing 25 ml of medium.

Within a month after placing a light green friable callus into liquid MS medium with 1.0 μM NAA and 1.0 μM BA on a platform shaker at 100 rpm, one culture produced an embryonic structure (unpublished results). More embryos were observed in other subcultures from this same callus clone after an additional month. Somatic embryos were separated from callus at the time of subculturing so that one subculture contained somatic embryos with as little callus as possible and the other subculture contained the light green callus with no embryos. Somatic embryos produced secondary somatic embryos that were broken apart at the time of subculturing.

White callus obtained from an immature seed was placed in liquid culture and within one week somatic embryos were observed. This second callus clone produced mainly embryos with little or no additional callus formation. Both callus clones produced embryos that had various developmental stages from globular to torpedo to cotyledonary stages.

Only a few normal shaped embryos were produced, many embryos exhibited abnormalities. The main abnormality observed was the fusion of several embryos together at their root apices. Smaller embryos often budded off at this point and the embryos could be broken apart with forceps. Several embryos exhibited precocious germination where the radicle would elongate. When this occurred, cotyledons that emerged appeared small or were absent. Multiple cotyledons and fused cotyledons were observed.

Growth of somatic embryos in liquid suspension cultures differed from agar solidified cultures. Multiplication rates and the number of somatic embryos were greater in liquid cultures compared to solidified cultures. Hundreds of somatic embryos were obtained by subculturing embryogenic callus in agitated liquid medium. In solidified medium, only a few somatic embryos were obtained.

When somatic embryos were removed from liquid cultures and placed onto

*Figure 4.* White ash seedling (left) and germinated somatic embryo (right) after 2 months *in vitro*.

solidified secondary medium with 0.5 μM BA and 0.5 μM NAA, numerous secondary somatic embryos developed, similar to those pictured in Fig. 2.

Liquid cultures on a shaker were excellent for rapid multiplication of white ash somatic embryos. This appeared to be the best method for production of secondary somatic embryos. On agar-solidified medium, there was a minimal number of embryos budding off secondary embryos. The embryos appeared to develop from proembryonic cells that may have begun the somatic embryo developmental pathway while still on the primary medium. However, in the liquid medium somatic embryos grew differently than on solidified medium and produced many abnormal secondary somatic embryos. This could be reduced by adding 1 μM glyphosate to the medium (unpublished observation). The glyphosate caused the embryos to enlarge and inhibited secondary embryogenesis. The embryos appeared uniform in response to glyphosate.

## 4. Plantlet conversion

We have had limited success converting somatic embryos into plantlets. Somatic embryo cotyledons have grown and expanded and radicles have elongated (Fig. 3), but only a few have developed into plants (Fig. 4).

*Dehydration*. Green somatic embryos and embryo masses were dehydrated for 0, 7, and 14 days in empty, non-sealed Petri dishes that were placed in a sealed desiccator and stored in the dark. Relative humidity was maintained at 70% by placing a saturated solution of 1 $NH_3Cl$:1 $KNO_3$ (by volume) at the bottom of the desiccator. Embryos were rehydrated by placing on an agar solidified medium containing 1 μM indolebutyric acid (IBA) and 1 μM BA. Embryo appearance was observed for 3 months.

Somatic embryos and embryo masses started to dehydrate visibly within a day after placing in the desiccator. The embryos appeared slightly shriveled. After 7 and 14 days, the embryos appeared quite shrunken and dried compared to nondehyrated embryos. When placed back on MS medium, embryos and embryo masses rehydrated. Embryos not desiccated enlarged without any cotyledon, stem, or root growth. Dehydrated embryos had the same response except embryos did not enlarge as much. No plant development was observed in any treatments.

*Plant growth regulators*. Various other treatments, such as chilling the embryos and incorporating $GA_3$ into the medium have been unsuccessful. Best plantlet conversion was obtained when 1, 3, or 10 μM TDZ was incorporated into liquid or agar-solidified medium. Because of the potency of TDZ, multiple axillary shoots elongated on plantlets from somatic embryos germinated on medium containing this cytokinin. This was not considered to be a disadvantage because the shoots can be excised and easily rooted, resulting in $> 1$ plantlet per converted somatic embryo.

*Root system*. Preece *et al.* (1989) reported that although some white ash somatic embryos germinated, many did not survive transplanting into a greenhouse medium. They attributed this to poor root growth. The plants were wilted in the greenhouse (unpublished observation). Bates *et al.* (1992, 1993) reported that rooted epicotyls from somatic embryos survived and grew in the greenhouse. They rooted them under intermittent mist in a greenhouse or in peat plugs in high humidity flats in the laboratory, similar to rooting adventitious shoots and epicotyls from *in vitro* germinated seeds.

It is suspected that the root that forms when the radicle elongates *in vitro* is not well adapted to the soil. We have found that seeds that germinate *in vitro* are also more difficult to establish *ex vitro* when the radicle is attached than when the epicotyl is excised and rooted *ex vitro*. Adventitious roots that formed on epicotyls from seeds or somatic embryos under mist or high relative humidity in peat plugs or greenhouse soil apparently were better adapted for growth *ex vitro*.

## 5. Acclimatization

Acclimatization began when elongated epicotyls were excised from germinated somatic embryos and placed under mist on a greenhouse bench or into preformed peat plugs in high humidity trays in the laboratory for rooting. After remaining under intermittent mist for about 25 days, rooted plantlets were transplanted into small pots containing a greenhouse potting medium (2 sphagnum moss peat: 1 vermiculite:1 perlite (by volume)). They remained on the mist bench for one week and were then moved to a nonmisted greenhouse bench. Survival has been excellent.

### 5.1. *Overcoming dormancy*

White ash plants that have been maintained in the greenhouse, grew until the terminal buds set. After bud set, stem caliper increased and the stems became more woody. Leaves remained green for at least six months. To overcome dormancy, plants were placed into a cooler (4°C in darkness). In the cooler, leaves began to turn a yellowish green color then abscised within the first or second week. After two to three months, the plants were removed from the cooler and placed on greenhouse benches. Buds began to grow after one week. White ash in the greenhouse had an episodic growth pattern in which a flush of growth occurred followed by bud set. Occasionally a second growth flush occurred later if the growing conditions were optimal.

## 6. Somaclonal variation

*In vitro* embryos have been variable with one to several cotyledons (Preece *et al.*, 1989). However, *ex vitro*, adventitiously regenerated white ash plants have been uniform (Fig. 5A). Only one plant with obvious somaclonal variation has been observed (Fig. 5B). White ash has opposite leaf phyllotaxy. This particular plant from a somatic embryo had sub-opposite to alternate phyllotaxy. After one chilling cycle to overcome dormancy, the apical shoot continued to grow with the alternate phyllotaxy, but the side shoots had the normal opposite phyllotaxy.

## 7. Conclusions

Somatic embryogenesis is a viable method for micropropagating white ash. Unlike other woody plants where there is a limited window of time when adventitious regeneration is possible, white ash forms somatic embryos relatively easily from mature seeds. Ash seeds remained viable for years when stored properly, therefore a batch of seeds could be kept until needed to produce somatic embryos. However, at best, we have only been able to

*Figure 5.* (A) (Left to right) Rooted adventitious shoot, and epicotyls from a seed germinated *in vitro* and a somatic embryo. (B) Rooted epicotyl from a somatic embryo showing sub-opposite to alternate phyllotaxy.

stimulate up to 20% of the seeds to form somatic embryos. Additionally there was a strong genotypic effect that resulted in batches of seeds that would not form somatic embryos (tabular data not presented).

Our "standard" protocol for obtaining somatic embryogenesis from white ash seeds has begun with selection of a tree that produces embryogenic seeds. We have placed disinfested cut seeds onto MS medium with 10 μM 2,4-D and 1 μM TDZ for four weeks. Seedlings were then transferred to MS secondary medium containing 0.5 μM BA and 0.5 μM NAA. Incubation has been under fluorescent lamps that provide a 16 h photoperiod. Explants were then transferred to fresh medium every two to three months.

Shoot organogenesis is an alternative route for adventitious regeneration of white ash (Bates *et al.*, 1992). There is also a strong genotypic effect on shoot organogenesis, similar to that reported here for somatic embryogenesis. However, the percentage of seeds that are organogenic ranged up to 60%. This was considerably higher than with somatic embryogenesis. A problem with shoot organogenesis of white ash was stimulating elongation of bud-like structures into shoots that could be rooted. Shoot elongation could be obtained if the correct medium was used.

Whether regeneration of white ash was via somatic embryogenesis or shoot organogenesis, clearly this tree species can be regenerated adventitiously. This may have important implications for the use of biotechnology for tree improvement.

## 8. Acknowledgements

We thank NOR-AM Chemical Co., Wilmington, DE, for providing thidiazuron and J.W. Van Sambeek for help providing our trees with NCFES numbers and for seeds.

## 9. References

Bates, S.A., J.E. Preece and J.H. Yopp, 1993. Secondary somatic embryogenesis and plantlet conversion of white ash. HortSci. 28: 405. (Abstr.).

Bates, S., J.E. Preece, N.E. Navarrete, J.W. Van Sambeek and G.R. Gaffney, 1992. Thidiazuron stimulates shoot organogenesis and somatic embryogenesis in white ash (*Fraxinus americana* L.). Plant Cell Tiss. Org. Cult. 31: 21–29.

Clausen, K.E., F.H. Kung, C.F. Bey and R.A. Daniels, 1981. Variation in white ash. Silvae Genetica 30: 93–97.

Driver, J.A. and A.H. Kuniyuki, 1984. *In vitro* propagation of paradox walnut rootstock. HortSci. 19: 507–509.

Elias, T.S., 1980. The Complete Trees of North American Field Guide and Natural History. VonNostrand Reinhold Co., New York.

Gautheret, R.J., 1934. Culture du tissu cambial. C.R. Acad. Sci. Paris. 198: 2195–2196.

Huetteman, C.A. and J.E. Preece, 1993. Thidiazuron: A potent cytokinin for woody plant tissue culture. Plant Cell Tiss. Org. Cult. 33: 105–119.

Lloyd, G. and B. McCown, 1980. Commercially-feasible micropropagation of mountain laurel, *Kalmia latifolia*, by use of shoot tip culture. Comb. Proc. Intl. Plant Prop. Soc. 30: 421–427.

Murashige, T. and F. Skoog, 1962. A revised medium for rapid growth and bioassays with tobacco tissue cultures. Physiol. Plant. 15: 473–497.

Merkle S,A., H.Y. Wetzstein and H.E. Sommer, 1987. Somatic embryogenesis in tissue cultures of pecan. HortSci. 22: 128–130.

Navarrete, N.E., J.W. Van Sambeek, J.E. Preece and G.R. Gaffney, 1989. Improved micropropagation of white ash (*Fraxinus americana* L.). In: G. Rink and C.A. Budelsky, (Eds), Proceedings of the Seventh Central Hardwood Forest Conference; 1989 March 5–8; Carbondale, IL, pp. 146–149

Neuman, M.C., J.E. Preece, J.W. Van Sambeek and G.R. Gaffney, 1993. Somatic embryogenesis and callus production from cotyledon explants of Eastern black walnut (*Juglans nigra* L.). Plant Cell Tiss. Org. Cult. 32: 9–18.

Preece, J.E. and S. Bates, 1991. *In vitro* studies with white ash (*Fraxinus americana*) nodules. In: M.R. Ahuja, (Ed.), Woody Plant Biotechnology, NATO ASI Series 210, pp. 37–44. Plenum Press, New York.

Preece, J.E., S. Bates and J.W. Van Sambeek, 1993. Variability in germination of ash (*Fraxinus* spp.) cut seeds *in vitro*. Seed Sci. Technol (submitted).

Preece, J.E., J. Zhao and F.H. Kung, 1989. Callus production and somatic embryogenesis from white ash (*Fraxinus americana* L.). HortSci. 24: 377–380.

Preece, J.E., J. Zhao and F.H. Kung, 1987a. *In vitro* callus production and somatic embryogenesis of ash (*Fraxinus*). HortSci. 22(5): 675 (Abstr.).

Preece, J.E., P.H. Christ, L. Ensenberger and J. Zhao, 1987b. Micropropagation of ash (*Fraxinus*). Comb. Proc. Intl. Plant Prop. Soc. 37: 366–372.

Tulecke W. and G. McGranahan, 1985. Somatic embryogenesis and plant regeneration from cotyledons of walnut, *Juglans regia* L. Plant Sci. 40: 57–63.

Wolter, K.E. and F. Skoog, 1966. Nutritional requirements of *Fraxinus* callus cultures. Amer. J. Bot. 53: 263–269.

Young, J.A. and C.G. Young, 1992. Seeds of woody plants in North America. Dioscorides Press, Portland, OR.

# 18. Somatic embryogenesis in coconut (*Cocos nucifera* L.)

Jennet Blake and Roland Hornung

**Contents**



## 1. Introduction

Coconut (*Cocos nucifera* L.) is a woody, monocotyledonous species producing an unbranched stem without a cambium and which therefore has no secondary growth. New leaves are initiated at the terminal apex continuously throughout the year, usually at the rate of one or two per month. Each axillary bud produces an inflorescence which consists of 20–30 branches, or rachillae, bearing large numbers of male flowers with a few female flowers at the base of each rachilla. Pollination is by wind or insects and the nut is ripe about 12 months after fertilisation. Propagation is entirely by seed.

The coconut palm is an important subsistence crop in all tropical areas of the world, and is less commercialised than, for example, the oil palm (*Elaeis guineensis* Jacq.). Most of the production is used for domestic consumption in the country of origin; few countries have a surplus for export. Even in countries such as the Philippines, which is the largest world exporter of coconut, the average yield is very low. There is, however, great potential for increased production as high-yielding palms may have ten times as many nuts per year as low-yielding palms. A comprehensive survey of coconut production has been published by the FAO (Ohler, 1984) and Green (1991) has edited a series of contributions on the present status of coconut production and priorities for research.

Each area where the palm is grown has its own selection of native varieties which are predominantly of the Tall type. These may continue in bearing for 80 years or more, so that there is little incentive for new plantings of improved varieties. As Tall varieties do not come into bearing for 7–8 years, breeding programmes are slow to make improved varieties available.

Fewer Dwarf varieties are cultivated as they give smaller nuts which are less acceptable to farmers. However, they have been utilised for crossing with Tall varieties and there are considerable areas now planted with hybrids which generally yield better than either of their parents. Yield improvements

*S. Jain, P. Gupta & R. Newton (eds.), Somatic Embryogenesis in Woody Plants, Vol. 2*, 327–340.

have been obtained with the development of Dwarf × Tall hybrids such as the Maypan (Malayan Dwarf × Panama Tall) in Jamaica and the PB 121 or MAWA (Malayan Dwarf × West African Tall) in the Ivory Coast. However, the widespread plantings of PB 121 have been less successful in areas of the world other than West Africa, and most countries now wish to produce their own hybrids from indigenous stock.

Although increased yield is often of primary concern to the coconut grower, disease-resistance or tolerance is even more important in areas where lethal diseases can decimate large areas in a short time. For example, Lethal Yellowing which killed the majority of Tall palms in Jamaica is now seriously affecting the Yucatan peninsula in Mexico.

Coconut stocks are innately heterogeneous and selection of uniform parental lines for hybrid production is therefore often difficult. The initial aim for the tissue culture propagation of coconut is to clonally propagate selected, elite, parental lines so that coconut breeders will be able to speed up their breeding programmes. Later, if clonal plants are acceptable to farmers, propagation of hybrids could be undertaken.

Encouragement to propagate the coconut palm clonally has been provided by achievements with other palm species, including the commercially important oil palm in which early success enabled the establishment of considerable areas of clonal plantations (Corley *et al.*, 1981). The date palm (*Phoenix dactylifera* L.) has also been propagated successfully by tissue culture (see review by Branton and Blake, 1988) with subsequent field plantings. However, the coconut has proved to be much more difficult to regenerate in culture than either oil or date palm; although at least two laboratories have claimed production of clonal plants (Branton and Blake, 1986; Verdeil *et al.*, 1992), the numbers are relatively few with, as yet, no clonal field plantings. There are a number of reports of somatic embryogenesis with the production of the occasional plantlet (see review by Blake, 1990), but improved success has not been reported from these laboratories.

The method of clonal propagation in all three of these commercially important palms has been by somatic embryogenesis, although axillary bud proliferation in date palm (see review: Branton and Blake, 1988) has also been reported. Despite the prolonged callus phase in oil palm, the early clonal plants showed little variation within the clones and appeared to be true to type (Corley, 1982). However after some years, abnormalities in early flowering of some clones were observed which were thought to be epigenetic and attributable to the culture conditions (see review: Wooi, 1990). There have been no similar abnormalities reported with date palm, but the callus phase in this species is generally much shorter. As all published methods of clonal propagation of coconut also appear to be through somatic embryogenesis, it will be very important to keep the period in culture as short as possible and, subsequently, to carry out stringent field trials, particularly as Armstrong and Francis (1987) reported that long-term cultures of one line of coconut callus maintained on a 2,4-D medium produced some chromosomal abnormality.

Although it has been suggested that the coconut might be propagated also by vegetative proliferation of the flower meristems, considerable research was unable to obtain true reversion of floral primordia to the vegetative state (Blake and Eeuwens, 1982). As more information becomes available on the molecular control of flowering, it may eventually be possible to obtain true reversion. This could prove to be the micropropagation method of choice because each coconut inflorescence contains over 3,000 flower primordia and the chances of somatic mutations would be reduced to almost zero in comparison with those which might arise during a long callus phase. The disadvantage would be a limitation to the degree of multiplication achievable. However, this may not be so critical if clonal propagation is utilised primarily for the multiplication of elite palms for parental breeding stocks which are then used to produce hybrid nuts. Seed nuts may be more acceptable to farmers than clonal plants, and would be less expensive.

Normal dicotyledonous systems of somatic embryogenesis have an initial callus phase during which cell proliferation is random. In the coconut, the initial callus is only partially dedifferentiated before reorganisation occurs. The initial tissue which is developed may be referred to as callus (Branton and Blake, 1983; Pannetier and Buffard-Morel, 1986) or as calloid (Brackpool *et al.*, 1986). The two terms describe the same type of tissue and both are used in this review.

In many embryogenic systems, differentiation occurs within the callus to produce somatic embryos which develop (or germinate) into normal plants, often by a developmental pathway showing similarities to the germination of zygotic embryos. In the coconut, good quality somatic embryos are rare and normal development is difficult to achieve. A range of abnormal structures are commonly observed: a comparison of zygotic and somatic embryo development will be made in this review in an attempt to suggest how such structures arise. Finally suggestions will be made for future research which could lead to the production of greater numbers of normal plantlets.

## 2. Culture initiation

Most parts of the adult palm have been used as starting material, including roots (Justin, 1978) young leaves (Pannetier and Buffard-Morel, 1986) and inflorescences (Pannetier and Buffard-Morel, 1986; Blake, 1990). Juvenile tissues such as young leaves (Karunaratne *et al.*, 1991) and embryos have been used also (Karunaratne and Periyapperuma, 1989) and recent work has indicated the advantages of excising the plumule from the embryo and using this as starting material for somatic embryogenesis (Hornung, unpublished).

Calloid arises on the very young leaves in a manner similar to that on older leaves.

As with many other monocotyledons, the initial coconut explants generally produce callus most successfully when cultured with 2,4-D, usually at about 100 μM concentration in the presence of activated charcoal. The latter is required to maintain an active level of the auxin which will induce dedifferentiation of the tissues, without having toxic effects. Other auxins have been tested (unpublished data) but have shown no clear advantage over 2,4-D, though NAA (without charcoal) was preferred for root explants (Justin, 1978). The presence of charcoal means that the concentration of auxin available to the tissues cannot be defined. Investigations of the auxin adsorption pattern of charcoal have shown that the charcoal in the culture medium continues adsorption over a period of at least 2 weeks (Ebert and Taylor, 1990) so that the concentration of auxin is reduced over this period after the preparation of the medium.

In addition to the problem of defining actual auxin levels in the medium, there is the difficulty of recognising the state and stage of development of the initial explants, which may come from organs of different ages, and will vary according to the explant position in the organ, as well as with the type of organ i.e., root, leaf or inflorescence. The size of the initial explant may also be important, whether in actual dimensions or in relation to the number of cells, or potential cells, which are present. In general, younger tissues can tolerate, and need, a higher level of auxin to achieve dedifferentiation.

Cytokinins are usually incorporated in the culture media, but a specific requirement has not been consistently demonstrated. The pattern of cytokinin adsorption by activated charcoal is similar to that already shown for 2,4-D (Ebert *et al.*, 1993).

Leaf explants from adult palms have been studied in several laboratories. Highly meristematic nodular structures have been developed within leaf tissues and these may form roots or cease development (Buffard-Morel *et al.*, 1992; Kendurkar *et al.*, personal communication). When a cambial-type layer is formed, these nodules can produce embryos which may give secondary embryos (Buffard-Morel *et al.*, 1992). Alternatively a granular callus of secondary origin may be produced which appears to lead to improved embryo formation (Buffard-Morel *et al.*, 1992). Leaf explants from seedlings were most embryogenic when taken from 10–20 cm-long leaves of 12–24-month-old seedlings rather than younger or older plants (Karunaratne *et al.*, 1991). The maximum embryogenic response was obtained from the leaf base with diminishing activity towards the apex.

Inflorescence explants also produce nodular structures, thought to originate either from the rachis (Verdeil *et al.*, 1992) or from the meristematic dome of the early flower primordium (Fig. 1A). These nodules have a different structure to those in the leaf. They are formed superficially and consist of files of cells produced as though from a cambium, although such a layer is not evident (Fig. 1B). They are similar in structure to the granular

*Figure 1.* Early developmental stages of embryogenic tissue. (A) Nodular calloid produced from coconut inflorescence (Wye College). (B) Section of nodular calloid shown in A. (C) Production of somatic embryos from calloid shown in A. (D) Embryogenic callus of unicellular origin (reproduced from Verdeil *et al.*, 1992).

callus developed from leaf nodules (Buffard-Morel *et al.*, 1992; Fig. 2). Since this type of structure is not produced by random cell proliferation as in the callus of most dicotyledons, but shows considerable organisation, it was termed calloid by Brackpool *et al.* (1986).

### 3. Embryo development

The early nodules, whether produced in the leaf or from inflorescence tissue, can be regarded as pre-embryogenic structures (Fig. 1A). Further development of these nodules to embryos (Fig. 1C) has been achieved by a reduction in the concentration of 2,4-D (Blake, 1990) or by an increase followed by a

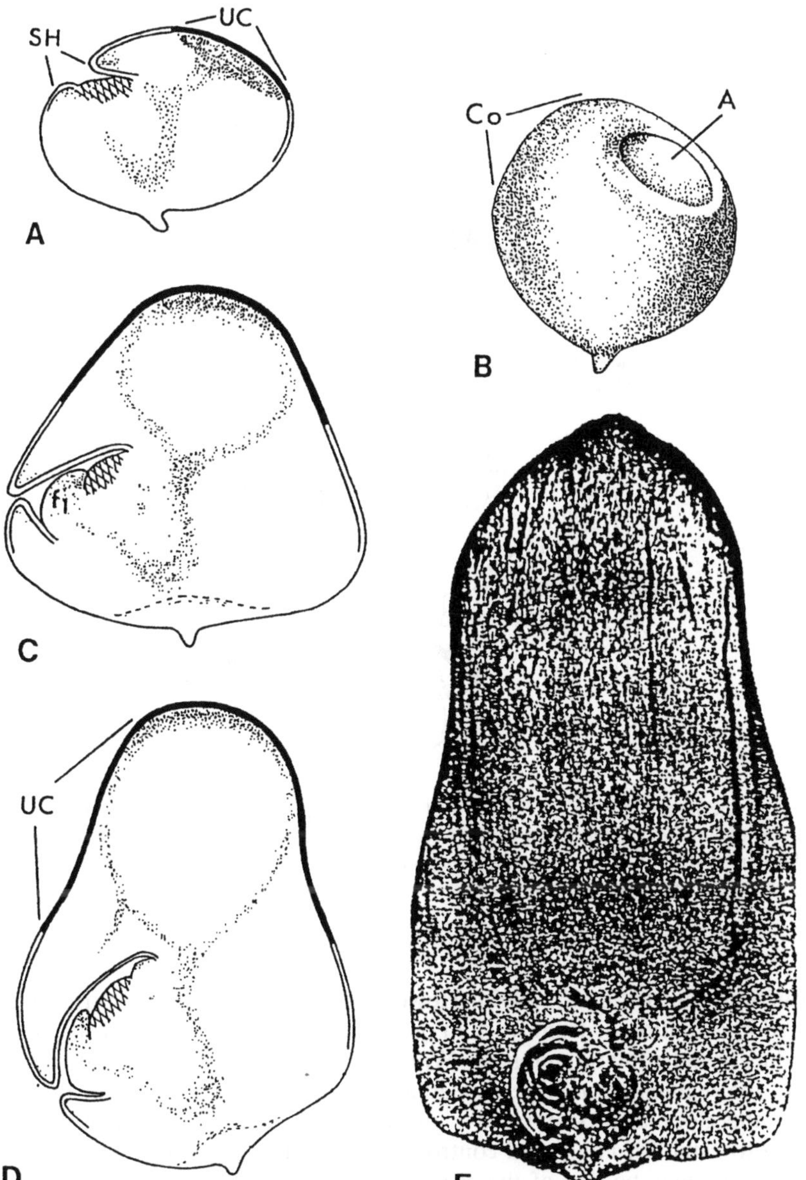


*Figure 2.* Zygotic embryo development in *Cocos nucifera* L. (reproduced from Haccius and Philip, 1979). (A) Early stage showing apex (hatched) with sheathing base (SH) and upper part (UC) of the cotyledon. (B) Early stage showing the cotyledonary primordium (Co) surrounding the apex (A). (C,D) Successively later stages of the developing embryo. (E) Section of mature embryo showing shoot and root apices at base with large cotyledon above. Single outline – hypocotyl and root pole; double outline – sheathing base; thick outline – upper part of cotyledon; f1 – first plumular leaf.

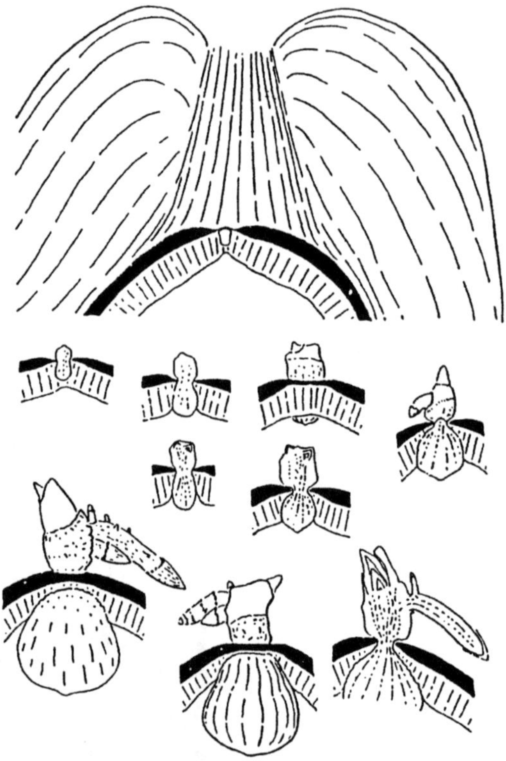

*Figure 3.* Germination of zygotic embryo of coconut (reproduced from Corner, 1966). Top: position of embryo behind soft eye (germ-pore) showing husk, woody shell (in black) and endosperm (hatched). Below: successive stages in growth of the embryo as the cotyledon swells into and digests the endosperm.

reduction (Verdeil *et al.*, 1992). However, it is not clear, either from published information or from work in our own laboratory, how such structures produce good somatic embryos. Williams and Maheswaran (1986) suggest that "the segmentation patterns leading to formation of a meristematic nodule – the globular proembryoid – are independent of the ability of this cell group to establish integrated control and to initiate bipolar embryo formation". This may be one of the reasons why, although many embryoid-type nodules are formed, the majority produce abnormal developments such as fused embryos, fused leafy structures with or without roots (Fig. 5A) and haustorial (cotyledonary) type tissues (Sugimura and Salvaña, 1989; Blake, 1990; Verdeil *et al.*, 1992).

However it is important to consider that the zygotic embryo is not actually a bipolar structure with root and shoot meristems, but also has a cotyledonary

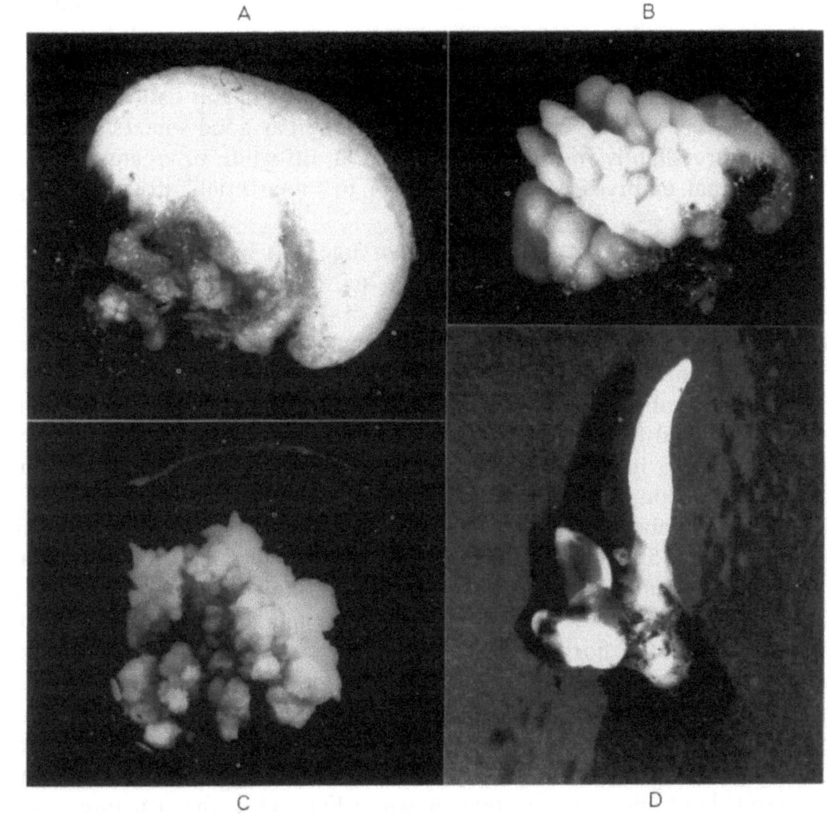


*Figure 4.* Abnormal developments from embryogenic tissue. (A) Somatic "haustorium". (B,C) Embryo-shaped haustorial tissue. (D) Haustorial tissue forming elongated embryo-like structure (right) with adjacent green shoot (left) which developed into a plantlet.

meristem. This is clearly evident in the work of Haccius and Philip (1979). Fig. 2 shows the early development when the positions of shoot and root apices can be identified (Fig. 2A), clearly surrounded by the cotyledon (Fig. 2B). As the embryo develops, the proportions change so that the cotyledon becomes the dominant tissue (Fig. 2C,D). Finally, at maturity, the volume of the cotyledon is so increased that the shoot and root primordia are proportionately insignificant (Fig. 2E). The continuing increase in size of the cotyledon of the zygotic embryo reflects the potential for growth of the cotyledonary pole at the stage of embryo initiation. Therefore, it may not be surprising that cotyledonary type tissue can become dominant during the process of somatic embryogenesis.

In the germination of the zygotic embryo, the cotyledon enlarges into a white spongy structure (Fig. 3) composed primarily of parenchyma-type cells with some vascular tissue. After somatic embryogenesis, the cotyledonary meristem can produce a range of formations, often disorganised, but sometimes organised to resemble the zygotic haustorium (Fig. 4A) or embryo-like structures (Fig. 4B,C,D; also Verdeil *et al.*, 1989). This tissue is recognisable by its very white, spongy appearance whereas good somatic embryos (Fig. 2C) are slightly translucent and have an off-white or creamy appearance. Normal embryos can arise adjacent to "haustorial" structures (Fig. 4D).

Cotyledonary tissue from somatic embryogenesis has great potential for growth, particularly when auxin is very low or omitted. However, when recultured with a very high concentration of 2,4-D (400 μM with activated charcoal) such tissue will regenerate calloid which has been shown to repeat the embryogenic process and produce a plantlet (Blake and Wilson, 1990). The re-cycled calloid can be regularly subcultured and maintained on media without charcoal for long periods of time when 2,4-D or NAA is supplied at 0.1–0.01 μM. It is considered to be in an early embryogenic phase, and is identified as long-term calloid.

As indicated above, the so-called bipolar embryo is really "tripolar". This is shown in Fig. 5B where a somatic plantlet with shoot and root also has an haustorium. In culture the early somatic embryo rarely manages to maintain a balanced or integrated development of the three meristems and although the shoot may be quite normal (Fig. 5C), the root system may be poorly developed. Often cotyledonary development predominates, but leafy structures or roots may also be formed (Branton and Blake, 1986). Sometimes two of the three poles may develop and many cultures are seen with cotyledon and roots (e.g., Blake, 1990; Fig. 1G). Although leaves are often formed, an organised shoot apex, such as that shown in Fig. 5D, is the structure which most often fails to develop.

In attempts to alter the balance of embryo development towards the shoot meristem, long-term calloid has been cultured on media in which the concentrations of ammonium and nitrate nitrogen have been varied. After two consecutive periods of culture on media with 0, 10, 20 or 40 m eq. of ammonium or nitrate in all combinations, calloid became very haustorial in the absence of ammonium, particularly in the presence of high nitrate (Table 1). Variations in pH (from 3.6–5.5) were associated with the ratios of ammonium to nitrate in the medium, but apparently were not related to the production of haustorial tissue. There were no suggestions that any of the treatments promoted the development of somatic embryos.

Obtaining good quality somatic embryos is the first necessity for obtaining well-balanced plantlet development (Fig. 5B). So far, it is not clear how this is to be achieved but Verdeil *et al.* (1992) suggest there are two alternative routes which may be taken during embryogenic development: (1) the most commonly observed is where embryos are of multicellular origin, formed from superficial meristematic structures with a clear epidermis which often leads to the abnormal structures described above. These embryos arise from

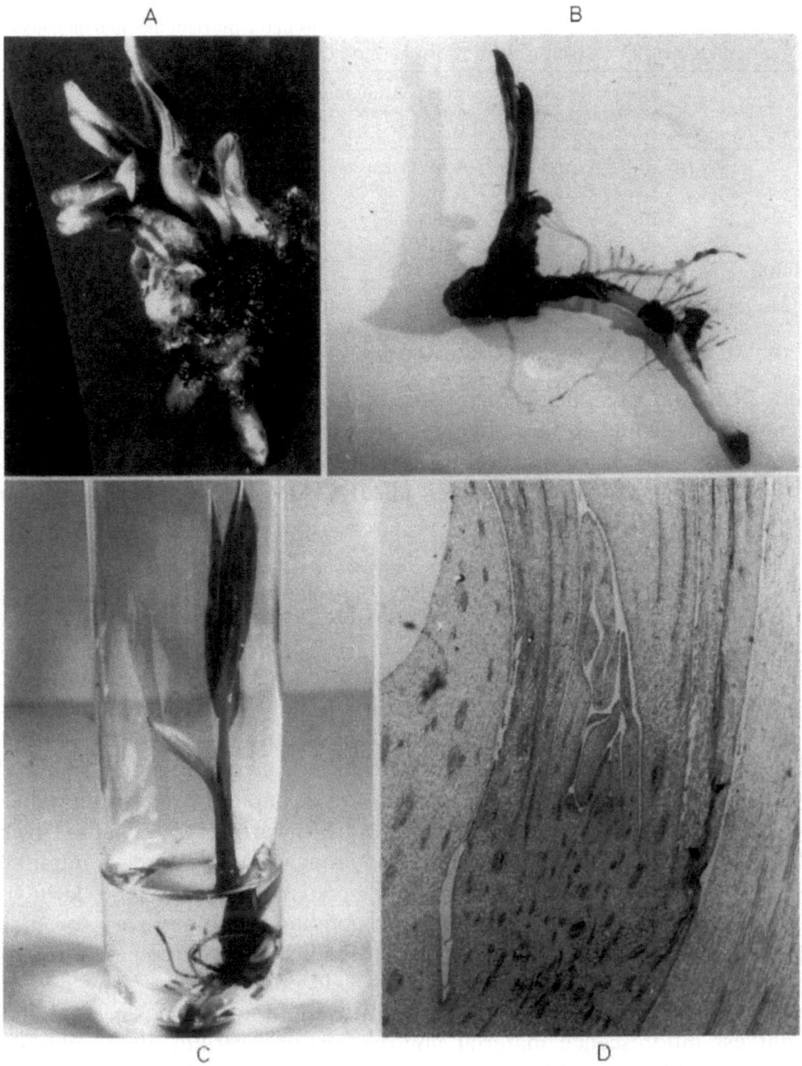


*Figure 5.* Somatic plantlets. (A) Fused leaves and roots. (B) Balanced development of shoot and root with "haustorium". (C) Good shoot development with roots inadequately developed. (D) Production of leaf primordia from the apex of a somatic plantlet.

the type of calloid shown in Fig. 1A. (2) The more recently observed case where embryos are of unicellular origin and embryogenic groups of cells are separated from each other by thickening of their walls (Fig. 1D). Such cell groups resemble the embryogenic cells described for oil palm (Schwendiman

*Table 1.* The effect of ammonium and nitrate concentrations in the nutrient medium on haustorial-type development in coconut calloid. Mean haustorial score of 10 replicates per treatment.

| | m eq. Nitrate | m eq. Ammonium | | | |
|---|---|---|---|---|---|
| | | 0 | 10 | 20 | 40 |
| | 0 | 3.6 | 3.3 | 2.7 | 2.7 |
| | 10 | 4.5 | 2.7 | 1.9 | 1.8 |
| | 20 | 4.4 | 2.3 | 2.3 | 1.7 |
| | 40 | 5.0 | 4.1 | 1.7 | 1.0 |
| Control | 40 | | | 0.5 | |

No haustorial development = 0; maximum development = 5.0.

*et al.*, 1990) and have the characteristics of the first stages of zygotic embryos. Such material would be more likely to produce good cell suspensions, as has already been shown for oil palm (de Touchet *et al.*, 1991). A more detailed consideration of the recent research at ORSTOM, Montpellier is given by Verdeil and Buffard-Morel (1994).

## 4. Future directions for research

The importance of the coconut as a subsistence crop in all parts of the tropical world justifies the continued efforts to develop a reliable method of clonal propagation. Breeders are anxious, not only to be able to multiply selected parental lines, but also to have homogeneous breeding material. This could be achieved with the regeneration of haploid plants from pollen: early results indicated good possibilities (Monfort, 1985; Thanh-Tuyen, 1990), but the work is slow and difficult and much more research is required. Even with haploid plants available to the breeder, heterosis could occur.

The difficulties of finding genotypes responding reliably *in vitro* to obtain reproducible production of embryogenic callus are compounded by the necessity to obtain explants from a mixed field population of palms and the destructive methods of harvesting generally used. However, it is now possible to harvest inflorescence material non-destructively (Rillo, 1989) and ultimately it could be possible to use roots as source material.

In order to achieve embryogenesis reproducibly it will be necessary to evaluate the right developmental stage of the explant tissue and find the best "window for regeneration" by screening tissues at different developmental stages. This can be done by monitoring metabolic changes (such as the changes in lipid content as carried out by Ceniza *et al.*, 1992), by using molecular techniques to detect early protein markers for embryogenesis using monoclonal antibodies (Altherr *et al.*, 1993), and by time-consuming studies of the cell cycle (Jesty and Francis, 1992). Investigations of endogenous hormone concentrations would also add to the knowledge of regenerating systems (as with the oil palm; Jones, 1990). Studies with protoplasts from highly embryogenic tissues could provide useful information, but embryogenic cell suspensions from selected callus lines would greatly facilitate many of these areas of future research.

## 5. Conclusion

The coconut is an important crop which is propagated entirely by seed. A method of clonal propagation is needed urgently so that improved breeding stocks can be obtained more rapidly.

Although the coconut has been propagated through somatic embryogenesis, the process is not well understood and few clonal plants have been produced. A good regenerative callus, or calloid, is difficult to induce and it is rare to obtain good quality somatic embryos which give normal plants. A range of abnormal embryogenic structures are commonly observed. Cotyledonary (or haustorial) tissues often predominate indicating that the process of somatic embryogenesis has occurred, but that shoot and root meristems have not developed. Empirical experimentation has not been successful in developing a reliable protocol for coconut micropropagation and more fundamental investigations are now required.

## 6. Acknowledgements

We are grateful to the Deutsche Gesellschaft für Technische Zusammenarbeit (GTZ) for supporting coconut research at Wye College over many years. R.H. has also received funding from the International Scientific Co-operation programme of the European Community. We thank all colleagues in the Department who have helped in the preparation of this review. Mrs. M. Wrigley contributed Fig. 5D.

## 7. References

Altherr, S., S. Stirn and H.J. Jacobsen, 1993. Immunobiochemical analysis of a nuclear protein marker for regeneration potential in higher plants. J. Plant Physiol. 141: 415–422.

Armstrong, S.W. and D. Francis, 1987. Cell and nuclear area, RNA and protein content and chromosome counts in meristematic cells of secondary roots and callus of *Cocos nucifera*. Can. J. Bot. 65: 547–552.

Blake, J., 1990. Coconut (*Cocos nucifera* L.): Micropropagation. In Y.P.S. Bajaj (Ed.), Biotechnology in Agriculture and Forestry, Vol. 10, Legumes and Oilseed Crops I, pp. 538–554. Springer-Verlag, Berlin.

Blake, J. and C.J. Eeuwens, 1982. Culture of coconut palm tissues with a view to vegetative propagation. In: A.N. Rao (Ed.), Tissue Culture of Economically Important Plants: Proc. Int. Symp. Singapore, pp. 145–148. COSTED/ANBS.

Blake, J. and N. Wilson, 1990. Problems of somatic embryogenesis in coconut (*Cocos nucifera* L.). In: Abstr. VIIth Int. Congr. Plant Tissue Cell Culture, p. 245.

Brackpool, A.L., R.L. Branton and J. Blake, 1986. Regeneration in palms. In: I.K. Vasil (Ed.), Cell Culture and Somatic Cell Genetics of Plants Vol. 3, pp. 207–222. Academic Press, New York.

Branton, R.L. and J. Blake, 1983. Development of organised structures in callus derived from explants of *Cocos nucifera* L. Ann. Bot. Lond. 52: 673–678.

Branton, R.L. and J. Blake, 1986. Clonal propagation of coconut. In: E. Pushparajah and Chew Poh Soon (Eds.), Cocoa and Coconuts: Progress and Outlook, pp. 771–780. Incorporated Society of Planters, Kuala Lumpur.

Branton, R.L. and J. Blake, 1988. Date palm (*Phoenix dactylifera* L.). In: Y.P.S. Bajaj (Ed.) Biotechnology in Agriculture and Forestry. Vol. 5 Trees II, pp. 161–175. Springer-Verlag, Berlin.

Buffard-Morel, J., J.L. Verdeil and C. Pannetier, 1992. Embryogenèse somatique du cocotier (*Cocos nucifera* L.) a partir d'explants foliaires: étude histologique. Can. J. Bot. 70: 735–741.

Ceniza, M.S., S. Ueda and Y. Sugimura, 1992. *In vitro* culture of coconut endosperm: callus induction and its fatty acids. Plant Cell Rep. 11: 546–549.

Corley, R.H.V., 1982. Clonal planting material for the oil palm industry. The Planter 58: 515–528.

Corley, R.H.V., C.Y. Wong, K.C. Wooi and L.H. Jones, 1981. Early results from the first oil palm clone trials. Paper presented at Int. Conf. The Oil Palm in Agriculture in the Eighties, Kuala Lumpur.

Corner, E.J.H., 1966. The Natural History of Palms. Weidenfeld and Nicolson, London.

Ebert, A. and H.F. Taylor, 1990. Assessment of the changes of 2,4–dichlorophenoxyacetic acid concentrations in plant tissue culture media in the presence of activated charcoal. Plant Cell Tiss. Org. Cult. 20: 165–172.

Ebert, A., F. Taylor and J. Blake, 1993. Changes of 6–benzylaminopurine and 2,4–dichlorophenoxyacetic acid concentrations in plant tissue culture media in the presence of activated charcoal. Plant Cell Tiss. Org. Cult. 33: 157–162.

Green, A.H. (Ed.), 1991. Coconut Production. Present Status and Priorities for Research. World Bank Technical Paper 136.

Haccius, B. and U.J. Philip, 1979. Embryo development in *Cocos nucifera* L. – critical contribution to a general understanding of palm embryogenesis. Plant Syst. Evol. 132: 91–106.

Jesty, J.H.F. and D. Francis, 1992. Cellular responses of leaf explants of *Cocos nucifera* L. *in vitro*. Plant Cell Tiss. Org. Cult. 28: 235–244.

Jones, L.H., 1990. Endogenous cytokinins in oil palm (*Elaeis guineensis* L.) callus, embryoids and regenerant plants measured by radioimmunoassay. Plant Cell Tiss. Org. Cult. 20: 201–209.

Justin, S.H.F.W., 1978. Vegetative propagation of coconuts. Rep. East Malling Res. Stn. 1977, pp. 175–176.

Karunaratne, S. and K. Periyapperuma, 1989. Culture of immature embryos of coconut (*Cocos nucifera* L.) callus proliferation and somatic embryogenesis. Plant Sci. 62: 247–253.

Karunaratne, S., C. Gamage and A. Kovoor, A. 1991. Leaf maturity, a critical factor in embryogenesis. J. Plant Physiol. 139: 27–31.

Monfort, S., 1985. Androgenesis of coconut – embryos from anther culture. Z. Pflanzenzücht. 94: 251–254.

Ohler, J.G., 1984. Coconut, Tree of Life. FAO Plant Production and Protection Paper 57.

Pannetier, C. and J. Buffard-Morel, 1986. Coconut Palm (*Cocos nucifera* L.) In: Y.P.S. Bajaj (Ed.), Biotechnology in Agriculture and Forestry Vol. 1. Trees, pp. 430–450. Springer-Verlag, Berlin.

Rillo, E.P., 1989. A nondestructive technique for collecting immature coconut inflorescence for tissue culture. Philipp. J. Coconut Stud. 14: 16–17.

Schwendiman, J., C. Pannetier and N. Michaux-Ferrière, 1990. Histology of embryogenic formations during *in vitro* somatic embryogenesis of oil palm *Elaeis guineensis* Jacq. Oléagineux 45: 409–418.

Sugimura, Y. and M.J. Salvaña, 1989. Induction and growth of callus derived from rachilla explants of young inflorescences of coconut palm. Can. J. Bot. 67: 272–274.

Thanh-Tuyen, N.T., 1990. Coconut (*Cocos nucifera* L.): anther culture. Y.P.S. Bajaj (Ed.), In: Biotechnology in Agriculture and Forestry, Vol. 10, pp. 555–568. Springer-Verlag, Berlin.

Touchet, B. de, Y. Duval and C. Pannetier, C. 1991. Plant regeneration from embryogenic suspension cultures of oil palm (*Elaeis guineensis*) Jacq.). Plant Cell Rep. 10: 529–532.

Verdeil, J.L. and J. Buffard-Morel, 1994. Somatic embryogenesis in coconut (*Cocos nucifera* L.). In: Y.P.S. Bajaj (Ed.), Biotechnology in Agriculture and Forestry. Somatic Embryogenesis. Springer-Verlag, Berlin (in press).

Verdeil, J.L., J. Buffard-Morel and C. Pannetier, 1989. Somatic embryogenesis of coconut (*Cocos nucifera* L.) from leaf and inflorescence tissue. Research findings and prospects. Oleagineux 44: 403–411.

Verdeil, J.L., C. Huet, F. Grosdemanges, A. Rival and J. Buffard-Morel, 1992. Coconut (*Cocos nucifera* L.) somatic embryogenesis: obtention of several clone ramets. Oléagineux 47: 465–469.

Williams, E.G. and G. Maheswaran, 1986. Somatic embryogenesis: factors influencing co-ordinated behaviour of cells as an embryogenic group. Ann. Bot. Lond. 57: 443–462.

Wooi, K.C., 1990. Oil palm (*Elaeis guineensis* Jacq.): tissue culture and micropropagation. In: Y.P.S. Bajaj (Ed.), Biotechnology in Agriculture and Forestry, Vol. 10, pp. 569–592. Springer-Verlag, Berlin.

# 19. Somatic embryogenesis in mango (*Mangifera indica* L.)

Richard E. Litz, Pamela A. Moon, Helena Mathews, Subramanian Jayasankar, Mary Joy Monsalud and Fernando Pliego-Alfaro

**Contents**



## 1. Introduction

### 1.1. *Importance and history*

The mango *Mangifera indica* L. is considered to be one of the most important fruit crops of the world (Fig. 1). Its annual production is exceeded only by *Musa* (bananas and plantains), citrus, grapes and apples (FAO Production Yearbook, 1990). Currently, it is the most important fruit crop of Asia. Mango is in the Anacardiaceae family, which includes several other tropical and subtropical tree species of economic importance, e.g., cashew *Anacardium occidentale*, pistachio *Pistachia vera*, and several *Spondias* species, including *S. cytherea*, *S. mombin* and *S. purpurea*. The center of diversity for the genus *Mangifera* is in southeast Asia, with the greatest number of species being found on the island of Borneo. According to Mukherjee (1985), there are 39 currently recognized *Mangifera* species. The mango is the most widely cultivated species within the genus, and has a natural distribution throughout southeast Asia. It is found in forests as far west as the Indo-Burman region (Assam). There appear to be two distinct geographical races of the species: a polyembryonic race that occurs in the tropical rainforest of southeast Asia and a monoembryonic race that is associated with India. Some taxonomists have suggested that the mango originated in India, because of the ancient association between Indian culture and religion with this fruit. It has also been noted that the word for mango used throughout Asia is derived from the Tamil "maanga". Indian culture permeated southeast Asia 1–2 millenia

*S. Jain, P. Gupta & R. Newton (eds.), Somatic Embryogenesis in Woody Plants, Vol. 2,* 341–356.

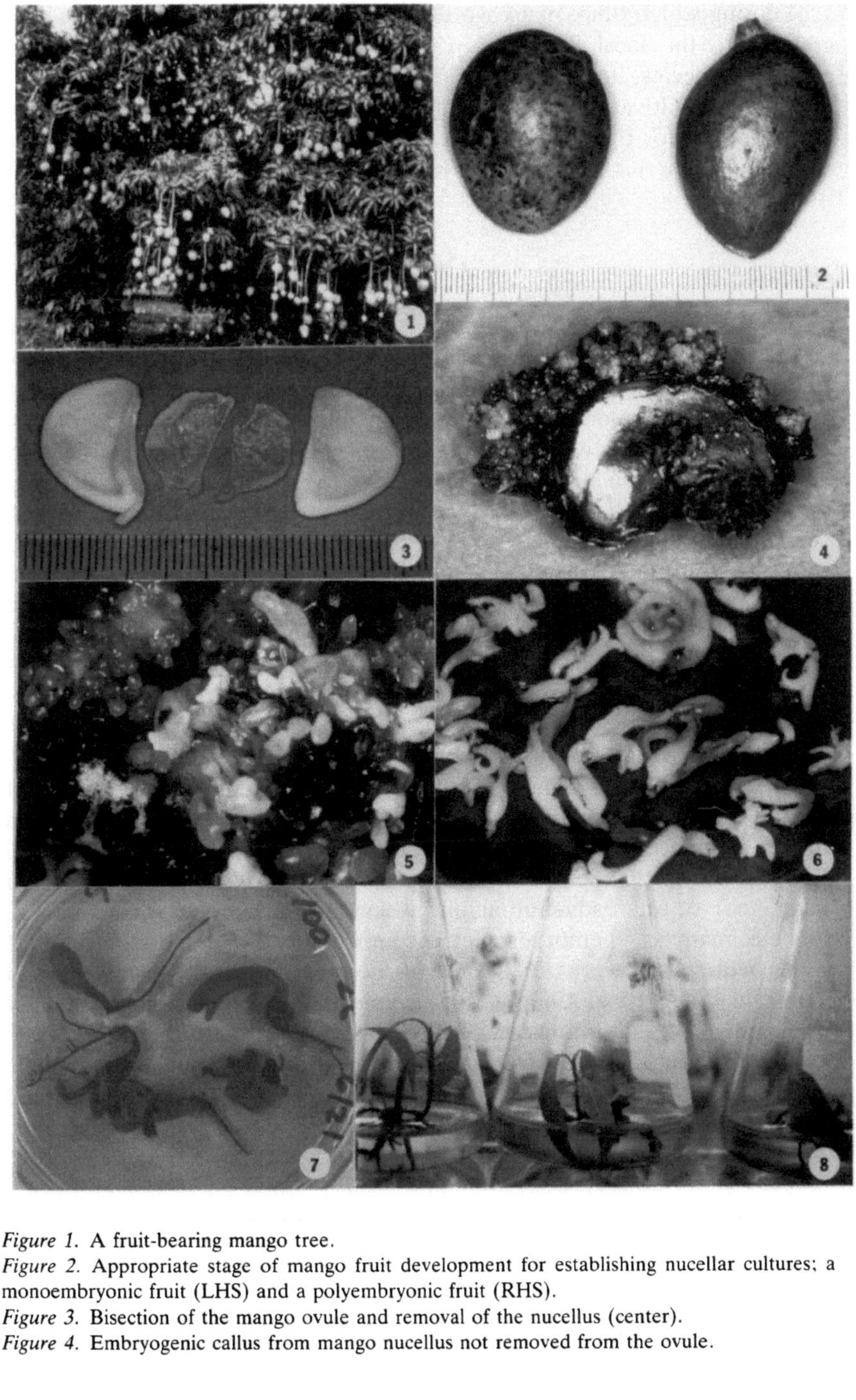

*Figure 1.* A fruit-bearing mango tree.
*Figure 2.* Appropriate stage of mango fruit development for establishing nucellar cultures; a monoembryonic fruit (LHS) and a polyembryonic fruit (RHS).
*Figure 3.* Bisection of the mango ovule and removal of the nucellus (center).
*Figure 4.* Embryogenic callus from mango nucellus not removed from the ovule.

ago, during which time, many words from Sanskrit and Tamil became integrated into the local languages. On the basis of the distribution of the genus and species, however, it is more probable that the mango was already available for cultivation throughout the region. Other *Mangifera* species that are cultivated for their fruit include *M. caesia*, *M. foetida* and *M. odorata*, although Mukherjee (1985) has indicated that at least 14 species bear edible fruit.

### 1.2. *Polyembryony*

In polyembryonic mangos, adventitious embryos arise from the nucellus at the micropylar end of the ovule. Embryogenic cells within the nucellus of polyembryonic mangos are normally densely cytoplasmic, and are usually found within the epidermis of the nucellus, but have also been observed in the hypodermis of the nucellus (Juliano, 1937). Occasionally, but particularly in certain varieties such as "Strawberry", the zygotic embryo fails to develop. The failure of fertilization and the abortion of the zygote due to adverse weather conditions are usually responsible for this anomaly. Although fertilization may not be an essential stimulus for induction of embryogenic competence in cells of the nucellus, the growth and maturation of nucellar and zygotic embryos embryos are dependent on the presence of the endosperm, i.e, fertilization (Sachar and Chopra, 1957). It is also possible that adventitious embryos can be formed by secondary inductions from the cotyledons and hypocotyls of developing embryos (Singh, 1969).

The number of adventitious embryos within the developing mango polyembryonic ovule varies with the cultivar and environmental factors that affect seed development. No suspensor has ever been observed to be associated with an adventitious embryo within a polyembryonic mango ovule (Juliano, 1934; Sachar and Chopra, 1957). In some polyembryonic cultivars, only a single embryo, often adventitious, is present. In the cultivar "Olour", there can be as many as 40 embryogenically competent cells within the nucellus of young ovules; however, only 3 or 4 of these can develop to maturity and germinate (Sachar and Chopra, 1957). Polyembryony in mango is thought to be mediated by 1 or more recessive genes (Leroy, 1947), although Sturrock (1968) considers that the trait is more likely to be determined by a single recessive gene. The progeny of hybridizations between monoembryonic and polyembryonic cultivars are often polyembryonic (Juliano, 1937; Wester,

---

*Figure 5.* Somatic embryogenesis from mango nucellar callus. Maturation of somatic embryos is occurring from callus regions removed from the induction medium.

*Figure 6.* Maturation of mango somatic embryos.

*Figure 7.* Mature, green mango somatic embryos. Germination has occurred.

*Figure 8.* Plant recovery from mango somatic embryos in an atmosphere of high carbon dioxide levels with high illumination.

1920). Because mangos are considered to be allotetraploids ($2n = 4x = 40$) (Mukherjee, 1950), determining the nature of inheritance of a polygenically mediated trait such as polyembryony is complex.

Mango embryos require 4–5 months to develop to maturity, and can attain a length (including the cotyledons) of 3–6 cm after this time. In some polyembryonic cultivars, e.g., "Cambodiana" and "Sabre", the embryo mass sometimes elongates so rapidly that it grows through the ovule wall in immature fruit before the seed coat has formed. The large cotyledons are rich in starch. Mango embryos are recalcitrant, and cannot be stored for more than 1 or 2 weeks at room temperature. When stored at 5°C, the seeds lose viability almost immediately (Parisot, 1988); between 10°C and 20°C , embryos retain their viability for 1 and a half to 3 weeks; at 30°C, mango seeds are viable for 5 days only. Mango seeds cannot withstand dehydration.

## 2. Morphogenesis

Somatic embryogenesis in mango has been reviewed previously by Litz (1986a,b), Litz *et al.* (1993) and by Mathews and Litz (1992). Other relevant reviews that have discussed somatic embryogenesis in mango in the context of other tropical fruit include Litz *et al.* (1985), Litz (1987) and Litz and Jaiswal (1991).

### 2.1. *Establishment of embryogenic cultures*

In a historically interesting paper, Maheshwari and Rangaswamy (1958) attempted to stimulate the proliferation of nucellar tissue from mango and *Citrus microcarpa*. Although the induction of embryogenic cultures was successful with *C. microcarpa*, they were unable to stimulate adventitious embryony in mango ovules *in vivo*. The earliest attempt to regenerate mango was reported by Rao *et al.* (1982), who described the initiation of callus from explanted cotyledon pieces on MS (Murashige and Skoog, 1962) medium containing 15% coconut water, 40 g/l sucrose, 11.5 μM kinetin and 27 μM NAA. Callus was formed within 2 weeks after explanting, and was described as being dark, globular and compact. Although adventitious roots developed from the callus after a few weeks, shoot formation was never observed. Rao *et al.* (1982) did not indicate that the mango used in this study was from a monoembryonic or a polyembryonic cultivar.

Litz *et al.* (1982) reported that somatic embryogenesis could occur from the nucellus of explanted ovules of polyembryonic mango cultivars "Chino", "Heart", "Ono", "Sabre" and "Turpentine" on modified MS medium containing 400 mg/l glutamine, 100 mg/l ascorbic acid, 60 g/l sucrose and either 20% coconut water or BA at 4.4 and 8.8 μM. They indicated that the efficiency of somatic embryogenesis could be greatly enhanced if the cultures were maintained in liquid medium of the same formulation. In a later report,

Litz *et al.* (1984) described the initiation of embryogenic cultures of mango cultivars "Chino" and "Ono" from explanted nucellus on modified MS medium containing 20% coconut water. Embryogenic cultures were established by explanting the excised nucellus on plant growth medium. The optimum response for establishing embryogenic cultures was obtained when ovules were used 30–45 days after pollination/fertilization. Litz (1984) was able to induce a transient embryogenic callus from the explanted nucellus of 3 monoembryonic mango cultivars, "Irwin", "Ruby" and "Tommy Atkins" on modified MS medium (Litz *et al.*, 1982) with 4.5 μM 2,4-D, and observed that the embryogenic response was cultivar-related. In later reports, embryogenic cultures of monoembryonic "Lippens" (Litz, 1987) and "Keitt" (Litz and Schaffer, 1987) were induced.

In contrast with the response of polyembryonic mango nucellar explants on medium containing 2,4-D, monoembryonic mango nucellus produces a transient embryogenic callus, which is typically friable and yellow – to – light brown in color. Very rapidly, however, this callus differentiates entirely on medium with or without 2,4-D. On medium with 2,4-D, proembryos differentiate throughout the callus, and continue to proliferate by the formation of secondary proembryos from the protoderm; whereas, on medium containing no plant growth regulators, the proembryos can develop to maturity after a brief sequence of proliferation. Proembryos develop directly from the nucellar explants of polyembryonic mango without the formation of an intermediate callus. Moreover, they can develop on medium that does not contain any plant growth regulators. When subcultured on growth media with and without 2,4-D, the developmental sequence is the same as that described for monoembryonic mango. It is apparent that an inductive event is essential in order to establish embryogenic cultures of monoembryonic mango. Embryogenic competent cells are already present in the explanted nucellus of polyembryonic mangos, and therefore induction is unnecessary; the *in vitro* response is permissive.

The stage of fruit and ovule development can be critical for establishing embryogenic cultures from mango nucellus (Fig. 2), and usually is optimum when the embryo:ovule ratio is not greater than 0.5, although both the optimum ratio and the efficiency of establishment of embryogenic cultures is cultivar-related (Litz, 1987). The nucellus in mango ovules of either the polyembryonic or monoembryonic type is enlarged, and is normally quite thick and easily removed from the ovule at its micropylar end (Fig. 3). Although differentiation most commonly occurs from the micropylar end of nucellar explants, Dewald (1987) demonstrated that the entire tissue can produce embryogenic cultures. The current protocol for establishing embryogenic cultures involves the use of plant growth medium consisting of B5 (Gamborg *et al.*, 1968) major salts, MS minor salts and organics, 400 mg/l glutamine, 60 g/l sucrose, 4.5 μM 2,4-D and 2.0 g/l gellan gum (after DeWald *et al.*, 1989a). In order to avoid damaging the nucellus during its excission, we routinely place into culture bisected ovules from which the embryo mass has been removed, so that the nucellus is in direct contact with the plant growth medium (Fig. 4). Mango explants are incubated on induction medium in sterile plastic disposable petri dishes sealed with Parafilm in darkness at 25°C.

### 2.2. *Maintenance of embryogenic cultures*

When embryogenic cultures are maintained on semi-solid plant growth medium containing 2,4-D, the cultures normally become dark brown within a few days, although this is not necessarily an indication of mortality. Newly formed tissue is pale yellow or light brown, but rapidly becomes discolored. The growth rate is low. Somatic embryo maturation can occur in embryogenic cultures maintained on semi-solid medium containing 2,4-D, but usually only if sectors of the cultures are not in direct contact with the medium (Fig. 5). For ease of handling, embryogenic cultures are usually maintained in liquid medium containing 4.5–9 μM 2,4-D (Litz *et. al.*, 1984; Dewald *et al.*, 1989a). Some cultivars, including "Chino", "Hindi", "Ono" (all polyembryonic) and "Lippens" (monoembryonic) can be maintained easily as embryogenic suspension cultures; whereas, nucellar cultures of other cultivars of mango: 1) form large spherical structures that fail to produce secondary somatic embryos, e.g., polyembryonic "Sabre", 2) produce rapidly proliferating cultures that appear to be embryogenic but fail to differentiate following subculture into plant growth regulator free-medium, e.g., polyembryonic "Carabao", and 3) fail to proliferate at all, e.g., monoembryonic "Keitt". Embryogenic cultures consist of modified somatic proembryos that proliferate by the formation of secondary somatic proembryos. They can be routinely maintained in semi-darkness in Erlenmeyer flasks (125 ml) containing 50 ml plant growth medium (with 2,4-D) on a rotary shaker at 100 r.p.m. Subculture into fresh liquid medium is essential at 4–5-day intervals in order to prevent the rapid darkening of the tissue.

At the time of subculture, the embryogenic suspensions of mango cultivars such as "Hindi", "Lippens" and "Tommy Atkins" consist of proembryonic masses of different sizes. The larger proembryonic masses produce secondary somatic embryos continuously from the protoderm; however, the larger masses have lost their integrative ability, and cannot form singulated somatic embryos following subculture into medium without 2,4-D. The smaller proembryos have the ability to form single somatic embryos in most mango cultivars. Hence, to initiate the process of somatic embryo maturation, the embryogenic cultures are passed through filtration fabric, and only the smaller proembryos are retained. The large proembryonic masses are recultured in medium containing 2,4-D, and continue to proliferate.

Prolonged maintenance of embryogenic cultures of mango can result in the loss of morphogenic potential over time. This response is highly cultivar-dependent, and "Hindi" embryogenic cultures have been maintained in suspension for up to 5 years without loss of morphogenic potential. Another implication of long-term maintenance of embryogenic cultures is the increased frequency over time of developmental anomalies, i.e., fasciation, polycotyly, fused cotyledons and the loss of bipolarity in differentiated somatic embryos (Litz *et al.*, 1993). Hyperhydricity is a serious problem during maturation of somatic embryos of certain mango cultivars. Affected somatic embryos are characteristically glassy in appearance, and cannot develop beyond an intermediate stage of maturation. They become necrotic; however, secondary somatic embryogenesis often occurs from their hypocotyls, and these somatic embryos appear to be normal. This problem appears to be associated with rapid proliferation in suspension culture, rapid maturation and the very large size of these somatic embryos. Mathews *et al.* (1992) indicated that "Hindi" somatic embryos appear to be particularly prone to this disorder. Hyperhydricity of somatic embryos can be reversed by subculturing them on hard growth medium (6 g/liter gellan gum) or by partially dehydrating them (Monsalud, 1994).

### 2.3. *Maturation of somatic embryos*

Cotyledonary stage mango somatic embryos usually develop asynchronously and inefficiently from embryogenic cultures that have been maintained for long periods without subculture on semi-solid medium with 2,4-D. Maturation of somatic embryos occurs from sectors of these cultures that are physically isolated from the plant growth medium. Subculture onto semi-solid plant growth medium that is free of 2,4-D is essential for the efficient production of heart stage somatic embryos, although these cultures are not synchronized due to the presence of proembryos of different developmental stages. Synchronization of embryogenic cultures is much more practical using suspension cultures; however, many mango cultivars, e.g., Cambodiana", "Keitt", "Sabre", etc., do not proliferate as suspension cultures. If suspension cultures consisting of proembryos are passed through nylon filtration fabric, a degree of synchronization can be achieved. Proembryos between 520 and 860 μm can be collected and subcultured in liquid embryogenesis medium (no 2,4-D). After 15 days, proembryos between 860 and 1910 μm are collected and recultured in the same medium formulation. At each subculture thereafter, proembryos of the same size (860–1910 μm) are recultured in liquid embryogenesis medium until somatic embryos of the heart stage of development are present. The time required for the appearance of heart stage mango somatic embryos following the subculture of proembryos into liquid medium is cultivar-dependent, and varies from 30 days for "Hindi" to 65 days for "Lippens" among cultivars in which this character has been measured (Litz *et al.*, 1993). The presence of 0.22 μM BA in liquid embryogenesis medium can reduce this time by stimulating differentiation of the cotyledons (Fig. 6). It is necessary for "Hindi" somatic embryos that have reached the heart stage of development to be transferred onto semi-solid medium in order for normal embryo maturation to occur. If "Hindi" embryogenic cultures in liquid growth medium with 2,4-D are transferred to semi-solid medium without 2,4-D, the cultures very quickly become dark brown, and somatic embryo differentiation cannot occur. Proembryos of "Carabao", on the other hand, can be subcultured from 2,4-D-containing liquid medium onto semi-solid medium without 2,4-D to induce maturation.

The medium that has been developed for maturation of mango somatic embryos consists of half strength B5 major salts with the MS minor elements and organic components of maintenance and embryogenesis media, 40 g/l sucrose, 400 mg/l glutamine and 20% (v/v) filter-sterilized coconut water (DeWald *et al.*, 1989b). This medium, rich in organic composition, is used because of the long maturation period of mango embryos, i.e., 120–160 days from pollination and/or fertilization and the extremely large size of mature mango somatic embryos (3–6 cm). This medium enables the somatic embryos to attain a large size without precocious germination. Somatic embryo maturation occurs entirely in darkness. The somatic embryos remain white until they are mature, at which time they become green in the absence of light. Maturation and germination are continuous, and there is no period of developmental arrest. Germination is evident at the time that the hypocotyl elongates. Soon afterwards, the radicle emerges, and rapidly grows into the medium (Fig. 7). At the earliest indication of germination, mango somatic embryos are transferred from dark to light conditions (60 $\mu$mol/m$^2$/s$^1$) with a 16 h photoperiod. The apical meristem does not organize until after germination begins; however, approximately 1–2 weeks after germination, the shoot elongates rapidly.

Maturation of mango somatic embryos was also obtained in liquid basal medium, and germination occurred following the transfer of apparently mature somatic embryos onto semi-solid medium supplemented with 4.4 $\mu$M BA. Plantlets could not be recovered, and secondary somatic embryogenesis was reported to occur from the transitional zone between the hypocotyl and radicle in germinating somatic embryos (Litz *et al.*, 1984).

## 2.4. *Plant recovery*

When mango somatic embryo-derived plantlets are grown under the conditions described above, they invariably do not survive transfer to soil. An apical necrosis develops that gradually results in mortality. By adapting a protocol that was originally developed by Kozai and his coworkers (Kozai and Iwanami, 1988; Fujiwara *et al.*, 1988; etc.) for stimulating autotrophic growth in micropropagated nursery plants and later modified for recovering *Theobroma cacao* plantlets from somatic embryos that could survive the transfer to soil (Figueira *et al.*, in press), it has been possible to recover plantlets with a much greater efficiency (Litz *et al.*, 1993; Mathews and Litz, 1992). *In vitro* germinated mango plantlets on semi-solid medium in Erlenmeyer flasks or G7 containers that have been sealed with gas permeable membranes are transferred to an incubator that has been supplied with

*Figure 9.* Possible somaclonal variant derived from a "Keitt" mango somatic embryo. Regenerant shows precocious branching.

20,000 ppm $CO_2$ in nitrogen as a carrier. The light intensity is 180–200 $\mu mol/m^2/s^1$ (16 h photoperiod) and is provided by cool white fluorescent lamps. The plantlets that grow under these conditions have thicker leaves, and do not decline as rapidly from apical necrosis (Fig. 8).

Another approach that has been adopted for improving the survival of plantlets derived from somatic embryos of mango has involved the partial desiccation of heart stage somatic embryos of cultivars such as "Hindi" that have a strong tendency to become vitrified (Monsalud, 1994). Accordingly, highly synchronized cultures of heart stage somatic embryos (3–5 mm length) can be partially desiccated by transferring them from liquid embryogenesis medium to dry sterile filter paper within petri dishes. After a few minutes, the petri dishes are sealed with parafilm, and the somatic embryos can remain viable for up to 9 days if stored in the dark at 25°C. This treatment results in the loss of approximately 18% of the fresh weight of the somatic embryos. When they are recultured on semi-solid maturation medium (see above), most of the somatic embryos resume development. The quality of these somatic embryos is enhanced, and they germinate efficiently *in vitro*.

Plants that have been regenerated from embryogenic nucellar cultures of mango appear to be normal, although precocious branching has been observed in "Keitt" mango plants derived from cultures maintained in 2,4-D – containing medium for only 6 months (Fig. 9). Litz *et al.* (1993) compared isozymes extracted from leaves and roots of nucellar seedlings of the polyembryonic mango cultivar "Philippine" with the same isozymes extracted from leaves and roots of plantlets derived from somatic embryos of the same cultivar. They found that there was considerable variation (≥20%) in the banding pattern of leaf esterase (EST) and alcohol dehydrogenase (ADH) of somatic embryo-derived plantlets. Although variation in isozyme banding patterns was also observed in nucellar seedlings, the frequency was much less (≤2%). This indicates that genetic variation occurs naturally in nucellar seedlings. However, variation is much greater in plants that have been regenerated from somatic embryos. It is possible that off-types have a selective advantage in rapidly proliferating embryogenic cultures.

## 3. Applications

### 3.1. *Micropropagation*

Somatic embryogenesis has certain obvious implications for the propagation and improvement of mango. Mango cultivars are routinely propagated by cleft grafting, bud grafting and inarching. All of these procedures are fairly routine and efficient with standard mango cultivars. However, rootstocks can significantly affect the vigor of grafted clones, since monoembryonic mango seedling rootstocks can be very heterogeneous. In certain mango-producing regions, such as India, there is a scarcity of uniform, i.e., polyembryonic, mango seedling rootstocks. This shortage could be addressed through the production of superior rootstocks by somatic embryogenesis.

One of the greatest achievements of conventional mango breeding programs in recent years has been the identification of dwarf selections with fruit of acceptable quality. In heavily populated areas of the world, arable land is needed for the production of staple food crops. Mango production in the future will probably be in densely planted blocks of dwarf, high yielding cultivars. Unfortunately, because of the slow growth of these selections, they cannot be routinely propagated vegetatively in numbers that would be meaningful due to the shortage of grafting bud wood. Somatic embryogenesis could be useful for propagating sufficient material for the rapid expansion of this new industry.

The apparently high incidence of somaclonal variation in plantlets derived from somatic embryos of mango must be a limiting factor in considering whether somatic embryogenesis is appropriate for micropropagating mango cultivars (Litz *et al.*, 1993). Similarly, the medium- to long-term preservation of mango cultivar germplasm in the form of embryogenic cultures or as early developmental stages of somatic embryos may be inappropriate because of somaclonal variation.

### 3.2. *Cultivar improvement*

Mango cultivars represent trees having unique gene combinations that have been selected for their superior fruit quality and yield within a particular agroecosystem. With a few exceptions, all of the existing mango cultivars have been clonally propagated from dooryard selections that have been made over many hundreds of years. Conventional breeding has had relatively little impact on mango cultivar development, with the exception of "Mallika", "Amrapali", "Au-Rumani", "Manjera" and "Ratna" that have emerged from breeding programs in India during the past few years (Iyer, 1989). The disappointing impact of conventional breeding on mango improvement is due to several factors. Mango is extremely heterogeneous, which is complicated by its allotetraploidy. Thus, the simplest trait is mediated by at least 4 genes at each locus. Since most horticulturally important traits are conferred by complexes of genes, the probability of securing a superior seedling tree as a result of controlled pollinations is very remote. Furthermore, mango has a relatively long juvenile period, i.e., 7 years, and seedling trees normally require another 10–15 years before all of the important horticultural parameters can be evaluated. For these reasons, mango cultivar improvement via somatic cell genetics and by the production of transgenic plants has great attraction.

#### 3.2.1. In vitro *selection*

Litz *et al.* (1989) described preliminary attempts to recover somaclonal variants of mango that had greater resistance to anthracnose, the most serious disease, worldwide, affecting mango production. Anthracnose is caused by the fungal pathogen *Colletotrichum gloeosporioides* Penz. (Fig. 10). There is strong supportive evidence that anthracnose symptom development is associated with the presence of one or more phytotoxins produced by the pathogen. Colletotin has been isolated from cultures of *Colletotrichum* sp. (Goodman, 1960); another phytotoxin, β-lycomarasmic acid, has also been isolated and characterized from *C. gloeosporioides* cultures (Ballio *et al.*, 1969). By co-culturing embryogenic cultures of mango with partially purified culture filtrates of *C. gloeosporioides* that had been isolated from anthracnose-affected mango leaves and fruit, Litz *et al.* (1989) indicated that it was possible to differentiate the interaction responses of resistant and susceptible cultivars. It was also possible to demonstrate that suspension cultures could be challenged successfully with partially purified culture filtrate. After recurrent selection with culture filtrate, embryogenic cultures of susceptible culti-

*Figure 10.* "Lippens" somatic embryos that have been challenged with partially purified *Colletotrichum gloeosporioides* culture filtrate.

vars showed greater tolerance to the culture filtrate. The response of regenerated plants to controlled inoculations by the pathogen was not reported.

### 3.2.2. *Genetic transformation*

Mathews and Litz (1990) reported that the growth of embryogenic "Parris" and "Tommy Atkins" suspension cultures is inhibited by 12.5 μg/ml kanamycin; whereas the maturation of somatic embryos was inhibited on semi-solid medium by 200 μg/ml kanamycin. Growth inhibition of the latter was incomplete because regions of cultures on semi-solid medium were not in complete contact with the growth medium. Litz *et al.* (1990) also described the regeneration of putatively genetically transformed plantlets derived from embryogenic cultures of a "Keitt" zygotic embryo. *Agrobacterium tumefaciens* strain C58C1 harboring the disarmed cointegrate plasmid pGV 3850::1103 that conferred resistance to kanamycin was utilized as the vector. Molecular confirmation for the stable integration of the neophosphate transferase gene (NPT II) into mango was ambiguous, and it is now considered that the plantlets were not solid transformants, but were more likely to be chimeras.

Subsequently, Mathews *et al.* (1992), using highly embryogenic "Hindi"

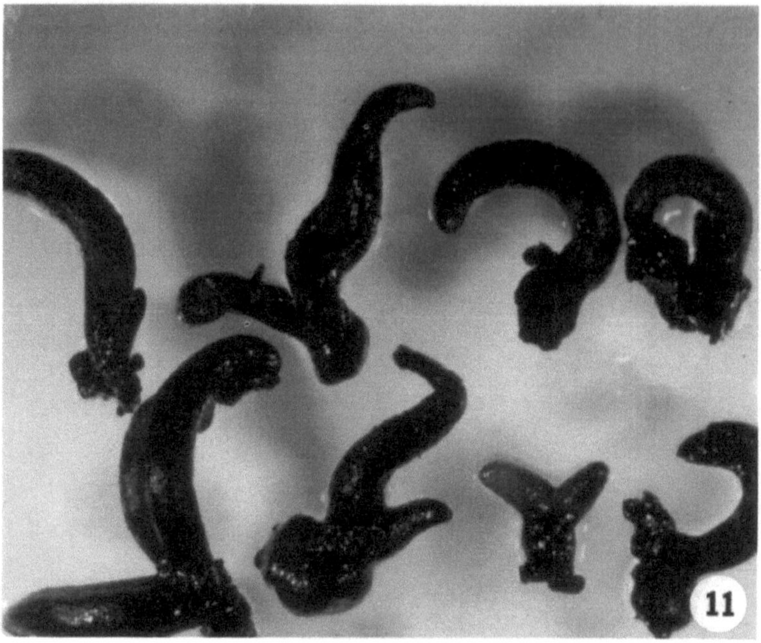


*Figure 11.* X GLUC reaction of "Hindi" somatic embryos indicating the presence of GUS.

suspension cultures, demonstrated that solid transformants for the NPT II and $\beta$-glucuronidase (GUS) genes could be recovered. The vector that was utilized was disarmed, genetically engineered *A. tumefaciens* containing both of these genes [strain A208 harboring pTiT37–SE::pMON 9749 (9749 ASE)]. In order to eliminate chimeras from the cultures, a step-wise selection strategy was adopted: 1) initial screening at 200 μg/ml kanamycin, 2) screening at 400 μg/ml kanamycin, and 3) recovery of purely transformed proembryos in liquid selection medium containing 100 μg/ml kanamycin. The somatic embryos that were recovered stained intensely and completely blue in the X-GLUC reaction (Jefferson, 1987), indicating the incorporation of GUS into the mango genome, and grew normally in growth medium containing very high concentrations of kanamycin, indicating the stable incorporation of the NPT II gene into mango (Fig. 11). Although "Hindi" is highly embryogenic, and is in many ways an ideal model for many *in vitro* studies, its somatic embryos become vitrified during maturation and deteriorate. Consequently, plantlets were never recovered from the genetically transformed "Hindi" somatic embryos.

## 4. Conclusion

Many of the important parameters that influence the induction of embryogenic cultures of mango from nucellar explants, the maintenance of these cultures, the maturation of somatic embryos and the regeneration of plants from somatic embryos have been described. The potential for utilizing these protocols to improve existing mango cultivars is evident; however, there has been almost no attempt to isolate and characterize genes that confer horticulturally interesting traits in mango. Gomez Lim *et al.* (1993) have identified proteins that are associated with fruit ripening in mango. These have been identified as enzymes that degrade the cell wall, e.g., polygalacturonase, invertase, amylase and cellulase. These same proteins are also detectable when mRNA extracted from unripe and ripe fruit is translated *in vitro*. They also reported the preparation of a cDNA bank from ripe mango fruit, and isolated several genes that are differentially expressed during fruit development. The implication of this work for mango improvement is that certain mango fruit ripening genes, e.g., for polygalacturonase or one of the rate limiting genes involved in ethylene synthesis, could be inserted into the mango genome in the antisense configuration, thereby blocking or impeding ripening (Oeller *et al.*, 1991; Smith *et al.*, 1990). This would be of great benefit for mango producers in developing countries, as it would extend the shelf life of mango and extend the market for this fruit in temperate countries. Other molecular strategies that could result in improvement of existing cultivars of mango involve the genetic transformation of mango with antimicrobial and pesticidal genes; however, because mango is a perennial species, there remain unanswered questions about this approach. Will pest resistance develop in the field in a similar manner that has been observed with virtually all pesticides? Since somatic embryogenesis rejuvenates clonal mango, regenerated plants require several years in order to reach maturity. Will the genetically engineered pest resistance in mango break down before maturity and production are reached? These and other questions must be addressed before we can rely on some contemporary genetic approaches at the expense of traditional plant breeding.

## 5. Acknowledgements

The authors would like to express their gratitude to those who made much of this work possible, including the Florida Mango Forum, Monsanto, and the International Board for Plant Genetic Resources. Florida Agricultural Experiment Station Journal Series No. R-03334.

## 6. References


Ballio, A., A. Bottalico, A. Carilli, V. DiVittorio and A. Graniti, 1969. Production and isolation of aspergillomarasmin $\beta$ (lycomarasmic acid) from cultures of *Colletotrichum gloeosporioides* Penz. (*Gloeosporioides olivarum* Alm.). Phytopath. Medit. 8: 187–196.

DeWald, S.G., 1987. *In vitro* somatic embryogenesis and plant regeneration from mango (*Mangifera indica* L.) nucellar callus. Ph.D. Thesis, University of Florida, Gainesville.

DeWald, S.G., R.E. Litz and G.A. Moore, 1989a. Optimizing somatic embryo production in mango. J. Amer. Soc. Hort. Sci. 114: 712–716.

DeWald, S.G., R.E. Litz and G.A. Moore, 1989b. Maturation and germination of mango somatic embryos. J. Amer. Soc. Hort. Sci. 114: 837–841.

FAO Production Yearbook, 1990. Food and Agricultural Organization of the United Nations, Rome.

Figueira, A., A. Whipkey and J. Janick. Elevated $CO_2$ facilitates micropropagation of *Theobroma cacao* L. In: Proc. Int. Cocoa Conf., Kuala Lumpur (in press).

Fujiwara, K., T. Kozai and I. Watanabe, 1988. Development of a photoautotrophic tissue culture system for shoots and/or plantlets at rooting and acclimatization stages. Acta Hort. 230: 153–158.

Gamborg, O.L., R.A. Miller and K. Ojima, 1968. Plant cell cultures. 1. Nutrient requirements of suspension cultures of soybean root cells. Exp. Cell Res. 50: 151–158.

Gomez Lim, M., L.J. Saucedo and R. Lopez Gomez, 1993. The study of mango fruit ripening at the molecular level. Acta Hort. (in press).

Goodman, R.N., 1960. Colletotin, a toxin produced by *C. fuscin* Laub. Phytopathology 50: 325–327.

Iyer, C.P.A., 1989. Recent advances in varietal improvement in mango. Acta Hort. 291: 109–132.

Jefferson, R.A., 1987. Assaying chimeric genes in plants: the gus gene fusion system. Plant Mol. Biol. 5: 387–405.

Juliano, J.B., 1934. Origin of embryos in the strawberry mango. Philip. Agric. 25: 553–563.

Juliano, J.B., 1937. Embryos of Carabao mango (*Mangifera indica* Linn.). Philip. Agric. 25: 749–760.

Kozai, T. and Y. Iwanami, 1988. Effects of $CO_2$ enrichment and sucrose concentration under high photon fluxes on plantlet growth of carnation (*Dianthus caryophyllus* L.) in tissue culture during the preparation stage. J. Jap. Soc. Hort. Sci. 57: 279–288.

Leroy, J.F., 1947. La polyembryonie chez les *Citrus*. Son intérêt dans la culture et l'amélioration. Rev. Int. Bot. Appl. (Paris) 27: 483–495.

Litz, R.E., 1984. *In vitro* somatic embryogenesis from nucellar callus of monoembryonic *Mangifera indica* L. HortSci. 19: 715–717.

Litz, R.E., 1986a. Mango. In: D.A. Evans, W.R. Sharp and P.V. Ammirato (Eds.), Handbook of Plant Cell Culture, Vol. 4. Techniques and Applications, pp. 612–625. Macmillan, New York.

Litz, R.E., 1986b. Mango (*Magifera indica* L.). In: Y.P.S. Bajaj (Ed.), Biotechnology in Agriculture and Forestry, Trees I, pp. 220–232. Springer-Verlag, Heidelberg.

Litz, R.E., 1987. Application of tissue culture to tropical fruit trees. In: C. E. Green, D.A. Somers, W.P. Hackett and D.D. Biesboer (Eds.), Plant Tissue and Cell Culture, pp. 407–418. Alan R. Liss, New York.

Litz, R.E. and V.S. Jaiswal, 1991. Micropropagation of tropical and subtropical fruits. In: R.H. Zimmerman and P. Debergh (Eds.), Micropropagation Technology and Application, pp. 247–264. Kluwer Academic Publishers, Dordrecht.

Litz, R.E., R.J. Knight and S. Gazit, 1982. Somatic embryos from cultured ovules of polyembryonic *Mangifera indica* L. Plant Cell Rep. 1: 264–266.

Litz, R.E., R.J. Knight and S. Gazit, 1984. *In vitro* somatic embryogenesis from *Mangifera indica* L. callus. Sci. Hort. 22: 233–240.

Litz, R.E., H. Mathews, N. Bharathan and K.R. Narayanan, 1990. Transformation of somatic embryos of mango. In: VII Int. Cong. Plant Tiss. and Cell Cult., Amsterdam, p. 67 (Abstr.).
Litz, R.E., V.H. Mathews, R.C. Hendrix and C.Y. Yurgalevitch, 1989. Mango somatic cell genetics. Acta Hort. 291: 133–140.
Litz, R.E., V.H. Mathews, P.A. Moon, F. Pliego-Alfaro, C.Y. Yurgalevitch and S.G. DeWald, 1993. Somatic embryos of mango (*Mangifera indica* L.). In: K. Redenbaugh (Ed.), Synseeds Applications of Synthetic Seeds to Crop Improvement, pp. 409–425. CRC Press, Boca Raton.
Litz, R.E., G.A. Moore and C. Srinivasaan, 1985. Tissue culture for propagation and improvement of tropical fruits and palms. Hort. Rev. 7: 157–200.
Litz, R.E. and B. Schaffer, 1987. Polyamines in adventitious and somatic embryos of mango. J. Plant Physiol. 128: 251–258.
Maheshwari, P. and N.S. Rangaswamy, 1958. Polyembryony and *in vitro* culture of embryos of *Citrus* and *Mangifera*. Indian J. Hort. 15: 272–282.
Mathews, H. and R.E. Litz, 1990. Kanamycin sensitivity of mango somatic embryos. HortSci. 25: 965–966.
Mathews, H. and R.E. Litz, 1992. Mango. In: F.A. Hammerschlag and R.E. Litz (Eds.), Biotechnology of Perennial Fruit Crops, pp. 433–448. CAB International, Wallingford, Oxon.
Mathews, H., R.E. Litz, H.D. Wilde, S.A. Merkle and H.Y. Wetzstein, 1992. Stable integration and expression of $\beta$-glucuronidase and NPT II genes in mango somatic embryos. In Vitro Cell. Dev. Biol. 28P: 172–178.
Monsalud, M.-J.R., 1994. Control of hyperhydricity in 'Hindi' mango somatic embryos. M.S. Thesis, University of Florida, Gainesville.
Mukherjee, S.K., 1950. Mango: its allopolyploid nature. Nature 166: 196–197.
Mukherjee, S.K., 1985. Systematic and Ecogeographical Studies of Crop Genepools: 1. *Mangifera* L. International Board for Plant Genetic Resources, Rome.
Murashige, T. and F. Skoog, 1962. A revised medium for growth and bioassays of tobacco tissue cultures. Physiol. Plant. 15: 473–497.
Oeller, P.W., M.-W. Lu, L.P. Taylor, D.A. Pike and A. Theologis, 1991. Reversible inhibition of tomato fruit senescence by antisense RNA. Science 254: 437–439.
Parisot, E., 1988. Étude de la croissance rhythmique chez de jeune manguière (*Mangifera indica* L.). I. Description, germination et conservation des graines polyembryonnées de manguier. Fruits 43: 97–105.
Rao, A.N., Y.M. Sin, N. Kathagoda and J.F. Hutchinson, 1982. Cotyledon tissue culture of some tropical plants. In: A.N. Rao (Ed.), Tissue Culture of Economically Important Plants, pp. 124–137. COSTED, Singapore.
Sachar, R.C. and R.N. Chopra, 1957. A study of the endosperm and embryo in *Mangifera* L. Indian J. Agric. Sci. 27: 219–228.
Singh, L.B., 1969. Mango. In: F.P. Ferwerda and F. Wit (Eds.), Outlines of Perennial Crop Breeding in the Tropics, pp. 309–327. H. Veenen and Zonen, Wageningen.
Smith, C.J.S., C.F. Watson, P.C. Morris, C.R. Bird, G.B. Seymour, J.E. Gray, C. Arnold, G.A. Tucker, W. Schuch, S. Harding and D. Grierson, 1990. Inheritance and effect on ripening of antisense polygalacturonase genes in transgenic tomatoes. Plant Mol. Biol. 14: 369–374.
Sturrock, T.T., 1968. Genetics of mango polyembryony. Proc. Fla. State Hort. Soc. 81: 311–314.
Wester, P.J., 1920. The mango. Philip. Bur. Agric. Bull. 18: 3–70.

# 20. Somatic embryogenesis in neem (*Azadirachta indica*)

Shubhada Thengane, Medha Joshi and A.F. Mascarenhas

**Contents**



## 1. Introduction

*Azadirachta indica* A. Juss is commonly known as Neem or "Margosa", and also Indian Lilac (synonym *Melia azadirachta* L.; *Melia indica* A. Juss Brand or *Melia parviflora* Moon; family Meliaceae). It is a native to Burma and dry arid regions of the Indian subcontinent. During the last century, it was introduced into the arid zones of Africa and is now being widely planted in Australia, Fiji, Mauritius, Central and South America, The Caribbean, Puerto Rico, Haiti, Philippines and other countries.

*A. indica* is an evergreen, tall, fast growing tree, which can grow up to a height of 25 meters. It has a spreading habit with a broad crown. It thrives on nutrient-deficient soils and is tolerant to high temperatures but is susceptible to excessive cold or frost (Radwanski, 1977a). The native habitat occurs at altitudes between 50 and 100 meters. Rainfall of 130 mm per annum is sufficient for its normal growth. Its deep root system has the capability to extract nutrients and moisture from highly leached, sandy soils and it can withstand long periods of drought.

Although the tree grows well in soil at pH 5, its leaf litter gradually brings the surface soil pH to neutrality (Radwanski, 1969).

*A. indica* can be easily propagated by seed. The reproductive phase normally begins after 5 years. It flowers from January through April and the fruits mature from May to August. The fruits are produced in drooping panicles (Fig. 1). Each fruit is about 2 cm long and when ripe, has a yellow fleshy pericarp, a white hard shell and a brown oil-rich kernel. Yield ranges between 30–100 kg per tree depending upon the ecosystem, genotype and habitat.

*S. Jain, P. Gupta & R. Newton (eds.), Somatic Embryogenesis in Woody Plants, Vol. 2*, 357–374.

*Figure 1.* Drooping panicles of Neem.

For centuries, the derivatives of this tree have found uses in agriculture, health, medicine, cosmetics and livestock production. The recent spurt of interest in this tree is due to its spectacular biological activity against a wide range of insects and nematodes. Because of its legendary insect repellent and medicinal properties, *A. indica* is being identified as "the most promising of all plants", by the National Research Council, USA. It is now being identified as "nature's gift to mankind", "the wonder tree", "the tree for solving global problems", or "Global tree".

### 1.1. *Chemistry*

The pioneering work on the isolation and identification of this tree's constituents was initiated in India in 1942 and has continued in various parts of the world.

The bitterness of this tree is due to the presence of a wide array of biologically active compounds which are chemically diverse and structurally complex. In addition to general types of constituents such as carbohydrates, fatty acids, amino acids, sugars and polysaccharides, a number of novel chemical compounds have been isolated from various parts of the tree. These have been well reviewed by Mitra (1970), Kraus (1983), Schmutterer *et al.* (1981), Schmutterer and Ascher (1984) and Siddiqui *et al.* (1988).

The first two crystalline compounds to be isolated were nimbin and nimbinin along with an amorphous bitter substance nimbidin (Siddiqui *et al.*, 1988). Since then, nearly 100 more constituents have been isolated and their structures elucidated (Jones *et al.*, 1989). These include protolimonoids,

*Table 1.* List of some of the chemical constituents of *A. indica*.

| Plant part | Chemical constituents |
|---|---|
| Bark | Nimbin, Nimbinin, Nimbidin, Nimbindiol, tannins |
| Flowers | Flavonoids, Nimbosterol, Quercetin |
| Heart-wood | Nimbatone |
| Leaves | Quercetin, Nimbosterol, gedunin |
| Seed | Azadirachtin, deacetylnimbin, meliacin, melantriol, salanin |

limonoids or tetranortriterpenoids, pentanortriterpenoids, hexanor-triterpenoids, and nontriterpenoidal constituents.

A major breakthrough in *A. indica* research was the isolation of Azadirachtin (AZ), a limonoid , in the early 1970s by Butterworth and Morgan of the University of Keele, UK (Butterworth and Morgan, 1968). The structure of the compound was established in 1980s by Kraus of the University of Hobenhein in Germany (Kraus *et al.*, 1985).

It is considered as the most important active principle in *A. indica* seeds. The percentage of azadirachtin in the seed varies from 0.3% in India to 0.6% in Africa. Azadirachtin works at a very low concentration of 1–10 parts per million by blocking ecdysone's (moulting hormone) activity, thereby preventing the larvae from shedding their external skeletons and maturing. It also prevents feeding in about 200 insects at a concentration of 10–100 parts per million.

Some of the important constituents and their location in the plant are listed in Table 1.

### 1.2. *Uses*

The commercial uses of *A. indica* were known from the Vedic period in India over 4000 years ago and the domestic uses have been mentioned by Kautilya in his "Arthashastra", as early as 4th century B.C. *A. indica* has multifarious properties and uses in agriculture, as an insecticide and in the pharmaceutical and soap industries (Koul *et al.*, 1990).

#### 1.2.1. *As a forest tree*

Unlike other multipurpose forest tree species, *A. indica* has its own advantages. The wood resembles *Tectona grandis* (teak) in its strength and properties (Table 2).

The wood is aromatic and moderately heavy (Sp.gr=0.56–0.85) (mean= 0.68, National Academy of Sciences, USA, 1980). It is however, straight grained and more resistant to shock. The wood seasons well except for end splitting. It is resistant to fungi and insect attack and immune to termite attack. It is durable even out of doors, but does not polish well (Mitra,

*Table 2.* Properties of *A. indica* timber in comparison to *Tectona grandis*.

| Properties | *T. grandis* (quality index) | *A. indica* (quality index) |
|---|---|---|
| Weight | 100 | 124 |
| Strength (as beam) | 100 | 87 |
| Stiffness (as beam) | 100 | 81 |
| Suitiability (as post) | 100 | 82 |
| Shock resistance | 100 | 105 |
| Shape retention | 100 | 77 |
| Shear | 100 | 129 |
| Surface hardness | 100 | 131 |
| Nail holding ability | 100 | 144 |

Source: CSIR, 1985.

1963). *A. indica* timber is used for house building, furniture, bullock carts, wooden ploughs and other agriculture implements.

### 1.2.2. *As pesticide*

The first scientific report of the repellent action of *A. indica* against storage pests dates back to early 1930s. It has been an age-old practice in rural India to mix dried leaves with stored grain to repel insects. Today, numerous reports are available describing the activities of *A. indica* against a broad spectrum of insects. Amongst a host of constituents, a large number of triterpenoids, diterpenoids, sterols and sulphur containing heterocyclic compounds have been isolated from various parts of the tree. The best documented bioactivities involve both feeding deterrence and insect growth regulation.

When applied, the constituents are absorbed by the seedling through the roots and the active components are transported to the leaves. It has been noted, however, that the level of this systemic activity differs from plant to plant and formulation to formulation. Extracts without oil or with different concentrations of oil, exhibit different levels of systemic action. This action differs with insects as well.

In terms of individual compounds, azadirachtin stands out and is the only constituent to show all the diverse activities such as insecticidal, antifeedant, growth inhibitory, oviposition deterring, antihormonal and antifertility.

Azadirachtin and other triterpenoids are natural, environmentally compatible products which are promising candidates for pest control. The relatively short residual life of the active principles of *A. indica* derivatives may be considered as a disadvantage from an economic point of view. But, such products will not disturb the ecosystem and will not cause outbreak of new pests.

### 1.2.3. *As fertilizer*

With the current thrust on sustainable agriculture and organic farming, the use of natural resources like *A. indica*, has a greater practical significance especially in augmenting nitrogen use efficiency.

The seed cake, a residue of the oil extract, has a greater concentration of nitrogen, phosphorus, potassium, calcium and magnesium than farm yard manure and thus it is a potential source of organic manure.

When alcoholic extracts of the cake are mixed with urea and ammonium sulphate fertilizers, high nitrate levels can be achieved. A slow rate of nitrification follows application of the cake extracts and oil. By arresting the development of grazing invertebrates as well as nitrifying bacteria, biological fixation can be increased. Addition of the cake in soil reduces the population of ammonium oxidizers (*Nitromonas* sp.) without affecting nitrite oxidizers (*Nitrobacter* sp.). When urea treated with cake is added to soil, the urease activity (urea hydrolysis reaction) is reduced. Moreover ammonia volatilization loss is only 4% with cake coated urea fertilizer. It also reduces leaching loss of nitrogen.

### 1.2.4. *As therapeutic agent*

The tree has been used as a traditional remedy in ayurvedic medicine in India since antiquity. The medicinal properties have been ascribed to leaves, fruits and bark. Seed oil has been used for antimalarial, antihelmenthic, vermifuge, febrifuge, antiseptic and antimicrobial purposes, for bronchitis control and as a healing agent against various skin disorders (Chopra *et al.*, 1952; Radwanski, 1977b; Sinniah and Bhaskaran, 1981).

Certain organosulphur compounds of *A. indica*, such as sulfone dapsone, have been found to be antileprotic agents. These sulfones (Gupta and Joshi, 1983) along with aqueous leaf extracts also act as antimalarial agents. The analgesic and antipyretic effects of the components are encouraging. These have been shown to reduce eosinophil count by 75% in eosinophilia patients (Gopinath, 1983).

Another widely studied effect of *A. indica* has been its antifertility properties. The oil administered orally to female albino rats has prevented pregnancy (Khare *et al.*, 1984; Lal *et al.*, 1987) with normal restoration of fertility after 30 days. Its contraceptive effect has been tried through oral, vaginal and subcutaneous routes of applications.

Another important pharmacological use is as a dentrifice. It is reputed to produce remarkable healing of gum inflammations and pardontosia.

Table 3 enumerates some biological properties of *A. indica* preparations.

### 1.2.5. *As air purifier*

Researchers at the Indian Toxicological Research Institute and National Botanical Research Institute, Lucknow, India, have jointly claimed that *A. indica* is capable of absorbing dust and gas from polluted atmosphere, thus purifying the atmospheric air. Even though *A. indica* is indigenous and has

*Table 3.* Preparations of *A. indica* and their biological activity.

| Preparation | Biological activity | | |
|---|---|---|---|
| | As pesticide | Therapeutic use | Others |
| Oil | Antifeedant growth regulator | Antifertility antimicrobial | Soap making |
| Seed kernel extract | Oviposition deterrent, antifeedant | Antiinflammatory | – |
| Leaf extract | – | Blood purifier, antileprotic | – |
| Deoiled seed cake | – | – | Organic manure, cattle feed |
| Root extract | – | – | Preparation of toddy |
| Bark/stem extract | – | Antipyretic, antiseptic | – |

many reputable properties, it has been generally neglected by plant scientists. As this tree survives varied climatic and edaphic conditions, it is presumably heterozygous. No serious attempts seem to have been made to collect its germplasm.

Selection of plus trees would require a thorough knowledge of its breeding system, pollination biology and reproductive strategies (Mohan Ram *et al.*, 1993).

### 1.3. In vitro *approaches*

*In vitro* techniques have assumed special importance in tree improvement and propagation programs due to several constraints and limitations of the conventional methods. *In vitro* techniques have been utilized with considerable success in many tree species (Kendurkar and Mascarenhas, 1987). To combat ecological and socio- economic needs of an ever increasing population, the necessity for large scale plantation of multi-purpose trees is evident. Plantlet regeneration is best achieved with adventitious shoots which are then rooted or through somatic embryogenesis. The advantages of somatic embryogenesis over the conventional two stage method has drawn a lot of attention (Gupta *et al.*, 1991).

There are numerous potential applications of *in vitro* culture techniques towards improvement of *A. indica*. Its mass propagation on a commercial scale has limitations of seed viability. Moreover, vegetative propagation of *A. indica* using conventional methods is not possible (Narayan and Jaiswal, 1985). The interest is also on developing high yielding lines with improved seed and quality with high oil content. Tissue culture offers fascinating methods for large scale production of plants in shorter duration devoid of seasonal constraints.

Several workers have reported successful regeneration of plantlets in *A. indica* by using a variety of explants like leaves (Jaiswal and Narayan, 1983; Narayan and Jaiswal, 1985; Schulz, 1984), stem (Sanyal *et al.*, 1981), axillary buds (Jaiswal and Narayan, 1983; Drew, 1993; Joshi and Thengane, 1993), bark (Sanyal and Datta, 1986), inflorescences, cotyledons (Muralidharan and Mascarenhas, 1989; Shrikhande *et al.*, 1993; Nirmala Kumari *et al.*, 1993) and mature and immature embryos (Rangaswamy and Promila, 1972; Thiagarajan and Murali, 1993).

In most cases, proliferation of shoot and root meristems either directly or via callus has been reported. The first report of morphogenesis in *A. indica* was from Delhi University by Rangaswamy and Promila (1972), where development of normal seedling and differentiation of 10–20 shoot buds from plumular callus was observed from mature embryos of *A. indica* cultured on minimal medium containing Hoagland's solution in distilled water. However, no root formation was noticed.

Recently, world wide interest has been generated in this tree which is evident by an ever increasing number of publications and four International Conferences on *A. indica* held in 1980, 1983, 1987 and 1993. Plant regeneration has been achieved on Murashige and Skoog's MS (1962), basal medium supplemented with different growth hormones like 2,4-dichlorophenoxy acetic acid (2,4-D), 3–indoleacetic acid (IAA), 3-indolebutyric acid (IBA), naphthalene acetic acid (NAA), 6–benzylamino purine (BA), kinetin (KN) and N-isopenylamino purine (2-iP), individually or in various combinations and concentrations (0.05–2.0 mg/l).

In all the reports, differentiation of shoot or root meristems was via callus. Sanyal *et al.* (1981, 1983, 1986, 1988) conducted a series of experiments to study the effect of hormonal regime on regeneration and histochemical differentiation in *A. indica*. Basal MS medium supplemented with 50 g/l sucrose and phytohormones BA and NAA (0.1, 0.25 and 0.5 mg/l), individually and in combination or NAA (0.1–1.0 mg/l) and KN and BA (0.5 mg/l) were used for plantlet regeneration via callus. Among the auxins (IAA, NAA, IBA and 2,4–D) and cytokinins (KN and BA) tried singly, IAA (0.5 mg/l) and BA (0.5 mg/l) produced healthier and higher number of shoot buds. Addition of adenine sulphate (10 mg/l) gave best results with development of leafy shoots.

However, the main emphasis was on the production of active principles like nimbin and $\beta$-sitosterol from callus tissue. Regeneration studies were mainly confined to the relationship of chemical and histological differentiation in *A. indica* (Sanyal *et al.*, 1981).

Recently induction of adventitious shoots has been observed from immature embryos, shoot tips and axillary buds (Gurumurthi and Jagadees, 1992; Thiagarajan and Murali, 1993; Joshi and Thengane, 1993; Joarder *et al.*, 1993; Nirmala Kumari *et al.*, 1993; Drew, 1993). About 5–10 shoots per explant were regenerated using MS basal medium supplemented with cytokinins.

*Table 4*. Tissue culture studies on *Azadirachta indica*.

| Explant | Response | Results | References |
|---|---|---|---|
| Mature embryo | Callus | Organogenesis | Rangaswamy and Promila, 1972 |
| Stem bark | Callus | – | Sanyal *et al.*, 1981 |
| Stem, leaf, axillary buds | Callus | Organogenesis | Jaiswal and Narayan, 1983 |
| Cotyledon | Callus | Organogenesis | Sanyal *et al.*, 1983 |
| Leaf | Callus | Organogenesis | Schulz, 1984 |
| Leaf | Callus | Organogenesis | Narayan and Jaiswal, 1985 |
| Stem bark | Callus | Organogenesis | Sanyal and Datta, 1986 |
| Cotyledon | Callus | – | Sanyal *et al.*, 1988 |
| Cotyledon | Bipolar structures | Morphogenesis | Muralidharan and Mascarenhas, 1989 |
| Cotyledon | Callus | Embryogenesis | Shrikhande *et al.*, 1993 |
| Immature embryo | Germination | Plantlets | Thiagarajan and Murali, 1993 |
| Cotyledons | Shoot induction | Morphogenesis | Nirmala Kumari *et al.*, 1993 |
| Nucellar tissue and cotyledon | Callus | Embryogenesis | Joarder *et al.*, 1993 |
| Axillary buds | Shoot induction | Plantlets | Joshi and Thengane, 1993 |
| Axillary buds | Shoot induction | Plantlets | Drew, 1993 |

## 2. Culture initiation

Callus induction using different explants has been observed by many workers (Rangaswamy and Promila, 1972; Sanyal *et al.*, 1981, 1983; Jaiswal and Narayan, 1983; Narayan and Jaiswal, 1985). In most of the reports callus induction followed by shoot proliferation was observed (Table 4).

The first attempt to induce somatic embryogenesis in *A. indica* was made by Muralidharan and Mascarenhas (1989), where induction of bipolar structures from cotyledon induced callus was observed. Cotyledons from unripe green fruits were used for induction of callus. The excised cotyledonary pieces were inoculated on Gamborg *et al.*'s B5 (1968) medium supplemented with 6–benzylamino purine (BA) and kinetin (KN) or N-isopenylamino purine (2iP), (0.5–5.0 mg/l), with 20 g/l sucrose. The suspension cultures were initiated by incubating the cultures in dark at 27°C. No studies were conducted to confirm the embryonic nature of the bipolar structures.

The first conclusive evidence of induction of somatic embryogenesis in *A. indica* was demonstrated by Shrikhande *et al.* (1993) from National Chemical Laboratory, Pune, India.

Joarder *et al.* (1993) obtained induction of plant regeneration via somatic embryogenesis from nucellar tissue of immature seeds and from mature and immature cotyledons of *A. indica*. The cultures were initiated on MS and LS medium (Linsmaier and Skoog, 1965), supplemented with different con-

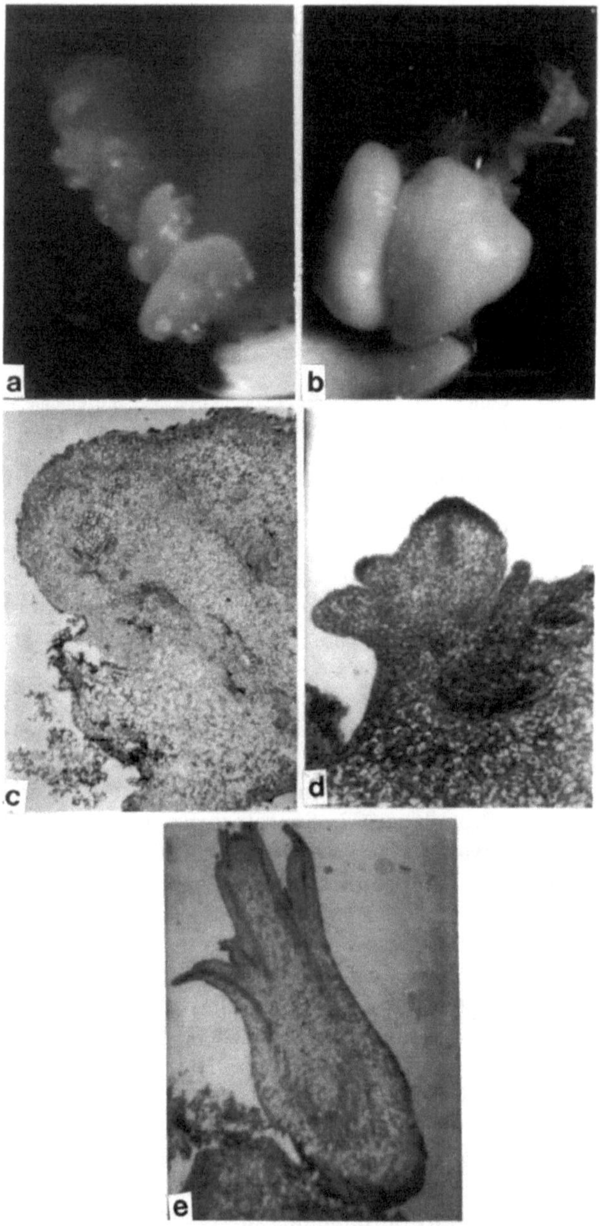


*Figure 2.* Direct somatic embryogenesis. (a) Embryo induction directly on cotyledon edge. (b) Germinating embryoid still attached to cotyledon. Histological examinations of indirect somatic embryogenesis. (c) Callus showing meristematic zone (25×). (d) Proembryoid emerging from callus (50×). (e) Germinating embryo still attached to callus (25×).

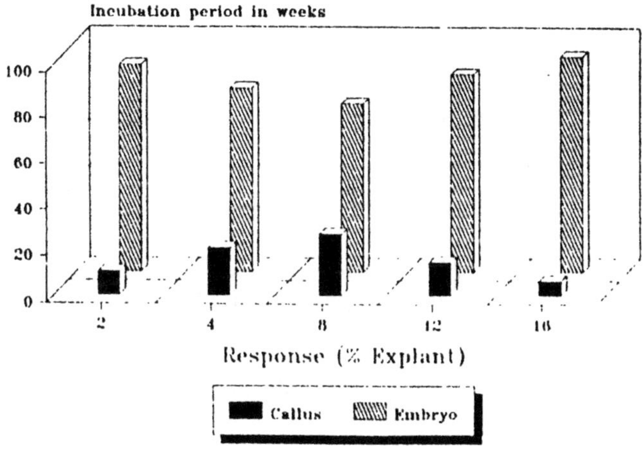


*Figure 3.* Embryogenic response in relation to incubation period.

centrations of cytokinins and auxins, either singly or in combinations. Nucellus, 40–45 days after fertilization, was found to be the most responsive. Basal MS medium with 2.0 mg/l BA and 0.2 mg/l NAA induce callus. According to them callus induction was genotypic dependent.

Direct somatic embryogenesis in *A. indica* was reported by Joshi and Thengane (1993) and Joarder *et al.* (1993). According to Joshi and Thengane (1993), type of auxin was found to be a deciding factor in the embryo induction pathway. When 2,4–D was used in combination with BA, somatic embryos were induced directly on the cotyledonary explant (Fig. 2a). However, when IAA was used callus induction was followed by somatic embryogenesis.

Joarder *et al.* (1993) claimed that, the basal medium was the deciding factor. According to them, LS medium was found to be effective for indirect somatic embryogenesis via callus stage.

### 3. Growth maintenance

Muralidharan and Mascarenhas (1989) observed bipolar structures irrespective of light or dark incubation at 27°C. Suspension cultures were raised on rotary shakers by using liquid media of the same composition.

Shrikhande *et al.* (1993) studied the effect of IAA (0.5 to 5.0 mg/l) alone and in combination with BA or KN (0.2 to 3.0 mg/l) on callus induction. According to them, all combinations of IAA and KN induce callus and rhizogenesis while combination of IAA and BA induced compact nodular callus.

They also reported the effect of incubation period on embryo induction. According to them, the embryogenic response increased (26%) by enhancing the incubation period up to 8th week (Fig. 3). Further increase in incubation duration had an adverse effect and a gradual decline in embryogenic response was noted. After the 16th week, cultures did not produce any embryos.

## 4. Embryo development

The promotory effect of reduced nitrogen especially at the induction phase of somatic embryogenesis has been emphasized (Rangaswamy, 1986). Different sources of reduced nitrogen such as coconut milk (CM), casein hydrolysate (CH), mixture of amino acids or a single amino acid such as L-glutamine, L-alanine has been used (Ammirato, 1983). Casein hydrolysate has been shown to allow vigorous organ development over a broad range of auxins (Murashige and Skoog, 1962).

Shrikhande *et al.* (1993) observed induction of defined embryos with distinct bipolar structures on MS medium with IAA (0.5 mg/l), BA (1.0 mg/l), CH (1000 mg/l) and 50 g/l sucrose Other concentrations of BA and IAA did not show somatic embryogenesis. Transfer of callus to a 16/8 h photoperiod was, however, found to be necessary for embryo induction.

Histological studies were carried out to confirm the origin of somatic embryos from callus tissue (Shrikhande *et al.*, 1993). They observed that differentiation of meristematic pockets occurred in the inner cortical paranchymatous tissue of cotyledons which are pushed towards the periphery due to swelling and thus formed proembryogenic masses (Fig. 2c,d,e). As the process was continuous and non-synchronous, initiation and development of all stages of embryo was observed.

According to Muralidharan and Mascarenhas (1989), bipolar structures were induced irrespective of light or dark incubation within 1–3 weeks. Beyond this period on the same medium in dark, no further differentiation was observed. Joarder *et al.* (1993) observed induction of up to 500 embryos per culture from nucellar tissue in the presence of low concentrations (less than 0.1 mg/l) of BA. With immature cotyledons, the highest percentage of embryogenesis was induced on MS medium supplemented with 1.0 mg/l NAA, 0.1 mg/l $GA_3$.

## 5. Embryo germination

The bipolar structures did not show germination if left on the same medium (Muralidharan and Mascarenhas, 1989). These structures were transferred to a hormone-free solid medium under continuous light (1000 lux) condition and resulted in the elongation and proliferation of leafy appendages. No root induction was observed from these structures, However, a tap root with leaf

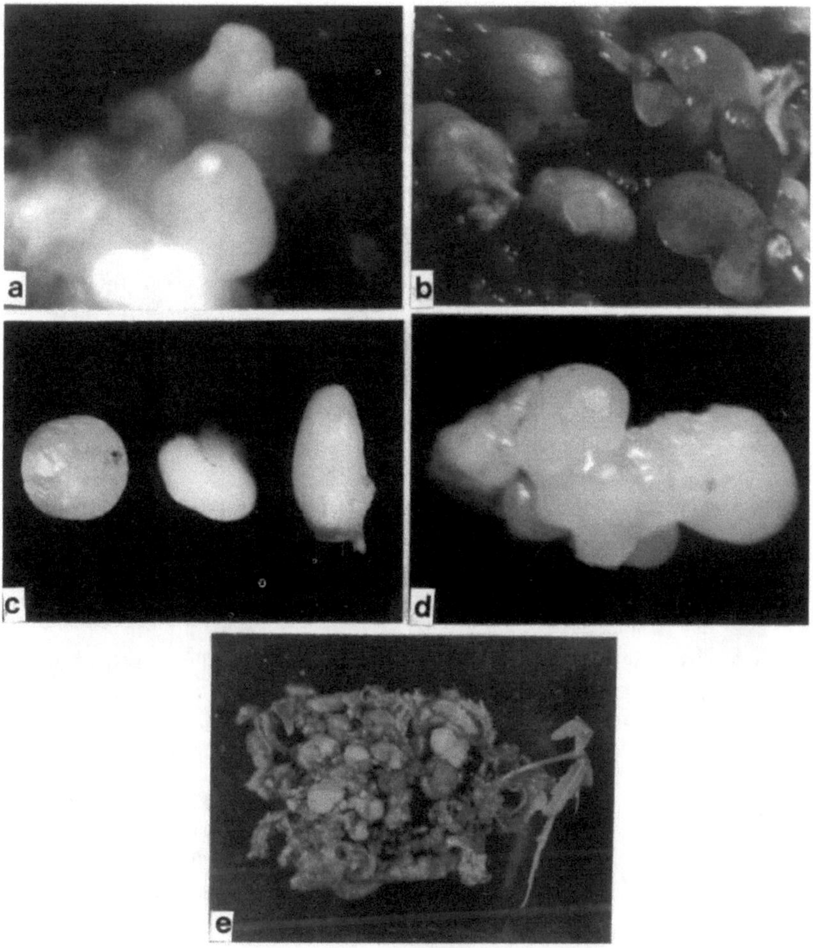


*Figure 4.* Somatic embryogenesis via callus. (a) Callus showing shiny globular masses (10×). (b) Embryogenic mass (10×). (c) Different stages of embryo development (16×). (d) Abnormal embryo. (e) Germinating embryo mass.

like appendages developed on transfer to a hormone-free liquid medium. The detailed studies on embryo development or plantlet formation were not conducted.

Shrikhande *et al.* (1993) tested different media for maturation and germination of embryos. According to them, increase in BA concentration to 2.0 mg/l, increases the number of cotyledonary stage embryos. Some of the embryos still attached to underlying callus tissue showed elongation of shoot axis and development of rudimentary roots (Fig. 5a). About 5% abnormal

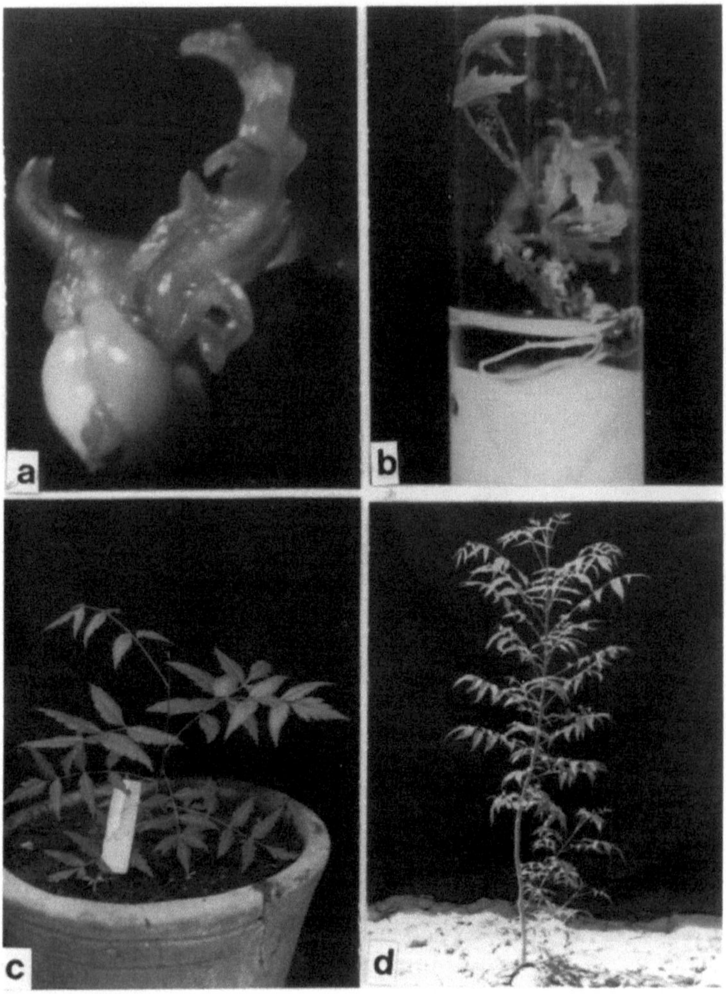


*Figure 5.* Plantlets obtained via indirect somatic embryogenesis. (a) Single germinating embryo showing elongated shoot and rudimentary root. (b) Complete plantlet on MS half medium. (c) One year old well established plant. (d) Plant in field (age 2 years).

embryos were noted in these experiments. The abnormal embryos were formed by the union of two or three embryos during early developmental stages (Fig. 4d). Such embryos did not germinate but later showed callusing. Improvement in the percentage of normal embryos was observed when the concentration of CH was reduced to 200 mg/l. However, about 70% of the embryos germinated on a medium containing 2.0 mg/l BA with the complete omission of CH.

Somatic embryos directly induced on the explant tissue, rarely germinated (2–5%) (Fig. 2b). The resulting plantlets were, however, necrotic and fasciated (Joshi and Thengane, 1993).

According to Joarder *et al.* (1993), germination of embryoids and development of the root and shoot axis was a rare phenomenon. Only 1% embryos of the nucellar origin germinated into complete plantlets. Transfer of embryogenic tissue to half strength MS medium containing 0.1 mg/l BA led to greening and further development of embryoids.

## 6. Growth in the field

The only report on transfer of embryogenically derived plantlets to soil or green house conditions has been by Shrikhande *et al.* (1993). They transferred 87 embryogenically derived plantlets to sterilized sand:soil mixture (1:1). 90% survival was obtained. The plants showed normal development and morphology.

## 7. Protocol

A detailed step-by-step procedure developed to obtain plantlets from cotyledonary callus of *A. indica* is as follows:

1. Unripe mature green fruits collected in July and August were first washed thoroughly under running tap water to remove latex. These were then washed with liquid detergent (10% savlon), followed by thorough washing under running tap water. Fruits were then sterilized in an inoculation chamber with 70% alcohol for one min and 1.5 g/l mercuric chloride for 10 min. They were finally washed 5 to 6 times with sterile distilled water and transferred to a sterile Petri dish.
2. Fruits were then cut transversely into two halves and embryo axis excised. The cotyledons were scooped out and inoculated in culture tube with its abaxial surface in contact with the medium.

The initiation medium consists of:

MS basal medium supplemented with MS vitamins, 1000 mg/l CH, 0.5 mg/l IAA, 1.0 mg/l BA, 50 g/l sucrose and 4 g/l agar.

3. The cultures were incubated in dark at 25±1°C.
4. Callus initiation with swelling of cotyledons was observed after 1 to 2 weeks of incubation.
5. Callus was recultured after 20 day interval. Initiation of embryogenic globular masses was evident after two subcultures (Fig. 4a).
6. To enhance adventitious embryo induction, the calli were recultured on the same medium, where callus mass with embryos of different developmental stages was observed (Fig. 4b,c).

7. Transfer of these embryos on to the medium containing MS basal medium supplemented with MS vitamins, 0.5 mg/l IAA, 2.0 mg/l BA, 50 g/l sucrose and 4 g/l agar, resulted in germination of embryos within 20 days (Fig. 5a). The cultures were incubated in 16/8 h photoperiod at 3000 lux light intensity.
8. Germinated embryos were then transferred to MS half strength liquid medium with 20 g/l sucrose, where complete plantlets with a well developed root and shoot were obtained in 10 days (Fig. 5b).
9. Transfer of plantlets to greenhouse conditions: Plantlets were carefully removed and thoroughly washed under running tap water and then transferred to pots filled with a mixture of sand:soil (1:1) (Fig. 5c). The potted plantlets were then transferred to shaded green house conditions for further growth.
10. 90% survival of the plants was observed. After two months, the plants were transferred to field conditions (Fig. 5d).

The above protocol was tested three times for its repeatability.

## 8. Conclusion and prospects

The application of tissue culture techniques to semi-tropical and tropical trees is not common and success in plant production through organogenesis or embryogenesis in these tree species is limited.

During the last two decades, *A. indica* has attracted the attention of the scientific community worldwide which is evident with the publication of a large number of articles and papers.

The commercial value of *A. indica* has been known to human for ages. Every part of the tree has been in use and traded for various purposes. Today, its products are favourably used worldwide due to their environmental friendliness. In the field of pest control, they are revolutionalizing the total concept by shifting emphasis from instant pest mortality to a gradual and nearly invisible pest extinction (Parmar and Ketkar, 1993).

There are a large number of industries world wide which are using *A. indica* preparations. The use of *A. indica* based products for medicinals, toileteries and other purposes has a large consumer acceptability.

However, propagation of *A. indica* using conventional methods has limitations of:

1. Keeping quality due to high moisture and oil content of seed, 30–35% and 40–45% respectively.
2. Poor seed viability.
3. Late fruiting, therefore seedlings available for planting late in the season.
4. Slow growth of the tree with long gestation – economic life starts after 10 years.
5. Fruiting not uniform.

6. Tremendous variation in germplasm.

Somatic embryogenesis in *A. indica* can hasten the process of increasing genetic diversity, thereby, increasing the chances of isolating plants with desirable traits.

What is needed today, is large scale research and development efforts for sustainable neem-based technology. The ever increasing demand for its products needs major emphasis on planned propagation of elite germplasm.

## 9. References


Ammirato, P.V., 1983. Embryogenesis. In: D.A. Evans, W.R. Sharp, P.V. Ammirato *et al.* (Eds.), Handbook of Plant Cell Cultures. Techniques for Propagation and Breeding, pp. 82–113. Macmillan Publishing Co., New York.

Butterworth, J.H. and E.D. Morgan, 1968. Isolation of a substance that suppresses feeding in locusts. Chem. Commun. Vol. 1: 23–24.

Chopra, I.C., K.C. Gupta and B.N. Nazir, 1952. Preliminary study of antibacterial substances from *Melia azadirachta*. Indian J. Med. Res. 40: 511–515.

CSIR, 1985. The Wealth of India-I: Raw Materials, pp. 504–511. Government of India, New Delhi.

Drew, R.A., 1993. Clonal propagation of Neem by tissue culture. Paper presented at the World Neem Conf., February 24–28, Bangalore.

Gamborg, O.L., R.A. Miller and K. Ojima, 1968. Nutrient requirements of suspension cultures of soybean root cells. Exp. Cell Res. 50: 151–158.

Gopinath, G., 1983. A clinical study of the therapeutic value of nimbin (from *Melia azadirachta*) on tropical eosinophilia. M.D. Thesis, University of Kerala, Kerala.

Gupta, D. and B.C. Joshi, 1983. Electron impact studies of some sulphur compounds. In: A. Frigeria (Ed.), Recent Developments in Mass Spectrometry in Biochemistry, Medicine and Environmental Research, Part 8, pp. 143–157. Elsevier Scientific, Amsterdam.

Gupta, P.K., R. Timmis, J. Pullman *et al.*, 1991. In: I.K. Vasil (Ed.), Cell Culture and Somatic Cell Genetics of Plants, Vol. 8., pp. 75–92. Academic Press, New York.

Gurumurthi, K. and S. Stanley Jagadees, 1992. Micropropagation of *Azadirachta indica* A. Juss from buds of coppice shoots. In: Proceedings of Workshop on Vegetative Propagation and Biotechnologies for Tree Improvement, Tirupathi (AP), pp. 137–140.

Jaiswal, V.S. and P. Narayan, 1983. Regeneration of plantlets from the somatic tissue of some Indian trees. J. Exp. Biol. 22: 138–139.

Joarder, N., R. Islam and O.I Joarder, 1993. Micropropagation of *Azadirachta indica* through axillary bud culture. Paper presented at the World Neem Conf., February 24–28, Bangalore.

Jones, P.S., S.V. Ley, E.D. Morgan and D. Santafianos, 1989. The chemistry of the neem tree. In: M. Jacobson (Ed.), Focus on Phytochemical Pesticides, Vol. 1, The Neem Tree, p. 19. CRC Press, Boca Raton.

Joshi, M.S. and S.R. Thengane, 1993. Potential application of in vitro methods for propagation of Neem. Paper presented at the World Neem Conf., February 24–28, Bangalore.

Kendurkar, S.V. and A.F. Mascarenhas, 1987. Micropropagation of forest trees through tissue culture. In: G.M. Reddy (Ed.), Plant Cell and Tissue Culture of Economically Important Plants, pp. 181–190. Hyderabad, India.

Khare, A.K., M.C. Srivastava, M.K. Sharma and J.D. Tewari, 1984. Antifertility activity of neem oil in rabbits and rats. Probe 23: 90–93.

Koul, Opender., B.I. Murray and C.M. Ketkar, 1990. Properties and uses of neem, *Azadirachta indica*. Can. J. Bot. 68: 1–11.

Kraus, W., 1983. Biologically active compounds from Meliaceae. In: Proceedings of the 2nd

International Conference on the Chemistry and Biotechnology of Biologically Active Natural Products, Budapest, pp. 331–345.

Kraus, W., M. Bokel, A. Klenk and H. Pohnl, 1985. The structure of azadirachtin and 22, 23-dihydro-23-$\beta$-methoxyazadirachtin. Tetrahedron Lett. 26: 6435–35.

Lal, R., M. Gandhi, A. Sandaranarayan , V.S. Mathur and P.L. Sharma, 1987. Antifertility effect of *Azadirachta indica* oil administered to female albino rats on selected days of pregnancy. Fitoterapia 58: 239–242.

Linsmaier, E.M. and F. Skoog, 1965. Organic growth factor requirements of Tobacco tissue cultures. Physiol. Plant. 18:100–127.

Mitra, C.R., 1963. Neem. Indian Council Oil Seed Committee, Himayatnagar, Hyderabad, pp. 15–68.

Mitra, C.R., 1970. Studies of certain plants (inclusion of neem) extracts and isolates having pesticidal properties – final report. PL 480 Project No. A-T-ENT.

Mohan Ram, H.Y. and N.B. Nair, 1993. Botany. In: N.S. Randhawa and B.S. Parmar (Ed.), Neem – Research and Development, pp. 6–26. SPS Publications, New Delhi.

Muralidharan, E.M. and A.F. Mascarenhas, 1989. *In vitro* morphogenesis in *Azadirachta indica* A. Juss and *Eucalyptus* citriodora Hook. In: A.K. Kukreja, A.K. Mathur, P.S. Ahuja and R.S. Thakur (Ed.), Tissue Culture and Biotechnology of Medicinal and Aromatic Plants, pp. 49–55. Paramount Publ., New Delhi.

Murashige, T. and F. Skoog, 1962. A revised medium for rapid growth and bioassays with tobacco tissue cultures. Physiol. Plant. 15: 473–497.

Narayan, P. and V.S. Jaiswal, 1985. Plantlet regeneration from leaflet callus of *Azadirachta indica* A. Juss. J. Tree Sci. 4: 65–68.

Nirmala Kumari, A., N.M. Rangaswamy and S.R. Sree Rangaswamy, 1993. Tissue Culture Studies in Neem. Paper presented at the World Neem Conf., February 24–28, Bangalore.

Parmar, B.S. and C.M. Ketkar, 1993. Commercialization. In: N.S. Randhawa and B.S. Parmar (Eds.), Neem – Research and Development, pp. 270–283. SPS Publications, New Delhi.

Radwanski, S., 1969. Improvement of red acid sand by neem tree (*Azadirachta indica*) in Sokoto, North Western State of Nigeria. J. Appl. Ecol. 6: 507–511.

Radwanski, S., 1977a. Neem Tree. 1. Commercial potential characteristics and distribution. World Crops Livest. 29: 62–63, 65–66.

Radwanski, S., 1977b. Neem Tree. 2. Uses and Potential uses. World Crops Livest. 29: 111–113.

Rangaswamy, N.S., 1986. Somatic embryogenesis in angiosperm cell tissue and organ cultures. Proc. Indian Acad. Sci. Plant Sci. 96: 247–271.

Rangaswamy, N.S. and Promila, 1972. Morphogenesis of the adult embryos of *Azadirachta indica*. Z. Pflanz. Physiol. Bd. 67s: 377–379.

Sanyal, M., A. Das, M. Bannerjee and P.C. Datta, 1981. *In vitro* hormone induced chemical and histological differentiation in stem callus of neem, *Azadirachta indica* A. Juss. J. Exp. Biol. 19: 1067–1068.

Sanyal (Sarkar), M. and P.C. Datta, 1986. Nutritional environment and redifferentiation from cultured tissues of *Azadirachta indica*. Acta Hort. 188: 99–103.

Sanyal (Sarkar), M., A. Mukherjee and P.C. Datta, 1988. Glycine on *in vitro* biosynthesis of Nimbin and $\beta$-sitosterol in tissues of *Azadirachta indica*. Curr. Sci. 57: 40–41.

Sanyal, M., S. Tikadar and P.C. Datta, 1983. Nimbin and $\beta$-sitosterol in cotyledon and cultured tissues of *Azadirachta indica*. Indian Drugs 20: 479–481.

Schmutterer, H., K.R.S Ascher and H. Rembold, 1981. Natural pesticides from the neem tree (*Azadirachta indica*) In: Proceedings of the Ist Int. Neem Conf., Roltach-Egern, p. 297.

Schmutterer, H. and K.R.S. Ascher, 1984. Natural pesticides from the neem tree and other tropical plants. In: Proceedings of the 2nd Int. Neem Conf., Rauischholzhaussen, p. 43.

Schulz, F.A., 1984. Tissue culture of *Azadirachta indica* A Juss. In: Proceedings of the 2nd Int. Neem Conf., Rauischholzhaussen, p. 83.

Shrikhande, Medha., S.R. Thengane and A.F. Mascarenhas, 1993. Somatic embryogenesis and plant regeneration in *Azadirachta indica* A. Juss. In Vitro Cell. Dev. Biol. 29P: 38–42.

Siddiqui, S., B.S. Siddiqui S. Faizi and T. Mahmood, 1988. Tetracyclic triterpenoids and their derivatives from *Azadirachta indica*. J. Nat. Prod. (Llyodia) 51: 30–43.
Sinniah, D. and G. Bhaskaran, 1981. Margosa oil poisoning as a cause of Reye's syndrome. Lancet 1: 487–489.
Thiagarajan, N. and P.M. Murali, 1993. High frequency regeneration of neem *in vitro* from immature embryo explants. Paper presented at the World Neem Conf., February 24–28, Bangalore.

# 21. Somatic embryogenesis in chestnut

F.J. Vieitez

**Contents**



## 1. Introduction

The chestnut tree (genus *Castanea*) belongs to the Fagaceae, along with beeches (*Fagus*), oaks (*Quercus*), evergreen chinkapins (*Castanopsis*) and tanoaks (*Lithocarpus*). Chestnut trees became distinct from other members of their family as early as 43 million years ago, and probably earlier (Rutter, 1987).

*Castanea* is a small genus with 13 species native to the temperate zones of Asia, Europe and Eastern United States (Camus, 1929). The species that are most representative and of greatest economic importance are *C. crenata* Sieb. and Zucc (Japanese chestnut), *C. dentata* Borkl. (American chestnut), *C. mollisima* Bl. (Chinese chestnut) and *C. sativa* Mill. (European chestnut). These four species constitute the true chestnut tree section of the genus. The other species are small trees or shrubs of importance only for breeding, as rootstocks or for special uses. All the species are closely related and interhybridize very readily. The normal somatic number of the species and interspecific hybrids in the genus is 24 (Jaynes, 1962).

Chestnut is a deciduous tree with alternate, simple, 20–30 cm long oblong to lanceolate straight-veined leaves with a serrate margin and an acute to acuminate apex. Chestnuts are monoecious, the flowers being produced in two types of catkin borne on the current season's shoots. The staminate flowers are borne in long, slender male catkins located on the lower portion of the shoot, while the female flowers are clustered at the base of bisexual catkins near the terminal portion of the shoot. Chestnuts are self-sterile and

*S. Jain, P. Gupta & R. Newton (eds.), Somatic Embryogenesis in Woody Plants, Vol. 2,* 375–407.

at least two varieties are required for pollination. They are, predominantly, a wind-pollinated species. The fruits are large dark brown nuts enclosed in a prickly hard involucre or bur. True chestnut trees produce edible nuts and are in no way related to the "horse chestnut" (*Aesculus hippocastanum*) or the "water chestnut" (*Eleocharis dulcis*).

Chestnuts are exceptionally long-lived trees, with a natural life expectancy in the range of 500–1000 years. Some extant European trees are reputed to be over 2000 years old. Chestnuts are shade tolerant, fast growing, and possess long taproot trees which are slow to start but, once established grow rapidly under good growing conditions and may reach up to 30–35 metres in height and/or several metres in diameter. The famous "100 horses chestnut" located at Sant'Alfio (Catania, Italy), reputed to be 3600–4000 years old, was over 15 metres in diameter. Chestnuts require a moderately humid, warm climate and preferably deep, well drained, and slightly acid soil (pH 4.5–6), but have been known to thrive on a wide variety of sites and soils in association with other hardwoods.

Root rot (ink disease) caused by the fungus *Phytophthora cinnamomi*, and blight caused by the bark fungus *Cryphonectria parasitica*, are the two major disease problems of chestnut. Between them, they have destroyed most of the chestnut populations that existed in Europe and America 100 years ago. In 40 years, over 3.5 billion of American trees were killed by blight, which meant the virtual obliteration of *C. dentata* as a species. Fortunately, root-rot-resistant hybrids have been developed in both Europe and America by crossing the susceptible native species with the full resistant Asian species *C. crenata* or *C. mollisima*. Both Asian species are fully resistant to root rot and also have high natural resistance to blight. In America, some *C. dentata* × *C. mollisima* hybrids, such as all the Dunstan hybrid varieties, have resisted blight during more than 25 years of breeding and evaluation (Wallace, 1987).

### 1.1. *Importance*

Archaeologists report that chestnut trees were cultivated at least 6000 years ago in China. About 2000 years ago, they were being bred, propagated asexually, and widely planted in most temperate climates of the world. True chestnuts and their interspecific hybrids (*C. crenata* × *C. mollisima*; *C. sativa* × *C. crenata*; *C.sativa* × *C. mollisima* and *C. dentata* × *C. mollisima*) are commercially important species, producing edible nuts and/or a highly valuable and versatile timber which is used for everything from plywood cores to heavy construction. Chestnut extracts provide tannin for the leather industry. Additionally, chestnuts supported large wildlife populations, including squirrels, deer, pheasant, bears, and turkeys.

Chestnut trees have outstanding characteristics as nut trees. They are annual bearers and start to bear at a younger age than most other nut trees. Some hybrids begin production when they are around 4–5 years old. Chestnut orchards usually start yielding a positive cash flow in 7–10 years. At maturity, a single tree can produce over 100 kg of nuts annually. Nuts ripen in September and October, and most fall free from the burs, making easy for harvesting.

The chestnuts (the nuts of the chestnut tree) are highly nutritious, and have been called "the grain that grows on a tree". Nutritionally, European chestnuts closely resemble wheat in composition, but with less starch and more dextrose and other water-soluble carbohydrates (about 40%). Chestnuts also contain a very high quality protein (10–15% dry weight), with an amino acid balance similar to an egg but with no cholesterol. Chestnuts are low in fat (2–3%) compared with almonds, walnuts, hazelnuts or pecans, which are over 50% fat.

Chestnuts are a versatile food. They can be eaten raw, boiled or roasted as a snack. Peeled chestnuts can be dried and ground into a flour. Chestnut puree and paste are used for baking pastries and making ice cream in Europe and Asia. Marron glacé is a famous gourmet dessert of Europe. Commercially important chestnut industries have been established in most chestnut producing countries.

The annual world production of chestnuts is about half a million metric tons, which does not cover the great current demand. The two major production areas are Asia (China, Japan and Korea) and Europe (Italy, Spain, France, Portugal and Turkey). China is the world's largest producer and consumer of chestnut. The United Nations lists China as producing 240,000 metric tons annually.

### 1.2. *Propagation*

Chestnut trees, like most large seeded hardwoods, are very difficult to propagate vegetatively. Only limited success has been achieved by grafting methods (Graves and Nienstadt, 1953; Moore, 1963; Jaynes, 1965; Shafer, 1966; Park, 1967; Beck, 1970; Eltkins *et al.*, 1980; Lagerstedt, 1981; Vieitez and Vieitez, 1981a,b), stooling (Urquijo, 1952; Schad *et al.*, 1952; Vieitez, 1952, 1953, 1955; Nienstadt and Graves, 1955; Solygnat, 1964), cutting techniques (Urquijo, 1952; Pease, 1953; Vieitez, 1956, 1963, 1974; Jaynes, 1961; Solygnat, 1964; Jaynes and Messner, 1967) and micropropagation (Vieitez *et al.*, 1986).

The review on chestnut tissue culture lists works on 8 *Castanea* species and hybrids (Vieitez *et al.*, 1986). Generally, micropropagation has been achieved by organogenesis via development of axillary buds from zygotic embryos or seedling material (Vieitez and Vieitez, 1980a,b, 1982/83; Vieitez *et al.*, 1981; Keys and Cech, 1981, 1982; Rodriguez, 1982; Chevre *et al.*, 1983) and occasionally via adventitious buds (Skirvin, 1981; San-José, 1983; San-José *et al.*, 1984). Micropropagation from adult trees has also been reported (Biondi *et al.*, 1981; Vieitez *et al.*, 1983; Read *et al.*, 1985; Sanchez and Vieitez, 1991). The application of rejuvenation methods (Ballester *et al.*,

1989, 1990; Sanchez, 1991) has improved the micropropagation rates attained with culture material from mature chestnut trees.

### 1.3. *Biotechnology of somatic embryogenesis in chestnut*

Interest in growing chestnuts is now gradually extending to countries such as New Zealand, Australia, Argentina or Chile, with no tradition in chestnut culture. The propagation methods cited above cannot supply enough plants to satisfy the current demand. Biotechnology based on somatic embryogenesis may be an alternative approach of overcoming these constraints, since micropropagation via somatic embryogenesis is, theoretically, the most efficient method of propagation and offers the possibility of mass cloning of chestnut on a scale far greater than any other method. As a rapid vegetative propagation method, somatic embryogenesis can also accelerate breeding programs, and thus facilitate efforts to develop chestnut blight resistance.

To date, however, somatic embryogenesis in chestnut has met with only limited success. Skirvin (1981), in a review of fruit-crops grown in tissue culture, stated that McPheeters *et al.* (1980) observed embryoids on cotyledon pieces from seedlings of a *C. mollisima* × *C. dentata* hybrid cultured on a modified MS (Murashige and Skoog, 1962) medium containing 8.87 μM $N_6$-benzyladenine (BA) and 0.54 μM naphthaleneacetic acid (NAA), though this information was not specified in the original paper of McPheeters *et al.* (1980). González *et al.* (1985) reported somatic embryogenesis on slices of cotyledon of a *C. sativa* × *C. crenata* hybrid; most of explants had formed embryogenic calli after 8 weeks of culture on media supplemented with 0.45–45.2 μM 2,4-dichlorophenoxy acetic acid (2,4-D) alone or combined with a cytokinin. Histological sections showed globular, heart-shaped and torpedo-shaped embryoid-like structures differentiating within the explant tissue. After transfer to auxin-free medium, such structures neither proliferated nor developed further. Vieitez *et al.* (1990) and Merckle *et al.* (1991, 1992) obtained true somatic embryos from immature zygotic embryos of two *C. sativa* × *C. crenata* hybrids and from *C. dentata* respectively, but they did not regenerate plants. At the International Chestnut Conference held at Morgantown (West Virginia), Vieitez *et al.* (1992 a) reported the regeneration of chestnut plantlets (emblings) from somatic embryos developed from both callus and cell suspension cultures. The procedures employed in our laboratory for the establishment and maintenance of embryogenic cultures of chestnut and for the conversion of embryos into emblings, will be described in the remainder of this chapter.

## 2. Culture initiation

For several years, we have attempted to induce the production of chestnut somatic embryos on explants of both juvenile and mature tissues, taken from various specimens of ink-disease resistant *C. sativa* × *C. crenata* clones. The juvenile explants included zygotic embryos excised at developmental stages ranging from fertilized ovules to mature embryos, and various plant parts excised from micropropagated juvenile shoots. Mature explants included developing amments, whole stamens, filaments, and anthers. All explants were cultured on an induction medium consisting of basal medium formulation (MS, Heller's (1953) or woody plant medium (Lloyd and McCown, 1980) WPM) supplemented with a reduced nitrogen source (casein hydrolysate or a mixture of amino acids) plus a wide range of auxin-based hormone treatments, which included 2,4-D, indole-3-butyric acid (IBA) or NAA at concentrations between 0.45 and 53 μM, and Picloram (4-amino-trichloropicolinic acid) at 0.02–8.2 μM, alone or in combination with 0.44–9.2 μM BA, kinetin (K) or zeatin (Z) (Vieitez *et al.* 1990). Cultures were maintained either under a 16 h photoperiod at 30 μmol $m^{-2}$ $s^{-1}$ (from cool white fluorescent lamps) or in darkness with 25°C day and 20°C night temperatures, respectively.

Despite intensive work, to date somatic embryogenesis has only been achieved on explants obtained from immature zygotic embryos excised from 15–20 mm immature nuts (with white to pale green shells) collected 10–12 weeks post fertilization from clones 431 and HV. Our current embryogenic cultures were initiated in the summer of 1988 from a small proportion of a group of these explants that had been cultured on a modified MS medium containing amino acids and supplemented with either 0.45 μM 2,4-D plus 0.46 μM Z, or 2.26 μM 2,4-D with or without 4.4–9 μM BA or Z (Vieitez *et al.*, 1990). After two months in darkness, cultures were transferred to 2,4-D free, half strength MS induction medium supplemented with combinations of IBA or NAA with Z or BA and maintained under a 16 h photoperiod (30 μmol $m^{-2}$ $s^{-1}$). After gradual reduction of hormone levels, a few whitish embryo-like structures were observed protruding from the surface of callus on 6–8% of the initial cultures. After isolation and subculture on the same medium, these embryogenic structures did not develop any further, but did proliferate to form a yellowish embryogenic callus with a nodular appearance.

Initially, these embryogenic calli could be maintained by regular subculture on either $\frac{1}{2}$ MS or WPM media supplemented with amino acids plus combinations of BA or Z with IBA or NAA. After 2–3 months, however, only cultures derived from two Clone 431 nuts and one from Clone HV nut remained embryogenic. These viable embryogenic cultures consisted of friable clusters of globular structures (Figs. 1 and 2) which upon subculture proliferated further and, occasionally, developed somatic embryos (Fig. 3). The other callus cultures became brown and died, or degenerated into green compact, non-embryogenic calli. These losses meant that fewer than 2% of initial explants originated sustainable embryogenic cultures.

Merckle *et al.* (1991) reported similar induction frequencies for somatic embryogenesis on ovules and immature zygotic embryos of American chestnut (*C. dentata*) cultured on solid WPM medium supplemented with 32.2 μM NAA or 18.1 μM 2,4-D plus 1.11 μM BA and, subsequently, transferred to

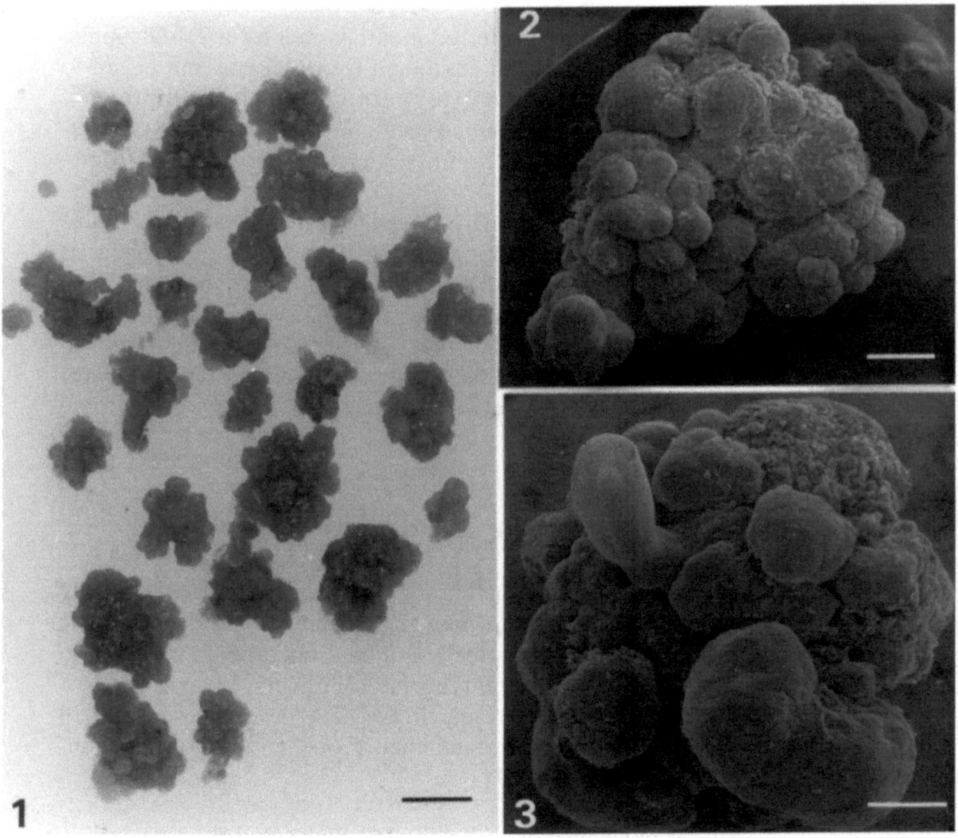

*Figure 1.* Embryogenic callus of chestnut consisting of friable clusters of yellowish globular structures referred to as globular embryogenic callus (GEC) (bar = 5 mm).
*Figure 2.* Scanning electron micrograph of a sample of GEC (bar = 1 mm).
*Figure 3.* Scanning electron micrograph of a somatic embryo developing on GEC from a single globular structure after 4 weeks of culture on WPM containing Z and NAA (bar = 1 mm).

a hormone-free or BA-supplemented medium. They obtained two types of embryogenic cultures depending on the length of time the explant remained on an auxin-supplemented medium. The first type consisted of somatic embryos that formed directly from the radicle of zygotic embryos which remained on NAA- or 2,4-D-supplemented medium for 1 or 2 weeks before transfer to a hormone-free or BA-supplemented medium. Primary somatic embryos and one or two cycles of secondary somatic embryos were produced on both media. The second type of embryogenic cultures consisted of ovules that received a continuous exposure to NAA or 2,4-D for 5 weeks and formed a cream-coloured nodular growth that resembled proembryogenic masses (Halperin, 1966); when transferred monthly on a fresh 2,4-D-supplemented medium this nodular growth continued to proliferate and began to produce globular and heart-stage embryos, even in the continued presence of auxin. Transfer to NAA-supplemented medium failed to sustain proliferation of proembryogenic masses or repetitive somatic embryogenesis (Merckle *et al.*, 1991).

Although explants from the *C.sativa* × *C. crenata* hybrid used by Vieitez *et al.* (1990) did not produce proembryogenic masses, somatic embryogenesis in American chestnut was similar to that observed in the hybrid material, most notably as regards the appearance of the embryogenic calli and of cotyledonary embryos. Merckle *et al.* (1991) suggested that, in addition to interspecies differences, interactions between medium and growth regulator might account for the differences between the responses observed by them and by Vieitez *et al.* (1990).

## 3. Culture maintenance

The low somatic embryogenesis induction frequency and the losses occurring during the first few months' culture, along with the low yield of somatic embryos, compelled us to seek maximally optional operational methods to achieve efficient proliferation of embryogenic callus and efficient embryo production. After several years of research, we have developed a protocol that has proved to allow long-term maintenance of the embryogenic competence of chestnut cultures (Fig. 4).

### 3.1. *Maintenance of embryogenic callus cultures*

In a first series of experiments to optimize embryogenic callus proliferation, calli were subcultured onto more than 60 media differing in mineral salt, amino acid, sucrose, and growth regulator contents. Cultures were kept under a 16 h photoperiod at various irradiance levels (6–30 $\mu$mol $m^{-2}$ $s^{-1}$) with 25°C day and 20°C night temperatures, respectively; or in darkness with either the same temperature regime; or with a constant temperature of 25°C. Five pieces of callus ca. 0.5 cm in diameter were cultured in each culture vessel (Petri dishes or glass jars containing, respectively, 18 ml and 50 ml of solid medium) and at least 6 vessels of each type per culture medium and growth condition were assayed. Due to the great variation in callus responses even within the same culture vessel, quantitative comparisons among treatments failed to achieve statistical significance. It was nevertheless noted that both high-salt media such as MS and high hormone levels resulted in compact calli and abnormal embryo development (thickening of cotyledons, overgrowth, fused embryos), whereas media with lower salt contents, such as WPM or $\frac{1}{2}$ MS (MS with half strength major salts), supplemented with low concentrations ($\leqslant$ 2.2 $\mu$M) of a cytokinin plus an auxin at lower concentra-

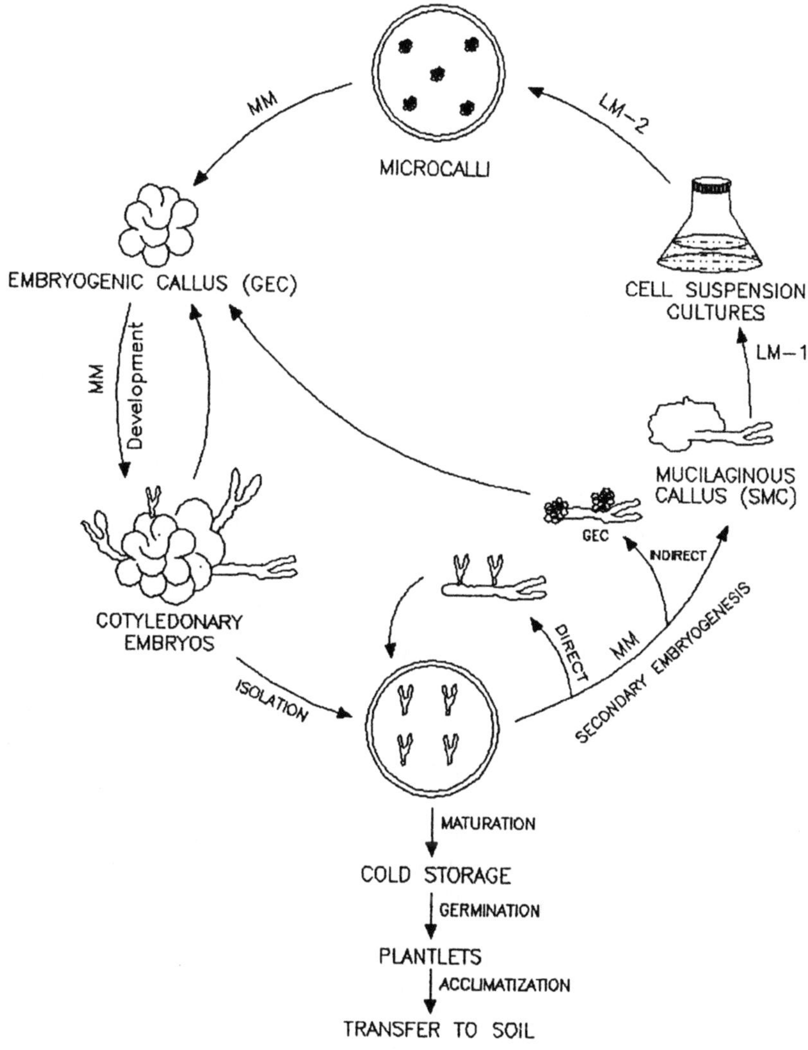


*Figure 4.* Protocol for maintenance of embryogenic competence of chestnut cultures and conversion of somatic embryos into plants.

tion (0.05–0.25 μM) resulted in the proliferation of embryogenic callus as masses of yellowish globular structures from which well-formed embryos developed. The most suitable combination of growth regulators was 0.92 μM Z and 0.25 μM IBA. Glutamine was found to be necessary for the maintenance of embryogenic cultures, since in its absence callus proliferation and embryo development ceased after 4–5 successive subcultures. Increasing the concentration of sucrose from 3 to 4–5% had no effect on the behaviour of cultures, while decreasing it to 1% resulted in the reduction of cotyledonary stage embryo production. The most appropriate environmental conditions were found to be a 16 h photoperiod of diffuse light (6–9 μmol $m^{-2}$ $s^{-1}$). The type of culture vessel did not appear to affect the proliferation rates of cultures, but the number and mean size of somatic embryos with normal morphology were always greater in dishes than in glass jars.

To optimize the production of cotyledonary stage embryos, three mineral salt formulations were tested: Heller's, WPM and $\frac{1}{2}$ MS. Five samples of callus (ca. 100 mg fresh weight) were subcultured per 9 mm plastic dish containing 18 ml of medium, with 4 dishes per mineral medium tried. All three media were supplemented with 100 μM Fe-Na-EDTA, MS vitamins and m-inositol, 3 mM glutamine, 0.92 μM Z, 0.25 μM IBA and 30 g/l sucrose and were solidified with 8 g/l agar (pH 5.8). Cultures were kept under a 16 h photoperiod of diffuse light. The experiments was repeated during three successive subcultures at 4 week intervals. The effect of mineral media was assessed in terms of the increase in the fresh weight of calli and the number of cotyledonary embryos formed per piece of callus after 4 weeks of culture.

The mineral composition of the medium did not affect the proliferation rates of cultures. In all three media tested, embryogenic callus multiplied its fresh weight three- to four-fold over 4 weeks' culture. Cotyledonary embryos with normal morphology were formed in all three media, as were atypical embryos (embryos that were swollen, callused or fused, or had only one or more than two cotyledons). The embryos were translucent or white to pale yellow in colour, and ranged between 1.5 and 10 mm in length. Although the smallest embryos were generally translucent, there was no correlation between embryo size and colour.

Embryo production data in the three media are expressed in Table 1 as the mean numbers (± se) of cotyledonary embryos formed per sample of callus after 4 weeks of culture, using 3 replicates of 20 samples each. The highest yield of cotyledonary embryos was always achieved on $\frac{1}{2}$ MS medium, with a mean production of 10.5 embryos per 100 mg of callus; Heller's medium, which has been used in some chestnut tissue cultures (Vieitez *et al.*, 1986), gave the lowest yield. Although the production of embryos with abnormal morphology was also higher on $\frac{1}{2}$ MS than on the other two media, abnormal embryos turned out to retain their embryogenic competence (their potential to produce new normal somatic embryos upon reculture), and $\frac{1}{2}$ MS was adopted for the maintenance of embryogenic callus cultures. After more than 4 years of repeated subculture on this medium, embryogenic callus continue to thrive as clusters of globular structures, and the production of cotyledonary embryos has remained undiminished (Fig. 5). In fact, in experiments conducted 3 years after $\frac{1}{2}$ MS plus 0.92 μM Z and 0.25 μM IBA was adopted as the standard maintenance medium (hereinafter designated MM), up to 200 cotyledonary embryos longer than 2 mm were recorded per gram of subcultured callus.

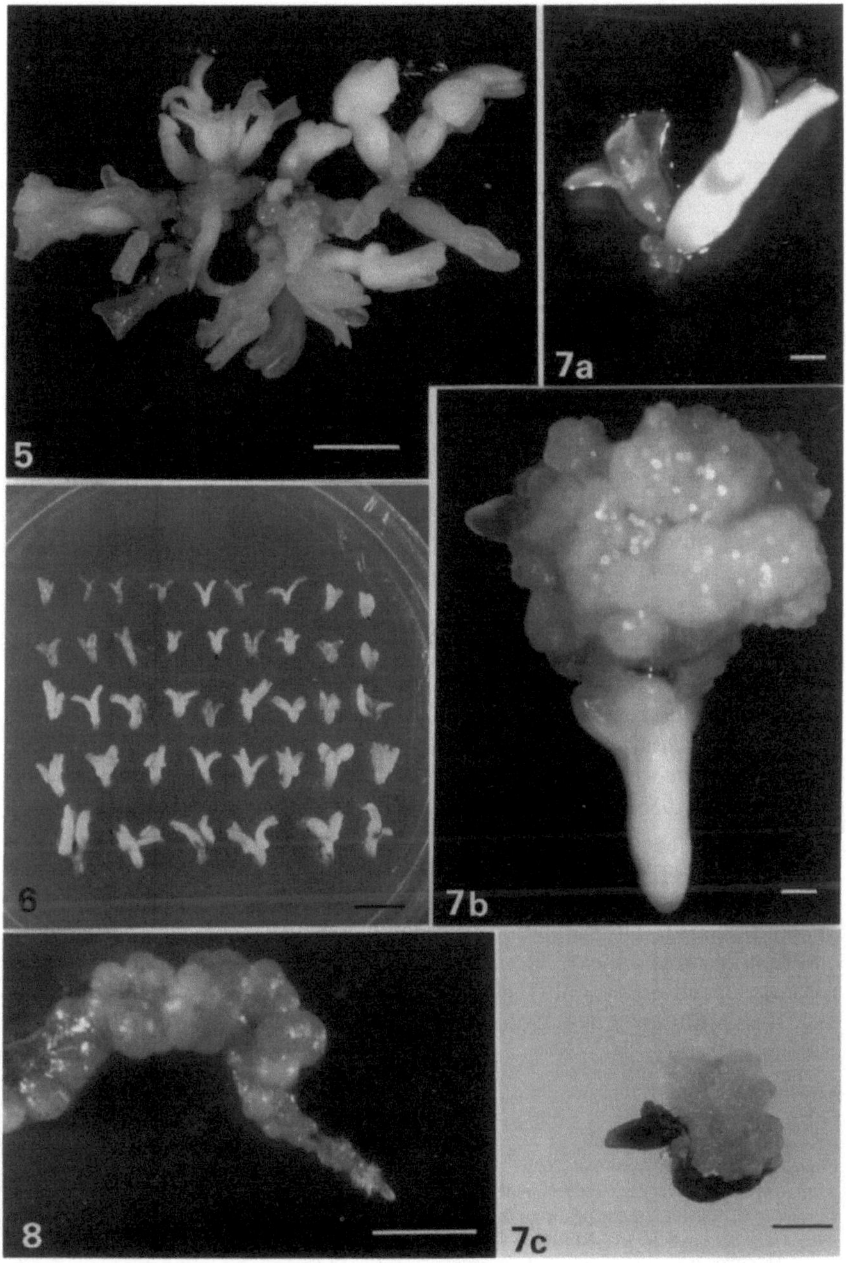


*Figure 5.* Development of cotyledonary embryos from GEC after 4 week's culture on MM ($\frac{1}{2}$ MS supplemented with 3 mM glutamine, 0.92 μM Z and 0.25 μM IBA, bar = 5 mm).

*Table 1.* Effects of three mineral media on production of somatic embryo from callus. Values are mean numbers (± se) of cotyledonary embryos formed per sample of callus (ca. 100 mg fresh weight) after 4 weeks of culture.

| Medium[a] | Total no. embryos | No. normal embryos | | | No. abnormal embryos |
|---|---|---|---|---|---|
| | | <3 mm | 3–5 mm | >5 mm | |
| Heller | 6.42 ± 0.23 | 2.95 ± 0.33 | 1.01 ± 0.14 | 0.78 ± 0.07 | 1.70 ± 0.16 |
| WPM | 8.80 ± 0.32 | 3.41 ± 0.21 | 1.83 ± 0.15 | 0.89 ± 0.10 | 2.58 ± 0.23 |
| $\frac{1}{2}$ MS | 10.57 ± 0.66 | 4.24 ± 0.36 | 2.32 ± 0.20 | 1.16 ± 0.15 | 2.84 ± 0.28 |

[a] All media supplemented with 3 mM glutamine, m-inositol and MS vitamins, 0.92 μM Z, 0.25 μM IBA and 3% sucrose.

In our current protocol for maintenance of the embryogenic competence of chestnut cultures, cotyledonary embryos (Fig. 6) are removed from calli and then either transferred to fresh MM for increasing the number of embryogenic cultures by secondary somatic embryogenesis, or used in experiments on maturation and subsequent conversion into plants.

### 3.2. *Secondary somatic embryogenesis*

In the literature, the term repetitive, or secondary, somatic embryogenesis has been used to refer to the formation of new somatic embryos on cultures of primary somatic embryos. Secondary embryogenesis has been reported in most species in which somatic embryos have been obtained (Williams and Maheswaran, 1986; Wann, 1989), including the chestnut (Vieitez *et al.*, 1990, 1992a; Merckle *et al.*, 1991).

The process of secondary embryogenesis has been associated with the repression of normal development of primary embryos (Maheswaran and Williams, 1985). The formation of secondary embryos in our embryogenic chestnut cultures appears to occur in keeping with this hypothesis: during routine subculture of embryogenic callus on MM, embryos detached from calli frequently form new embryos instead of developing further to mature

*Figure 6.* Isolated cotyledonary embryos of various sizes and morphologies for use in experiments on either secondary embryogenesis or conversion into plants (bar = 10 mm).

*Figure 7.* (a) Development of direct secondary embryos at the root pole of a primary embryo after 4 weeks of culture in MM (bar = 1 mm). (b) Indirect secondary embryogenesis via development of GEC (bar = 1 mm). (c) Indirect secondary embryogenesis via development of soft mucilaginous callus (SMC) from which embryogenic cell suspension cultures are initiated (bar = 5 mm).

*Figure 8.* Secondary embryogenesis. Development of GEC on the radicle of a germinating somatic embryo after 6 weeks of culture in MM lacking glutamine (bar = 5 mm).

stages. Isolated cotyledonary embryos behave in a similar manner in germination experiments, regardless of mineral medium and hormone treatment.

Secondary embryos develop either directly or indirectly (with prior callus formation) from the surface of primary embryos. Direct secondary embryos are, generally, formed on the hypocotyl region and/or near the root pole of primary embryos, and reach the cotyledonary stage within the following 4–6 weeks of culture (Fig. 7a). Indirect secondary embryos develop from two different types of callus: a) The first one is morphologically identical to the globular embryogenic callus (GEC; Fig. 7b) described above, on which embryos form upon subculture on fresh MM medium, and b) the other type is a soft-textured, friable callus with a mucilaginous appearance (SMC; Fig. 7c) resembling the calli developed in embryogenic cultures of *Fagus sylvatica* (Vieitez *et al.*, 1992b). These three different secondary embryogenic responses observed in embryos, maintained under identical culture conditions, suggest the presence of physiologically different embryogenic cell populations which respond differently to the signal for embryogenesis.

The secondary embryogenic capacity of both translucent and white somatic embryos with normal morphology was evaluated in four media: 1) MM; 2) MM containing 2.69 μM NAA instead of IBA (M-NAA); 3) hormone-free MM, pH 5.8 (MM-1); and 4) hormone free MM, pH 5.0 (MM-2). Groups of six cotyledonary embryos about 5 mm in length were placed in plastic dishes containing 18 ml of medium. For each combination of embryo type and medium, 54 embryos were used (three replicates of 18 embryos each). The percentages of embryos developing direct secondary embryos, GEC and SMC were recorded after 6 weeks of culture, and for each of the three type of response, these percentage data were subjected to an arcsine square root transformation for to two-way analysis of variance (Table 2).

Although translucent embryos are, apparently, more immature than white ones, there were no significant differences between them as regards the frequency of any of the three secondary embryogenesis responses. The responses to the three media did not depend on embryo type either. However, there were significant differences ($p = 0.01$) among the media as regards the frequency of SMC formation, which was significantly higher on M-NAA than on the other three media for both translucent embryos (12.9% as against values between 1.8 and 4.8%) and white embryos (20.3% as against values between 2.4 and 7.4%).

In all cases, the highest incidence of secondary embryogenesis occurred through the development of GEC. The frequency of GEC formation by translucent embryos ranged between 30.8% on hormone-free MM of pH 5.8 and 44.2% on MM medium, and the frequency for white embryos ranged between 30% on MM and 38.8% on M-NAA.

All four media supported direct secondary embryogenesis on both types of embryos. The highest frequency (16.9%) was achieved with translucent embryos cultured on hormone-free MM of pH 5.0, and the lowest frequency (3.7%) with white embryos cultured on MM.

*Table 2.* Frequency of direct and indirect secondary somatic embryogenesis (the latter via formation of either globular embryogenic callus (GEC) or soft, mucilaginous callus (SMC)), on translucent and white somatic embryos of chestnut after 6 weeks of culture on four media. Values are mean percentages (± se) of three replicates of at least 18 embryos each.

| Medium | Direct | | Indirect | | | |
|---|---|---|---|---|---|---|
| | Translucent | White | GEC | | SMC | |
| | | | Translucent | White | Translucent | White |
| MM | 10.55 ± 2.73 | 3.70 ± 1.85 | 44.25 ± 6.25 | 30.05 ± 4.57 | 1.85 ± 1.85 | 7.40 ± 4.89 |
| M-NAA | 12.03 ± 0.92 | 7.40 ± 3.70 | 36.10 ± 2.77 | 38.88 ± 5.55 | 12.96 ± 4.90* | 20.36 ± 8.07* |
| MM-1 | 9.87 ± 1.23 | 8.64 ± 2.47 | 30.86 ± 6.53 | 32.08 ± 8.08 | 1.23 ± 1.23 | 2.46 ± 1.23 |
| MM-2 | 16.97 ± 1.23 | 13.48 ± 3.34 | 35.79 ± 3.26 | 36.46 ± 7.30 | 4.84 ± 1.14 | 2.41 ± 1.20 |

* Significantly different from all other values in the same column at $p = 0.01$, according to 2-way ANOVA test. MM – Maintenance medium; M-NAA – MM supplemented with 2.69 μM NAA instead of IBA; MM-1 – Hormone free MM of pH 5.8; MM-2 – Hormone free MM of pH 5.0.

Direct secondary embryos and GEC were also, occasionally, observed on the radicle of the germinating somatic embryos (Fig. 8) and on plantlets regenerated from somatic embryos. Carman (1990) used the term "memory effect" to describe the high levels of embryogenic competence often observed in plants regenerated from embryogenic cultures, and stated that this carry-over effect might be due to the occurrence of somatic embryogenesis in the presence of exogenous auxins. In chestnut a "memory effect" of this kind might explain the repetitive embryogenesis observed on primary embryos cultured on hormone-free medium, since embryos cultured on hormone-free media did not usually regenerate plants.

Direct secondary embryos are now being used to perform experiments on germination in our laboratory. GEC is subcultured on MM to increase the number of embryogenic cultures and to maintain a continuous supply of somatic embryos, while SMC is dispersed in liquid medium to initiate cell suspension cultures following a protocol for the establishment and maintenance of embryogenic cell suspension cultures of *Fagus sylvatica* (Vieitez *et al.*, 1992b).

### 3.3. *Cell suspension cultures*

New cycles of somatic chestnut embryos can be obtained from suspension cultures established from SMC masses (Figs. 4 and 7c), which are made up of both free cells and loose cell aggregates composed of large, rounded, highly vacuolated cells together with a minority of cells with morphological characteristics associated with embryogenic cells (such as small size, oval or rounded shape, dense cytoplasm, voluminous nuclei and minimal vacuolization). The presence of embryogenic cells was confirmed by the acetocarmine stain (Gupta and Durzan, 1987). Suspension cultures are initiated by dispersing yellowish SMC masses (ca. 250–600 mg fresh weight) in 100–250 ml Erlenmeyer flasks containing 25–60 ml of liquid medium (LM-1) consisting of half strength major salts, microelements, m-inositol and MS vitamins, 100 μM Fe-Na-EDTA, 2 mM glutamine, 1.13 μM 2,4-D, 0.45 μM BA and 2% sucrose (pH 5.8). Suspensions are kept on a shaker at 90 rpm either in diffuse light or in darkness.

During the first 2–3 weeks of culture, prolific growth of embryogenic cells is achieved by renewing the medium almost completely every 3 days. Since most vacuolated cells seem not to divide, these quick subcultures increase the proportion of embryogenic cells, whose mean cell size decreases with successive subcultures. It must, however, be pointed out that there is a great interclonal variation in the ability to proliferate in suspension cultures. The proliferation rates of cell suspensions derived from the embryogenic line E-431 were always higher than those of suspensions from line E-HV.

Thereafter, suspensions were subcultured every 8 days following the procedures employed for the maintenance of cell suspension cultures of *F. sylvatica* (Vieitez *et al.*, 1992b). Suspensions are sieved through 104 μm nylon screens or allowed to settle for about 30 min, after which the filtrate or the suspended fraction, which contain only floating free cells and the smaller cell aggregates, are subcultured by 1:1–5 dilution with fresh LM-1, making a final liquid volume of 60 ml. The culture fraction retained on the 104 μm screens or which settle down at the bottom of the flasks are transferred to LM-2 medium, which has similar composition to LM-1 except for having no glutamine and having 0.46 μM Z in place of 2,4-D and BA.

Microscopic examination of cultures after one or two 8-day passages through LM-1 showed numerous preglobular proembryos (Halperin and Wetherell, 1965) consisting of a few closely packed cells (Figs. 11–17) which upon transfer to LM-2 gave rise to more-or-less spherical, compact cell aggregates (Fig. 18), or proembryogenic masses (PEMs, Halperin, 1966). Apparently, PEMs proliferated in a repetitive manner by fragmentation of the relatively large PEMs and subsequent proliferation of their fragments. This suggestion is based on the fact that PEMs larger than 2 mm were never observed, while their settled volume increased upon each subculture. Within 2–4 passages, depending on the cell line, PEMs acquire a slight yellowish colour, and subsequently multiple globular proembryos develop by cleavage polyembryony (Halperin, 1966). We called these yellowish PEMs bearing globular embryos (Fig. 19) "microcalli" due to their resemblance to the globular embryogenic callus formed on the solid medium. Suspensions derived from the embryogenic line E-431 again showed higher capacity for the formation of PEMs and microcalli than suspensions derived from line E-HV, which suggests that genotype is an important factor for the establishment of productive suspension cultures of chestnut. However, we have no explanation for the great variation in embryogenic capacity among different replicates of the same cell suspension lines maintained under identical conditions. Some replicates showed great profusion of globular proembryos, indicating the great potential of suspension cultures for large-scale production of somatic embryos, whereas in others hardly any proembryos were formed.

Once globular embryos are formed, they usually start to proliferate, producing clusters of new globular embryos instead of progressing to further stages of development (for the latter to occur requires to plate microcalli on solid MM; see below, late embryo development). Proliferating microcallus suspensions can be maintained for 3–4 months by subculturing 5 ml of settled culture into 60 ml of LM-2 every 6–8 days. If the subculture period is longer than 8 days, the culture medium becomes yellowish brown and necrotic areas appear on the early-formed embryos, which is followed by browning of all plant material.

Although suspensions lose their capacity to produce somatic embryos after 6–8 months of culture, our results show that the liquid media LM-1 and LM-2 are suitable for the establishment of embryogenic cell suspension cultures and for the production of globular somatic embryos. Thus, the potential for large-scale production of chestnut somatic embryos exists. However, in our cultures, embryos developed asynchronously because the presence of mixed

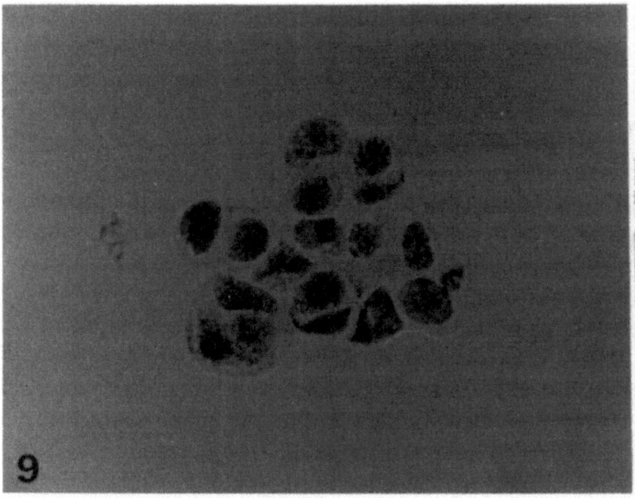


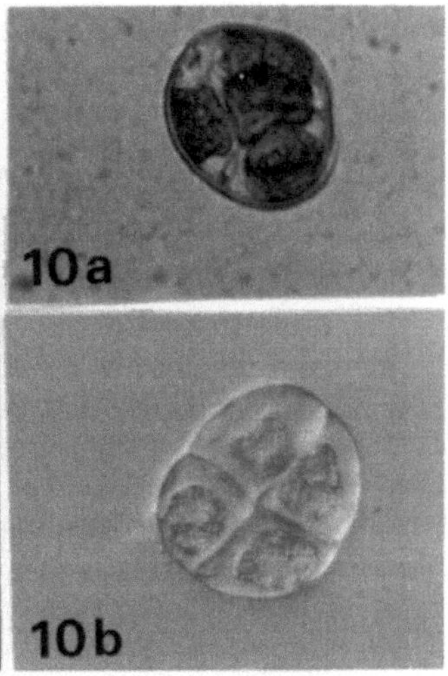


*Figure 9.* Large round pollen-like cells observed in 12-day old cell suspension cultures (hematoxylin stain after Feulgen reaction, ×270).
*Figure 10.* Tetrad-like structures floating free in LM-1 (×285). (a) Acetocarmine stain. (b) Observed with differential interference contrast optics.

populations of whitish PEMs and microcalli was unavoidable. Studies focused on achieving a high degree of synchronization in embryo formation are required before our system can be employed for practical purposes.

It is worth pointing out a number of cytological abnormalities observed in our embryogenic cell suspensions, including coenocytes or large cells with two or more nuclei resulting from acytokinetic mitoses (D'Amato, 1989) and cells that appeared to be in a meiotic stage. The latter were large, round cells (Fig. 9) that closely resembled the pollen-like cells described in two carrot cell lines (Nuti Ronchi *et al.*, 1992a,b), and tetrad-like structures (Fig. 10) and others like those drawn by Nuti Ronchi *et al.* (1992b) in their diagrammatic model of the parasexual cycle of two carrot cell lines.

Several reports in the literature (Halperin, 1966; Smith and Street, 1974; Reinert *et al.*, 1977) indicate that the tendency to cytogenetic alterations is a widespread phenomenon during *in vitro* culture, and it is normally correlated with the loss of morphogenetic potential due to prolonged subculture. However, in some cell lines, we have observed that pollen-like cells and tetrads were numerous 6–9 days after the establishment of suspensions but declined in number after several subcultures (though it should be pointed out that these suspension cultures were established from SMC masses developed either from primary somatic embryos or from embryogenic callus cultures maintained for 2–4 years by regular subculture, without loosing their embryogenic potential). The genetic stability of somatic embryos and plants derived from them clearly needs to be fully examined.

## 4. Embryo development

The ontogeny of somatic embryos has been paid little attention in the literature. In few a cases, however, it is well documented that in suspension cultures the first step in the differentiation of embryos is the occurrence of an asymmetric first division in single cells leading to the formation of two daughter cells of different size and morphology (Backs-Hüsemann and Reinert, 1970; Konar *et al.*, 1972; Maheswaran and Williams, 1985; Nomura and Komamine, 1986). Subsequently, one or both daughter cells undergo a series of cell divisions which do not give rise to embryos directly, but to characteristic cell aggregates referred to as PEMs (Halperin, 1966), proembryonal complexes (Haccius, 1978) or embryogenic cell clusters (Nomura and Komamine, 1985, 1986); it is from these aggregates that somatic embryos develop. Haccius (1978) reported these embryos to be of unicellular origin.

Although we have not been able to follow sequentially all the stages from a single cell to the globular embryo, the formation of chestnut somatic embryos appears to confirm the above mentioned pattern. This assertion is based on the observations: 1) somatic embryos only developed from PEMs; 2) in cell suspension cultures with high rates of PEM and globular embryo production, oblique to transverse, asymmetric first divisions were often observed in single cells floating free in the medium; and 3) asymmetric transverse first divisions were also observed in single cells located at the periphery of PEMs developing globular embryos.

### 4.1. *Early embryo development*

After isolated single cells had divided asymmetrically (Fig. 11), their unequal daughter cells followed a variety of different patterns of early division to form PEMs. In some cases, the smaller cell underwent a longitudinal division, and its derivative cells then divided repeatedly to form a morulloid mass, while the larger cell either did not divide any further (Fig. 12) or, occasionally, it underwent an oblique division before the smaller cell began dividing (Fig. 13). In most cases, however, both daughter cells contributed to the formation of 4-celled preglobular proembryos, though two division patterns were observed. In one pattern, both nuclei divide simultaneously (Fig. 14a) and the formation of new cell walls between their daughter nuclei leads to a 4-celled preglobular proembryos (Fig. 14b); in the other pattern, the smaller cells undergo a longitudinal division, giving rise to a 3-celled preglobular structures, and then the larger cells also divide longitudinally, forming a 4-celled preglobular proembryos (Fig. 15). In spite of intensive observations of suspensions with an inverted microscope, their further divisions

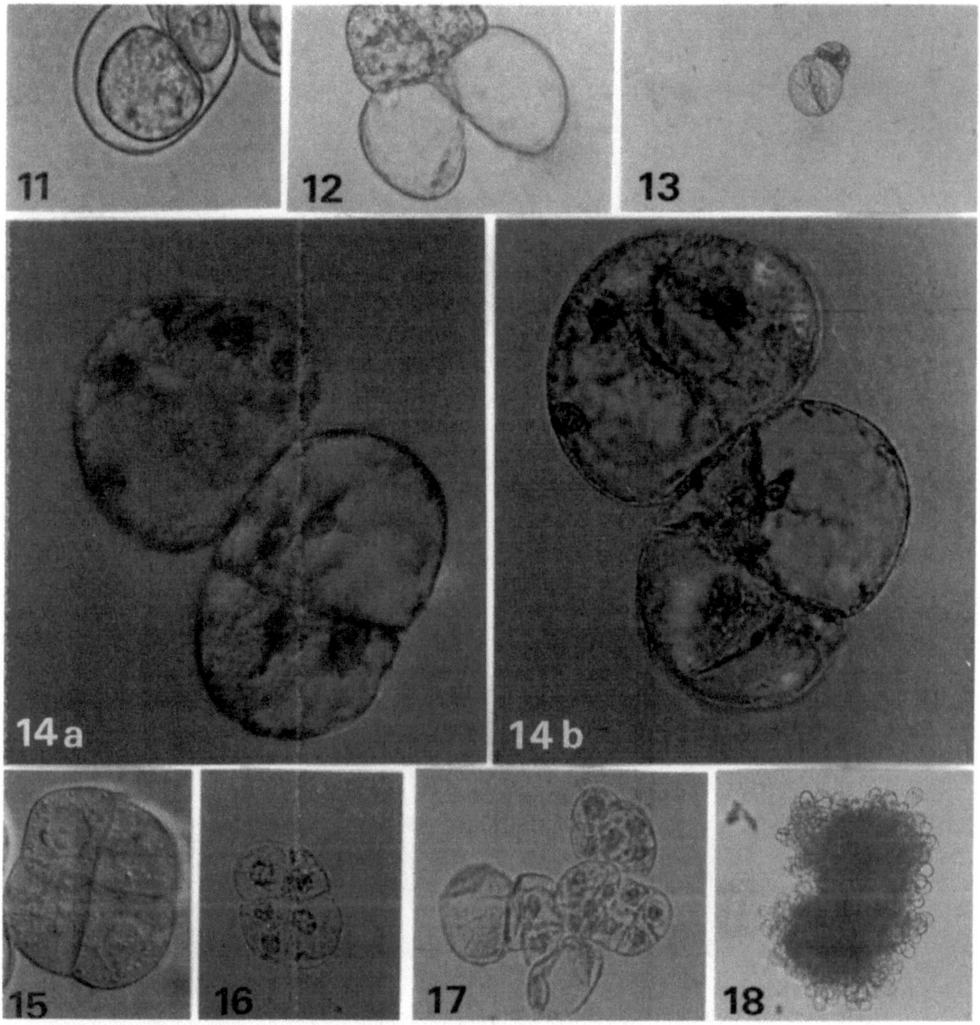


*Figure 11–18.* Development of proembryogenic masses (PEMs). *Fig.* 11. First asymmetric division in a free floating single cell leading to two unequal daughter cells (stained with Toluidine Blue 0 and observed on 1:1 glycerol-liquid medium, ×525). *Fig.* 12. Further divisions of the smaller daughter cells (×380). *Fig.* 13. Oblique division of the larger daughter cell (×145). *Fig.* 14. Simultaneous divisions of both daughter cells, and formation of cell walls between their daughter nuclei leading to 4-celled preglobular proembryo (acetocarmine stain, ×585). (14a) Differential interference contrast optics. (14b) Phase contrast optics. *Fig.* 15. Four-celled preglobular proembryo (×285). *Figs.* 16–17. Further divisions in a 4-celled preglobular proembryo (×285). *Fig.* 18. Proembryogenic mass developed after two or three 8-day passages in LM-2 (×90).

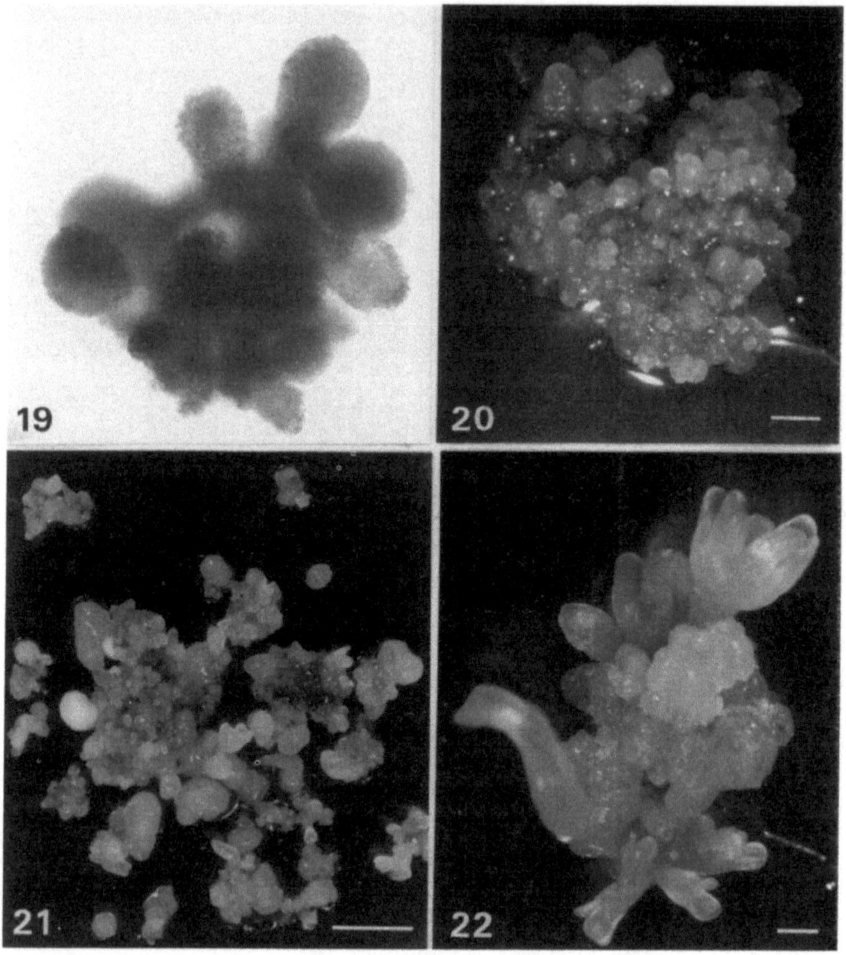


*Figure 19.* Microcalli. Development of multiple globular proembryos from PEMs (×90).
*Figure 20.* Microcalli one week after plating on solid MM (bar = 1 mm).
*Figure 21.* Proliferation of microcalli and embryo development within 2–3 weeks after plating (bar = 5 mm).
*Figure 22.* Development of cotyledonary embryos upon microcalli subcultured on MM (bar = 1 mm). Note the similarity with the GEC of Fig. 5.

to give rise to a more-celled preglobular proembryos (Figs. 16 and 17) and PEMs (Fig. 18) could not be traced in detail. Attempts to culture isolated embryogenic cells and 4-celled structures were unsuccessful. Following Nomura and Komamine (1985), isolated embryogenic cells and 4-celled structures were removed from suspensions manually using a micropipette under an inverted microscope and were singly transferred to thin layers of a variety of solid media, but they did not divide any further.

*Figure 23.* Conversion of somatic embryos into plants. *Fig.* 23. Plantlet regeneration from somatic embryos of different sizes after 8 weeks of culture on MS supplemented with 150 μM Fe-Na-EDTA and 0.92 μM Z (bar = 10 mm). *Fig.* 24. Emblings regenerated from embryos stored at 4°C for 16 weeks after 4 weeks of preculture on MS plus 0.92 μM Z (bar = 10 mm). *Fig.* 25. E-HV chestnut plant after 8 weeks of acclimatization (bar = 10 mm). (26) E-431 somatic embryo-derived chestnut during the second year's growth in soil (bar = 25 cm).

Once PEMs had formed, the development of globular embryos of chestnut appeared to proceed, in keeping with the literature (McWilliam *et al.*, 1974; Philip and Schraudolf, 1987), by transverse, unequal first division of single cells located at the periphery of PEMs. The next division was longitudinal in the apical daughter cell, but further divisions could not be followed in detail. Halperin (1966) reported that the segmentation patterns of young proembryos as they grow outward from PEMs are extremely irregular, indicating that there may be no definite sequence of cell lineage associated with proembryo growth, which has little effect on subsequent embryo development.

According to Haccius (1978) somatic chestnut embryos may be considered to be of unicellular origin, because they develop from PEMs, whose formation is strongly correlated with the occurrence of asymmetric first divisions in isolated single cells. Single-cell origin is necessary if somatic embryogenesis is to be employed for plantlet regeneration after the application of gene transfer techniques, because it cannot be assumed that regenerated plants are completely transformed unless their origin is unicellular (Michaux-Ferrière *et al.*, 1992).

### 4.2. *Late embryo development*

Embryos remained arrested at the globular stage in the liquid medium, and their further development required microcalli (Fig. 19) to be plated on solid MM. When samples of suspensions containing both whitish PEMs and microcalli were plated, only the proliferation of PEMs occurred; embryos did not develop any further on either microcalli or PEMs. To avoid this, the suspensions were allowed to settle for 1 min, the suspended fraction was discarded, and the settled fraction was resuspended in the liquid medium and filtered through a 40 mesh stainless steel screen. The retained fraction was then washed several times with liquid medium to remove remaining whitish PEMs, and microcalli were collected singly and transferred to solid MM. Within the first week of culture, numerous larger globular embryos were observed protruding from the microcalli (Fig. 20). Subsequently, microcalli became friable as they continued proliferating and producing new embryos. Again, we have no explanation for the variation among replicates of the same cultures as regards their ability to produce embryos. In some replicates all microcalli proliferated and developed new globular embryos, while in others none did. Embryos at all stages of development were observed together with abnormal structures in the same cultures after 3–4 weeks of culture on MM (Fig. 21). It was noted that globular embryos either developed further to give rise to single cotyledonary embryos as shown in Fig. 3, or grew and acted as proembryonal complexes (Haccius, 1978) from which more than one cotyledonary embryo developed. In some cultures up to 80% of microcalli had developed cotyledonary embryos within 4–5 weeks after plating. Some cotyledonary embryos germinated or elongated their radicles even while attached to microcalli. After subculture on MM, the morphology and behaviour of microcalli (Fig. 22) were identical to those of the initial embryogenic calli, thus closing the cyclic culture system developed for the maintenance of embryogenic competence in chestnut cultures.

*Table 3.* Effect of illumination regime on the conversion of somatic embryos into plants. Results recorded after 6 weeks of culture on solid $\frac{1}{2}$ MS containing 0.92 μM Z.

| Photoperiod | Irradiance μmol $m^{-2}\ s^{-1}$ | No. embryos tested | No. plantlets regenerated | Conversion% |
|---|---|---|---|---|
| 0 h | 0 | 54 | 0 | 0.00 |
| 12 h | 9 | 54 | 1 | 1.85 |
| 12 h | 30 | 54 | 3 | 5.55 |
| 12 h | 48 | 54 | 2 | 3.70 |
| 16 h | 9 | 54 | 2 | 3.70 |
| 16 h | 30 | 54 | 7 | 12.96 |
| 16 h | 48 | 54 | 5 | 9.25 |
| 24 h | 30 | 48 | 0 | 0.00 |
| 24 h | 48 | 48 | 1 | 2.08 |

## 5. Embryo germination

After 4 weeks on solid MM, somatic embryos, 3–10 mm in length, appearing mature and well formed with two cotyledons and a dense yellow root pole, were removed from the calli and used in germination experiments. Initially, isolated embryos were sorted by size into small (<5 mm), medium (5–8 mm) and large (>8 mm) classes and were cultured in hormone-free medium. This generally resulted in the growth of the radicle, but no shoots. In all the mineral medium formulations tested, large embryos developed their radicle more readily than smaller embryos. Similar germination results were obtained when embryos were cultured on media containing low levels (<1 μM) of growth regulators, but the frequency of secondary embryogenesis increased. On media supplemented with higher levels of growth regulators, the embryos showed general signs of overgrowth and, in particular very thickened cotyledons; though when the growth regulator formulation was 7.2 μM gibberellic acid ($GA_3$) and 5.7–11.4 μM indole-3-acetic acid (IAA), 100% of the embryos developed a strong root system. Though in many cases hypocotyl grew to 2–3 cm in length, subsequently secondary embryos or new cotyledons developed instead of true leaves, which is characteristic of immature embryos or precocious germination of embryos (De Wald *et al.*, 1989); shoots never developed. There were no differences in response between embryos cultured in darkness and those cultured with a 16 h photoperiod, nor between embryos cultured on solid medium or on paper bridges wetted with liquid media (but culture on vermiculite impregnated with liquid medium resulted in the necrosis of all embryos regardless of the composition of the medium).

Paradoxically, after we had tried a great variety of media, hormone treatments and environmental conditions, the first regenerated plantlet was obtained from an embryo that had been cultured on MM to induce secondary embryogenesis about 14 months after the initiation of experiments. To our surprise, since then the percentage of embryos germinated into whole plantlets (emblings) has increased gradually from experiment to experiment, and emblings have been obtained on most of the tested media, including those that had previously been tried without success. Although trials of different media and environmental conditions are still in progress with a view to enhance the conversion rate, we describe below our preliminary results and tentative conclusions.

In an experiment conducted after achieving a slight but sustainable enhancement of conversion by removing both IBA and glutamine from MM, somatic embryos 5–8 mm long were cultured on this $\frac{1}{2}$ MS medium containing 0.92 μM Z and kept under nine different light conditions. Best conversion into emblings was achieved with a 16 h photoperiod at irradiances of 30 and 48 μmol $m^{-2}$ $s^{-1}$, which afforded 12.96 and 9.25% conversion rates, respectively (Table 3). No emblings were obtained either in darkness or under continuous irradiance with 30 μmol $m^{-2}$ $s^{-1}$. Although these trials used a small number of embryos, further experiments confirmed that a 16 h photoperiod at these irradiance levels is the most suitable illumination regime for the conversion of embryos. Consequently, only these conditions have since been employed.

### 5.1. *Effect of mineral content of medium*

In rather different experiment, Misoo and Matsubayashi (1977, 1978) and Heberle-Bors (1980) reported that the addition of Fe-EDTA or $Na_2$-EDTA to the medium increased the frequency of formation of green plantlets in cultures of tobacco anthers. Similarly, Saito and Ide (1985), Sato *et al.* (1986) and Iliev and Chavdarov (1990) have reported enhanced regeneration of plantlets from adventitious buds of *Betula* species after increasing the concentrations of iron and chelates in combination with cytokinins. With the aim of improving shoot development from somatic embryos of chestnut, we investigated the effect of increasing the concentration of Fe-Na-EDTA from 100 to 150 μM in both half and full strength MS media supplemented with 0.92 μM Z. After 4 weeks, embryos were transferred to the fresh medium, and after 8 weeks of culture the number of embryos converted into plants was recorded. The experiment was repeated four times using at least 9 plastic dishes (6 embryos each) per medium tested.

In all four media, embryos of very different sizes and morphologies were able to regenerate plants (Fig. 23). On $\frac{1}{2}$ MS, the increase in Fe-Na-EDTA level did not affect the conversion frequency of embryos, defined as the mean percentage of embryos showing emergence of both shoot and root after 8 weeks of culture (Table 4). However, on full MS medium containing 150 μM Fe-Na-EDTA the conversion frequency, 23.34% (99 plantlets from 424 embryos), was significantly higher than that achieved on the correspond-

*Table 4.* Conversion of somatic embryos of chestnut Clone E-431: effect of increasing the concentration of Fe-Na-EDTA in half and full strength MS media supplemented with 0.92 μM Z. Conversion frequency expressed as mean percentage (± se) of embryos developing both shoot and root after 8 weeks of culture.

| Medium | [Fe-Na-EDTA] μM | No. embryos cultured | No. plantlets regenerated | Conversion frequency (%) |
|---|---|---|---|---|
| Half MS | 100 | 252 | 30 | 11.87 ± 4.25 |
| Half MS | 150 | 274 | 31 | 11.31 ± 2.94 |
| Full MS | 100 | 380 | 46 | 12.10 ± 3.92 |
| Full MS | 150 | 424 | 99 | 23.34 ± 6.62* |

* Value significantly higher than the others in the same column ($p = 0.05$) according to the chi-square test.

ing control that contained the standard concentration (100 μM) of Fe-Na-EDTA. The conversion frequency of the latter, 12.1% (46 plantlets from 380 embryos), did not differ significantly from those of the $\frac{1}{2}$ MS control, 11.87% (30 plantlets from 252 embryos).

Two other striking results should also be pointed out. Firstly, on full MS plus 150 μM Fe-Na-EDTA, there occurred the first conversion of E-HV embryos into plants, though the conversion rate (5–7%) was lower than for E-431 embryos. No plants were regenerated on the other three media in this experiment, but as in the case of E-431 embryos (see above), the conversion frequency of E-HV embryos also increased afterwards and plants were regenerated on most of the media tried. Secondly, adventitious buds developed from unconverted green embryos cultured on $\frac{1}{2}$ MS containing 150 μM Fe-Na-EDTA and on both full MS media.

From these experiments it appears that neither increasing the mineral salt content of MS medium from half to full strength, nor the increasing Fe-Na-EDTA concentration, by themselves had any effect on the embryo conversion frequency. The enhancement of conversion achieved on full MS plus 150 μM Fe-Na-EDTA can, at present, only be attributed to joint effects of increasing both Fe-Na-EDTA level and mineral salt content. It has been suggested that EDTA may change the net concentration of free calcium and magnesium ions in the medium (Kyo, 1990), which might result in a more balanced medium for embryo conversion. However attempts to optimize the composition of the mineral medium by modifying the concentrations of calcium chloride and/or magnesium sulphate including their elimination were unsuccessful.

### 5.2. *Effect of abscisic acid*

The variety of responses observed during germination experiments suggests that embryos were in a variety of different maturation states at time of the experiments, probably as a consequence of the lack of synchronization during embryo development. In other embryogenic cultures, such problems have been solved by supplementing the medium with abscisic acid (ABA), which may function as an osmoregulator causing cell water content to decrease, and so lead to more normal development (Gray and Purohit, 1991). The addition of ABA has proved effective for controlling repetitive embryogenesis (Ammirato, 1974, 1977, 1983; Kamada and Harada, 1981; Vasil and Vasil, 1981), for improving synchronization of development and maturation, and for enhancing conversion of somatic embryos of angiosperms (Ammirato, 1974, 1977, 1983; Vasil and Vasil, 1981; Lutz *et al.*, 1985; Higuchi and Maeda, 1990) and conifers (Durzan and Gupta, 1987; Dunstan *et al.*, 1988, 1991; Hakman and Von Arnold, 1988; Krogstrup *et al.*, 1988; Von Arnold and Hakman, 1988; Thompson and Von Aderkas, 1992) into plantlets.

Although Merckle *et al.* (1993) reported that synchronization itself did not appear to improve the maturation of somatic embryos of yellow poplar (*Liriodendrom tulilpifera*), we initially tried to induce synchronization by culturing embryogenic calli (GEC) on MM supplemented with ABA instead of Z and IBA. The result was total failure: no cotyledonary embryos developed, and when calli were transferred back to standard MM, embryos again reverted to asynchronous development.

Similarly, the culture of isolated embryos on media containing ABA at concentrations of 0.38, 1.89, 3.78 and 7.57 μM failed to prevent secondary embryogenesis and had no effect on either the morphology of the embryos or on their subsequent conversion on MS medium containing 0.92 μM Z and 150 μM Fe-Na-EDTA or on MS with $GA_3$ at various concentrations.

### 5.3. *Effect of cold treatment*

Like zygotic embryos, somatic chestnut embryos appear to adopt a state of epicotyl dormancy, and may thus require a period of low temperature to break dormancy and stimulate the development of shoots. To test the effect of cold storage on conversion, cotyledonary embryos larger than 5 mm were isolated from calli and transferred to plastic dishes containing 15 ml of solid MS medium supplemented with 0.92 μM Z. In a first series of experiments, the dishes were immediately stored at 4°C in pharmaceutical refrigerators for up to 36 weeks; while in a second series, embryos were maintained in the growth chamber (at 25°C day and 20°C night temperatures, respectively) for 4 weeks prior to cold storage in the same dishes. In both cases, after various periods of cold treatment, those embryos that remained healthy (with no signs of degeneration such as browning or soft texture) were selected for conversion experiments on solid MS medium containing either 100 or 150 μM Fe-Na-EDTA, and 0.92 μM Z. Experiments were repeated at least four times using at least 4 dishes (24 embryos) per cold treatment and medium assayed. Embryos were transferred to fresh medium after 4 weeks of culture,

*Table 5.* Conversion of E-431 somatic embryos stored at 4°C immediately after their isolation from calli. Conversion frequencies (mean percentage ± se) were calculated with respect to the number embryos surviving after 8 weeks of culture on MS plus 0.92 μM Z and 100 or 150 μM Fe-Na-EDTA.

| Cold storage (weeks) | [Fe-Na-EDTA] (μM) | No. embryos cultured | Surviving embryos | No. plantlets regenerated | Conversion frequency (%) |
|---|---|---|---|---|---|
| 6–8 | 100 | 180 | 118 | 24 | 20.29 ± 4.81 |
| 6–8 | 150 | 180 | 131 | 37 | 27.88 ± 9.46 |
| 12–16 | 100 | 180 | 104 | 19 | 17.27 ± 8.65 |
| 12–16 | 150 | 180 | 83 | 12 | 14.26 ± 4.79 |
| 20–26 | 100 | 156 | 109 | 17 | 14.89 ± 7.63 |
| 20–26 | 150 | 144 | 121 | 26 | 21.03 ± 9.19 |
| 32–36 | 100 | 132 | 96 | 21 | 21.23 ± 5.19 |
| 32–36 | 150 | 120 | 87 | 22 | 24.48 ± 6.03 |

*Table 6.* Conversion of E-431 somatic embryos stored at 4°C after 4 weeks of preculture on MS plus 0,92 μM Z. Conversion frequencies (mean percentage ± se) were calculated with respect to the number embryos surviving after 8 weeks of culture on MS plus 0.92 μM Z and 100 or 150 μM Fe-Na-EDTA.

| Cold storage (weeks) | [Fe-Na-EDTA] (μM) | No. embryos cultured | Surviving embryos | No. plantlets regenerated | Conversion frequency (%) |
|---|---|---|---|---|---|
| 6–8 | 100 | 156 | 132 | 30 | 22.09 ± 9.23 |
| 6–8 | 150 | 204 | 179 | 52 | 28.53 ± 5.97 |
| 12–16 | 100 | 120 | 105 | 31 | 29.42 ± 7.67 |
| 12–16 | 150 | 136 | 111 | 36 | 32.37 ± 10.8 |
| 20–26 | 100 | 126 | 94 | 29 | 30.03 ± 12.6 |
| 20–26 | 150 | 132 | 114 | 32 | 27.13 ± 12.4 |
| 32–36 | 100 | 96 | 73 | 18 | 24.15 ± 7.24 |
| 32–36 | 150 | 96 | 71 | 17 | 23.55 ± 6.16 |

and conversion frequencies were calculated with respect to the number or embryos surviving after 8 weeks of culture.

The results of the first and second series of experiments are listed in Tables 5 and 6, respectively. In general, cold treatment resulted in an overall enhancement of conversion with respect to previous experiments and decreased the frequency of secondary embryogenesis; and the conversion frequencies of embryos precultured (in the growth chamber with 25°C day and 20°C nights temperatures, respectively) for 4 weeks prior to cold storage were higher than those of embryos stored in the cold immediately after isolation from callus. However, the effects of cold and Fe-Na-EDTA concentration were not statistically significant because conversion frequencies showed great variability both between and within the treatments. This variability may have been due to employment of embryos from different harvests, and also to not all experiments having been conducted simultaneously. Furthermore, our ability to select high quality embryos increase with experience. The highest conversion frequencies (up to 46%) were obtained with embryos possessing thin cotyledons, while embryos bearing large, thick cotyledons, like those of mature zygotic chestnut embryos, did not regenerate plants at all.

Although the embryo survival rate decreased with increasing cold storage time under both storage regimes, preculture favoured survival. After 20–24 weeks in the cold, over 70% of precultured embryos appeared to be healthy, though some died upon reculture. Thus, criteria other than appearance must be employed for checking embryo viability. Apparently, droplet condensation on dishes was the main cause of embryo mortality during storage. This problem was partially overcome by placing dishes vertically in the refrigerator, which resulted in water accumulating at the bottom of the dishes instead of on the embryos.

Besides enhancing conversion, cold treatment also affected both the pattern of embryo development and embling morphology. After cold storage many embryos developed roots and shoots almost simultaneously (Fig. 24); whereas non-cold-treated embryos generally developed shoots only after the radicle had already somewhat grown, and in the final weeks of culture shoot growth halted and secondary roots developed. Emblings regenerated from cold-treated embryos showed more vigorous growth than those from non-treated embryos.

Since desiccation is believed to be important in the transition between embryo maturation and germination (Kermode and Bewley, 1985), we also investigated the effect of combining desiccation and cold storage on somatic embryo conversion. Embryos were kept in empty dishes for 24 h at 25°C and then stored for 4 weeks at 4°C in plastic dishes that were either empty or contained a thin layer of solid medium for 4 weeks. Upon reculture on conversion media, most of the embryos remained colourless and did not convert to plants, though in a few cases small radicles were developed.

### 6. Transfer to soil

Plantlets regenerated from both E-431 and E-HV embryos were transferred from dishes to glass jars containing 50–60 ml of solid MS medium (containing or not 0.46 μM Z) and were allowed to grow to about 4–5 cm in height. By then, they usually had 4–5 well-formed leaves and a 2–3 cm taproot with several secondary roots. When shoot growth required stimulation, the addition of liquid medium proved better than transfer to a fresh solid medium, which generally injured roots. Plantlets with swollen radicle and plantlets with only a few thin secondary roots (e.g., those derived from embryos that were damaged at the root pole during isolation from callus) were treated with a quick dip in 490 μM IBA solution to hasten development of strong root systems.

After washing with tap water to remove agar from their roots, emblings were transferred to 12 cm pots containing a 1:1 mixture of peat moss and quartz sand, and were kept in the greenhouse inside a plastic acclimatization tunnel, equipped with basal heating and a fog and mist system for hardening. Some 70–80% of emblings (116 of 147 E-431 emblings, and 38 of 52 E-HV emblings) survived and resumed growth within 4–8 weeks of transplantation. Losses were apparently due to excessive moisture in pots. Surviving plants (Fig. 25) were moved to normal greenhouse conditions and allowed to grow in the same pots for 2–6 months. Their morphology and behaviour were similar to those of seedlings grown in the same pots, though the latter grew more vigorously. Both formed resting buds and lost their leaves by mid November, by when *in vitro* plants were about 20 cm in height. Plants acclimatized in spring and summer were transferred to 30 cm pots containing a 1:1 mixture of soil and peat-quartz and kept in shade to avoid direct sun. A sample of acclimatized plants (30 E-431 and 12 E-HV plants) were transferred to field at the beginning of the following March and all resumed growth in spring. After one year in soil their heights ranged between 70 and 110 cm. In the second growing season they reached heights exceeding 2 m (Fig. 26).

## 7. Concluding remarks

The results of the limited research on somatic embryogenesis in chestnut suggest that the problems involved are similar to those encountered with most other woody species. Somatic embryogenesis has only been induced on immature zygotic embryo explants and with low initial success rates. Once initiated, however, the embryogenic cultures can be efficiently multiplied and maintained without loss of embryogenic potential, either by secondary embryogenesis or by regular subculture of embryogenic callus on media supplemented with glutamine, low levels of a cytokinin and an auxin. Embryogenic cell suspension cultures have also been established, and the formation of globular somatic embryos have been observed. Our observations suggest that somatic embryo development is initiated from single cells floating free in the liquid medium. This single-cell origin of embryos holds out the possibility of improving chestnut through gene transfer techniques.

There remain problems to be overcome. Although high yields of cotyledonary somatic embryos are obtained from both callus and plated microcallus suspensions, embryo development is asynchronous and germination rates low, embryo-to-plant conversion being limited by poor shoot development.

Conversion frequencies can be increased by cold storage of embryos, but even so remain too low in absolute terms. Currently, only a small fraction of the cotyledonary embryos produced reach maturity and show capacity for germinating into emblings. Embryos of very different sizes and morphologies are capable of producing emblings, but embryos possessing large, thick cotyledons, like those of mature zygotic embryos, do not regenerate plants at all. Conversion is generally preceded by slight enlargement, and always by progressive greening of the embryos. A short tap root always develops prior to shoot emergence, and after shoot growth secondary roots develop. It should be stressed that although it is now possible to regenerate chestnuts from somatic embryos, somatic embryos have not yet been obtained from mature material. The regenerated plants can, therefore, be clones of a sexually recombined genotype, but not of a single donor tree.

In order to realize the potential of this technique, further studies are necessary: firstly, to induce somatic embryo production from mature material; secondly, to achieve synchronized embryo development, and thirdly, to optimize medium composition so as better to satisfy the physiological requirements of embryos, thus enabling more embryos to mature and enhancing conversion rates.

## 8. Acknowledgements

Thanks are given to Prof. Dr. R. Salema (University of Porto, Portugal) and Dr. M[a]. L. Vieitez (University of Santiago de Compostela, Spain) for their assistance with electron microscopy. This study was partially supported by the EEC ECLAIR Programme contract AGRE-CT91–0067.

## 9. References

Ammirato, P.V., 1974. The effects of abscisic acid on the development of somatic embryos from cells of caraway (*Carum carvi* L.). Bot. Gaz. 135: 328–337.

Ammirato, P.V., 1977. Hormonal control of somatic embryo development from cultured cells of caraway: Interactions of abscisic acid, zeatin and gibberellic acid. Plant Physiol. 59: 579–586.

Ammirato, P.V., 1983. The regulation of somatic embryo development in plant cell cultures: Suspension culture techniques and hormonal requirements. Biotechnology 1: 68–74.

Backs-Hüsemann, D. and J. Reinert, 1970. Embryobildung durch isolierte Einzelzellen aus Gewebekulturen von *Daucus carota*. Protoplasma 70: 49–60.

Ballester, A., M.C. Sanchez and A.M. Vieitez, 1989. Etiolation as a pretreatment for *in vitro* establishment and multiplication of mature chestnut. Physiol. Plant 77: 395–400.

Ballester, A., M.C. Sanchez, M.C. San-José, F.J. Vieitez and A.M. Vieitez, 1990. Development of rejuvenation methods for *in vitro* establishment, multiplication and rooting of mature trees. In: R. Rodriguez, R. Sanchez and D.J. Durzan (Eds.), Plant Aging: Basic and Applied Approaches, pp. 43–49. Plenum Press, New York.

Beck, J., 1970. Nurse nut grafting. Ann. Rep. Northern Nut Growers Assoc. 61: 84–90.

Biondi, S., L. Canciani, G. De Paoli and N. Bagni, 1981. Shoot formation from bud cultures of mature chestnut. In: AFOCEL (Ed.), Proc. Coll. Int. Cult. In Vitro Essences For., pp. 180–185. IUFRO, Fontainebleau.

Camus, A., 1929. Chataigniers. Monographie des genres *Castanea* and *Castanopsis*. In: P. Le Chevalier (Ed.), Enciclopédie Economique de Sylviculture. Paul Le Chevalier, Paris.

Carman, J.G., 1990. Embryogenic cells in plant tissue cultures: Occurrence and behavior. In Vitro Cell. Dev. Biol. 26: 746–753.

Chevre, A.M., S.S. Gill, A. Mouras and G. Saleses, 1983. *In vitro* vegetative multiplication of chestnut. J. Hort. Sci. 58: 23–29.

D'Amato, F., 1989. Polyploidy in cell differentiation. Caryologia 42(3–4): 183–211.

De Wald, S.G., R.E. Litz and G.A. Moore, 1989. Maturation and germination of mango somatic embryos. J. Amer. Soc. Hort. Sci. 114: 837–841.

Dunstan, D.I., F. Bekkaoui, M. Pilon, L.C. Fowke and S.R. Abrams, 1988. Effects of abscisic acid and analogues on the maturation of white spruce (*Picea glauca*) somatic embryos. Plant Sci. 58: 77–84.

Dunstan, D.I., T.D. Bethune and S.R. Abrams, 1991. Racemic abscisic acid and abscisyl alcohol promote maturation of white spruce (*Picea glauca*) somatic embryos. Plant Sci. 76: 219–228.

Durzan, D.J. and P.K. Gupta, 1987. Somatic embryogenesis and polyembryogenesis in Douglas-fir cell suspension cultures. Plant Sci. 52: 229–235.

Eltkins, J.R., B.J. Given, E. Vieitez, G. Bazzigher and G. Griffin, 1980. Vegetative propagation of large, surviving American chestnut trees. Ann. Rep. Northern Nut Growers Assoc. 71: 56–71.

González, M.L., A.M. Vieitez and E. Vieitez, 1985. Somatic embryogenesis from chestnut cotyledon tissue cultured *in vitro*. Sci. Hort. 27: 97–103.

Graves, H and H. Nienstadt, 1953. Chestnut breeding. Ann. Rep. Northern Nut Growers Assoc. 44: 136–144.

Gray, D.J. and A. Purohit, 1991. Somatic embryogenesis and development of synthetic seed technology. Crit. Rev. Plant Sci. 10(1): 33–61.

Gupta, P.K. and D.J. Durzan, 1987. Somatic embryos from protoplasts of loblolly pine proembryonal cells. Biotechnology 5: 710–712.

Haccius, B., 1978. Question of unicellular origin of nonzygotic embryos in callus cultures. Phytomorphology 28: 74–81.

Hakman, I. and S. von Arnold, 1988. Somatic embryogenesis and plant regeneration from suspension cultures of *Picea glauca* (White spruce). Physiol. Plant 72: 579–587.

Halperin, W. and D. Wetherell, 1965. Ontogeny of adventive embryos of wild carrot. Science 147: 756–758.

Halperin, W., 1966. Alternative morphogenetic events in cell suspensions. Am. J. Bot. 53(5): 443–453.

Heberle-Bors, E., 1980. Interaction of activated charcoal and iron chelates in anther cultures of *Nicotiana* and *Atropa belladonna*. Z. Pflanzenphysiol. 99: 339–347.

Heller, R., 1953. Reserches sur la nutrition minérale des tissus végétaux cultivés *in vitro*. Ann. Sci. Nat. Bot. Biol. Veg. 14: 1–223.

Higuchi, N. and E. Maeda, 1990. Enhanced plant regeneration in rice callus cultures following abscisic acid treatment. Jpn. J. Crop. Sci. 59(2): 359–368.

Iliev, I. and J. Chavdarov, 1990. *In vitro* propagation of silver birch (*Betula pendula* Roth.) through induced adventitious bud formation from leaf cutting. Int. Plant Prop. Soc. News. Autumn 1990.

Jaynes, R., 1961. Buried-inarch technique for rooting chestnut cuttings. Ann. Rep. Northern Nut Growers Assoc. 52: 37–39.

Jaynes, R., 1962. Chestnut chromosomes. For. Sci. 8(4): 372–377.

Jaynes, R., 1965. Nurse seed grafts of chestnut species and hybrids. Proc. Amer. Soc. Hort. Sci. 86: 178–182.

Jaynes, R. and G.A. Messner, 1967. Four years of nut grafting chestnut. Int. Plant Prop. Soc. Comb. Proc. 17: 305–310.

Kamada, H. and H. Harada, 1981. Changes in the endogenous levels and effects of abscisic acid during somatic embryogenesis in *Daucus carota* L. Plant Cell Physiol. 22: 1423–1429.

Kermode, A.R. and J.D. Bewley, 1985. The role of maturation drying in the transition from seed development to germination. J. Exp. Bot. 36: 1906–1915.

Keys, R.N. and F.C. Cech, 1981. Plantlet formation in American chestnut embryonic tissue *in vitro*. Proc. 2nd North Central Tree Imp. Conf., pp. 189–194.

Keys, R.N. and F.C. Cech, 1982. Propagation of American chestnut *in vitro*. In: H.C. Smith and W.L. MacDonald (Eds.), Proc. 2nd Meeting USDA Forest Service American Chestnut Cooperators, Morgantown, WV, Jan 5–7, pp. 106–110. USDA Forest Service and College of Agriculture and Forestry, West Virginia University.

Konar, R.N., E. Thomas and H.E. Street, 1972. The diversity of morphogenesis in suspension cultures of *Atropa belladonna* L. Ann. Bot. 36: 249–258.

Krogstrup, P., E.N. Eriksen, J.D. Moller and H. Roulund, 1988. Somatic embryogenesis in sitka spruce (*Picea sitchensis* (Bong.) Carr.). Plant Cell Rep. 7: 594–597.

Kyo, M. 1990. Effects of EDTA and acidified medium on the differentiation of immature pollen in a tobacco pollen culture. Plant Cell Physiol. 31(8): 1249–1251.

Lagerstedt, H.B., 1981. A new device for hot-callusing graft unions. HortSci. 16: 529–530.

Lloyd, G. and B.H. McCown, 1980. Commercially feasible micropropagation of mountain laurel, *Kalmia latifolia*, by use of shoot-tip culture. Int. Plant. Prop. Soc. Comb. Proc. 30: 421–428.

Lutz, J.D., J.R. Wong, J. Rowe, D.M. Tricoli and R.H. Lawrence Jr., 1985. Somatic embryogenesis for mass cloning of crops plants. In: R.R. Hencke, K.W. Hughes, M.J. Constantin and A. Hollaender (Eds.), Tissue Culture in Forestry and Agriculture, pp. 105–116. Plenum Publishing Corp., New York.

Maheswaran, G. and E.G. Williams, 1985. Origin and development of somatic embryos formed directly on immature embryos of *Trifolium repens in vitro*. Ann. Bot. 56: 619–630.

McPheeters, K., R.M. Skirvin and C.A. Bly-Monnen, 1980. Culture of chestnut (*Castanea* spp.) *in vitro*. HortSci. 15: 417–418 (Abstr.).

McWilliam, A., S.M. Smith and H.E. Street, 1974. The origin and development of embryoids in suspension cultures of carrot (*Daucus carota*). Ann. Bot. 38: 243–250.

Merckle, S.A., D.T. Carraway and B.A. Watson-Pauley, 1992. Somatic embryogenesis and gene transfer in American chestnut. International Chestnut Conference, Morgantown, WV, July 10–14, p. 8 (Abstr.).

Merckle, S.A., A.T. Wietcko and B.A. Watson-Pauley, 1991. Somatic embryogenesis in American chestnut. Can. J. For. Res. 21: 1698–1701.

Merckle, S.A., H.D. Wilde and H.E. Sommer, 1993. *In vitro* culture of *Liriodendron tulipifera*. In: M.R. Ahuja (Ed.), Micropropagation of Woody Plants, pp. 281–302. Kluwer Academic Publishers, Dordrecht.

Michaux-Ferrière, N., H. Grout and M.P. Carron, 1992. Origin and ontogenesis of somatic embryos in *Hevea brasiliensis* (Euphorbiaceae). Am. J. Bot. 79(2): 174–180.

Misoo, S. and M. Matsubayashi, 1977. The effect of $Na_2$-EDTA on the induction of green embryoids in tobacco anther culture. Jpn. J. Breed. 2: 84–85.

Misoo, S. and M. Matsubayashi, 1978. Interaction of $Na_2$-EDTA and mineral salts in their effects on green embryoid formation in tobacco anther culture. Jpn. J. Breed. 28(1): 116–117.

Moore, J.C. 1963. Propagation of chestnuts and *Camellia* by nurse seed grafts. Int. Plant Prop. Soc. Comb. Proc. 13: 141–143.

Murashige, S. and F. Skoog, 1962. A revised medium for rapid growth and bioassay with tobacco tissue cultures. Physiol. Plant 15: 473–497.

Nienstadt, H. and A.H. Graves, 1955. Blight resistant chestnut culture and care. Conn. Agr. Exp. Sta. Circ. No. 192.

Nomura, K. and A. Komamine, 1985. Identification and isolation of single cells that produce somatic embryos at high frequency in a carrot suspension culture. Plant Physiol. 79: 988–991.

Nomura, K. and A. Komamine, 1986. Polarized DNA synthesis and cell division in cell clusters during somatic embryogenesis from single carrot cells. New Phytol. 104: 25–32.

Nuti Ronchi, V., L. Giorgetti, M. Tonelli and G. Martini, 1992a. Ploidy reduction and genome segregation in carrot cell lines. I. Prophase chromosome reduction. Plant Cell Tiss. Org. Cult. 30: 107–114.

Nuti Ronchi, V., L. Giorgetti, M. Tonelli and G. Martini, 1992b. Ploidy reduction and genome segregation in cultured carrot cell lines. Plant Cell Tiss. Org. Cult. 30: 115–119.

Park, K.S., 1967. New method of juvenile tissue grafting on some special-use trees. I. Studies on juvenile tissue grafting on some crop-tree species (walnut, chestnut, ginkgo and oak). Res. Rep. Inst. For. Gen. Korea 5: 75–84.

Pease, R.W., 1953. Rooting chestnuts from softwood cuttings. Amer. Hort. 32: 259–264.

Philip, V.J. and H. Schraudolf, 1987. Origin and development of embryos in suspension cultures of rose apple (*Eugenia jambos* L.). Betr. Biol. Pflanzen. 62: 9–21.

Read, P.E., M.A. Hosier and Y. Quiguang, 1985. Tissue culture of chestnuts. Ann. Rep. Northern Nut Growers Assoc. 76: 142–145.

Reinert, J., Y.P.S. Bajaj and B. Zbell, 1977. Aspects of organization-organogenesis, embryogenesis, cytodifferentiation. In: H.E. Street (Ed.), Plant Tissue and Cell Culture, Vol. II, pp. 389–427. Blackwell, Oxford.

Rodriguez, R., 1982. Multiple shoot-bud formation and plantlet regeneration on *Castanea sativa* Mill. seeds in culture. Plant Cell Rep. 1: 161–164.

Rutter, P.A., 1987. Chestnut ecology and the developing orchading industry. In: M.S. Burnett and R.D. Wallace (Ed.), Chestnuts and Creating a Commercial Chestnut Industry. Proc. 2nd Northwest Chestnut Cong, pp. 33–48. Chestnut Growers Exchange, Portland, OR.

Saito, A. and Y. Ide, 1986. *In vitro* plantlet regeneration from adventitious buds induced by petiole culture in Japanese white birch. J. Jap. For. Soc. 67(9): 373–375.

Sanchez, M.C., 1991. Propagación *in vitro* de material adulto de castaño y roble: Desarrollo de metodos de rejuvenecimiento. Doctoral Thesis, Univ. Santiago de Compostela.

Sanchez, M.C. and A.M. Vieitez, 1991. *In vitro* morphogenetic competence of basal sprouts and crown branches of mature chestnut. Tree Physiol. 8: 59–70.

San-José, M.C., 1983. Organogenesis y regeneración de plantas de castaño mediante cultivo *in vitro*. Doctoral Thesis, Univ. Santiago de Compostela.

San-José, M.C., A.M. Vieitez and E. Vieitez, 1984. *In vitro* plantlet regeneration from adventitious buds of chestnut. J. Hort. Sci. 59(3): 359–365.

Sato, T., Y. Ide and A. Saito, 1986. Tissue culture technology in rapid clonal propagation of Japanese white birch. J. Jap. For. Soc. 68(8): 343–346.

Schad, C., G. Solygnat, J. Grente and P. Venot, 1952. Recherches sur le chataignier à la station de Brive. Ann. Amèl. Plantes 2: 369–458.

Shafer, J., 1966. A chestnut challenge. Ann. Rep. Northern Nut Growers Assoc. 57: 32–35.

Skirvin, R.M., 1981. Fruit crops. In: B.V. Conger (Ed.), Cloning Agricultural Plants via *In Vitro* Techniques, pp. 51–139. CRS Press, Boca Raton, FL.

Smith, S.M. and H.E. Street, 1974. The decline of embryogenic potential as callus and cell suspension cultures of carrot (*Daucus carota* L.) are serially subcultured. Ann. Bot. 38: 223–241.

Solygnat, G., 1964. Rooting chestnut trees. Ann. Rep. Northern Nut Growers Assoc. 55: 33–36.

Thompson, R.G. and P. von Aderkas, 1992. Somatic embryogenesis and plant regeneration from immature embryos of western larch. Plant Cell Rep. 11: 379–385.

Urquijo, P., 1952. Multiplicación asexual de castaños. An. Inst. Nac. Invest. Agron. 1: 317–323.

Vasil, V. and I. Vasil, 1981. Somatic embryogenesis and plant regeneration from suspension cultures of pear millet (*Pennisetum americanum*). Ann. Bot. 47: 669–678.

Vieitez, A.M., A. Ballester, M.L. Vieitez and E. Vieitez, 1983. *In vitro* plantlet regeneration of mature chestnut. J. Hort. Sci. 58: 457–463.

Vieitez, A.M. and E. Vieitez, 1980a. Plantlet formation from embryogenic tissue of chestnut grown *in vitro*. Physiol. Plant 50: 127–130.
Vieitez, A.M. and M.L. Vieitez, 1980b. Culture of chestnut shoots from buds *in vitro*. J. Hort. Sci. 55: 83–84.
Vieitez, A.M. and M.L. Vieitez, 1982/83. *Castanea sativa* plantlets proliferated from axillary buds cultivated *in vitro*. Sci. Hort. 18: 343–351.
Vieitez, A., M.L. Vieitez and A. Ballester, 1981. *In vitro* chestnut regeneration, anatomical and chemical changes during the rooting proccess. In: AFOCEL (Ed.), Proc. Coll. Int. Cult. In Vitro Essences For., pp. 148–152. IUFRO, Fontaineblau.
Vieitez, A.M., M.L. Vieitez and E. Vieitez, 1986. Chestnut (*Castanea* spp.). In: Y.P.S. Bajaj (Ed.), Biotechnology in Agriculture and Forestry, Vol. 1, Trees I, pp. 391–314. Springer-Verlag, Berlin.
Vieitez, E., 1952. Ensayos de reproducción vegetativa de híbridos de castaño *C. sativa* × *C. crenata*. An. Edafol. Fisiol. Veg. 11: 185–209.
Vieitez, E., 1953. Estudios sobre la reproducción vegetativa del castaño. I. Enraizamiento en el acodo alto mediante el empleo de fitohormonas. An. Edafol. Fisiol. Veg. 12(4): 337–373.
Vieitez, E., 1955. El empleo de sustancias de acción hormonal en el enraizamiento del castaño por acodo alto. An. Edafol. Fisiol. Veg. 14(9/10): 483–518.
Vieitez, E., 1956. Problemas que plantea el estaquillado del castaño. An. Edafol. Fisiol. Veg. 15: 629–659.
Vieitez, E., 1963. Enraizamiento de estaquillas juveniles de castaño. Montes 12: 331–333.
Vieitez, E., 1974. Vegetative propagation of chestnut. N.Z. J. For. Sci. 4(2): 242–252.
Vieitez, F.J., M.C. San-José, A. Ballester and A.M. Vieitez, 1990. Somatic embryogenesis in cultured immature zygotic embryos in chestnut. J. Plant Physiol. 136: 253–256.
Vieitez, F.J., A. Ballester and A. M. Vieitez, 1992a. Regeneration of chestnut via somatic embryogenesis. International Chestnut Conference, Morgantown, WV, July 10–14, p. 11 (Abstr.).
Vieitez, F.J., A. Ballester and A.M. Vieitez, 1992b. Somatic embryogenesis and plantlet regeneration from cell suspension cultures of *Fagus sylvatica* L. Plant Cell Rep. 11: 609–613.
Vieitez M.L. and A.M. Vieitez, 1981a. Injerto sobre castañas germinadas. An. Edafol. Agrobiol. 40: 317–326.
Vieitez, M.L. and A.M. Vieitez, 1981b. Injerto en hipocotilo de plántulas de castaño. An. Edafol. Agrobiol. 40: 647–655.
Von Arnold, S. and I. Hakman, 1988. Regulation of somatic embryo development in *Picea abies* by abscisic acid (ABA). J. Plant Physiol. 132: 164–169.
Wallace, R.D., 1987. History of breeding and production of hybrid chestnut cultivars. In: M.S. Burnett and R.D. Wallace (Eds.), Chestnuts and Creating a Commercial Chestnut Industry. Proc. 2nd Northwest Chestnut Congr., pp. 24–29. Chestnut Growers Exchange, Portland, OR.
Wann, S.R., 1989. Somatic embryogenesis in woody plants. In: J. Janick (Ed.), Horticultural Reviews, Vol. 10, pp. 153–181. Timber Press, Portland, OR.
Williams, E.G. and G. Maheswaran, 1986. Somatic embryogenesis: Factors influencing coordinated behaviour of cells as an embryogenic group. Ann. Bot. 57: 443–462.

# 22. Somatic embryogenesis in horse chestnut (*Aesculus hippocastanum* L.)

Ljiljana Radojević

**Contents**



## 1. Introduction

### 1.1. *Distinguishing features*

Horse chestnut (*A. hippocastanum* L., 2n = 40) forest tree belongs to the genus *Aesculus* and family *Hippocastanaceae*. It can grow as a solitary tree, attains a height up to 25 m, and the diameter of the trunk can reach up to 1 m (Fig. 1). According to Tutin *et al.* (1968) its main morphological characteristics are as follows: the trunk is covered with thin, dark brown, smooth bark which becomes darker and regularly fissured into rectangular scales during ageing. Young twigs are thick, brownish and slender with ovoid leaf scars. The buds are large and glabrous, covered with numerous viscid scales (Fig. 3a). Cylindrical panicles (15–30 cm long) develop from 2.5 cm long apical buds. The large, dark-green leaves are palmate, with 5–7 cm long leaflets which are obovate, cuneate, usually irregularly accumulated, crenate-

*S. Jain, P. Gupta & R. Newton (eds.), Somatic Embryogenesis in Woody Plants, Vol. 2*, 409–444.

*Figure 1. Aesculus hipocastanum* L. growing in "Jevremovac" Botanical Garden, Belgrade. The tree is 100 years old and 30 m high (Photo M. Radojević, March, 1993).
*Figure 2.* Ripe horse chestnut seeds.

serrate, glabrous and tomentose above, and glabrous beneath (Fig. 3b,$b_1$). Zygomorphic bisexual flowers on large, terminal, erect panicles appear from April till May. The calyx encloses five white-yellow petals (size up to 1 cm) with a pink spot at the base (Fig. 3c,$c_1$). Orange-yellow, hairy stamens surround the style. The ovary is superior, consisting of three fused carpels and three locules (seldom two or one by abortion), each locule containing two ovules (Fig. 3d). Some flowers of the inflorescence are sterile, i.e., only the flowers situated near the base get pollinated to develop green, and spiny burs (Fig. 3e). The burs (about 5 cm long) usually contain two (2–4 cm long) brown shiny nuts with a large white hilum (Fig. 3$e_1$), which continue to ripen till August or September. Horse chestnut achieves maturity relatively rapidly and forms fertile nuts after 10 to 15 years. The tree grows rapidly and

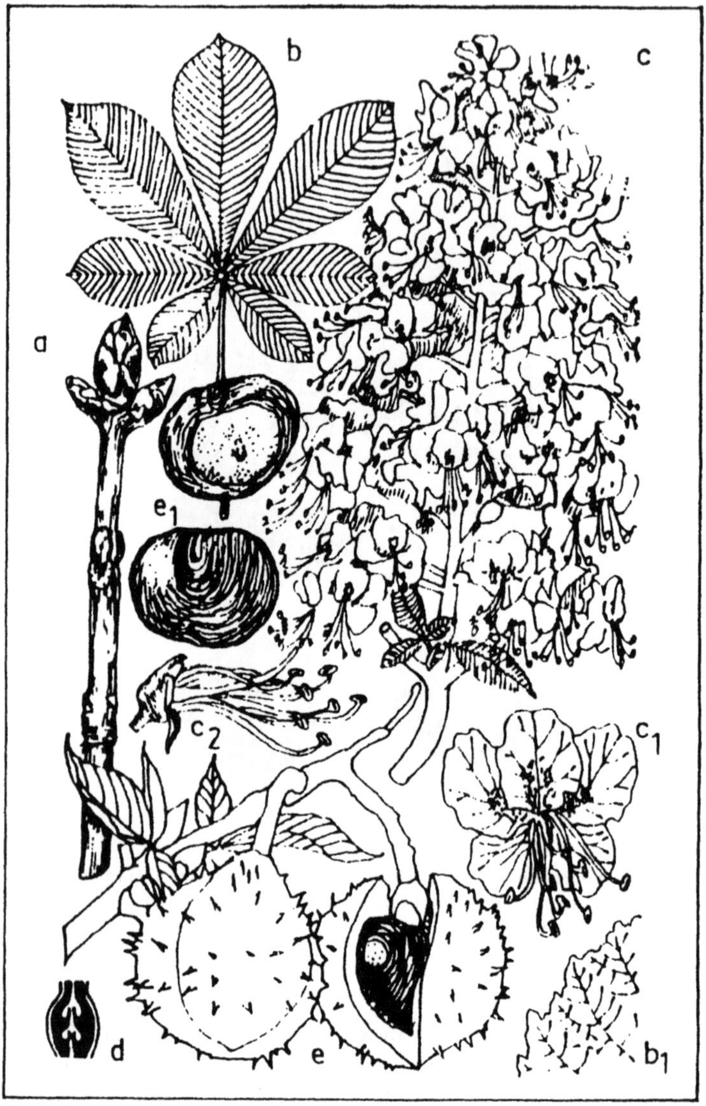


*Figure 3.* Morphology of vegetative and generative organs in *A. hippocastanum* L. (a) A winter twig. (b-$b_1$) Palmate leaf. (c) Inflorescence. ($c_1$) Flowers in native size. ($c_2$) Calyx with stamens. (d) vertical section of the ovary. (e) Twig with dehiscing bur. ($e_1$) Seed with large hilum (Radojević, 1977; drawing M. Radojević).

develops an ovoid and ramified crown. The trunk is relatively short (Fig. 1), and the root system is extensively branched and shallow.

### 1.2. *Distribution and cultivation area*

Horse chestnut represents a relict species of the tertiary flora. Košanin (1926) found horse chestnut in south-west Macedonia, Galičica Mt., gorges of Crni Drim river, and at the mountain foot of the Bistra. This tree grows in natural habitats together with *Acer platanoides*, *Alnus glutinosa*, *Fraxinus oxycarpa*, *Juglans regia*, and *Platanus orientalis*. Em (1959) found horse chestnut at several spots in south-west Macedonia. Its present range of distribution was confirmed only on Balkan peninsula in south-east Albania, Greece (Epirus), north-east Bulgaria and south-west Macedonia (Markgraf, 1932; Procter, 1977). On the Greek mountains, it appears in the lower zone of *Abies cephalonica*, altitude 1000–1300 m above sea level, whereas in Bulgaria horse chestnut grows in coastal forests and ravines of the eastern Balkans. Its range of distribution is depicted in Fig. 4 (Stevanović and Radojević, 1993).

Horse chestnut represents a tree of humid forests, sandy river alluvia and humus clay being the best substrates in its natural habitats. Moderately warm climate is suitable for its growth, but it can very well be grown successfully in the zones of cool temperate climate.

### 1.3. *Diseases*

Numerous fungal species developing on *A. hippocastanum* have been identified so far. Both the visible changes and the plant damage upon infestation depend on the fungal species. As early as 1929, Seymour identified some thirty genera of the fungi, mainly representatives of micromycetes attacking *A. hippocastanum*. We will list here the most frequently recorded fungal diseases of this tree.

*Guignardia aesculi* (Pk.) V.B. Stewart and its anamorphic stage *Phyllosticta sphaeropsoidea* Ell. and Ev. causes brown spots on the chestnut leaves. This disease appears most frequently on the transplants in the nurseries and more seldom in the forests and parks.

*Mycrosphaerella maculiformis* (Pers.) Schraet with its anamorphic stages *Cylindrosporium castaneocolum* (Desm.) Berl. and *Phyllosticta maculiformis* Sacc. causes the leaf spots which appear as minute polygonal gray spots bordered by dark lines. This fungus is very harmful to the horse chestnut and humid summers support its development. As a result of the infestation, premature leaves fall.

*Mycrosphaerella hyppocastani* (Saap.) Kleb, causes tiny pink spots in the beginning and later become brown with a gray spot in the central part. Heavy infestation with this fungus causes wilting of the leaves which eventually fall down during the spring. In the gray part of the spot, anamorphic stage of *Septoria aesculicola* (Fr.) Sacc. develops.

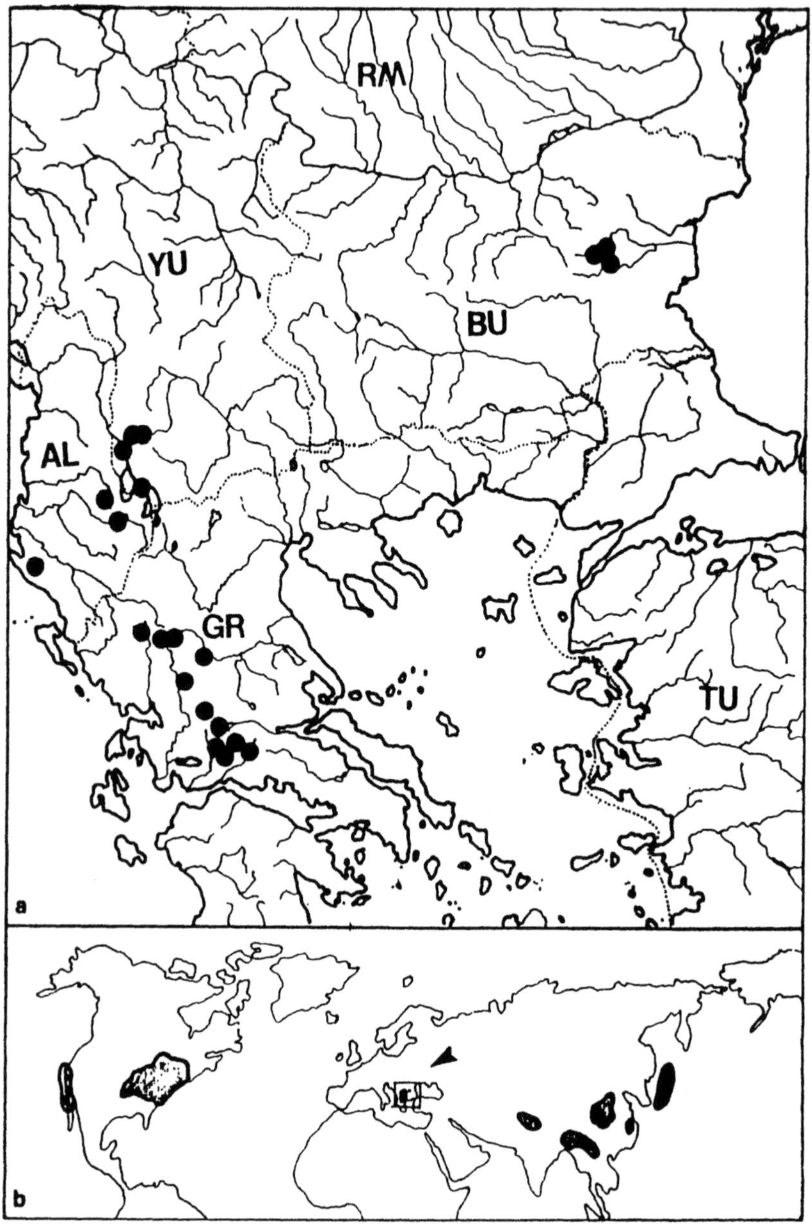


*Figure 4.* (a) General range of distribution of *Aesculus hippocastnum* on Balkan peninsula (Stevanović and Radojević, 1993). (b) Disjunct distribution of genus *Aesculus* according to Tachtadzhyana (1980). The part of the Balkan peninsula presented on Map a is designated by arrow.

*Nectria cinnabarina* (Tode) Fr. and its anamorph *Tubercularia vulgaris* Tode ex Fr., a wound parasite, attacks branches and trunk of the tree. Severe infestation leads to the death of the trees.

*Collybia velutipes* Curt. causes soft, porous yellowish-white putrefaction of both vegetative plants and cut down trees.

*Fomes applanatus* (P. ex Wallr.) Gill., *Ganoderma applanatum* (P. ex. Wallr.) Pat. and *Laetiporus sulphureus* Fr. often cause the putrefaction of older trees. They attack the core but do not prevent tree growth (Tucović, 1973).

### 1.4. *Importance*

Horse chestnut can be grown in the warmer habitats of the beech (*Fagetum montanum*) forests. Because of its beautiful and dense crown and inflorescence, both *A. hippocastanum* and its hybrid *A. carnea* Hayne (*A. hippocastanum* L. × *A. pavia* L.) are widely used as ornamental trees. Great variability with regard to the shape of the crown, appearance and size of the leaves, and flower colour makes horse chestnut highly suitable for producing new ornamental types such as f. *baumanni* Schned., f. *laciniata* Jacq., f. *luteovariegata* West., f. *pendula* Pavill., etc. Since horse chestnut is resistant to different pollutants urban environmental pollutants, it is often planted as a solitary tree in the parks and avenues.

Horse chestnut wood, so-called "blindholz", is not durable and it is used in carpentry and woodcarving (Richardson, 1978).

### 1.5. *Conventional practices for the propagation*

Both *A. hippocastanum* and its hybrid *A. carnea* Hayne can be propagated *via* seeds (Fig. 2) which should be slightly dried and seeded immediately (50 nuts/$m^2$) or exposed to stratification till spring because of the embryo dormancy.

However, this tree species can also be propagated vegetatively and cover sprouts with soil in the spring or in the autumn, using root cuttings in the autumn or by grafting. In order to obtain the hybrids, suitable for horticulture (e.g., *A. carnea* Hayne and *A. glabra*), top or root graftings are applied usually by occulation. For root grafting *A. hippocastanum*, *A. carnea* and *A. octandra* grafts are used, whereas for top grafting of *A. pavia*, *A. carnea* grafts are more suitable due to devoid of callosities formation at the graft site.

### 1.6. In vitro *propagation*

In order to improve and/or to preserve elite characteristics of woody plant species, different biological programmes including tissue culture, molecular genetics and genetic engineering have been employed (Cheliak and Rogers,

1990). Various aspects of tree improvement with biotechnologies have been discussed at length in several review articles (Sommer and Brown, 1979; Radojević and Kovoor, 1985; Bajaj, 1986; Gunn and Day, 1986; Hasnain and Cheliak, 1986; James, 1987; Cheliak and Rogers, 1990; Radojević, 1991). Their common theme is to increase the number of regenerants. Only Hasnain *et al.* (1989) discussed the economic aspects of tissue culture applications for rapid forest improvement. The data on the characterization of DNA from woody species, and isolation and characterization of individual genes are still scarce (Cheliak and Rogers, 1990). Only recently, extensive research has been conducted to improve trees with genetic engineering technologies (Dandekar *et al.*, 1987; Mackay *et al.*, 1988; McGranahan *et al.*, 1988; Vahala *et al.*, 1989; Vardi *et al.*, 1990). However, genetic manipulation of trees through *in vitro* cultures of protoplasts, cells, different plant tissues and organs has a vast potential to play an important role in forest and orchard biotechnology programmes for the production of desirable woody plant species.

In order to improve classical vegetative propagation of *A. hippocastanum*, several different methods have been used such as: 1) anther culture for haploid plant production (Radojević, 1978); 2) leaf explant (Dameri *et al.*, 1986), filament (Jörgensen, 1989; 1991a), and zygotic embryo cultures (Radojević, 1988; Profumo *et al.*, 1991a) for the production of somatic embryos; and 3) shoot and meristem cultures for rapid micropropagation (Radojević *et al.*, 1987; Radojević, in preparation). The procedures 1 and 2 can give gametoclonal and somaclonal variants which could be applied in breeding programmes for the horse chestnut and/or for aescin production (Profumo *et al.*, 1987a, 1991a, 1992).

## 2. Culture initiation and maintenance

Different explants were excised from *in vitro*-grown *A. hippocastanum* plants and cultured to produce friable or compact calli to fully differentiated plantlets capable of growing in the field conditions. The results of several authors were applied not only to various explants of this species but also to the culture media of different composition containing hormones and other plant growth regulators (see Table 1).

### 2.1. *Cotyledon culture*

The first callus cultures of the horse chestnut were established using either different parts of cotyledons devoid of embryogenic axis or explants isolated from near the embryogenic axis. The explants (500–600 mg f.w.) taken from ripe seeds, which are closer to the embryogenic axis, provided a better proliferation and then they were transferred into the stationary liquid medium. Mineral solution supplemented with macro-inorganic nutrients after Bertossi (1960) and micro-inorganic Heller's nutrients after Gautheret

*Table 1. A. hippocastanum* explants used for calli, somatic embryos and plantlet production by *in vitro* culture in different media.

| Inoculum | Medium (mg $L^{-1}$) | Growth response | Reference |
|---|---|---|---|
| Cotyledon explants | MS** + 2,4-D (0.5) or NAA (0.5) + Kin (0.5) + CM (20%) + YE (10.0–20.0) | Calli | Profumo *et al.*, 1976, 1980 |
| | MS** + NAA (0.1) + $GA_3$ (0.1) | Friable calli with solid areas, Roots | ″ |
| Cotyledon-embryonal axis | MS** + 2,4-D (2.0) + NAA (2.0) + Kin (2.0) | Embryogenic calli, Somatic embryos | Profumo *et al.*, 1989 |
| Cotyledon explants of ripe seeds | MS* or MS** + 2,4-D (1.0) | Embryogenic calli, Somatic embryos, Secondary somatic embryos | Profumo *et al.*, 1990 |
| ″ | MS* or MS** + 2,4-D (0.1–2.0) + IAA (0.1–1.0) + NAA (0.1–2.0) + $GA_3$ (0.1–1.0) | Embryogenic calli, Somatic embryos | ″ |
| ″ | MS* + 2,4-D (0.1) + Kin (0.1) | Somatic embryos | Profumo *et al.*, 1991a |
| Leaf explants | MS* + 2,4-D (2.0) + NAA (2.0) + Kin (2.0) | Non-embryogenic and embryogenic calli, Somatic embryos | Dameri *et al.*, 1986 |
| Immature embryos | MS* + 2,4-D (3.0) + Kin (1.0) + CH (250) + Pro (250) | Non-embryogenic, Embryogenic calli, Organogenic calli | Radojević, 1988 |
| ″ | MS* + 2,4-D (1.0) + Kin (1.0) + CH (250) + Pro (250) | Embryogenic calli, Somatic embryos | ″ |
| ″ | MS*** + IAA (1.0) + $GA_3$ (1.0) + Glu (400) | Plantlets | ″ |
| ″ | MS*** + IAA (1.0) + $GA_3$ | ″ | ″ |

| | | | |
|---|---|---|---|
| | (1.0) + AC (1%) + Glu (400) | | |
| '' | MS*** + Kin (3.0) + Glu (400) | Germinated embryos | '' |
| '' | MS*** + IAA (1.0) + $GA_3$ (1.0) + Kin (3.0) + Glu (400) | '' | '' |
| '' | MS*** + IBA (1.0) + $GA_3$ (1.0) + Kin (3.0) + Glu (400) | '' | '' |
| '' | MS*** + IAA (1.0) + $GA_3$ (1.0) + BA (3.0) + Glu (400) | '' | '' |
| '' | MS*** + IAA (1.0) + $GA_3$ (1.0) + Zeatin (3.0) + Glu (400) | '' | '' |
| Filaments | WPM + 2,4-D (1.10) + BA (0.56) | Non-embryogenic calli | Jörgensen, 1989, 1991a |
| '' | WPM + BAP (0.17) | Embryogenic calli, Somatic embryos | '' |
| Immature ovules | MS* + 2,4-D (1.0) + Kin (1.0) + CH (200) + Pro (200) | Embryogenic calli, Somatic embryos | Radojević and Marinković, 1993 |
| Meristem tips from somatic embryos | WPM + 2,4-D (0.001) + BA (1.0) + $GA_3$ (0.1) | Leaf rosettes | Radojević *et al.*, 1988 |
| '' | WPM* + BA (1.0) + $GA_3$ (0.1) | Shoot multiplication | '' |
| '' | WPM + IBA (2.0) + vitamin mixture (Walkey, 1972) | Roots, Plantlets | '' |

MS – Murashige and Skoog basal medium (1962); MS* – agarized MS; MS** – stationary MS liquid medium; MS*** – liquid medium on a horizontal shaker; WPM – medium after Lloyd and McCown (1981); WPM* – "double-phase" (1/2 solid + 1/2 liquid) medium after Lloyd and McCown (1981). Ingredients: 2,4-D – dichlorophenoxyacetic acid; IAA – indoleacetic acid; IBA – indole-3-butyric acid; NAA – naphtaleneacetic acid; $GA_3$ – gibberellic acid; BA – 6-benzylaminopurine; Kin – 6-furfurylaminopurine; AC – activated charcoal; CH – casein hydrolysate; CM – coconut milk; Glu – glutamine; Pro – L-proline; YE – yeast extract.

(1959), or those after Murashige and Skoog (1962) with sucrose or glucose (5%), coconut milk (20%) and (in mg $L^{-1}$): $B_1$ (1.25), $B_2$ (0.375), $B_5$ (1.25), $B_6$ (0.375), $B_{12}$ (0.003) and PP (12.5) vitamins, cysteine (10.0), yeast extract (10.0–20.0), 2,4–D (0.5), NAA (0.5) or $GA_3$ (0.1) appeared insufficient for a proper proliferation of calli and rhizogenesis. Addition of NAA and $GA_3$ (0.1 mg $L^{-1}$, each) induced the friable calli with roots (Profumo *et al.*, 1976). Auxins (2,4-D or NAA) enhanced the growth of the newly-formed callus tissue, while $GA_3$ completely inhibited proliferation. The lowering of cell multiplication rate and the use of yeast extract indicated the formation of less exuberant calli (Profumo *et al.*, 1980).

Profumo *et al.* (1989, 1990) established embryogenic callus in cotyledonary explants. They failed to induce embryogenesis in cultured explants excised from ripe seeds without embryogenic axes on MS (Murashige and Skoog, 1962) nutrient liquid and agarized media supplemented with IAA, NAA, Kin and $GA_3$. When the media were supplemented with low levels (0.1–1.0 mg $L^{-1}$) of Kin, NAA and 2,4-D (in combination) or with 2,4-D alone, somatic embryogenesis was induced. This led to a conclusion that cotyledonary explants can produce embryogenic callus (EC) only if both liquid and solid media were supplemented with plant growth regulators. It is of interest to underline that MS liquid medium containing Kin, NAA and 2,4-D (2 mg $L^{-1}$, each) reduced callogenesis. Auxins (IAA, NAA), Kin or $GA_3$ applied in varying concentrations (0.1, 0.5 or 1.0 mg $L^{-1}$) as well as hormone-free control MS liquid or solid medium did not induce somatic embryogenesis. Embryogenic calli with numerous well-developed somatic embryos were formed in MS solid medium supplemented with 2,4-D (0.1–1.0 mg $L^{-1}$). Somatic embryogenesis reached 100% when embryogenic calli were subcultured in fresh control medium. In MS liquid medium, the embryogenic calli formed few somatic embryos, however, their number increased upon transfer to the fresh MS solid medium (Profumo *et al.*, 1990). These data suggest that combined action of auxins and cytokinins is not indispensable for somatic embryogenesis induction in cotyledonary explants, contrary to the results of Dameri *et al.* (1986) who used *A. hippocastanum* leaf explants and assumed that the balance of 2,4-D and kinetin is necessary for the induction of somatic embryogenesis. However, subculture of these initial explants in MS liquid or agarized media supplemented with 1.0–2.0 mg $L^{-1}$ 2,4-D for 1–3 weeks did not result in the formation of either embryogenic calli or somatic embryos and only tissue necrosis occurred (R.M. Dameri, personal communication). These findings could mean that chemical composition of the cotyledons is primarily connected with their storage function and thus the addition of exogenous auxins is absolutely required for the induction of embryogenesis. Our previous data have demonstrated that exogenously applied auxins are necessary for somatic embryogenesis induction in the cultures of horse chestnut immature embryos (Radojević, 1988).

## 2.2. *Leaf culture*

Leaf explants isolated from the primary leaves of *in vitro*-grown horse chestnut seedlings (seeds germinated in MS agar medium devoid of hormones) were cultured in MS medium containing KND mixture of growth hormones (Kin, NAA and 2,4-D, 2 mg $L^{-1}$, each). The explants swelled for ten days and then formed callus tissue after 1–2 months of culture. This callus was isolated and subcultured in MS liquid medium supplemented with the same additives as in the initial medium for 25 days. After that, callus was transferred to a solid MS medium containing KND and that resulted in the formation of whitish and friable callus incapable of somatic embryos formation. This callus was designated as non-embryogenic (NE) (Dameri *et al.*, 1986). Isolation of the initial callus and subculture in MS solid medium with KND led to the formation of white compact callus ($E_1$) which sometimes had yellow friable regions where somatic embryos were formed. Such a callus was designated as $E_2$ embryogenic callus (Dameri *et al.*, 1986). The same authors described somatic embryos as the structures morphologically similar to androgenic embryos (Radojević, 1978) and zygotic embryos (Radojević, 1988) of *A. hippocastanum*. Both leaf explants and $E_2$ callus formed somatic embryos with the same frequency after 60 days of subculture in MS medium containing KND. The capacity of embryogenic callus to produce embryos was directly dependent on hormones and it did not reduce even after repeated subcultures. Cell suspension cultures did not produce embryogenic calli in MS solid medium containing KND. In this case, shaking of calli in the liquid medium was inhibitory, although it was essential for somatic embryogenesis of *Cariaca candamarcensis* (Jordan, 1986), *Quercus serrata* (Sasamoto and Hosoi, 1989) and several other herbaceous plant species (Steward *et al.*, 1958; Evans *et al.*, 1984).

According to Dameri *et al.* (1986), cultures of *A. hippocastanum* leaf explants produced two types of embryogenic calli: $E_1$ (precursor of $E_2$), yellow and friable embryogenic callus, and a non-embryogenic callus.

## 2.3. *Embryo culture*

Somatic embryogenesis and plant formation of *A. hippocastanum* were achieved in immature (2–3 cm long) embryos on MS medium containing 2,4-D (3.0 mg $L^{-1}$) and Kin (1.0 mg $L^{-1}$) with or without proline (Radojević, 1988) (Fig. 5). Initially, only whitish compact callus was formed in the medium devoid of proline. Its growth was limited due to necrosis after six months of culture. In MS medium supplemented with proline and casein hydrolysate (250 mg $L^{-1}$, each), three types of calli were simultaneously formed: a) whitish and compact non-embryogenic callus (NEC), b) globular translucent embryogenic callus (EC), and c) compact and green organogenic callus (OC) (Plate 1: 1–3). After the first subculture, these calli were separated. Non-embryogenic callus and EC were cultured in the above mentioned

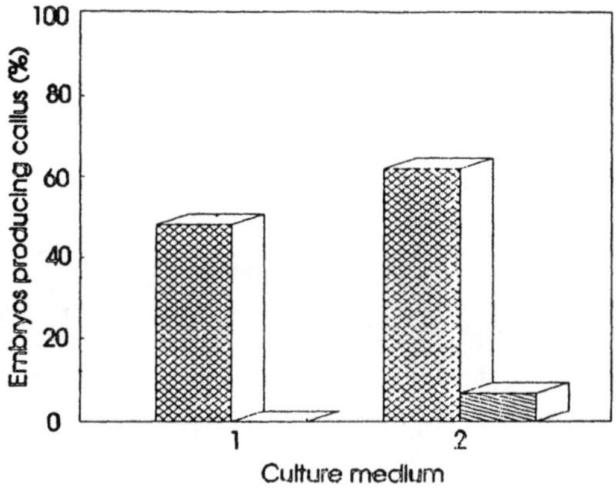


*Figure 5.* The effect of medium composition on embryogenic callus induction in culture of *A. hippocastanum* immature embryos. Media (ingredients in mg $L^{-1}$): (1) MS + 2,4-D (3) + Kin (1) + CH (250) and (2) MS + 2,4-D (3) + Kin (1) + L-proline (250) + CH (250). Fifty and 100 embryos were cultured in medium devoid of proline (1) and supplemented with proline (2), respectively. Percentage of embryos producing non-embryogenic callus (diamonds). Percentage of embryos producing embryogenic callus (bars).

medium, whereas for OC MS medium amended with BA and $GA_3$ (1.0 and 0.1 mg $L^{-1}$, respectively) was used.

In some regions of EC, somatic embryos were developed (Plate 1: 5). After the third passage, EC were transferred to MS medium supplemented with 2,4-D (1.0 mg $L^{-1}$) and that led to an increase in the callus mass and a number of SE appeared at the callus surface. Between 70 to 100 somatic embryos with different developmental stages including globular (Plate 1: 4) to cotyledonary stage (Plate 1: 6,7) were formed per culture.

In addition to embryos with normal morphology, a few irregularly shaped embryos developed, probably either due to high 2,4-D and proline levels in the medium (Radojević, 1985) or only due to high 2,4-D concentration. These results are similar to the observations with leaf segment (Dameri *et al.*, 1986) and filament (Jörgensen, 1989) cultures of horse chestnut.

Multiplication of the shoots originating from the OC type was achieved with the shoot multiplication procedure originating from meristem of horse chestnut androgenic embryos on WPM "double phase" medium (semi solid) amended with 1.0 mg $L^{-1}$ BA and 0.1 mg $L^{-1}$ $GA_3$ (Radojević *et al.*, 1987).

Auxin is an indispensable factor for somatic embryogenesis induction, as demonstrated earlier for callus originating either from leaf explants (Dameri *et al.*, 1986) or from immature embryo cultures of *A. hippocastnum* (Radojević, 1988), as well as most of the fruit and ornamental tree species (Radojević, 1980) studied so far. Recently, induction of OC together with that of

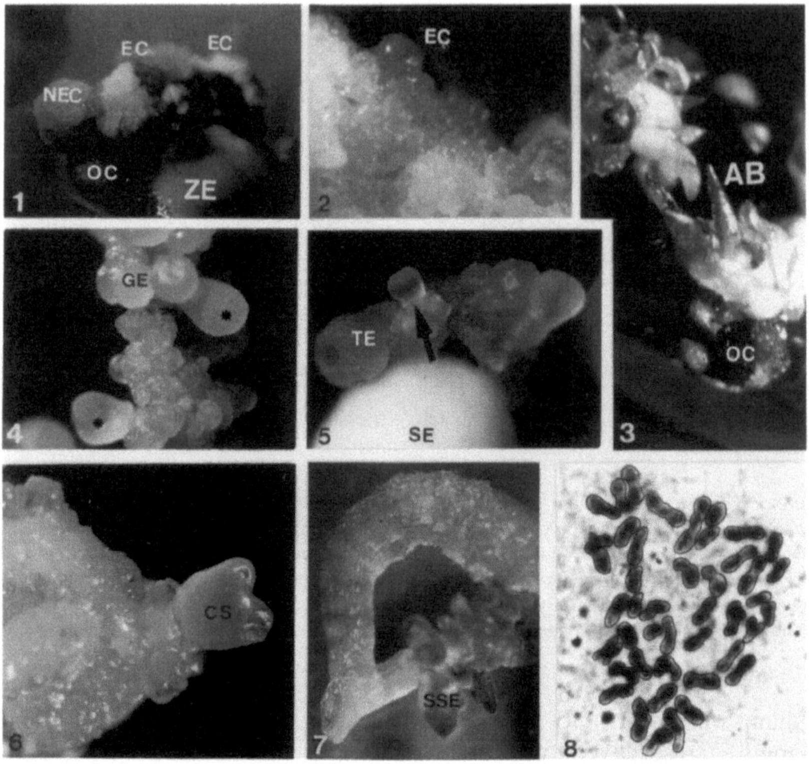


*Plate 1.* Somatic embryogenesis in the culture of immature zygotic *A. hippocastanum* L. embryos (Radojević, 1988). (1) Zygotic horse chestnut embryo (ZE) in MS medium enriched with (in mg $L^{-1}$): 2,4-D (3), Kin (1), Pro (250) and CH (250), which formed non-embryogenic (NEC), embryogenic (EC) and organogenic calli (OC); ×5.5. (2) A detail of embryogenic (EC), and (3) organogenic calli (OC) with adventive buds (AB); × 7. (4) Mass of globular embryos (GE) and torpedo-shaped embryos (asterisk; ×14). (5) Torpedo (TE) and heart-shaped embryoids (arrow) formed at the radicle of the somatic embryo (SE); ×14. (6) Somatic embryo in early cotyledonary stage (CS) and mass of globular embryos (GE); ×13. (7) Several secondary somatic embryos (SSE) formed on the cotyledons; ×14.5. (8) "Squash" preparation showing diploid chromosome number (2n = 40) of a regenerant; ×900.

EC was achieved for the first time (Radojević, in preparation). Simultaneous appearance and correlation of somatic embryogenesis and organogenesis was recently demonstrated in *Arachis hypogaea* cotyledon cultures (Ozias-Akins *et al.*, 1992). The number of shoots formed were highly correlated with each other in all examined genotypes, which were similar to the number of somatic embryos formed from cotyledons. They also confirmed that the maturity of the explants does not influence embryogenic response of the explant in the culture, although the formation of somatic embryos from older cotyledons was reduced.

The process of somatic embryogenesis in horse chestnut immature embryo cultures expressed a tendency of suppressing organogenesis (Radojević, in preparation) as demonstrated earlier in the cultures of *Quercus suber* (El Maâtaoui *et al.*, 1990), *Pisum sativum* (Tétu *et al.*, 1990) and *Iris setosa* (Radojević and Subotić, 1992).

Maturity of the zygotic embryos (Radojević, 1988) and cotyledons of the horse chestnut (Profumo *et al.*, 1990) did not affect somatic embryogenesis induction. However, opposite results were obtained with soybean cotyledon cultures which were competent for somatic embryogenesis only within a very short period of development (Ranch *et al.*, 1985; Sellars *et al.*, 1990).

### 2.4. *Filament culture*

Floral filaments collected from *A. hippocastanum* 10–100 year-old trees were cultured in WPM medium (Lloyd and McCown, 1981) supplemented with glutamine (1000 mg $L^{-1}$), glycine (50 mg $L^{-1}$), serine (50 mg $L^{-1}$), BA (0–0.5 μmol $L^{-1}$) and 2,4-D (0, 1.0, 2.5 or 5.0 μmol $L^{-1}$).

The filaments grown in WPM containing 2.5 μmol $L^{-1}$ of BA and 5.0 μmol $L^{-1}$ 2,4-D produced calli. Embryogenesis was induced by subculture of callus in the same medium devoid of 2,4-D. Decreasing levels of BA (0.75, 1.0 or 2.5 μmol $L^{-1}$) stimulated the number of embryo formation. In some cultures, callus turned yellow and became compact. Further reduction of BA concentration (below 0.75 μmol $L^{-1}$) resulted in the transformation of the whole callus into somatic embryos. In WPM medium supplemented with 0.75 μmol $L^{-1}$ BA, callus remained white and only a few somatic embryos were developed. According to Jörgensen (1989, 1991a), these somatic embryos could be cloned after addition of 2.5–5.0 μmol $L^{-1}$ BA to the culture medium similar to androgenic embryos of horse chestnut (Radojević, 1991).

### 2.5. *Immature ovule culture*

Immature ovules (1–2 mm long) of *A. hippocastanum* were isolated from the buds (4–5 mm long) excised from the basal, middle and apical parts of the inflorescence. They were cultured in MS solid medium amended with (in mg $L^{-1}$): 2,4-D (1.0), Kin (1.0), L-proline and casein hydrolysate (200 mg $L^{-1}$, each). After 2–3 weeks of culture, whitish compact callus was formed on individual ovules of all inflorescence segments. The highest number of calli was formed on the ovules taken from the basal flowers and only these calli produced somatic embryos (Table 2). Preliminary results showed that 21% ovules formed somatic embryos (Plate 2: 1,2). Immature horse chestnut ovules expressed a persistent embryogenic capacity when cultured in the presence of 2,4-D and kinetin. These data (Radojević and Marinković, 1993) resemble the results obtained during the studies of fertilized *Paulownia tomentosa* ovules cultured *in vitro* (Radojević, 1979).

*Table 2.* Influence of hormones and ovule origin on embryogenic callus induction in immature ovule cultures of *A. hippocastanum*.

| Medium | Inflorescence segment | No. of cultured ovules | No. of ovules forming NEC | Ovules forming EC with somatic embryos (SE) | |
|---|---|---|---|---|---|
| | | | | Number | % |
| | basal | 100 | – | – | – |
| 1 | middle | 100 | – | – | – |
| | apical | 100 | – | – | – |
| | basal | 191 | 11 | 40 | 21 |
| 2 | middle | 145 | 2 | – | – |
| | apical | 129 | 1 | – | – |

1. Control medium: MS + CH + L-proline (200 mg $L^{-1}$, each); 2. MS + 2,4-D + Kin (1 mg $L^{-1}$, each) + CH + L-proline (200 mg $L^{-1}$, each); Total inforescence length 22.0 cm; NEC – non-embryogenic callus; EC – embryogenic callus.

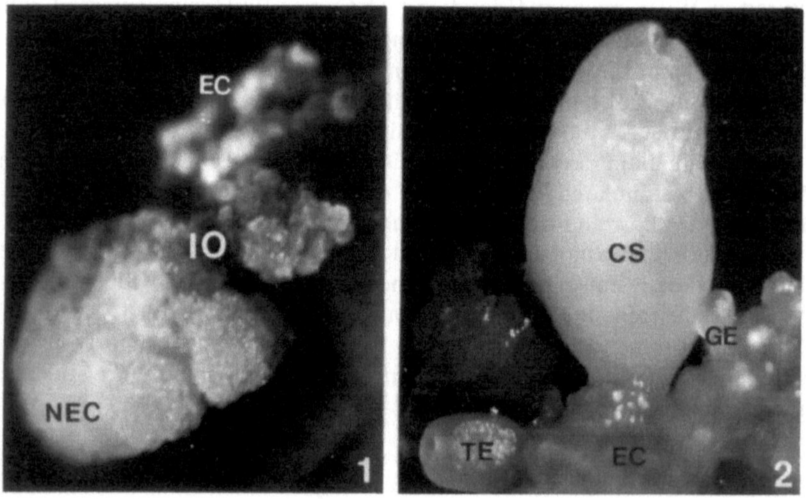


*Plate 2.* Somatic embryogenesis in the culture of immature ovules of *A. hippocastanum* (Radojević and Marinković, 1993). (1) Immature ovule (IO) after 2 weeks of culture in MS solid medium enriched with (in mg $L^{-1}$): 2,4-D (1), Kin (1), Pro and CH (200, each) formed non-embryogenic (NEC) and embryogenic calli (EC); ×6.5. (2) Embryogenic callus (EC) with somatic embryos in globular (GE), torpedo (TE) and early cotyledonary stage (CS); ×13.

## 2.6. *Meristem culture*

Somatic embryos originating from *A. hippocastanum* immature embryo cultures (Radojević, 1988) served as a suitable starting material for the isolation of meristem with two leaf primordia (0.2–0.3 cm long). Successive differentiation of the meristem into the leaf rosettes was achieved in WPM medium

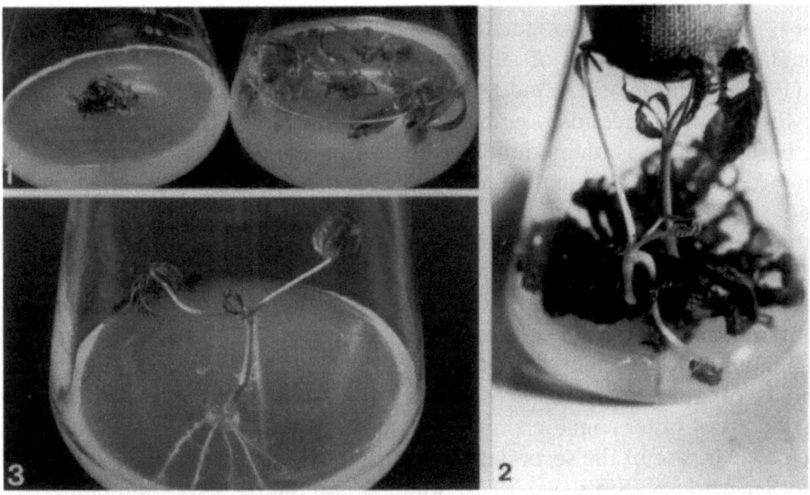


*Plate 3.* Meristem micropropagation from somatic *A. hippocastanum* embryos (Radojević, *et al.* 1988). (1) Branching and shoot multiplication: in WPM solid (left) and "double-phase" medium (right) containing (in mg $L^{-1}$): BA (1) and $GA_3$ (1). (2) Shoots during elongation phase in WPM "double-phase" medium with (in mg $L^{-1}$): BA (1) and $GA_3$ (10). (3) Rooting shoot in WPM solid medium with IBA (2 mg $L^{-1}$).

supplemented with 2,4-D, BA and $GA_3$ (0.001, 1.0 and 0.1 mg $L^{-1}$, respectively) for three months. Auxin 2,4-D was omitted from the medium during the first subculture.

Branching of the rosettes and satisfactory shoot multiplication (Plate 3: 1,2) was obtained in WPM "double-phase" medium containing vitamins (Walkey, 1972) after Jaquiot (Gautheret, 1959) and different amounts of $GA_3$ (1.0–10.0 mg $L^{-1}$). These results resembled those observed during the elongation of the shoots originating from the organogenic callus developed in horse chestnut immature embryo cultures (Radojević and Marinković, to be published). Shoots rooted satisfactorily on WPM solid medium supplemented with IBA (2.0 mg $L^{-1}$) and vitamin mixture after Walkey (1972) (Plate 3: 3).

## 3. Embryo development

### 3.1. *Secondary somatic embryogenesis*

Fragments of somatic embryos obtained from cotyledonary explants (Profumo *et al.*, 1990) formed secondary embryoids when transferred to an agarized culture medium supplemented with KND or only 2,4-D. Furthermore, somatic embryos originating from leaf cultures or immature embryos stimulated the formation of new somatic embryos. According to Dameri *et al.* (1986), addition of BA and IAA (1.0 mg $L^{-1}$ of each) appeared to impede the formation of secondary embryos. Addition of 2.0 g $L^{-1}$ yeast extract to the basal medium neither stimulated the formation of the new embryos, nor enhanced the development of the existing ones. For the formation of secondary embryos on the primary somatic embryos (originated from immature embryos), individual somatic embryos were grown in MS medium containing reduced 2,4-D levels (0.01–0.001 mg $L^{-1}$; Radojević, 1988). Increasing BA concentrations (0.5–2.5 μmol $L^{-1}$) resulted in the higher rate of secondary embryo formation on the surface of the cultured *A. hippocastanum* embryos originating from filaments (Jörgensen, 1989, 1991a). Somatic secondary embryoids of horse chestnut arose either from the basal region or on the cotyledons of other somatic embryos in clusters of 2, 4 or more (Plate 1: 7). Such a clear case of the secondary somatic embryogenesis was also described by other workers (Staritsky, 1970; Söndahl and Sharp, 1977; Söndahl *et al.*, 1980; Maheswaran and Williams, 1986; Ammirato, 1987; Rout *et al.*, 1991; Durham and Parrott, 1992). Maheswaran and Williams (1986) suggested that single-cell or multicellular initiation of somatic embryogenesis depends on the age or the degree of maturity of the parental tissues. Older tissues with some immature epidermal cells gave rise to somatic embryos originating from a single cell. It is, therefore, possible to use properly staged somatic embryos as explants to induce the development of the next generation of somatic embryos arising from a single cell. Recently, Liu *et al.* (1992) established an efficient somatic embryogenesis system in soybean and provided an evidence demonstrating the single-cell origin of somatic embryos through somatic embryo cycling. In this system a conspicuous increase of both the frequency of explant response and the number of embryos per responding explant was achieved when compared with those obtained from zygotic explants. Our previous cytological examinations of horse chestnut somatic embryos, originating from zygotic embryos, also indicated somatic embryo cycling which represents a promising system that could be used for genetic transformation (Radojević, 1988).

### 3.2. *Factors for the formation and maturation of somatic embryos*

Several culture media amended with different plant growth regulators, organic compounds and amino acids have been tried, depending on the initial explant used for the induction of somatic embryogenesis (Table 1). Addition of high auxin concentrations (2,4-D or NAA) was needed for the initiation of early embryogenic stages. After that, either reduce the auxin level or devoid auxin in the media for the differentiation and further development of cotyledonary somatic embryos. Most of the cotyledonary embryos were at the maturation stage.

Sequential changes in the levels of exogenously applied auxins stimulated the induction of somatic embryogenesis in the cultures of horse chestnut (Radojević, 1988, 1991) and other woody plant species (Radojević, 1980).

The balance of auxins and kinetin also influenced the induction of somatic embryogenesis. This was confirmed by the results of the authors who used the same auxin/kinetin ratios for the induction of somatic embryogenesis in cotyledonary (Profumo *et al.*, 1976, 1980, 1990, 1991a) and leaf segment explants (Dameri *et al.*, 1986). The balance of auxin and kinetin (3.0 mg $L^{-1}$ 2,4-D and 1.0 mg $L^{-1}$ Kin) also affected somatic embryogenesis induction in *A. hippocastanum* immature embryos. A slight decrease in auxin:kinetin ratio led to a decrease in the number of embryos (Radojević, 1991). On the contrary, the best yield of the somatic embryos was obtained in *Coffea arabica* leaf-derived callus when the level of Kin exceeded that of NAA (Söndahl and Sharp, 1977). These data suggest that species-specific requirement in the levels of exogenously applied auxins and Kin for the development of somatic embryos.

L-Proline and some other amino acids stimulated somatic embryogenesis in several plant species (Ammirato, 1986; Rangaswami, 1986; Radojević and Subotić, 1992).

Primary and secondary somatic embryogenesis in horse chestnut cultures can be maintained relatively for a longer period of time without losing the embryogenic potential. Jörgensen (1989) also observed no alterations in the embryogenic potential of the horse chestnut cultures originating from the filaments for ten months. Similar embryogenic potential of the *Prunus persica* embryogenic calli was reported by Raj Bhansali *et al.* (1990). They achieved a continuous multiplication of peach somatic embryos for more than ten months by a repetitive somatic embryogenesis process. In cultures originating from horse chestnut immmature embryos, somatic embryos were formed during 12 months with the same intensity on the medium having a reduced sucrose level (3% w/v) and 1.0 mg $L^{-1}$ of each 2,4-D and Kin (Radojević, 1988). Prominent embryogenic potential was observed and 70–100 somatic embryos per callus from the globular to the cotyledonary stage were formed (Plate 1: 6).

The rate of somatic embryo formation on the embryogenic calli differed depending on the horse chestnut explant used and composition of the culture medium. According to the results obtained with embryogenic calli originating from horse chestnut immature embryos (Radojević, 1988), it takes 3 months to form somatic embryos in MS medium supplemented with 2,4-D (1.0 mg $L^{-1}$), CH and Pro (200 mg $L^{-1}$, each). However, Dameri *et al.* (1986) obtained a cup shaped or slightly ovoid somatic embryos after two months of cultures of $E_2$ embryogenic callus, originating from horse chestnut leaf explants using KND medium. This morphological form of the somatic embryos was observed for the first time in androgenic embryos developed with the anther cultures of *A. hippocastanum* (Radojević, 1978).

When transferred to a hormone-free MS basal medium or to the same medium with reduced auxin level, somatic embryos developed two or more cotyledons which frequently varied in size, thickness and colour. Most of the somatic embryos formed were dicotyledonary in the cultures of different horse chestnut explants via embryogenesis and/or secondary embryogenesis (Plate 1: 7), and embryos with three or more cotyledons were also developed (Dameri *et al.*, 1986; Radojević, 1988; Profumo *et al.*, 1990). Earlier this phenomenon has been described in some woody plant species, such as apple (James *et al.*, 1984) and rose (Rout *et al.*, 1991).

In order to achieve a complete conversion of the somatic embryos into the plantlets, it is necessary to provide the optimum nutritive and environmental conditions.

Dameri *et al.* (1986) reported that on the $E_2$ embryogenic callus originating from the horse chestnut leaf explants, 150–200 somatic embryos were produced per each $cm^3$. About 80% of these embryos developed into plantlets upon transfer to the culture medium devoid of hormones. Shoot apices of these embryos grew rapidly and developed stem and leaves composed of 3–5 leaflets, while the root region grew much slower. The latter was characterized by a sheath similar to that which covers first the root tip and then the radicle (List and Steward, 1965). Embryos with poorly-developed cotyledons produced thin roots which were not covered with a sheath. This could mean that swollen cotyledons of somatic embryos inhibit premature root development. The same phenomenon was observed in the somatic embryos originating from immature embryos (Radojević, 1988) as well as in androgenic embryos originating from microspores of horse chestnut (Radojević, 1978). In order to prevent such inhibitory effects, we have removed cotyledons containing storage reserves and they may also contain some inhibitors preventing further development of the cotyledonary stage embryos into the plantlets, especially the root development. Furthermore, rooting did not improve. Dameri *et al.* (1986) cultured horse chestnut somatic embryos in the medium supplemented with 100 mg $L^{-1}$ D,L-ornithine monochlorhydrate for 2 weeks in the dark and again roots did not developed. Addition of BA and IAA (0.1 mg $L^{-1}$, each) or yeast extract (2 g $L^{-1}$) to the basal medium did not influence further development of the somatic embryos and inhibited formation of the new ones.

There are no data available in the literature on the application of abscisic acid (ABA) or other desiccants (e.g., increased sucrose concentration, sorbitol, polyethylene glycol (PEG), etc.) in order to enhance maturation and desiccation tolerance of *A. hippocastanum* somatic embryos. Such a treatment would probably lead to the complete conversion of somatic embryos into the plantlets, similar to recent results obtained with *Picea glauca* (Attree *et al.*, 1991; Dunstan *et al.*, 1991). Addition of PEG 4000 (5–7.5%) and ABA (16 μM) to the cultures of *P. glauca* for 28 days stimulated the maturation of the somatic embryos (Attree *et al.*, 1991). Somatic embryos that maturated in the presence of PEG remained quiescent during desiccation due to low level of moisture, but were quite capable of producing plantlets without change in morphology. Similar observations have been recorded with some herbaceous plant species, such as *Pisum sativum* (Barratt *et al.*, 1989) and *Medicago sativa* (Xu *et al.*, 1990; Senaratna *et al.*, 1990). Both ABA and PEG had ceased premature germination of the *P. glauca* somatic embryos, and accumulation of nutrients in the somatic embryos of some conifers (Feirer *et al.*, 1989), and *M. sativa* (Xu *et al.*, 1990).

The desiccation of embryos occurs in most of the seeds in the nature and this process plays a significant role during the developmental period between the maturation and germination of the embryos (Kermode, 1990). It was also reported that desiccation by 6% sorbitol improved germination of *Quercus robur* somatic embryos (Chalupa, 1990), and it is quite possible that similar effects could be achieved in *A. hippocastanum* somatic embryos.

### 3.3. *Cytological and anatomical studies of regenerants*

Ultrastructural studies demonstrated significant differences in different *A. hippocastanum* callus types (NE, $E_1$ and $E_2$) (Profumo *et al.*, 1987b). The cells of the $E_2$ calli expressed cytological characteristics reflecting very high metabolic activity. These cells divided rapidly, ribosomes were organized into the polysomes, and the number of mitochondrial cristae and the rate of dictyosomes proliferation were increased. Both $E_1$ and $E_2$ embryogenic calli formed somatic embryos (Profumo *et al.*, 1986). In $E_2$ calli, somatic embryoids were formed directly from meristemoids as described earlier for *Corylus avellana* (Radojević *et al.*, 1975), *Paulownia tomentosa* (Radojević, 1979), and *Theobroma cacao* (Kononowicz *et al.*, 1978). In some cultures of horse chestnut leaf explants, somatic embryos appeared without callogenesis (Profumo *et al.*, 1986). They were of single-cell origin similar to those formed on the calli. The embryogenic stages of the former were typical globular, heart-shaped, torpedo-shaped; and finally cotyledonary embryos were observed. Anatomical studies revealed that differentiated somatic embryos had a vascular continuum between provascular cylinder, cotyledons, apical shoots and radicular pole (Profumo *et al.*, 1986). The shoots differed in shape, depending on the age of the somatic embryos and the development of the root region covered with a sheath (List and Steward, 1965), completed bipolarity of the embryos. As demonstrated by light microscopy, the sheath derived from the junction of the cotyledonary base is born in a single structure of the vascular bundles (Profumo *et al.*, 1986). Inside the sheath, another structure belonging to the embryonal axis was formed. In the beginning, the axis had characteristics of the hypocotyl and further, it changed from the eustelic to the actinostelic form. Profumo *et al.* (1986) studied histological connections between the sheath and the real root. The root was developed after a much longer period than necessary for the formation of plantlet epigaeus parts.

Cytogenetic studies with the regenerants of horse chestnut, having different origin, revealed the genome with 40 chromosomes (Plate 1: 8), typical for the plants derived from ripe seeds (Darlington and Wylie, 1955).

## 4. Embryo germination

### 4.1. *Seed dormancy*

Similar to numerous woody plant species, seeds of the horse chestnut are dormant (Fig. 2). When seeds fall in the autumn from the tree, they are dormant (Hartmann and Kester, 1965; Bewley and Black, 1982; Profumo *et al.*, 1987b) and recalcitrant, losing their viability upon drying. If seeded in the soil immediately, they would germinate during the next spring after the winter chilling period. Regeneration of *A. hippocastanum* plantlets under *in vitro* conditions indicated the connection between a certain interruption of somatic embryo development and seed dormancy. It seems that the dormancy of horse chestnut seeds remains somehow "inscribed" into somatic embryos and secondary embryoids derived either from cotyledons (Profumo *et al.*, 1990) or zygotic immature embryos (Radojević, 1988). It has been shown, however, that horse chestnut plantlets can be obtained in the cultures of embryogenic axes isolated from the ripe seeds (Profumo *et al.*, 1989). This could be taken as a further evidence that the cause for the dormancy of embryos should be sought for in the cotyledons as already demonstrated in apple (Thevenot and Côme, 1973) and cherry (Aboud-Zeid and Neumann, 1973) seeds. However, it should be underlined that it is not always possible to suggest such a simple relationship to exist between the dormancy of somatic embryos and seeds (Profumo *et al.*, 1991a; Radojević, 1991). For instance, certainly dormancy was observed in: somatic embryos formed at the anther walls of *Vitis vinifera* (Rajasekaran and Mullins, 1979), somatic embryos of *Eschscholtzia* obtained in filament cultures (Kavathekar *et al.*, 1977), and horse chestnut androgenic embryos (Radojević, 1978; Jörgensen, 1991b). However, these cultures were not derived from the ripe seeds. These examples confirm that dormancy of *in vitro*-grown embryos is a rather complex phenomenon. It, nevertheless, creates big problems during the process of conversion of somatic embryos into the plantlets. These difficulties could be overcome by applying some plant growth regulators or by exposing them to low temperatures (4°C) together with a suitable light regime (light/-darkness).

### 4.2. *Germination of somatic embryos, plantlet formation and transfer to soil*

The germination of horse chestnut somatic embryos depends on normal morphology, embryo maturity, and concentration of auxins, cytokinins and gibberellins in the culture medium (Dameri *et al.*, 1986; Radojević, 1988, 1991; Profumo *et al.*, 1989, 1990). This has already been observed earlier in some herbaceous plant species (Lazzeri *et al.*, 1987). Gibberellins, such as $GA_3$, have greatly affected germination of *Corylus avellana* zygotic embryos (Bradbeer and Pinifield, 1967), somatic embryos of *Juglans regia* (Deng and

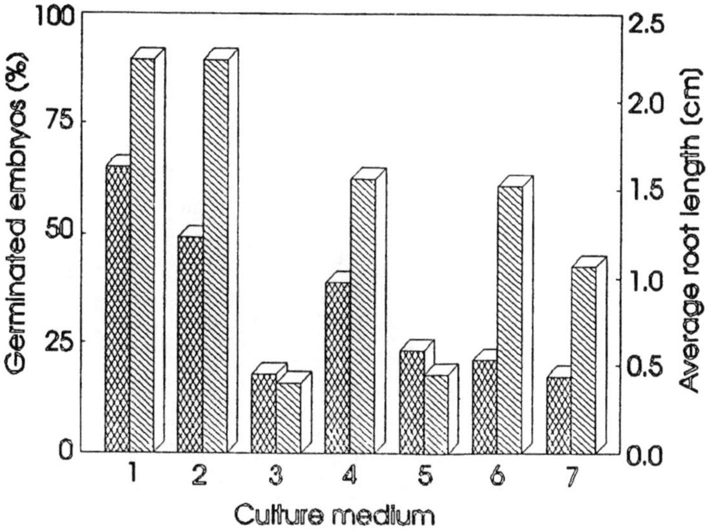


*Figure 6.* The effect of some growth promoting substances, glutamine and activated charcoal on the germination of somatic *Aesculus* embryos. Media (ingredients in mg $L^{-1}$): (1) MS + IAA (1) + $GA_3$ (1) + Glu (400); (2) MS + IAA (1) + $GA_3$ (1) + 1% AC + Glu (400); (3) MS + Kin (3) + Glu (400); (4) MS + IAA (1) + $GA_3$ (1) + Kin (3) + Glu (400); (5) MS + IBA (1) + $GA_3$ (1) + Kin (3) + Glu (400); (6) MS + IAA (1) + $GA_3$ (1) + BA (3) + Glu (400); (7) MS + IAA (1) + $GA_3$ (1) + Zeatin (3) + Glu (400). The number of embryos cultured in each medium varied from 80–100. Germinated embryos (%) (diamonds); Average root length (cm) (bars).

Cornu, 1992), somatic (Radojević, 1988) and androgenic embryos of *A. hippocastanum* (Radojević *et al.*, 1989; Marinković and Radojevic, 1992).

For the germination of horse chestnut cotyledonary somatic embryos, derived from immature embryos (Radojević, 1988), seven different liquid media supplemented with L-glutamine (400 mg $L^{-1}$) were used (Fig. 6). Induction of germination was not achieved in control medium devoid of glutamine and hormones. In MS medium containing kinetin (3.0 mg $L^{-1}$) and glutamine (400 mg $L^{-1}$), 18% of somatic embryos germinated. Addition of IAA and $GA_3$ to MS medium favoured germination and 65% of horse chestnut somatic embryos germinated (Fig. 7; medium 1). Addition of activated charcoal to the above medium slightly inhibited somatic embryo germination and 49% of somatic embryos germinated (Fig. 7; medium 2). They formed long and thin primary roots (Fig. 7; media 1 and 2). During prolonged growth as a result of 2–3 subcultures in MS liquid medium containing 1% activated charcoal (AC) and (in mg $L^{-1}$): IAA 1.0, $GA_3$ 1.0 and L-glutamine 400, most of the germinated embryos developed a primary root with secondary rootlets and a shoot with several leaves typical to the horse chestnut (Plate 4: 1). Somatic embryos were completley converted into plantlets in media 1 and 2 (Plate 4: 2,3).

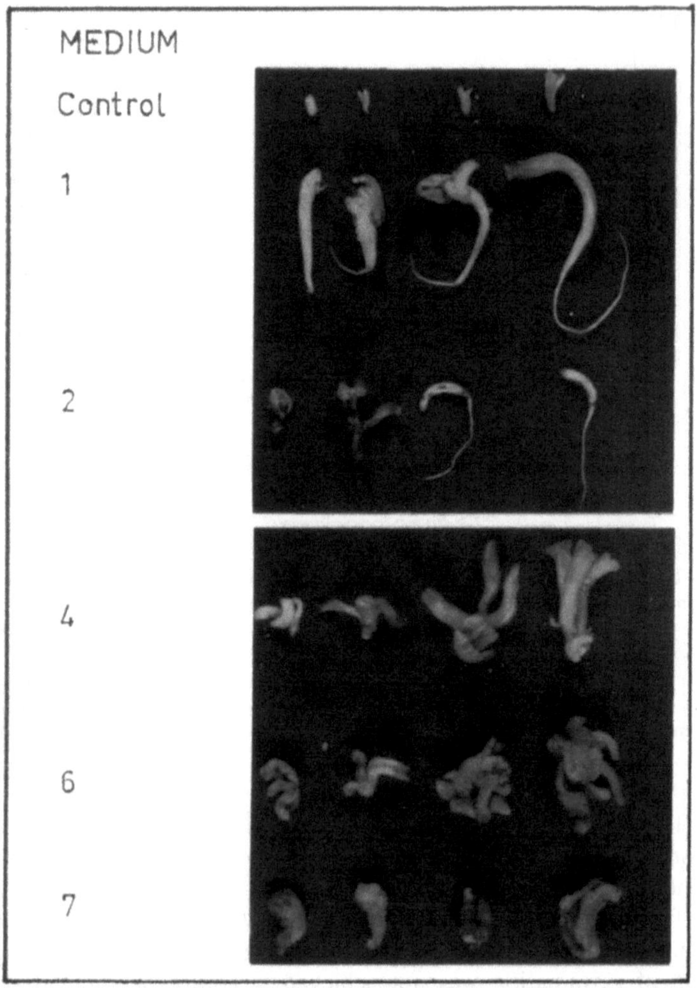


*Figure 7.* Germination of somatic embryos and plantlets formation of *A. hippocastanum*. Control: MS medium devoid of hormones; (1) Plantlets which formed primary root in MS medium containing IAA, $GA_3$ (1 mg $L^{-1}$, each) and L-Glu (400 mg $L^{-1}$); (2) Plantlets cultured in the medium given under 1 supplemented with 1% AC; (4) Germinated *Aesculus* embryos with thickened and elongated radicle and well-developed cotyledons in MS medium with IAA, $GA_3$ (1 mg $L^{-1}$, each), Kin (3 mg $L^{-1}$) and L-Glu (400 mg $L^{-1}$) and (6) embryos in the same medium containing BAP (3 mg $L^{-1}$); (7) Somatic embryos with elongated embryogenic radicle and root axes in the medium described under 6 and supplemented with zeatin (3 mg $L^{-1}$) instead of Kin (Radojević, 1988).

*Plate 4.* Plant recovery of *A. hippocastanum*. (1) Plantlet forming shoot and primary root with secondary rootlets in MS medium supplemented with (in mg $L^{-1}$): IAA (1), $GA_3$ (1) and Glu (400) with addition of 1% AC. (2) Early acclimatization phase. Rooted horse chestnut plantlet growing in perlite and watered with 1/2 strength liquid WPM medium. (3) Later acclimatization phase: horse chestnut plantlet growing in the pot filled with a perlite-peat (1:3) mixture (Radojević, unpublished).

When embryos were cultured in the medium containing kinetin or BA (3 mg $L^{-1}$), an increase in number and swelling of cotyledons was observed (Fig. 7; media 4 and 6). A uniform growth of cotyledons and radicle was noticed when embryos were cultured in the medium containing zeatin (Fig. 7; medium 7). Addition of zeatin or BA to the medium inhibited horse chestnut somatic embryo germination. These results are not in accordance with Gray (1992) who induced germination in 80–100% *Vitis rotundifolia* somatic embryos by adding 1 μM benzyl-adenine (BA) to MS medium. Also, BA (0.05 mg $L^{-1}$) in combination with IAA (5.0 mg $L^{-1}$) greatly stimulated germination of the somatic embryos in *Populus* spp.(Michler and Bauer, 1991), which were further successfully transferred *ex vitro* to the greenhouse conditions.

Only green cotyledonary horse chestnut embryos germinated, while malformed and translucent embryos were not capable of germination. According to Tulecke and McGranahan (1985), the translucency of the somatic embryos probably reflects the deficiency of starch and/or protein storage.

Tulecke and McGranahan (1985) and Tulecke *et al.* (1988) demonstrated that chilling for two months at 4°C improved the germination of the *Juglans regia* somatic embryos. It seems that such a treatment interrupted the dormancy of these embryos. It is known that the maturation of the zygotic embryos of *Juglans regia* is accompanied by abscisic acid (ABA) accumulation (Marx *et al.*, 1988). Chilling of the somatic embryos decreases endogenous ABA content and as a consequence the dormancy breaks enabling embryos to germinate including somatic embryos of *Corylus avellana* (Radojević, 1977), and *Vitis vinifera* (Rajasekaran and Millins 1979), androgenic (Radojević, 1991) and somatic (Profumo *et al.*, 1991a) embryos of *A. hippocastanum*.

Profumo *et al.* (1991a) induced somatic embryogenesis in the cultures of horse chestnut cotyledonary explants. However, these somatic embryos were dormant and incapable of producing plantlets. They interrupted the dormancy by culturing somatic embryos in the medium devoid of hormones and exposed them to chilling (6°C) for six months in the dark. Control embryos were grown in the same medium under normal environmental conditions (25°C, 12 h light/dark cycle). Several days after the transfer of chilled somatic embryos under growing conditions at 25°C, significant differences were observed between chilled and control embryos. In somatic embryos exposed to cold, active organogenesis was observed and the roots developed after six days, while during 12 days 100% of the somatic embryos formed roots. Examination of the regenerated roots demonstrated that this process is accompanied by the elongation of the cotyledonary basis, and the root axis appeared to be large and stumpy. After that, the axis developed and grew deeper into the solid medium and developed the secondary roots, while the apical part became more slender. On the 9th day, a plumule was formed and later the whole plant was derived. After 17th day, 40 chilled somatic embryos formed the plantlets. Control somatic embryos formed neither roots nor shoots. It is interesting to note that the embryos of cotyledonary origin exposed to cold first develop roots, while somatic embryos obtained from leaf explants develop the epigaean apparatus before the roots (Dameri *et al.*, 1986). In the latter case, the roots are thinner and more numerous than the former. Chilling treatment also leads to an enlarged root axis, similar to horse chestnut seed germination in the field. At present, there are no data available in the literature on the synchronous development of roots and shoots in the horse chestnut cultures.

Rooted plantlets were potted in a mixture of perlite-peat (1:3) (Plate 4: 3). After that, they were acclimatized under greenhouse conditions with aseptic precautions and optimum humidity in order to transfer them, eventually, in the field.

### 4.3. In vitro *culture of somatic horse chestnut embryos in the presence of mycorrhizal fungi metabolites*

Ectotrophic mycorrhizae are essential for the establishment of tree seedlings and for further improved growth and development, especially in poor soil with low levels of nutrients. The inability of tree seedlings to grow in mycorrhizal fungi free soil and their successful ability to grow in an inoculated soil with a suitable mycorrhizal fungus have been described several decades ago (Harley, 1959). The influence of mycorrhizal fungi on the above processes is directly connected to their metabolites. It is well documented that numerous plant growth regulators are present in the mixture of metabolites produced by certain microorganisms (Nitsch, 1956; Cutler, 1988). For example, metabolities of: a) *Lactarius piperatus* contain auxins, b) *Fusarium moniliforme* possess gibberellins, c) *Pseudomonas syrinae* pv. *savastanoi* contain cytokinins, and d) *Cercospora cruenta* have abscisic acid and related compounds.

As mentioned above, metabolites of *Lactarius piperatus* fungus contain high levels of auxins (IAA and IBA) and it has been widely applied for classical rooting of *Corylus*, *Fagus* and *Quercus* spp. seedlings. Metabolites of *Laccaria laccate* were found to favour rooting of *Betula*, *Pinus* and *Eucalyptus* seedlings (N. Gogala, personal communication). Marx (1969) described symbiosis of *Phytophtora cinnamomi* with *Pinus echinata* and *P. teada*, and antagonism of this mycorrhizal fungus to other fungal root pathogenes and soil bacteria. The fungus *Pisolithus tinctorius* also stimulated rooting of *Quercus* and *Eucalyptus* seedlings (N. Gogala, personal communication). It has been suggested that mycorrhizal fungi support selective absorption and accumulation of the ions by a host plant, i.e., they assist their host to absorb nutrients from insoluble soil minerals, organic components and forest litter which are not readily avaliable to non-mycorrhizal roots (Rosendahl, 1942; Routien and Dawson, 1943). It was also suggested that mycorrhizal roots are less sensitive to pathogenes than non-mycorrhizal ones.

There are no data available in the literature on the effect of mycorrhizal fungi on rooting of *A. hippocastanum* cuttings or on the symbiosis of this tree species with some other fungi. In this chapter, the effects of hormones and chilling on the dormancy and germination of the horse chestnut somatic embryos have been discussed. Here, we will briefly describe our results on the germination of horse chestnut somatic embryos in the presence of *Lactarius piperatus* metabolites (Radojević *et al.*, 1988). Water extract containing metabolites of this fungus was prepared according to the procedure of Marx (1969). Horse chestnut somatic embryos were grown in MMN solid (Marx, 1969) and MS liquid medium supplemented with (in mg $L^{-1}$): IAA 1.0, $GA_3$ 1.0 and Glu 400. Water extract of *L. piperatus* metabolites (WEM) was added to some cultures (100 ml $L^{-1}$). In MMN medium, most of the somatic embryos rapidly developed necrosis. In MS medium supplemented with hormones and devoid of WEM, 59% of the embryos germinated. The percentage of the germinated embryos decreased (46%) when WEM was added to the same medium. At the same time, 28% embryos germinated in the control MS medium devoid of hormones but supplemented with WEM. Embryos developed long and thin primary roots with well developed secondary rootlets. Although addition of WEM to the medium containing hormones decreased the number of the germinated embryos, the plantlets produced were greener than the controls and better synchronization in the development of roots and shoots was observed. Also, addition of WEM stimulated hardening of the plantlets and their survival rate was higher when compared with those produced in the medium devoid of WEM.

## 5. Secondary metabolites and other compounds of possible industrial significance in horse chestnut grown *in vivo* and *in vitro*

It is well documented that horse chestnut wood, bark and nuts contain a multitude of different chemical compounds. Some of these compounds have already been used for several purposes. In the near future, more substances occurring in different plant organs especially those produced by *in vitro* cultures will be found suitable for the industrial exploitation.

Aescin is one of the secondary metabolites of *A. hippocastanum* which has been pharmacologically well characterized. It is widely applied in the therapy of different peripheral vascular disorders. This compound is a mixture of triterpene glycosides and it occurs naturally, mostly, in horse chestnut seeds. However, even more extensive synthesis of this compound has been reported in tissue cultures of different plant organs (Profumo *et al.*, 1976, 1980, 1991a, 1992). High level of aescin was recorded in all analyzed horse chestnut tissues grown in MS medium supplemented with plant growth regulators or even in the absence of the growth regulators. Aescin content, synthesized by cotyledon-derived callus (Profumo *et al.*, 1980, 1992), was higher than that found in the cultures of leaf-derived embryogenic calli (Dameri *et al.*, 1986), and in both cases aescin level exceded the level in the horse chestnut ripe seeds. Profumo *et al.* (1991b) examined aescin synthesis in $E_1$ and $E_2$ embryogenic calli, and in leaf-derived somatic embryos using aescin content in ripe cotyledons *in vivo* as a reference. No significant differences in aescin level between $E_1$ callus and cotyledons were observed. However, aescin content in $E_2$ calli (in active proliferation phase and differentiation during 20-day-culture period) was higher than in the control. After 70 days of continuous culture without subculture into the fresh medium, aescin concentration decreased which was higher than in the control. In the cultures, subcultured every 25 days, aescin level was approximately three times higher than that in the control, whereas horse chestnut somatic embryos contained aescin slightly more than seven times than in the ripe cotyledons taken from the seeds (Profumo *et al.*, 1991a).

The above results indicate that both callus and somatic embryo of *A. hippocastnum* can be succesfully used for aescin production. There are many advantages of using tissue culture for aescin production. First of all, it is possible to obtain this compound in high quantity from uniform plant material (suspension cell lines and/or embryogenic calli) than the mature seeds, independent of the season. Profumo *et al.* (1991b) suggested that both $E_2$ calli and somatic embryos are highly suitable sources of aescin production when compared with other tissues. Maximum aescin content was obtained by transferring somatic globular embryos to MS medium devoid of hormones (Profumo *et al.*, 1991b). This increase of aescin level could be due to the transfer of horse chestnut tissue from hormone-containing MS medium to the MS medium without hormones.

Frischknecht *et al.* (1977) and Paupardin (1983) observed an antagonism between plant growth stimulation and accumulation of secondary metabolites, especially when auxins were applied. As far as horse chestnut and aescin production are concerned, growth intensity paralleled aescin production (Profumo *et al.*, 1991b), i.e., it seems that the same factors trigger cell division and multiplication, and somatic embryogenesis induction also increased aescin production. The presence of aescin content in different types of *in vitro*-grown *Aesculus* tissues is similar to the field-grown plants (Von Kartnig *et al.*, 1965, 1966, 1968). Data on aescin content in different *in vitro* and *in vivo*-grown *A. hippocastnum* explants are listed in Table 3.

In addition to aescin, the bark of horse chestnut tree contains certain amounts of gallic and tannic acids, which are used in dye and leather industries.

Horse chestnut seeds are rich in proteins (15% d.w.) and starch (34% d.w.) and may be used as fodder and for large-scale ethanol production (Tucović, 1973).

Some other biologically active compounds were also isolated from different *Aesculus* spp. Esters of indole-3-acetic acid were extracted and purified from the liquid endosperm of immature fruits of *A. parviflora*, *A. baumanni*, *A. pavia rubra* and *A. pavia humulis* (Domagalski *et al.*, 1987). Liquid endosperm contained at least 12 chromatographically distinct esters. One of them represented the ester of indole-3-acetic acid and myo-inositol, and the second

*Table 3.* Aescin content in different explants of *A. hippocastnum* grown *in vitro* and *in vivo*.

| Explants *In vivo* | *In vitro* | Reference |
|---|---|---|
| Petals | – | Von Kartnig *et al.*, 1965 |
| Flower stalks | – | '' |
| Seedling leaves | – | '' |
| Cotyledons | – | '' |
| Stem | – | Von Kartnig *et al.*, 1966 |
| Leaves | – | '' |
| Primary seedling leaves | – | Von Kartnig *et al.*, 1968 |
| '' | Cotyledon-derived calli | Profumo *et al.*, 1976, 1980 |
| '' | Embryogenic $E_1$ and $E_2$ calli; leaf-explant-derived somatic embryos | Profumo *et al.*, 1991a |
| Seeds | Cotyledon-explant-derived calli | Profumo *et al.*, 1992 |

one was identified as the ester of this acid and glucosyl-rhamnose, while the third one was the ester of indole-3 acetic acid and desoxyamino hexose.

Seeds of *A. californica* Nutt. contain hydroquinone and (-)-epicatechin (Kubo *et al.*, 1985). Hydroquinone acted by inhibiting seed germination when applied in high concentration (100 $\mu g\ ml^{-1}$) but stimulated growth of lettuce and rice seedlings at low concentration. (-)-Epicatechin had low effect on growth when applied alone, however, it suppressed growth inhibitory activity of hydroquinone when used in combination with hydroquinone.

During the recent years, the production of secondary metabolites by plant cell cultures *in vitro* attracted the greater attention of numerous researchers (Buitelaar and Tramper, 1992). The expectations from such approaches became high when first commercial product shikonin was produced by the cell cultures of *Lithospermum erythrorhizon* in Japan (Deno *et al.*, 1987). However, so far in spite of great efforts, only a few compounds have been obtained in large-scale by these techniques. Insufficient knowledge on the biosynthic pathways of secondary metabolites is the main problem. It is worth mentioning that Buitelaar and Tramper (1992) described several techniques including cell immobilization, application of a two-phase system for the improvement of secondary metabolites excretion, and choice of certain reactions for the stimulation of plant cell growth (elicitation, cell line selection, genetic manipulations, etc.) for the production and leaching of desired secondary metabolites.

Although some strategies described so far are promising, no general approach has been developed for a stable plant tissue culture system aimed at large-scale production of useful secondary metabolites. This opens a new area for research in horse chestnut and other plants.

## 6. Concluding remarks

Regeneration of *Aesculus hippocastnum* plants was achieved using the tissue (meristems from the somatic embryos) or different organs (leaves, filaments, ovules, zygotic embryos) via micropropagation, somatic embryogenesis, and organogenesis. Auxins (2,4-D, NAA) and cytokinins (Kin, BA) are indispensable factors for the induction of somatic embryogenesis and organogenesis.

During the early phase of embryogenic callus induction in the leaf and immature embryo cultures, auxin was necessary at much higher concentrations than cytokinin. For embryogenic callus growth and somatic embryo formation, auxin concentration was reduced without changing the cytokinin concentration. Sequential addition of auxins followed by $GA_3$ led to organogenic callus production and the appearance of adventitious buds.

For the further development of somatic embryos, auxins (IAA or IBA) and $GA_3$ were necessary. The stimulatory effects of $GA_3$ and/or cytokinins on the germination of somatic embryos and plantlet development have frequently been described for other plant species as well (Wang and Vasil, 1982; Ammirato, 1983; Ghazi *et al.*, 1986; Cruz *et al.*, 1990). These plant growth regulators could overcome some kind of "inscribed" or "acquired" dormancy, since it is known that dormant seeds can be germinated by exogenous $GA_3$ and/or cytokinins.

Chilling of somatic embryos at 4°C for six months greatly enhanced the germination of embryos (Profumo *et al.*, 1991a), and the rapid development of roots and shoots. At 25°C, chilled somatic embryos developed into plantlets within approximately 18 days. Therefore, this method could be applied either alone or in combination with hormone treatments as demonstrated previously in horse chestnut androgenic embryos (Radojević, 1977).

The stimulatory effects of metabolite extracts of *Lactarius piperatus* on hardening of the plantlets and their survival rate suggest that the future research could be directed towards the establishment of symbiosis between horse chestnut and some suitable fungus.

The yield of somatic embryos (150–200 embryos per culture) suggests that they can be successfully used for the production of artificial seeds of *A. hippocastanum* (Jörgensen, 1991a).

Since horse chestnut embryogenic calli and somatic embryos contain high amount of aescin (Profumo *et al.*, 1991b, 1992), they could be commercially exploited for large-scale aescin production. For this purpose, specific bioreactors should be introduced.

The combination of *in vitro* horse chestnut somatic embryo production and cryopreservation (Jörgensen, 1990, 1991c) will enable the long-term preservation of regenerants and/or meristem propagules in gene banks. In this way, gene loses resulting from an increasing environmental pollution would be avoided and gene preservation of elite horse chestnut trees would be provided (Jörgensen, 1990).

The production of horse chestnut regenerants by meristem micropropagation, organogenesis and somatic embryogenesis represents a promising approach for the propagation of other plants belonging to the genus *Aesculus*.

## 7. Acknowledgements

The experimental work presented in this chapter was supported by the Scientific Ministry of Serbia.

## 8. References

Aboud-Zeid, A. and K.H. Neumann, 1973. Preliminary investigations on the influence of cotyledons on the development of cherry embryos (*Prunus avium* L.) Z. Pflanzenphysiol. 69: 299–305.

Ammirato, P.V., 1983. Embryogenesis. In: D.A. Evans, W.R. Sharp, P.V. Ammirato and Y. Yamada (Eds.), Handbook of Plant Cell Culture, Vol. I, pp. 82–123. McMillan, New York.

Ammirato, P.V., 1986. Control and expression of morphogenesis in culture. In: L.A. Withers and P.G. Alderson (Eds.), Plant Tissue Culture and its Agricultural Applications, pp. 23–45. Butter-Worths, London.

Ammirato, P.V., 1987. Organizational events during somatic embryogenesis. In: Green, C.E., Sommers, D.A., W.P. Hackett and D.D. Biedboer (Eds.), Plant Tissue and Cell Culture, Plant Biol. 3, pp. 57–81. Alan R. Liss., Inc. New York.

Attree, S.M., Moore, D., V.K. Sawhney and C. Fowke, 1991. Enhanced maturation and desiccation tolerance of white spruce (*Picea glauca* Moench) somatic embryos: Effects of a non-plasmolysing water stress and abscisic acid. Ann. Bot. 68: 519–525.

Bajaj, Y.P.S., 1986. Biotechnology of tree improvement for rapid propagation and biomass energy production. In: Y.P.S. Bajaj (Ed.), Biotechnology in Agriculture and Forestry, Vol. 1. Tree I, pp. 1–23. Springer-Verlag, Berlin/Heidelberg/New York.

Barratt, D.H., Whitford, P.N., Cook, S.K., G. Butcher and T.L. Wang, 1989. Analysis of seed development in *Pisum sativum* L. VIII. Does abscisic acid prevent precoccious germination and control storage protein syntesis? J. Exp. Bot. 40: 1009–1014.

Bertossi, F., 1960. Ritmo autonomo di crescita del tessuto dilimone coltivato *in vitro*. Atti. Ist. Bot. Crittog. Univ. Pavia ser. V, 17: 210–221.

Bewley, J.D. and M. Black, 1982. Viability and Longevity. Springer-Verlag, Berlin/Heidelberg/-New York.

Bradbeer, J.W. and N.J. Pinifield, 1967. Studies in seed dormancy. III. The effects of the gibberellin on dormant seeds of *Corylus avellana*. New Phytol. 66: 515–523.

Buitelaar, R.M. and I. Tramper, 1992. Strategies to improve the production of secondary metabolites with plant cell cultures: a literature review. J. Biotechnol. 23: 111–141.

Chalupa, V., 1990. Plant regeneration by somatic embryogenesis from cultured immature embryo of oak (*Quercus robur* L.) and linden (*Tillia cordata* Mill.). Plant Cell Rep. 9: 398–401.

Cheliak, W.C. and D.L. Rogers, 1990. Integrating biotechnology into tree improvement programs. Can. J. For. Res. 20: 452–463.

Cruz, G.S., J.M. Canhoto and M.A.V. Abreu, 1990. Somatic embryogenesis and plant regeneration from zygotic embryos of *Feijoa sellowiana* Berg. Plant Sci. 66: 263–270.

Cutler, H.G., 1988. Unusual plant-growth regulators from microorganisms. CRC Crit. Rev. Plant Sci. 6: 323–343.

Dameri, R.M., Caffaro, L., P. Gastaldo and P. Profumo, 1986. Callus formation and embryogenesis with leaf explants of *Aesculus hippocastanum* L. J. Plant Physiol. 126: 93–96.

Dandekar, A.M., Gupta, P.K., D.J. Durzan and V. Knauf, 1987. Transformation and foreign gene expression in micropropagated douglas-fir (*Pseudotsuga menziesii*). Bio/Technol. 5: 587–588.

Darlington, C.D. and A.P. Wylie, 1955. In: Allen and Unwin Ltd. (Ed.), Chromosome Atlas of Flowering Plants 2., p. 197, London.

Deng, M.-D. and D. Cornu, 1992. Maturation and germination of walnut somatic embryos. Plant Cell Tiss. Org. Cult. 28: 195–202.

Deno, H., Suga, C., T. Morimoto and Y. Fujita, 1987. Production of shikonin derivatives by cell suspension cultures of *Lithospermum erythrorhizon*. IV. Production of shikonin derivatives by a two-layer culture containing an organic solvent. Plant Cell Rep. 6: 197–199.

Domagalski, W., A. Schulze and R.S. Bandurski, 1987. Isolation and characterization of esters of indole-3–acetic acid from the liquid endosperm of the horse chestnut (*Aesculus* species). Plant Physiol. 84: 1107–1113.

Dunstan, D.I., T.D. Bethune and S.R. Abrams, 1991. Recemic abscisic acid and abscisyl alcohol promote maturation of white spruce (*Picea glauca*) somatic embryos. Plant Sci. 76: 219–228.

Durham, R.E. and W.A. Parrott, 1992. Repetitive somatic embryogenesis from peanut cultures in liquid medium. Plant Cell Rep. 12: 12–25.

El Maâtaoui, M., H. Espagnac and N. Michaux-Ferrière, 1990. Hystology of callogenesis and somatic embryogenesis induced in stem fragments of cork oak (*Quercus suber*) cultured *in vitro*. Ann. Bot. 66: 183–190.

Em, H. 1959. Za diviot ili konjski kosten vo N.R. Makedonija. Godisen zbornik na Zemjodelsko-šumarski fakultet. Skoplje, Knjiga XII.

Evans, D.A., W.R. Sharp and J.E. Bravo, 1984. Cell culture methods for crop improvement. In: W.R. Sharp, D.A. Evans, P.V. Ammirato and Y. Yamada (Eds.), Handbook of Plant Cell Culture, Vol. 2, pp. 47–68, Collier Macmillan Publishers, London.

Feirer, T.P., J.H. Conkey and S.A. Verhagen, 1989. Triglycerides in embryogenic conifer calli: a comparison with zygotic embryos. Plant Cell Rep. 8: 207–209.

Frischknecht, P.M., T.W. Baumann and H. Wanner, 1977. Tissue culture of *Coffea arabica* (Coffee). Growth and caffein formation. Planta Med. 31: 344–350.

Gautheret, R.J., 1959. Mileux de culture. In: R.J. Gautheret (Ed.), La Culture des Tissus Végétaux, pp. 11–34. Masson et Cie Ed., Paris.

Ghazi, T.O., H.V. Cheema and M.W. Nabors, 1986. Somatic embryogenesis and plant regeneration from embryogenic callus of soybean, *Glycine max* L. Plant Cell Rep. 5: 452–456.

Gray, D.J., 1992. Somatic embryogenesis and plant regeneration from immature zygotic embryos of muscadine grape (*Vitis rotundifolia*) cultivars. Amer. J. Bot. 79: 542–546.

Gunn, R.E. and P.R. Day, 1986. *In vitro* culture in plant breeding. In: L.A. Winthers and P.G. Alderson (Eds.), Plant Tissue Culture and its Agricultural Applications, pp. 313–336. Butterworths, London.

Harley, J.L., 1959. The Biology of Mycorrhiza. Interscience Publishers, Inc., New York.

Hartmann, H.T. and D.E. Kester, 1965. Propagazione delle piante. Edagricole, Bologna, 3: 1–5.

Hasnain, S. and W. Chellak, 1986. Scientific and technical articles/Articles scientifiques et techniques. In: S. Hasnain (Ed.), Tissue Culture in Forestry: Economic and Genetic Potential, The Forestry Chronicle, pp. 219–225, Ottawa, Canada.

Hasnain, S., R. Pigeon and R.P. Overend, 1989. Economic analysis of the use tissue culture for rapid forest improvement. In: S. Hasnain (Ed.), Tissue Culture in Forestry: Economic and Genetic Potential, The Forestry Chronicle, pp. 240–245, Ottawa, Canada.

James, J.D., A.J. Passey and D.C. Deeming, 1984. Adventitious embryogenesis and the *in vitro* culture of apple seeds parts. Int. J. Plant Physiol. 115: 217–229.

James, J.D., 1987. Cell and tissue culture technology for genetic manipulation of temperate fruit trees. Biotechnol. Genetic Eng. Rev. 5: 33–490.

Jordan, M., 1986. Somatic embryogenesis from cell suspension cultures in *Cariaca candamarcensis*. Plant Cell Tiss. Org. Cult. 7: 257–261.

Jörgensen, J., 1989. Somatic embryogenesis in *Aesculus hippocastanum* L. J. Plant Physiol. 135: 240–241.

Jörgensen, J., 1990. Conservation of valuable gene resources by cryopreservation in some forest tree species. J. Plant Physiol. 136: 373–376.

Jörgensen, J., 1991a. Somatic embryogenesis in *Aesculus hippocastanum* and *Quercus petraea* from old trees (10 to 140 years). In: M.R. Ahuja (Ed.), Woody Plant Biotechnology, pp. 351–352. Plenum Press, New York.

Jörgensen, J., 1991b. Androgenesis in *Quercus petraea*, *Fagus sylvatica* and *Aesculus hippocastanum*. In: M.A. Ahuja (Ed.), Woody Plant Biotechnology, pp. 353–354. Plenum Press, New York.

Jörgensen, J., 1991c. Cryopreservation of haploid embryos of *Quercus petraea*, *Fagus sylvatica* and somatic embryos of *Aesculus hippocastanum*. In: M.R. Ahuja (Ed.), Woody Plant Biotechnology, pp. 355–356. Plenum Press, New York.

Kavathekar, A.K., P.S. Ganapathy and B.M. Johri, 1977. Chilling induces development of embryoids into plantlets in *Eschscholtzia*. Z. Pflanzenphysiol. 81: 358–363.

Kermode, A.R., 1990. Regulatory mechanisms involved in the transition from seed development to germination. CRC Crit. Rev. Plant Sci. 9: 155–195.

Kononowicz, H., A.K. Kononowicz and J. Janick, 1978. Asexual Embryogenesis via callus of *Theobroma cacao* L. Z. Pflanzenphysiol. 113: 347–358.

Košanin, N., 1926. Verbreitung einiger Baum- und Strauch-Arten in Südenserbien. Z. "Ungarische Botanische Blätter" 1/12: 116–123.

Kubo, I., A. Matsumoto, M. Tanguchi and W.F. Wood, 1985. Combined effect on plant growth of (-)-epicatechin and hidroquinone, compounds from *Aesculus californica* Nutt. (*Hippocastanaceae*). Chem. Pharm. Bull. 33(9): 3826–3828.

Lazzeri, P.A., D.F. Hildebrand and G.B. Coolins, 1987. Soybean somatic embryogenesis: effects of hormones and culture manipulations. Plant Cell Tiss. Org. Cult. 10: 197–207.

Lloyd, G. and B.H. McCown, 1981. Commercially-feasible micropropagation of mountain lawrel (*Kalmia latifolia*) by use of shoot-tip culture. Proc. Int. Plant. Prop. Soc. 30: 421–427.

List, A.J.R. and F.C. Steward, 1965. The nucellus, embryo sac, endosperm and embryo of *Aesculus* and their interdependence during growth. Ann. Bot. 29: 1–17.

Liu, W., P.J. Moore and G.B. Collins, 1992. Somatic embryogenesis in soybean via somatic embryo cycling. In Vitro Cell Dev. Biol. 28: 153–160.

Mackay, J., A. Sequin and M. Lalonde, 1988. Genetic transformation of 9 *in vitro* clones of *Alnus* and *Betula* by *Agrobacterium tumefaciens*. Plant Cell Rep. 7: 229–232.

McGranaham, G.H., C.A. Leslie, S.L. Uratsu, L.A. Martin and A.M. Dandekar, 1988. *Agrobacterium*-mediated transformation of walnut somatic embryos and regeneration of transgenic plants. Bio/Technol. 6: 800–804.

Maheswaran, G. and E.G. Williams, 1986. Direct secondary somatic embryogenesis from immature sexual embryos of *Trifolium repens* cultured *in vitro*. Ann. Bot. 57: 109–117.

Marinković, N. and Lj. Radojević, 1992. The influence of bud length, age of the tree and culture media on androgenesis induction in *Aesculus carnea* Hayne anther culture. Plant Cell Tiss. Org. Cult. 31: 51–59.

Markgraf, F., 1932. Pflanzengeographie von Albanien ihre bedentung für Vegetation und Flora der Mitteleneerländer. In: L. Diels (Ed.), Biblioteca Botanica (Original-Abhandlungen aus dem Gesamtgebiete der Botanik), Heft 105, p. 110. (Erwin Nägele), G.M.B.H., Stuttdgart.

Marx, D.H., 1969. The influence of ectotrophic mycorrhizal fungi on the resistence of *Pine* roots to pathogenic infections. I. Antagonism of mycorrhizal fungi to roots pathogenic fungi and soil bacteria. Phytopathol. 59: 153–163.

Marx, S., W. Grosse and H.A. Schneider-Poetsch, 1988. A combined HPLC-ELISA assay applied to measure abscisic acid (ABA) and abscisyl-β-D-glucopyranoside in ripening embryos of walnut (*Juglans regia* L.) J. Plant Physiol. 133: 475–479.

Michler, C.H. and E.O. Bauer, 1991. High frequency somatic embryogenesis from leaf tissue of *Populus* spp. Plant Sci. 77: 111–118.

Murashige, T and F. Skoog, 1962. A revised medium for rapid growth and bioassays with tobacco tissue cultures. Physiol. Plant 15: 473–497.

Nitsch, J.D., 1956. Methods for the investigation of natural auxins and growth inhibitors. In: R.L. Wain and F.W. Wightman (Eds.), The Chemistry and Mode of Action of Plant Growth Substances, pp. 3–31. Butterworth, London/Acad. Press, New York.

Ozias-Akins, P., W.F. Anderson and C.C. Holbrook, 1992. Somatic embryogenesis in *Arachis hypogaea* L.: genotype comparison. Plant Sci. 83: 103–111.

Paupardin, C., 1983. Elaboration of secondary metabolites by tree or shrub tissues culture *in vitro*. Bull. Soc. Bot. Fr. 130 Actual. Bot. 2: 61–68.

Procter, R., 1977. Trees of the World. Hamlyn, London.

Profumo, P., R.M. Dameri and C. Orsino, 1976. Frammenti cotiledonari di *Aesculus hippocastanum* L. coltivati *in vitro* primi data sul comportamento dell' amido e dell' escina. Giorn. Bot. Ital. 110: 155–171.

Profumo, P., R.M. Dameri, C. Orsino and P. Modensi, 1980. Aescin content in callus from explants of *Aesculus hippocastanum* cotyledons grown *in vitro*. Giorn. Bot. Ital. 114: 25–28.

Profumo, P., P. Gastaldo, R.M. Dameri and L. Caffaro, 1986. Histological study of calli and embryoids from leaf explants of *Aesculus hippocastanum* L. J. Plant Physiol. 126: 97–103.

Profumo, P., P. Gastaldo and M. Martinucci, 1987a. Variation in aescin content in *Aesculus hippocatanum* seeds during the year. Fitoterapia 58: 184–186.

Profumo, P., P. Gastaldo and N. Rascio, 1987b. Ultrastuctural study of different types of callus from leaf explants of *Aesculus hippocastanum* L. Protoplasma 138: 89–97.

Profumo, P., P. Gastaldo and R.M. Dameri, 1989. Studio preliminare sui rapporti cotiledoni-asse embrionale in *Aesculus hippocastanum* L. Boll. Soc. It. Biol. Sper. 65: 603–608.

Profumo, P., P. Gastaldo, A.M. Gaviglia and R.M. Dameri, 1990. Somatic embryogenesis from cotyledonary explants of *Aesculus hippocastanum* L. Acta Embryol. Exp. 11: 101–106.

Profumo, P., P. Gastaldo, A.M. Caviglia and R.M. Dameri, 1991a. Somatic embryogenesis from cotyledonary explants of *Aesculus hippocastanum* L. Acta Embryol. Sper. 7: 122–126.

Profumo, P., A.M. Caviglia and R. Dameri, 1991b. Aescin content in embryogenic callus and in embryoids from leaf explants of *Aesculus hippocastanum* L. Planta Med. 57: 50–52.

Profumo, P., A.M. Caviglia and P. Gastaldo, 1992. Formation of aescin glycosides by callus tissue from cotyledonary explants of *Aesculus hippocastanum* L. Plant Sci. 85: 161–164.

Radojević, Lj., 1977. Physiological and cytological investigations of embryogenesis in tissue culture of *Corylus avellana* L., *Paulownia tomentosa* Steud. and *Aesculus hippocastanum* L. Ph.D Thesis, Univ. Belgrade, Yugoslavia (in Serbian).

Radojević, Lj., 1978. *In vitro* induction of androgenic plantlets in *Aesculus hippocastanum* L. Protoplasma 96: 369–374.

Radojević, Lj., 1979. Somatic embryogenesis and plantlets from callus cultures of *Paulownia tomentosa* Steud. Z. Pflanzenphysiol. 91: 56–62.

Radojević, Lj., 1980. Embryogenèse somatique et androgénèse chez certaine espèces ligneuse. Bull. Soc. Bot. Fr. 127, Actuel. Bot. 3/4: 99–107.

Radojević, Lj., 1985. Tissue culture of *Zea mays* "Cudu". I. Somatic embryogenesis in callus tissue. J. Plant Physiol. 119: 435–441.

Radojević, Lj., 1988. Plant regeneration of *Aesculus hippocastanum* L. (horse chestnut) through somatic embryogenesis. J. Plant Physiol. 132: 322–326.

Radojević, Lj., 1991. Horse Chestnut (*Aesculus* spp.). In: Y.P.S. Bajaj (Ed.), Biotechnology in Agriculture and Forestry, Vol. 16. Tree III, pp. 111–141. Springer-Verlag, Berlin/Heidelberg.

Radojević, Lj. and A. Kovoor, 1985. Induction of haploids. In: Y.P.S. Bajaj (Ed.), Biotechnology in Agriculture and Forestry, Vol. 1. Tree I, pp. 65–86. Springer-Verlag, Berlin/Heidelberg/New York.

Radojević, Lj. and A. Subotić, 1992. Plant regeneration of *Iris setosa* Pall. through somatic embryogenesis and organogenesis. J. Plant Physiol. 139: 690–696.

Radojević, Lj. and N. Marinković, 1993. Induction of somatic embryogenesis in immature ovule culture of *Aesculus hippocastanum* L. Arch. Biol. Sci. 45: 13–14.

Radojević, Lj., R. Vujičić and M. Nešković, 1975. Embryogenesis in tissue culture of *Corylus avellena* L. Z. Pflanzenphysiol. 77: 33–41.

Radojević, Lj., P. Druart and P. Boxus, 1987. Vegetative propagation of androgenous embryos of horse chestnut by meristem culture *in vitro*. Acta Hort. 212: 531–537.

Radojević, Lj., N. Djordjević and N. Gogala, 1988. Influence of different factors on somatic and androgenic *Aesculus* spp. embryos development. 6[th] Congr. FESPP, Split, Abstr. 1430.

Radojević, Lj., N. Djordjević and B. Tucić, 1989. *In vitro* induction of pollen embryos and plantlets in *Aesculus carnea* through anther culture. Plant Cell Tiss. Org. Cult. 17: 21–26.

Raj Bhansali, R, J.A. Driver, and D.J. Durzan, 1990. Rapid multiplication of adventitious somatic embryos in peach and nectarine by secondary embryogenesis. Plant Cell Rep. 29: 280–284.

Rajasekaran, K. and M.G. Mullins, 1979. Embryos and plantlets from cultured anthers of hybrid grapevines. J. Exp. Bot. 30: 399–407.

Ranch, J.P., L. Oglesby and A.C. Zielinski, 1985. Plant regeneration from embryo-derived tissue cultures of soybeans. In Vitro 21: 653–658.

Rangaswamy, N.S., 1986. Somatic embryogenesis in angiosperm cell, tissue and organe culture. Proc. Indian Acad. Sci. 96: 247–271.

Richardson, I.B.K., 1978. Dicotyledon families. In: V.H. Heywood (Ed.), Flowering Plants of the World, pp. 194–195. Univ. Press, Oxford.

Rosendahl, R.O., 1942. The effect of mycorrhizal and non-mycorrhizal fungi on the availability of difficulty soluble potassium and phosphorus soil. Sci. Soc. Amer. Proc. 7: 477–479.

Rout, G.R., B.K. Debata and P. Das, 1991. Somatic embryogenesis in callus culture of *Rosa hybrida* L. ev. Landora. Plant Cell Tiss. Org. Cult. 27: 65–69.

Routien, J.B. and R.F. Dawson, 1943. Some interrelationships of growth, salt absorption, respiration and mycorrhizal development in *Pinus echinata*. Amer. J. Bot. 30: 440–449.

Sasamoto, H. and Y. Hosoi, 1989. Somatic embryogenesis in suspension cultures of *Quercus serrata* Thunb. J. Jpn. Soc. 71: 20–22.

Sellars, R.M., G.M. Southward and G.C. Phillips, 1990. Adventitious somatic embryogenesis from cultured immature zygotic embryos of peanut and soybean. Crop Sci. 30: 408–414.

Senaratna, T., B.D. McKersie and S.R. Bawley, 1990. Artificial seeds of alfalfa (*Medicago sativa* L.): induction of desiccation tolerance in somatic embryos. In Vitro Cell Dev. Biol. 26: 85–90.

Seymour, A.B., 1929. Host Index of the Fungi of North America. Harvard Univ. Press, Cambridge, Massachusetts.

Sommer, H.E. and C.L. Brown, 1979. Application of tissue culture to forest tree improvement. In: Sharp, W.R., Larsen, P.O. and V. Ragavan (Eds.), Cell and Tissue Culture, Principles and Applications, pp. 461–491. Ohio Univ. Press, Columbus, OH.

Söndahl, M.R. and W.R. Sharp, 1977. High frequency induction of somatic embryos in cultured leaf explants of *Coffea arabica* L. Z. Pflanzenphysiol. 81: 395–408.

Söndahl, M.R., D.A. Evans and W.R. Sharp, 1980. Coffee cell culture. Growth and somatic embryogenesis. Newslett. Int. Ass. Plant Tissue Culture 30: 2–7.

Staritsky, G., 1970. Embryos formation in callus tissues of coffee. Acta Bot. Neerl. 19: 509–514.

Stevanović, V. and Lj. Radojević, 1993. Distribution and area under cultivation of *Aesculus hippocastanum* L. Arch. Biol. Sci. 45: 15–16.

Steward, F.C., M.O. Mapes and J. Smith, 1958. Growth and organized development of cultured cells. I. Growth and division of freely suspended cells. Amer. J. Bot. 45: 693–703.

Tachtadzhyana, A.L., 1980. Flowering Plants. Prosveshchenie, Moscow (in Russian).

Tétu, T., R.S. Sangwan and B.S. Sangwan-Norreel, 1990. Direct somatic embryogenesis and organogenesis in cultured immature zygotic embryos of *Pisum sativum* L. J. Plant Physiol. 137: 102–109.

Thevenot, C. and D. Côme, 1973. Initiation de la germination a l'axe embryonaire par les cotyledons chez le pommier (*Prunus malus* L.). C.R. Acad. Sci. 277: 1873–1876.

Tucović, A., 1973. Fam. *Hippocastanaceae*. Flora Srbije 5: 103–110.

Tulecke, W. and G. H. McGranahan, 1985. Somatic embryogenesis and plant regeneration from cotyledons of walnut (*Juglans regia* L.). Plant Sci. 40: 57–63.

Tulecke, W., G. McGranahan and H. Ahmadi, 1988. Regeneration by somatic embryogenesis of triploid plants from endosperm of walnut (*Juglans regia* L. cv. Mangerian). Plant Cell Rep. 7: 301–304.

Tutin, T.G., V.H. Heywood, N.A. Burges, D.M. Moore, D.H. Valentine, S.M. Walters and D.A. Webb, 1968. Flora Europaea, *Rosaceae* to *Umbelliferae*. Cambridge Univ. Press, Cambridge.

Vahala, T., P. Stabel and T. Eriksson, 1989. Genetic transformation of willows (*Salix* spp) by *Agrobacterium tumefaciens*. Plant Cell Rep. 8: 55–58.

Vardi, A., S. Bleichman and D. Aviv, 1990. Genetic transformation of *Citrus* protoplasts and regeneration of transgenic plants. Plant Sci. 69: 199–206.

Von Kartnig, Th., R. Herbst and F.J. Graune, 1965. Zur Frage der Sapononverteilung bei *Aesculus hippocastanum* Während Verschiedener Keimungs- und Wachstumsstadien. 2. Mitteilung: Über den jahreszeitlichen Gehalt einiger Organe von *Aesculus hippocastanum* an Aescin. Planta Med. 1: 39–45.

Von Kartnig, Th., F.J. Graune and R. Herbst, 1966. Zur Frage der Sapononverteilung bei *Aesculus hippocastanum* Während Verschiedener Keimungs- und Wachstumsstadien. 3. Mitteilung: Die Bildung von Aescin bei *Aesculus hippocastnum* und *Aesculus parviflora* nach der Verabreichung von Mevalonsäurelacton. Planta Med. 2: 121–125.

Von Kartnig, Th. and A. Hiermann, 1968. Saponinverteilung bei *Aesculus hippocastnum* Während Verschiedener Entwicklungsstaiden. 4. Mitteilung: Beobachtungen über die Bildung von Aescin bei jungen Pflanzen nach Verabreichung von 2–$^{14}$C-Mevalonsäurelacton (2–$^{14}$C-MSL). Planta Med. 1: 109–113.

Walkey, D.G.A., 1972. Production of apple plantlets from axillary-bud meristem. Can. J. Plant Sci. 52: 1082–1085.

Wang, D.Y. and I.K. Vasil, 1982. Somatic embryogenesis and plant regeneration from inflorescence segments of *Pennisetum purpureum* Schum. (Napier of elephant grass). Plant Sci. Lett. 25: 147–154.

Xu, N., K.M. Coulter and D.J. Bewley, 1990. Abscisic acid and osmoticum prevent germinatin of developing alfalfa embryos, but only osmoticum mainteins the synthesis of developmental proteins. Planta 182: 382–390.

# 23. Somatic embryogenesis in apple (*Malus*)

A. Wallin, M. Nyman and M. Svensson

**Contents**



## 1. Introduction

### 1.1. *The species*

The apple is cultivated in temperate regions all over the world. The commercially produced scion cultivars belong to the species *Malus* × *domestica* Borkh. and most of the rootstocks to *Malus pumila* Mill. (Way *et al.*, 1990). Established apple cultivars as well as apple rootstocks are highly heterozygous. Most scions are difficult to root from shoot cuttings. In order to get new plants and to preserve the intact genome of a desired cultivar, scions are propagated by budding or grafting onto seedlings or vegetatively propagated rootstocks.

Most apple cultivars are diploid (2n = 34), but about 10% of the most common varieties are triploid (Way *et al.*, 1990). Triploids are often higher yielding than diploids, and they have a high amount of vitamin content (Isaev and Domracheva, 1981). All *Malus* cultivars together constitute a wide genetic base. Horticulturally desirable traits such as increased cold hardiness and drought tolerance, disease and insect resistance, late blooming for avoidance of late spring frost, long storage life of the fruits and self-fertile cultivars could be identified and transferred to the cultivated varieties from old or not widely-grown cultivars or from wild *Malus* species. For example of desirable horticultural traits in *Malus* (see Way *et al.*, 1990). Today, the main objective of plant breeders is to breed cost-effective cultivars with regularly heavy cropping cultivars with high quality and long storage (Alston, 1987). For the future, disease and pest resistance are of considerable importance. Many insecticides and fungicides that are frequently applied are considered unsafe

*S. Jain, P. Gupta & R. Newton (eds.), Somatic Embryogenesis in Woody Plants, Vol. 2*, 445–460.

and had to be substituted for by improved genetic resistance in varieties with great demand.

Most of the qualities that go into making a good quality apple cultivar are under polygenetic control, for example: yield, skin finish, fruit colour, fruit size, acidity, storage and texture (Alston, 1987). This means that when two cultivars are intercrossed, there will be a wide and continuous range of expression of all these characters (Brown, 1975). Besides high heterozygosity, the long juvenile period of the seedlings obstructs efficient conventional breeding. Genetic engineering and mutation breeding offer promising tools, especially in combination with other techniques which are available to the plant breeder, for example, shortening of the juvenile phase and *in vitro* selection of resistant individuals. In order to change single genes in established cultivars, special attention has been drawn to mutation breeding, somaclonal variation and gene transfer. A number of potentially useful clones have been obtained by irradiation (X-, gamma- and neutron irradiation) of dormant scions or by chemical treatment of growing branches (Broertjes and Van Harten, 1988). At Long Ashton, for example, compact-type Bramley's and self-fertile mutant clones of Cox have been produced (Lacey and Campbell, 1987). A somaclone of an apple clonal rootstock with improved resistance to *Phytophthora cactorum* was found by Ancherani *et al.* (1989). Marker genes have been introduced into *Malus* (James *et al.*, 1988, 1989, 1993a,b) using *Agrobacterium tumefaciens* mediated gene transfer. Apple trees expressing the CrylA (c) gene from *Bacillus thuringiensis* and conferring resistance to insects have been planted in field trials in California (James *et al.*, 1993a). The gene encodes for modified endotoxins produced as intracrystalline proteins which spoil the digestion in the gut of the insect.

### 1.2. *Regeneration of plants*

Apple plants can be regenerated *in vitro* from cultured meristems, or by formation of somatic embryos and adventitious buds. During the last decade, *in vitro* propagation of apple scions as well as rootstocks have been commercially established (Debergh and Zimmerman, 1990). In order to maintain the genetic stability, shoots from meristems are commonly used for mass propagation. *In vitro* shoots are induced to form axillary shoots by treatment with cytokinin (for review see Wilkins and Dodds, 1983; James, 1987) or chemicals with similar effect for example thidiazuron. Thidiazuron is highly efficient for the initiation of both axillary (Van Nieuwkerk *et al.*, 1986) and adventitious shoots (Fasolo *et al.*, 1989; Wallin and Johansson, 1989) in *Malus*. The axillary shoots are rooted *in vitro*. Normally, excised shoots are transferred to a root-inducing medium, a medium with an auxin often indole-3-butyric acid (IBA). Rooting of apple *in vitro* is not without problems. It varies greatly between cultivars but this insufficiency can often be overcome by treatments such as changed growth conditions, rejuvenation of shoots by repeated subcultures, addition of phloroglucinol, a root-stimulating chemical

(Zimmerman and Broome, 1981) or by excluding $NH_4NO_3$ in the rooting medium (Sriskandarajah *et al.*, 1990). Phloroglucinol, 1,3,5-trihydroxybenzoic acid, is a phenolic compound that has a positive synergistic effect on root initiation when combined with auxin (James, 1983; Alvarez *et al.*, 1989).

### 1.3. *Why somatic embryogenesis*?

The aims of somatic embryogenesis research in *Malus* are to:

a) get an alternative method for clonal propagation of intact plants, and
b) obtain a tool for breeding.

The genus *Malus* seems to have a morphogenic potential both via somatic embryogenesis and adventitious bud formation (for review see James, 1987). Somatic embryos of apple can originate from the seed, from sexual organs or from highly juvenile tissue in very young parts of the shoot. In seeds, adventitious embryos have been initiated from the nucellus (Eichholtz *et al.*, 1979; James *et al.*, 1984a; Saito *et al.*, 1989) and from the endosperm (Mu *et al.*, 1977; James *et al.*, 1984a; Svensson, 1986) (Table 1). Embryogenesis can be induced directly or via embryogenic calli.

Adventive embryogenesis has been obtained from flower buds and petals (Mehra and Sachdeva, 1984) (Table 1) and from leaves of *in vitro* grown seedlings (Liu *et al.*, 1983; Kromer, 1984; Mehra and Sachdeva, 1984). Welander (1988) also found embryoid-like structures on leaf explants from *in vitro* shoots of mature trees. Since repeated subculture treatment reverses plants from the adult growth phase to a juvenile one, these shoots had presumably been rejuvenated. Recently, structures resembling proembryos were also shown in callus induced from protoplasts of apple seedlings (Perales and Schieder, 1993).

Adventitious buds of apple have been regenerated into shoots from callus of stem and leaf tissue (James *et al.*, 1984b; Barbieri and Morini, 1987; Belaizi *et al.*, 1991) and directly from leaf-discs (James *et al.*, 1986a; Welander, 1988; Hanke *et al.*, 1991).

### 1.4. *The seed tissues and breeding aims*

The zygote in the apple seed is surrounded by the endosperm, the nucellus and by an inner and outer integument (Fig. 1). The genome in the nucellus cells is diploid and of maternal origin. Nucellular embryos might be a source for production of juvenile mother shoots. The endosperm originates from the fusion of two female synergids and one male sperm and is triploid. Most triploid apple varieties are results of open pollination of a normal egg cell in a diploid variety with unreduced gametes. It is believed that triploids derived from an unreduced diploid egg cell and fertilized by haploid pollen are superior to those derived from crossing of tetraploids × diploids (Brown, 1975). By *in vitro* culture the triploid endosperm cells can be the origin of

*Table 1.* Adventitious embryogenesis and callus formation from nucellus and endosperms and from flower buds in different apple cultivars. A review.

| Material | Basal medium | Growth regulators | Response | References |
|---|---|---|---|---|
| nucellus, 50d Golden Delicious | MS | no | adv. embryos | Eichholtz *et al*. 1979 |
| | | NAA (0.1 mg/l)<br>BA (3.0 mg/l) | callus | |
| | | NAA (3.0 mg/l)<br>BA (0.3 mg/l) | callus | |
| nucellus and/or endosperm, 50 d | LS | no | adv. embryos in all except in Cox and M26 | James *et al*. 1984a |
| Bramley, Cox, Greensleeves, Spartan, M.9, M.25, M.26 and M.27 | | | callus in all except in Bramley | |
| nucellus, 30 d | LS | no | no adv. embryos or callus | James *et al*. 1984a |
| nucellus, 30 d | LS | 2,4-D or NAA (1 mg/l) plus BA (1 or 5 mg/l) | callus | James *et al*. 1984a |
| nucellus, 50 d Fuji and CG 80 | MS | 1) aux + cyt for callus initiation<br>2) IAA (0.1 mg/l) and kinetin (0.1 mg/l) for induction of adv. embryos | callus in 1)<br>adv. embryos after transfer to 2) | Saito *et al*. 1989 |
| endosperm, 50 or 100 d<br>6 Chinese cultivars | MS | 1) aux + cyt for callus initiation<br>2) BA (0.1–1.0 mg/l) and NAA (0.01–0.05 mg/l) plus casein hydrolysate for induction of adv. embryos | callus in 1)<br>adv. embryos after transfer to 2) | Mu.*let al*. 1977 |
| endosperm, 50 d, Aroma | LS | no | adv. embryos and callus | Svensson 1986 |
| flower buds and excised petals, Golden Delicious | Nitsch's | NAA and BA (2 mg/l) | embryogenic, nodular, brownish-green calli with embryoids at various stages of development | Mehra & Sachdeva, 1984 |

adv. = adventitious, aux

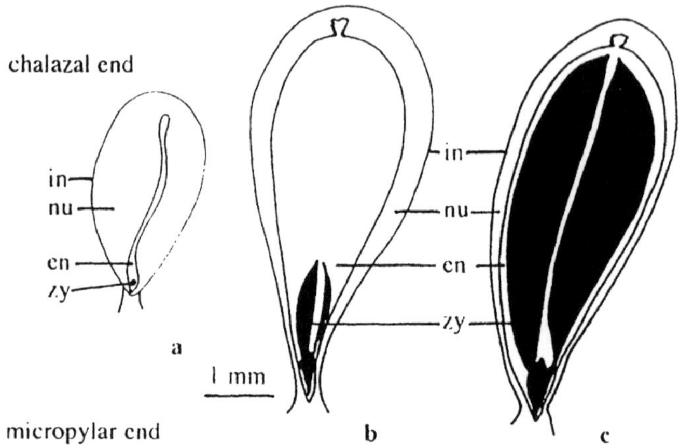


*Figure 1.* Apple seed longitudinal sections. (a) About 30 days post pollination. (b) About 50 days post pollination. (c) About 60 days post pollination. in – integuments (inner and outer); nu – nucellus; en – endosperm; zy – zygote.

somatic embryos and plants. This could thus become an alternative approach to produce new triploids. The endosperm has previously been *in vitro* cultured to produce shoots or embryos in a number of species (Bhojwani, 1984).

### 1.5. *Haploids*

Androgenesis has been reported from anther and pollen cultures (Milewska-Pawliczuk, 1990). Xue *et al.* (1990) successfully cultured pollen plantlets of 8 apple cultivars. Embryogenesis in anther cultures has been examined by Johansson (1986) and Zhang *et al.* (1987, 1992), and regenerated plants "thought" to be haploid were produced by the latter group. Introduction of haploids makes it possible to perform breeding at a reduced ploidy level.

In this chapter the most interesting results in somatic embryogenesis in *Malus* will be discussed and practical advice from experience gained in research with a Scandinavian cultivar will be presented.

## 2. Induction of adventitious embryos from nucellus and endosperm, directly or via embryogenic callus

### 2.1. *Material and methods*

Open-pollinated seeds of the cultivar Aroma, Balsgård, were used as a model system. Unripe apples were dissected 30–60 days after pollination to isolate the seed tissues. The apples were stored in darkness in plastic bags at 4°C until dissected (up to 8 weeks). Prolonged storage (3–4 months) resulted in drying in which case the tissues were difficult to separate and had a decreased viability. The apples were surface sterilized in 70% ethanol for 1 min, 10% Chlorine with 2 drops of Tween/50 ml for 5 min, and were rinsed three times in sterile, distilled water. The seeds were kept in a solution with 2% (w/v) polyvinylpyrrolidone (PVP) diluted in water and dissected under the binocular microscope.

In order to find a suitable seed age for the induction of somatic embryogenesis, 20-, 30-, 40-, 50- and 60-day-old fruits, were tested. Intact seeds from the 20-day-old fruits were cut longitudinally and placed on the medium (see below). At 30 days from anthesis, most of the immature seed is composed of the nucellus (Fig. 1a). In 30-day-old seeds, the integuments were removed and the inner part of the seed was cut longitudinally and placed with the cut in direct contact with the medium. Fifty days after pollination, the endosperm is cellular and has a dividing peripheral meristem, the zygote is easy to detach, and the nucellus has begun to degenerate (Fig. 1b). The integuments were removed from these seeds, and the nucellus and endosperm, were cut longitudinally, respectively. The endosperm parts were cultured with or without the presence of the zygote. In 60-day-old seeds, both the nucellus and endosperm were degenerating, and the zygote occupied the main part of the seed (Fig. 1c). Longitudinal halves of the nucellus and endosperm were cultured *in vitro*.

In accordance with Eichholtz *et al.* (1979) and James *et al.* (1984a), hormone-free MS medium (Murashige and Skoog, 1962) or LS medium (Linsmaier and Skoog, 1965) gelled with Gelrite Gellan Gum (0.3% w/v) and supplemented with 3% sucrose for somatic embryogenesis, was used.

Adventitious embryos were formed in endosperms from 40 and 50-day-old seeds (Fig. 2a; Table 2). In a single experiment, adventitious embryos from the nucellus in 50-day-old seeds, were also obtained (Fig. 2c). Adventitious embryos were always initiated at the micropylar end in accordance with Eichholz *et al.* (1979) and James *et al.* (1984a). Calli were also formed in endosperm halves from 40 to 60-day-old seeds. In tissues from young seeds (20 and 30-day-old fruits), no growth was observed.

On MS medium containing 2,4-D (1 mg/l) and BA (1 mg/l), no embryos were developed in tissues from 20 to 60-day-old fruits, but calli were formed in seeds of all tested ages, except for the nucellus halves. In 20-day-old seeds, the callus was very compact, and in older tissues it was more friable and often light-yellow in colour (Fig. 2g). Histological studies revealed that the friable callus was embryogenic (Fig. 2h). Embryogenic cells, rich in starch and with thick cell walls, were also observed in electron micrography (Fig. 3). Presumably, the older (40 to 60-day-old) endosperms contain enough endogenous plant hormones for callus growth without additional exogenous ones. The endosperm is known to be the main hormone producer in the seed. For indirect formation of adventitious embryos, embryogenic calli have

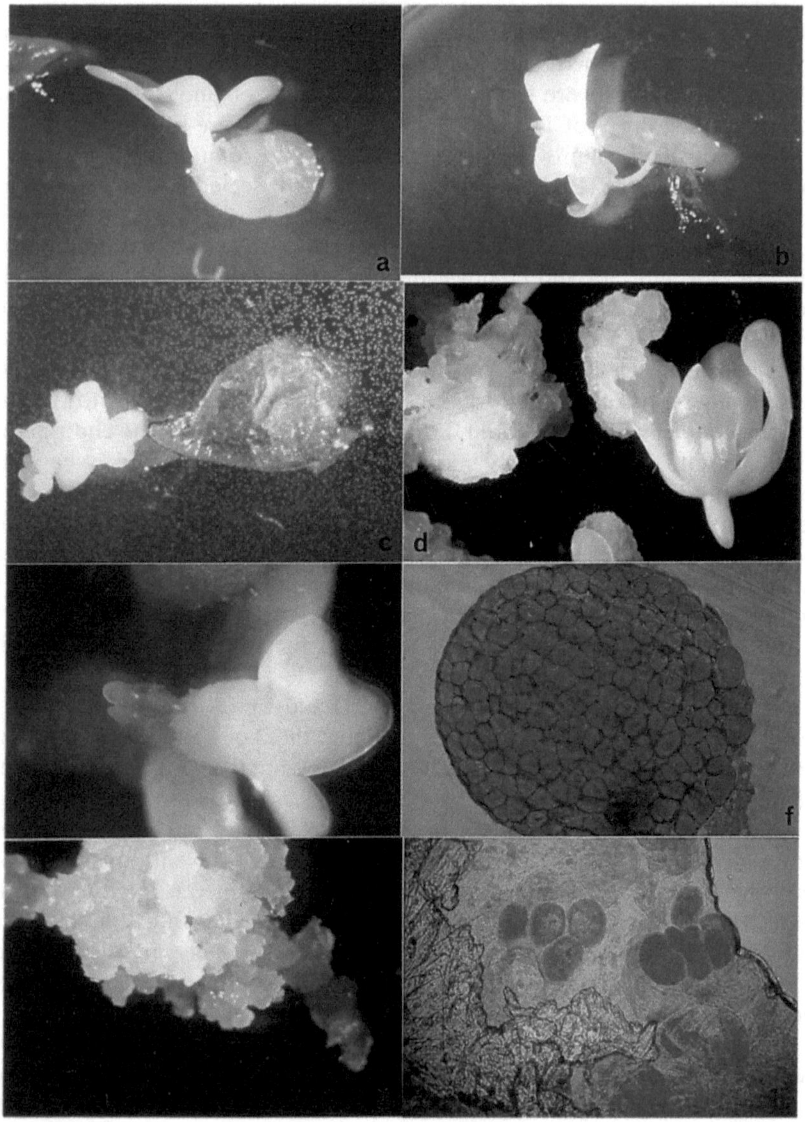


*Figure 2.* Embryos and embryogenic callus of the apple cultivar Aroma. The pictures are taken after 7–8 weeks in *in vitro* culture on MS medium without hormones. (a) An embryo from the endosperm in a 50-day-old seed. (b) A secondary embryo developed from the embryo in (a) some week later. (c) An embryo from the nucellus (micropylar end) in a 50-day old-seed. (d) Embryogenic callus. Growth stimulated by presence of a zygote. (e) Secondary embryos. (f) Longitudinal section of a globular secondary embryo. (g) Embryogenic callus. (h) Proembryos in embryogenic callus (as in (g)).

*Table 2.* Formation of adventitious embryos and embryogenic callus in tissues from Aroma seeds in MS medium without hormones (−h) or with 1 mg/l 2,4-D plus 1 mg/l BA (+h). In seeds 20 days from anthesis, integument, endosperm and nucellus were longitudinally cut and placed on medium. In 30-day-old apples the integuments were removed before cutting in longitudinal halves. In 40-, 50- and 60-day-old apples, longitudinally cut endosperm halves were tested as well as nucelli from 50- and 60-day-old. The zygote was excluded. Growth after 10 weeks.

| Material/Medium | Callus | Adventitious embryos |
|---|---|---|
| 20 days old, whole seed | | |
| MS − h | no | no |
| MS + h | yes, compact | no |
| 30 days old, seed − integument | | |
| MS − h | no | no |
| MS + h | no | no |
| 40 days old, endosperm | | |
| MS − h | yes, friable | yes |
| MS + h | yes, friable | no |
| 50 days old, endosperm | | |
| MS − h | yes, friable | yes |
| MS + h | yes, friable | no |
| 50 day old, nucellus | | |
| MS − h | no | no |
| MS + h | no | no |
| 60 day old, endosperm | | |
| MS − h | yes, friable | no |
| MS + h | yes, friable | no |
| 60 day old, nucellus | | |
| MS − h | no | no |
| MS + h | no | no |

been transferred to MS medium containing auxin and cytokinin (Mu *et al.*, 1977; Saito *et al.*, 1989).

The presence of the zygote stimulated both adventitious embryogenesis (not shown) and callus production (Fig. 2d; Table 3). The addition of activated charcoal (0.5%) inhibited callus formation (Table 3), and the growth of adventitious embryos started later on a medium with charcoal.

Other changes of the culture medium were also tested. Sorbitol was tested as a carbon source. Sorbitol is a major soluble carbohydrate in plants belonging to Rosaceae (Bieleski, 1982) and has been superior to sucrose for *in vitro* propagation of some *Malus* species (Pua and Chong, 1984). Sorbitol (3%) was not more efficient than sucrose in stimulation of embryogenesis.

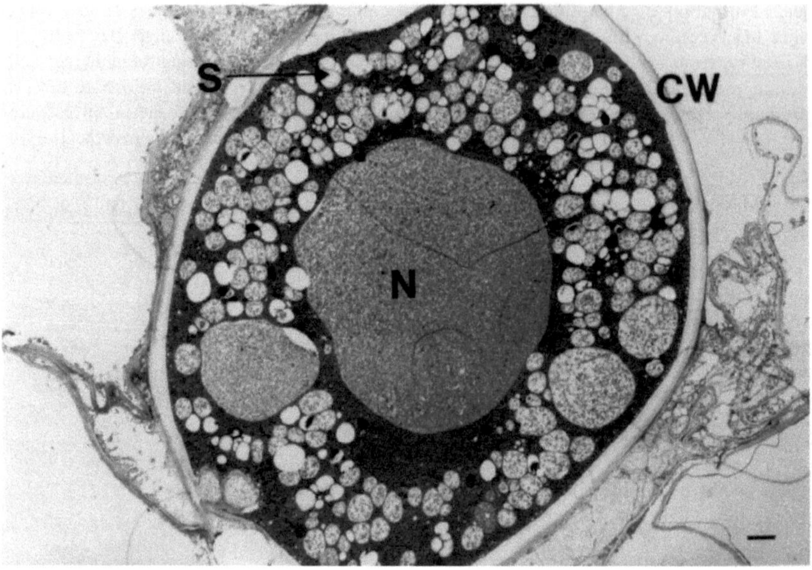


*Figure 3.* Cell from endosperm callus from the apple cultivar Aroma. Electron micrograph. Mark 1 μm. CW – cell wall; N – nucleus; S – starch grain.

*Table 3.* Growth of embryogenic calli from endosperm halves from Aroma seeds 40 and 50 days after pollination in absence and presence of the zygote. Callus growth was estimated in $mm^2$ after 10 weeks in culture. The values are mean values of $n > 20$ calli.

| Material | Callus growth | | |
|---|---|---|---|
| | MS – h | MS + h | MS – h + AC |
| En + zygote, 40 days | 0.71 | 0.78 | 0.35 |
| En, 40 days | 0.62 | 0.51 | 0.02 |
| En + zygote, 50 days | 1.09 | 1.02 | 0.64 |
| En, 50 days | 1.09 | 1.02 | 0.64 |
| En, 50 days | 0.69 | 0.52 | 0.23 |

En – endosperm halves; h – 2,4-D and BA (1 mg/l); AC – activated charcoal (0.5%).

In anther cultures of Anemone and Papaver an elevated $CO_2$ concentration stimulated embryogeneisis up to 5 times. We had indications that an increased level of $CO_2$ (5%) in the atmosphere and decreased nitrogen content (1/2 nitrogen) would increase the amount of embryogenesis (Table 4). The experiment was repeated the following year. The tendency was similar but the total capacity to form embryos was strongly reduced compared to earlier years due to the very bad year for apple fruit production. The immature fruits were dry and small for their age.

*Table 4.* Adventitious embryogenesis from endosperm of Aroma. The endosperms were isolated from seeds in fruits 50 days after pollination. The endosperms were cut longitudinally and cultured on hormone-free MS in air or on medium with normal or half nitrogen concentration in an atmosphere with 5% $CO_2$.

| Year | Normal N Air | | | Normal N $+CO_2$ | | | 1/2 N Air | | | 1/2 N $+CO_2$ | | |
|---|---|---|---|---|---|---|---|---|---|---|---|---|
| | a | b | c | a | b | c | a | b | c | a | b | c |
| 1985 | 12 | 100 | 35 | | | | | | | | | |
| 1986 | 23 | 50 | 5 | 7 | 60 | 24 | | | | | | |
| 1986 | 1 | 111 | 1 | | | | 5 | 90 | 5 | | | |
| 1987 | 0 | 100 | 0 | 1 | 100 | 4 | 1 | 100 | 3 | 4 | 100 | 22 |

a – Number of endosperm halves giving adventitious embryos after 10 weeks; b – Total number of endosperm halves; c – Total number of adventitious embryos after 10 weeks from all endosperm halves (primary + secondary).

Great differences between cultivars were reported by James *et al.* (1984a). In a comparable test between Aroma and Greensleeves cultivars, Aroma was observed to form less adventitious embryos and callus than was Greensleeves. This result support that Aroma is a recalcitrant *Malus* cultivar in *in vitro* culture. We have earlier had problems in establishing and cultivating Aroma shoots. For a model study, a cultivar that is easier to handle, is preferred.

### 3. Germinaton of adventitious and zygotic embryos

From the primary adventitious embryo initiated from Aroma seed tissues, new secondary embryos were frequently formed, most often from the hypocotyl-root axis (Fig. 2b,e). The development of the adventitious embryos was similar to that of the zygotic embryo, but the adventitious embryos were normally smaller and weaker and more transparent. Globular pro-embryos (Fig. 2f) became heart-shaped and the cotyledons enlarged (Fig. 2b). It was difficult to obtain hypocotyl elongation and to get vigorously growing shoot and root meristems. Futhermore, some adventitious embryos had more than two cotyledons. This occurs in zygotic embryos too (Fig. 2d).

Adventitious embryos of Golden Delicious developed embryo-like structures when placed in the dark, but embryos in light did not undergo morphogenesis (Eichholz *et al.*, 1979). Some of the adventitious embryos from the Fuji cultivar germinated and formed shoots, but only when they were transferred to light after being at 0°C in darkness for 6 months (Saito *et al.*, 1989). Zygotic embryos stored in darkness for 12–14 weeks germinated into seedlings after transfer to light and a medium with $GA_3$ (gibberellic acid); however, adventitious embryos did not respond (James *et al.*, 1984a).

In order to obtain maturation and germination of adventitious embryos a treatment involving the transfer to fresh hormone-free medium or a medium with BA (1 mg/l) under light (50 $W/m^2$, 16 h, 20°C), was tested. On hormone-free medium, 11% of the adventitious Aroma embryos grew into plantlets. On the BA medium, 34% of the embryos turned green and produced real leaves and could be cultured as shoot cultures. Such shoots had to undergo a rooting procedure. For details, see Svensson (1986).

In accordance with zygotic embryos of apple, adventitious embryos of most cultivars need to be stratificated to germinate. The level of abscisic acid (ABA) had to be reduced (Ryć and Lewak, 1980). A cold treatment (4–5 weeks) is normally sufficient (Dickie and Bowyer, 1985).

Reports about trees originating from adventitious embryos tested in the field have not been found.

## 4. Confirmation of genotype

In adventitious embryos initiated from seed tissue, the origin of the genome is not convincingly proved in all reports. The most comprehensive description of chromosome constitution of embryos is given by Mehra and Sachdeva (1984) in embryos initiated from calli of different explants. Mu *et al.* (1977) examined endosperm-initiated callus and obtained a chromosome number between 50–52 in most cells. Ploidy levels in plantlets were not examined. James *et al.* (1984a, 1986a,b) carried out controlled pollination with *Malus* baskatong pollen in order to distinguish the diploid genome of the maternal origin from that of the zygote. *Malus* baskatong pollen carries a dominant homozygous gene for purple pigmentation which is expressed in several plant parts. The purple colour was detected both in immature zygotes and in *in vitro* propagated shoots from the zygote as well as in plantlets developed from adventitious embryos and originating from nucellus and endosperm tissue. Svensson (1986) determined the ploidy level of the adventitious embryos by photofluorescentic DNA-measurements and found in preliminary results that some of the endosperm-derived adventitious embryos were most likely triploids. However, photofluoroscentic DNA measurements are statistically less significant than flow cytometric determinations.

A more thorough examination of the genotype of the germinated plants is necessary because there is a risk that single cells from the radicle of the zygote can penetrate the endosperm or the nucellus and can be the origin of the adventitious embryos. Hitherto, all reported adventitious embryos have been initiated at the micropylar end where the zygote has a close contact with other seed tissues.

For the determination of ploidy levels in apple, determination of the DNA content in nuclei can be determined with flow cytometric methods. Many nuclei can be rapidly counted and the coefficient of variation in apple is low (Wallin and Johansson, 1989). In a more detailed analysis of the genome, DNA based methods have to be utilized. Apple cultivars have been differentiated by M13 "DNA fingerprinting" (Nybom *et al.*, 1990) with the use of several restriction enzymes. Paternity could be provided in nearly 90% of seedlings with a similar analysis (Nybom and Schaal, 1990). Analysis of random amplified polymorphic DNA (RAPD) would be a still more efficient method for the future. Restriction fragment length polymorphisms (RFLP) can be used to detect chloroplast and mitochondrial DNA variations and trace maternal lineages in apple (Ishikawa *et al.*, 1992).

An alternative way for production of haploids, dihaploids or gynogenetic embryos could be the use of irradiated pollen (James *et al.*, 1986b). At low doses of ionizing radiation, the pollen can grow, penetrate, fertilize the egg and effect syngamy. At high doses the radiation results in an increased amount of embryo death and abnormality, penetration is obstructed and the egg is stimulated to gynogenetic development (parthenogenesis). This is called the "Hertwig effect" in plants (Pandey and Phung, 1982). Gamma-irradiated pollen of *Malus* baskatong were used for pollination of apple flowers by James *et al.* (1986a,b). Purple pigmentation was detected in the zygotic embryos of control pollen. If heavily irradiated pollen (30 kRad) was used, embryos that were "green-only" were derived from crosses (James *et al.*, 1986a,b).

## 5. Conclusion

Plant regeneration and breeding in apples with of somatic embryogenesis has not been as successful as it has been with *Citrus* species. *Citrus*, which has served as a model plant for many of the experiments above, has many similarities with apple including apomixis, which indicates that it can produce seeds from unfertilized eggs. In *Citrus*, nucellus-derived protoplasts can readily be induced to differentiate embryos which then can regenerate fruit-bearing trees (Vardi *et al.*, 1982). Intra- and interspecific somatic hybrids (Vardi and Galun, 1988; Grosser and Gmitter, 1990) have contributed to genetic variation obtained in *Citrus* scions and rootstocks, variation that was inaccessible previously.

In apple, plants of different ploidy level can be produced via embryogenesis. Most apple cultivars can directly (or via a callus phase), give rise to a great number of adventitious embryos. However, in most of the cultivars that have been tested, the embryogenic potential is not as high as in a species such as carrot. A cold treatment or other procedures to obtain embryo maturation and germination would be a major obstacle in mass propagation procedure. The technique is sufficient for: the production of plants with a genotype of specific interests, as an alternative in cultivars impossible to root, or to produce disease-free plants of maternal genotypes. Another specific niche would be cryopreservation of embryos for long-time storage of apple germplasm.

Hitherto, plants developed from adventitious embryos of nucellus or endosperm tissue have been incompletely assessed, and continued experimentation is needed in order to determine both ploidy level and origin of the genomes and to evaluate the plants in practical tests. From East Malling Research Station it has been confirmed that their adventitious embryos, originating from control seed tissues, had zygotic origin. In order to obtain diploids with the maternal genome, use of flower buds appears to be a more reliable material for initiation than from nucellus tissue.

One of the main interests in apple embryogenesis is the association with techniques for widening the gene pool. Unicellular somatic embryos are suitable for DNA transfer to form plants with new genes with a very low risk for chimeras. Protoplasts from plants or seed tissues of different ploidy levels could be utilized for fusion with the objective of combining interesting genomes into new diploids and/or triploids. Protoplasts of nucellular origin have already been fused with leaf protoplasts, but no plant regeneration was reported (Saito *et al.*, 1989). In Rosales, a somatic hybrid between sexually incompatible wild pear and Colt cherry, has been produced (Ochatt *et al.*, 1989).

Somatic embryogenesis in apple is a fascinating area which very likely will give new enthralling reports in the future. Regeneration of intact plants via adventitious shoots may in many cases be a better alternative.

## 6. Acknowledgements

This project was supported by the Swedish National Board for Technical Development. The authors are thankful to Elna Sjöström who patiently dissected the apple seeds during the first year.

## 7. References

Alston, F.H., 1987. Strategy for apple and pear breeding. In: A.J. Abbott and R.K. Atkin (Eds.), Improving Vegetatively Propagated Crops, pp. 113–123. Academic Press, Harcourt Jovanovich Publ., London, San Diego, New York.

Alvarez, R., S.J. Nissen and E.G. Sutter, 1989. Relationship between indole-3-acetic acid levels in apple (*Malus pumila* Mill.) rootstocks cultured *in vitro* and adventitious root formation in the presence of indole-3–butyric acid. Plant Physiol. 89: 439–443.

Ancherani, M., S. Predieri and P. Rosati, 1989. Adventitious shoots formation from *in vitro* leaves of MM106 apple clonal rootstock. In: International Symposium In Vitro Culture and Horticultural Breeding, Cesena, 1989, p. 28.

Barbieri, C. and S. Morini, 1987. *In vitro* regeneration from somatic tissues and seed explants of apple. Adv. Hort. Sci. 1(1): 8–10.

Belaizi, M., H. Paul, R.S. Sangwan and B.S. Sangwan-Norreel, 1991. Direct organogenesis from internodal segments of *in vitro* grown shoots of apple cv. Golden Delicious. Plant Cell Rep. 9: 471–474.

Bhojwani, S.S., 1984. Culture of endosperm. In: I.K. Vasil (Ed.), Cell Culture and Somatic Cell Genetics of Plants, Vol. I, pp. 258–268. Academic Press, New York.

Bieleski, R.L., 1982. Sugar Alcohols. Encyclopedia of Plant Physiology 13A: 158–192. Springer, Berlin.

Broertjes, C. and A.M. van Harten, 1988. Applied mutation breeding for vegetatively propagated crops, Development in Crop Science 12, pp. 229–245. Elsevier, Amsterdam.

Brown, A.G., 1975. Apples. In: J. Janick and J.N. Moore (Eds.), Advances in Fruit Breeding, pp. 3–37. Purdue Univ. Press, West Lafayette, Indiana, U.S.A.

Debergh, P. and R.H. Zimmerman (Eds.), 1990. Micropropagation: Technology and Application. Kluwer, Dordrecht.

Dickie, J.B. and J.T. Bowyer, 1985. Estimation of provisional seed viability constants for apple (*Malus domestica* Borkh. cv. Greensleeves). Ann. Bot. 56: 271–275.

Eichholtz, D.A., H.A. Robitaille and P.M. Hasegawa, 1979. Adventive embryony in apple. Hort. Sci. 14: 699–700.

Fasolo, F., R.H. Zimmerman and I. Fordham, 1989. Adventitious shoot formation on excised leaves of *in vitro* grown shoots of apple cultivars. Plant Cell Tiss. Org. Cult. 16: 75–87.

Grosser, J.W. and F.G. Gmitter Jr., 1990. Somatic hybridization of *Citrus* with wild relatives for germplasm enhancement and cultivar development. Hort. Sci. 25(2): 147–151.

Hanke, V., A. Rohde and C. Grafe, 1991. Studies on regeneration from somatic tissue *in vitro*: I. Adventitious shoot formation from apple leaf explant. Gartenbauwissenschaft 65(5): 214–220.

Isaev, S.I. and I.I. Domracheva, 1981. Use of polyploidy in apple breeding. Selektsiya yabloni v SSSR: 179–185.

Ishikawa, S., S. Kato, S. Imakawa, T. Mikami and Y. Shimamoto, 1992. Organelle DNA polymorphism in apple cultivars and rootstocks. Theor. Appl. Genet. 83: 963–967.

James, D.J., 1983. Adventitious root formation *in vitro* in apple rootstocks (*Malus pumila*) I. Factors affecting the length of the auxin-sensitive phase in M9. Physiol. Plant 57: 149–153.

James, D.J., 1987. Cell and tissue culture technology for the manipulation of temperate fruit trees. Biotech. Gen. Engin. Rev. 5: 33–79.

James, D.J., A.J. Passey and D.C. Deeming, 1984a. Adventitious embryogenesis and the *in vitro* culture of apple seed parts. J. Plant Physiol. 115(3): 217–229.

James, D.J., A.J. Passey and S.B. Malhotra, 1984b. Organogenesis in callus derived from stem and leaf tissues of apple and cherry rootstocks. Plant Cell Tiss. Org. Cult. 3: 333–341.

James, D.J., A.J. Passey, K.A.D. MacKenzie, O.P. Jones and E.C. Menhinick, 1986a. Regeneration of temperate fruit trees *in vitro* via organogenesis and embryogenesis. In: W. Horn, C.J. Jensen, W. Odenbach and O. Schieder (Eds.), Genetic Manipulation in Plant Breeding. Proceedings International Symposium organized by Eucarpia, September 8–13, 1985, Berlin, pp. 433–435. Walter de Gruyter, Berlin.

James, D.J., A.J. Passey, K.A.D. MacKenzie and E.C. Menhinick, 1986b. Protoplast culture and the use of apomixis *in vitro* for regeneration of fruit plants. In: L.A. Withers and P.G. Alderson (Eds.), Plant Tissue Culture and its Agricultural Applications. Proceedings at the Easter School in Agricultural Science, Vol. 41, pp. 349–357. Butterworths, London.

James, D.J., A.J. Passey, S. Predieri and E. Rugini, 1988. Regeneration and transformation of apple plants using wild-type and engineered plasmids in *Agrobacterium* spp. In: M.R. Ahuja (Ed.), Somatic Cell Genetics of Woody Plants. Proceedings of the IUFRO Working Party S2.04–07, held in Grosshansdorf, 10–13 August 1987, pp. 65–71. Kluwer Academic Publishers, Dordrecht.

James, D.J., A.J. Passey and D.J. Derek, 1989. Genetic transformation of apple and strawberry plants. In: International Symposium In Vitro Culture and Horticultural Breeding, Cesena, 1989. p. 112.

James, D.J., A.J. Passey, A.D. Webster, D.J. Barbara, P. Viss and A.M. Dandekar, 1993a. Transgenic apples and strawberries: Advances in transformation, introduction of genes for insect resistance and field studies of tissue cultured plants. Acta Hort. 336: 179–184.

James, D.J., S. Uratsu, J. Cheng, P. Negri, P. Viss and A.M. Dandekar, 1993b. Acetosyringone and osmoprotectants like betaine or proline synergistically enhance *Agrobacterium*-mediated transformation of apple. Plant Cell Rep. 12: 559–563.

Johansson, L.B., 1986. Effects of activated charcoal, cold treatment and elevated $CO_2$-concentrations on embryogenesis in anther cultures. In: W. Horn, C.J. Jensen W. Odenbach and O. Schieder (Eds.), Genetic Manipulation in Plant Breeding. Proceedings International Symposium organized by Eucarpia, September 8–13, 1985, Berlin, pp. 257–264. Walter de Gruyter, Berlin.

Kromer, K., 1984. Induction of somatic embryogenesis in apple culture. In: F.J. Novak, L. Havel and J. Dolezel (Eds.), Plant Tissue and Cell Culture Application to Crop Improvement, pp. 173–174. Czechoslovak Academy of Sciences, Inst. of Experimental Botany, Prague.

Lacey, C.N.D. and A.I. Campbell, 1987. Selection, stability and propagation of mutant apple. In: A.J. Abbott and R.K. Atkin (Eds.), Improving Vegetatively Propagated Crops, pp. 349–362. Academic Press, London.

Linsmaier, E.M. and F. Skoog, 1965. Organic growth factor requirements of tobacco tissue cultures. Physiol. Plant. 18: 100–107.

Liu, J.R., K.C. Sink and F.G. Dennis Jr., 1983. Adventive embryogenesis from leaf explants of apple seedings. HortSci. 18(8): 871–873.

Mehra, P.N. and S. Sachdeva, 1984. Embryogenesis in apple *in vitro*. Phytomorphology 34 (1/4): 26–36.

Milewska-Pawliczuk, E., 1990. Apple (*Malus domestica* Borkh.): *In vitro* induction of androgenesis. In: Y.P.S. Bajaj (Ed.), Haploids in Crop Improvement I, Biotechnology in Agriculture and Forestry, Vol. 12, pp. 250–263. Springer-Verlag, Berlin.

Mu, S.-K., S.-Q. Liu, Y.-K. Zhou, N.-F. Qian, P. Zhan, H.-X. Xie, F.-S. Zhang and Z.-L. Yan, 1977. Induction of callus from apple endosperm and differentiation of the endosperm plantlet. Scient. Sin. 20: 370–376.

Murashige, T. and F. Skoog, 1962. A revised medium for rapid growth and bioassay with tobacco tissue cultures. Physiol. Plant. 15: 473–497.

Nybom, H., S.H. Rogstad and B.A. Schaal, 1990. Genetic variation detected by use of the M13 "DNA fingerprint" probe in *Malus*, *Prunus*, and *Rubus* (Rosaceae). Theor. Appl. Genet. 79: 153–156.

Nybom, H. and B.A. Schaal, 1990. DNA "fingerprints" applied to paternity analysis in apples (*Malus* × *domestica*). Theor. Appl. Genet. 79: 763–768.

Ochatt, S.J., E.M. Patat-Ochatt, E.L. Rech, M.R. Davey and J.B. Power, 1989. Somatic hybridization of sexually incompatible top-fruit tree rootstocks, wild pear (*Pyrus communis* var. *pyraster* L.) and Colt cherry (*Prunus avium* × *pseudocerasus*). Theor. Appl. Genet. 78: 35–41.

Pandey, K.K. and M. Phung, 1982. "Hertwig Effect" in plants: Induced parthenogenesis through the use of irradiated pollen. Theor. Appl. Genet. 62: 295–300.

Perales, E.H. and O. Schieder, 1993. Plant regeneration from leaf protoplasts of apple. Plant Cell Tiss. Org. Cult. 34(1): 71–76.

Pua, E.-C. and C. Chong, 1984. Requirement for sorbitol (D-glucitol) as carbon source for *in vitro* propagation of *Malus robusta* No. 5. Can. J. Bot. 62: 1545–1549.

Ryć, M. and St. Lewak, 1980. The role of abscisic acid (ABA) in regulation of some photosynthetic enzyme activities in apple seedlings in relation to embryonal dormancy. Z. Pflanzenphysiol. 96: 195–202.

Saito, A., M. Niizeki and K. Saito, 1989. Organ formation from calli and protoplast isolation, culture, and fusion. J. Jap. Soc. Hort. Sci. 58(3): 483–490.

Sriskandarajah, S., R.M. Skirvin and H. Abu-Qaoud, 1990. The effect of some macronutrients on adventitious root development on scion apple cultivars *in vitro*. Plant Cell Tiss. Org. Cult. 21: 185–189.

Svensson, M., 1986. Somatic embryogenesis from the endosperm and nucellus of the apple seed. Växtförädling av frukt och bär. Verksamhetsberättelse 1984–1985. Försöksavdelningen för frukt och bär, Balsgård. Sveriges lantbruksuniversitet.

Van Nieuwkerk, J.P., R.H. Zimmerman and I. Fordham, 1986. Thidiazuron stimulation of apple shoot proliferation *in vitro*. Hort. Sci. 21(3): 516–518.

Vardi A. and E. Galun, 1988. Recent advances in protoplast of horticultural crops: *Citrus*. Sci. Hort. 37: 217–230.

Vardi, A., Spiegel-Roy, P. and E. Galun, 1982. Plant regeneration from *Citrus* protoplasts: Variability in methodological requirements among cultivars and species. Theor. Appl. Genet. 62: 171–176.

Wallin, A. and L. Johansson, 1989. Plant regeneration from leaf mesophyll protoplasts of *in vitro* cultured shoots of a columnar apple. J. Plant Physiol. 135: 565–570.

Way, R.D., H.S. Aldwinckle, R.C. Lamb, A. Rejman, S. Sansavini, T. Shen, R. Watkins, M.N. Westwood and Y. Yoshida, 1990. Apples (*Malus*). Genetic resources of temperate fruit and nut crops. 1. International Society for Horticultural Science. Acta Hort. 290: 3–62.

Welander, M., 1988. Plant regeneration from leaf and stem segments of shoots raised *in vitro* from mature apple trees. J. Plant Physiol. 132: 738–744.

Wilkins, C.P. and J.H. Dodds, 1983. Tissue culture propagation of temperate fruit trees. In: J.H. Dodds (Ed.), Tissue Culture of Trees, pp. 56–79. Croom Helm Ltd., Beckenham, Kent.

Xue, G.R., J.Z. Niu, Z.Y. Yang, Y.Z. Shi and K.W. Fei, 1990. Apple anther culture technique and the successful culture of pollen plantlets of 8 major apple cultivars. Scientia Agric. Sinica 23(3): 86.

Zhang, Y.X., J. Boccon-Gibod and Y. Lespinasse, 1987. Comptes Rendus de l'Academie des Science. III. Sciences de la Vie 305(11): 443–448.

Zhang, Y.X., L. Bouvier and Y. Lespinasse, 1992. Microspore embryogenesis induced by low gamma dose irradiation in apple. Plant Breed. 108(2): 173–176.

Zimmerman, R.H. and O.C. Broome, 1981. Phloroglucinol and *in vitro* rooting of apple cultivar cuttings. J. Amer. Soc. Hort. Sci. 106: 648–652.

# 24. Somatic embryogenesis in date palm (*Phoenix dactylifera* L.)

Shyamala Bhaskaran and Roberta H. Smith

**Contents**



## 1. Introduction

*Phoenix dactylifera* L., date palm, origins date back 4,000 to 5,000 years ago and are evident in the construction of a temple near Ur in Iraq (FAO, 1982). Current distribution of date palm is approximately 20° N and 35° N including dry, subtropical climates in northern and southern hemispheres of North and South America, Africa, and Asia. Recently it has been introduced into the Colorado Desert of North America, the Atacama Desert of South America, and the Kalahari Desert of South Africa, and into the Central Desert of Australia. The date palm requires hot summers without rain or high humidity for the five to seven months from pollination to harvest. Irrigation, however, is necessary.

The date palm grows to about 30 m, but it generally dies or is blown down before reaching 20 m. The fruit which are borne in large bunches have fleshy pericarps, and the pulp is a good source of calcium, phosphorus and carbohydrates (Martin, 1984). The moisture content of the fruit determines its classification as soft, semi-dry, or dry. It is a staple food in many Arab countries. The annual yield of the date is probably over 2.5 metric tons with Iraq, Iran, and Egypt being the main producers (Oudejans, 1979; FAO, 1982). It takes 5–7 years for a tree to begin to produce, and a single female tree can yield 30–100 kg of dates. In California fruit production can be 12,000, to 14,000 pounds of fruit per acre (Tisserat, 1983). Trees can live up to 150 years; the yield decreases as the tree ages, and they are generally cut down before they reach 100 years.

There are 12 species of *Phoenix* and they appear to be diploid (2n = 36). Interspecific hybrids are fertile. Male trees produce abundant pollen and are planted one male to 25–50 female trees. Pollination has been done by hand; however, since this is very labor intense, hand pollination is being replaced

*S. Jain, P. Gupta & R. Newton (eds.), Somatic Embryogenesis in Woody Plants, Vol. 2*, 461–470.

by blowing the pollen through long tubes (Oudejans, 1979). Mechanical methods are now replacing hand labor for dusting and harvesting particularly for the harder dates (Tisserat, 1983). Tree improvement via breeding has been limited, although, there is a need to develop disease resistance. A devestating disease is Bayoud which is caused by *Fusarium oxysporum* f. sp. *albedinis*. This is the most serious disease which has almost ruined the production in Morocco destroying 10 million palms (Tisserat, 1983). The disease is spread by contaminated soil and plant parts. In regions where humidity is high, graphiola leaf spot caused by *Graphiola phoenicis* (Mou;ng.) Poit. is a problem. Insect problems include those caused by scale, *Parlatoria blanchardi* (Targ.) and date mites, *Oligonychus pratensis* (Banks).

The diecious nature of the species *P. dactylifera*, the limited number of offshoots produced by each plant during its life span, and the heterozygosity of its seeds coupled with the unproven quality of seed-derived progeny have made date palm an ideal candidate for clonal propagation. A tree will produce offshoots from vegetative buds during the juvenile phase of growth (5–7 years) resulting in less than a dozen offshoots from a single tree (Tisserat, 1983).

The lack of adequate methods for rapid vegetative propagation of desirable female plants has spawned intensive efforts to clonally propagate this economically important species using *in vitro* techniques. These efforts have been successful, as evidenced by the number of reports published within the last 15 years. The earlier work on *in vitro* propagation of date palm has been adequately covered in a previous report by Tisserat (1983) and hence will not be described in detail here.

One of the successful methods for clonal propagation of date palm is via somatic embryogenesis, using soft, young, meristematic tissues as explants. Immature embryos, embryos from mature seeds, axillary buds, immature inflorescences and shoot tips from offshoots of established plants have been induced to form embryogenic callus *in vitro*. Plant regeneration from such embryogenic callus has been obtained from the cultivars Deglet Nour (Ammar and Benbadis, 1977), Halawy (Tisserat, 1979; Daguin and Letouze, 1988), Khadravi (Sharma *et al.*, 1984), Khadrawy (Kackar *et al.*, 1989), Bou Feggous, Bou Skri, Halawy, Kadrawy and Medjool (Daguin and Letouze, 1988) and Barhee (Bhaskaran and Smith, 1992) (Table 1). Asexual embryogenesis with commercial propagation in mind should be initiated with tissues from female plants of known quality because of the uncertainity associated with seed-derived material.

The basic technique for the induction of embryogenic callus, which makes use of Murashige and Skoog (MS) medium (Murashige and Skoog, 1962) supplemented with the plant growth regulators, auxins and cytokinins, has remained virtually unchanged over the course of years. The types, concentrations and combinations of growth regulators have varied with individual investigators. These variations may have been necessitated by the variety,

*Table 1.* A summary of reports on somatic embryogenesis in date palm, *P. dactylifera* L.

| Cultivar | Explant | Induction medium (conc. in mg/l) | Plants | Reference |
|---|---|---|---|---|
| Deglet Nour | Cotyledonary sheath | MS + $10^{-6}$ NAA, $10^{-7}$ kinetin, 15% coconut milk, 5% sucrose, 0.7% agar | Yes | Ammar and Benbadis, 1977 |
| Not indicated | Ovule sections | MS + 170 $NaH_2PO_4 \cdot H_2O$, 30000 sucrose, 40 adenine sulfate, 100 2,4-D, 1 2iP, 3000 charcoal, 8000 agar | Yes | Tisserat, 1979 Reynolds and Murashige, 1979 |
| Halawy | Embryo | MS + 100 i-inositol, 0.4 thiamine-HCl, 40 adenine sulfate, 30000 sucrose, 3 2iP, 10–100 NAA, 8000 agar, 3000 charcoal | Yes | Tisserat, 1979 |
| Khadravi | Axillary buds, shoot tips | MS + 3000 charcoal, 170 $NaH_2PO_4 \cdot 2H_2O$, 200 $KH_2PO_4 \cdot 2H_2O$, 100 2,4-D, 5 BA, 1 thiamine · HCl | Yes | Sharma *et al.*, 1984 |
| Bou, Feggous, Bou Skri, Halaway, Kadrawy, Medjool | Soft leaf base or rachis | MS + 3000 charcoal, 100 2,4-D, 3 2iP, 30000 sucrose, 7000 agar | Yes | Daguin and Letouze, 1988 |
| Khadrawy | Shoot tips | MS + 170 $NaH_2PO_4 \cdot 2H_2O$, 2 glycine, 100 2,4-D, 3 BA, 3000 charcoal, 30000 sucrose, 8000 agar | Yes | Kackar *et al.*, 1989 |
| Not indicated | Ovule segments | MS + 445 μM 2,4-D, 9.5 μM kinetin | Yes | Omar and Novak, 1990 |
| Barhee | Shoot tips, immature inflorescence | MS + 170 $NaH_2PO_4 \cdot 2H_2O$, 40 adenine sulfate, 0.4 thiamine · HCl, 100 2,4-D, 1 2iP, 30000 sucrose, 8000 agar, 3000 charcoal | Yes | Bhaskaran and Smith, 1992 |

age or growth conditions of the parent plant, as well as, the culture conditions in the individual laboratories which span many geographic locations.

Once the embryogenic callus is identified, a hormone-free medium is used in all experiments for further embryo development and maturation. A noteworthy improvement in recent years was the interpolation of a suspension culture phase (Daguin and Letouze, 1988; Bhaskaran and Smith, 1992). This step provides for synchronized development of a large number of embryos while allowing their individual growth and maturation.

With this background, a procedure for clonal propagation was developed in our laboratory using shoot tips and immature inflorescences of offshoots as explants. This procedure will be detailed in the following paragraphs. The main differences between our procedure and other published procedures will be indicated at the end of each section.

## 2. Culture initiation

### 2.1. *Plant material*

Immature tissues from offshoots of female plants are used. Since offshoots are generally obtained from field-grown plants, they may be heavily contaminated. It is preferable to remove the tough, outer leaves at a location away from the tissue culture facilities. Once the fibrous outer leaves are removed, the inner core of shoot material may be moved to a cleaner facility. Use clean knives or scalpels to cut off the soft inner leaves one at a time and collect any axillary bud or inflorescence present in the leaf axils. Intact inflorescences within their protective sheaths (spathe) may be saved in the refrigerator until use. After removing the leaves, collect the shoot tip consisting of the apical meristem and a few of the innermost leaves along with a piece of the subtending meristematic tissue and submerge it in an antioxidant solution comprised of 150 mg/l citric acid and 100 mg/l ascorbic acid in water. Any axillary bud obtained may also be stored in this antioxidant solution until sterilization. We have stored tissues for up to 4 h at room temperature in this solution, but refrigeration is recommended if overnight storage is needed.

Wipe the spathe with 70% ethanol before exposing the inflorescence. Cut the inflorescence into smaller segments for easier handling and place the segments in the antioxidant solution before sterilization.

### 2.2. *Tissue culture*

Sterilize the explants in a 1% solution of sodium hypochlorite containing a few drops of Tween-20 per liter for 15 to 30 min followed by three rinses with sterile water. Aseptically separate individual florets from the inflorescence. Cut the shoot tip into small pieces of approximately 5 mm. Dip each explant in a fresh solution of 1% sodium hypochlorite solution and place the explant on the callus induction medium.

The callus induction medium contains MS salts, and the following (in mg/l): $NaH_2PO_4 \cdot 2H_2O$ (170), adenine sulfate (40), thiamine-HCl (0.4), 2,4-dichlorophenoxyacetic acid (2,4-D) (100); isopentyl adenine (2-iP) (1); and (in g/l): sucrose (30); Difco Bacto Agar (8) and activated neutralized charcoal (3).

Adjust the pH of the medium to 5.7 before the addition of agar. After

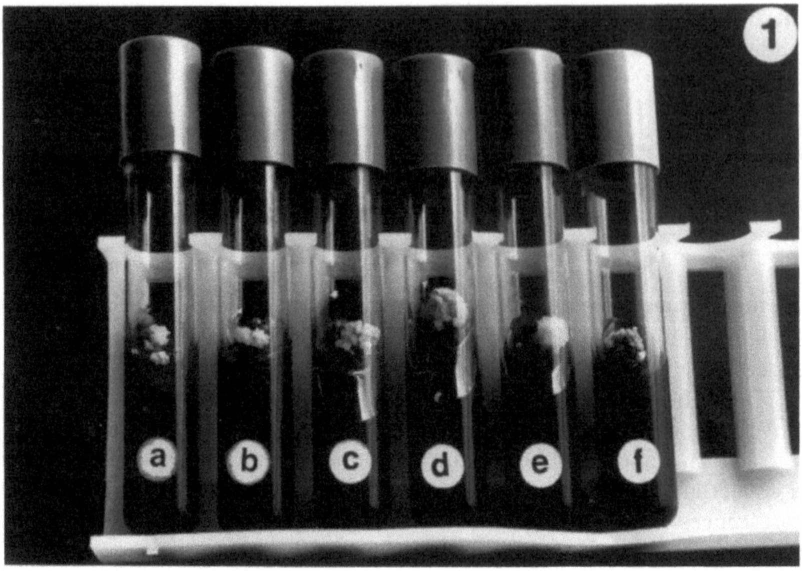


*Figure 1.* Callus from immature ovaries (a,b) and shoot tips (c,d) incubated in the dark and axillary buds (e) and shoot tips (f) incubated in the light.

solubilization of agar by heating, dispense the medium into culture tubes at 15–25 ml/tube and autoclave for 15 min.

Ammar and Benbadis (1977) obtained embryogenic callus from seedling tissues using MS medium supplemented with indole acetic acid (IAA), benzylaminopurine (BA), naphthalene acetic acid (NAA), kinetin and coconut milk and 5% sucrose. Tisserat (1979), Daguin and Letouze (1988) and Sharma *et al.* (1984) used 3 mg/l 2-ip. Kackar *et al.* (1989) substituted 2-ip with 5 mg/l BA and added 2 mg/l glycine and 7 or 8 g/l agar.

Cultures can be maintained in the dark or in the light at low light intensity (28 $\mu E\ m^{-2}\ s^{-1}$) with a photoperiod of 15–16 h at ambient temperature (24 to 28°C). They should be examined at weekly or monthly intervals until a white callus forms from the explant (Fig. 1). If no callus forms within two months, it is advisable to transfer the explants to fresh callus induction medium at 2 to 4 month intervals. It may take weeks (Kackar *et al.*, 1989) to months (Tisserat, 1979; Daguin and Letouze, 1988; Bhaskaran and Smith, 1992) before the characteristic nodular embryogenic callus forms from the explant. All explants should not be expected to form callus, especially pieces from the inflorescence. Some of the florets may turn tan in color and die, others may enlarge in size and yet others may turn green (those in the light).

*Figure 2.* Suspension culture with embryoids.

### 3. Callus maintenance

Once a well-defined nodular callus has formed from an explant, it should be transferred to a hormone-free medium. This medium is prepared by omitting 2,4-D and 2-ip from the callus induction medium. This procedure yielded the friable, white embryogenic callus in our experiments. Such callus could be maintaned for several months on this medium by subculture at monthly intervals.

An increase in phosphate concentration from 170 to 200 mg/l and of thiamine from 0.4 to 1 mg/l, and substitution of 2-ip by 5 mg/l BA was recommended by Sharma *et al.* (1984) for viable callus production, after which the cultures were placed on hormone-free medium for differentiation of somatic embryos. This intervening medium was not required for embryo differentiation in other studies.

## 4. Embryo development

For further multiplication and development of somatic embryos, approximately 200 mg of the friable embryogenic callus is placed in 25 ml of liquid medium in 50 ml DeLong flasks. This medium is prepared by omitting 2,4-D, 2-ip, agar and charcoal from the callus induction medium. Agitate the contents of the flask at 150 rpm in the light for one month, during which time hundreds of tiny individual embryos form in the flask (Fig. 2). After one month, take 2 ml aliquots of the embryos in suspension using a wide-mouth pipet and place in 25 ml of fresh medium of the same composition in 50 ml flasks and place on shakers at 150 rpm for an additional 3 weeks. Individual embryos develop into bipolar structures 0.5 to 1 cm in size during this phase. The yield from 200 mg of embryogenic callus is about 1000 embryos at the end of this period.

Daguin and Letouze (1988) deleted the solid hormone-free medium and replaced it with a liquid medium made up of half strength MS salts, 30 g/l sucrose and 3 g/l activated charcoal for individual development of embryos. Five grams of callus was placed in 100 ml of liquid medium in Erlenmeyer flasks at 100 rpm, and after 30 days in this medium, the suspension was sieved and embryos larger than 5 mm were placed on a different medium for embryo maturation, while embryos between 50 μm and 5 mm in size were returned to fresh suspension medium at 10 g/100 ml.

The embryogenic callus was left on solid hormone-free medium in all other studies. However, embryos differentiating on this medium matured at different rates and had to be individually separated.

## 5. Plant regeneration

The bipolar embryos are placed on the hormone-free medium described earlier in petri dishes (20 to 50 embryos per dish) (Fig. 3) and incubated in the light at low intensity with a 16 hr photoperiod. They enlarge into white opaque structures from which a root and a green shoot emerge from opposite ends. Approximately 40% of the embryos placed on this medium develop into such plantlets (Fig. 4), while others remain translucent or grow only roots or shoots, but not both.

Plantlets thus obtained are placed on another medium supplemented with vitamins for further plant development. This medium contains MS salts and in mg/l: thiamine-HCl (0.25); pyridoxine-HCl (0.25); calcium pantothenate (0.25), nicotinic acid (1.3); biotin (0.004), myo-inositol (50), glycine (7.7), and 40 g/l sucrose. The medium is solidified with 1.5% Gelrite (Kelco Div. of Merck and Co.).

Daguin and Letouze (1988) placed embryos larger than 5 mm on a medium containing half strength MS salts, 30 g/l sucrose, 3 g/l activated charcoal, solidified with 7 g/l agar for 45 days, after which plantlets were transferred

*Figure 3.* Embryos on solid medium in petri dishes.

to a second regeneration medium containing MS salts, 40 mg/l adenine, 5 mg/l kinetin, 2 mg/l BA, 0.5 mg/l NAA, 0.1 mg/l naphthoxyacetic acid (NOA), 40 g/l sucrose and 1 g/l activated charcoal.

Kackar *et al.* (1989) reported plantlet development on a medium supplemented with NAA, NOA and BA.

## 6. Establishment in soil

Successful establishment of plantlets in the soil has been reported from all laboratories. Treatment of plantlets with fungicides and antibiotics are recommended to prevent infections. The plantlets need to be covered initially

*Figure 4.* Plantlet in culture tube.

for a few weeks to prevent desiccation and to acclimatize them to free-living conditions.

## 7. Conclusion

Since conventional methods of clonal propagation of elite date palm cultivars are limited by the small number of off shoots produced by a tree in its life time, *in vitro* propagation is an attractive, practical alternative. There are at least six reports in the referred literature since 1977 on successful initiation of embryogenic callus and plant regeneration including nine different cultivars of date palm. The success of somatic embryogenesis, as well as, shoot organogenesis methods of propagation should result in increased commercial production of desirable date palm cultivars. When genetic engineering methodologies and more genes for disease resistance are more readily available, the tissue culture methodologies for date palm are at a stage where direct applications can be made.

## 8. Acknowledgements

This is technical article No. 31184 from the Texas Agricultural Experiment Station.

## 9. References

Ammar, S. and A. Benbadis, 1977. Multiplication végétative du Palmier-dattier (*Phoenix dactylifera* L.) par la culture de tissus jeunes plantes issues de semis. C.R. Acad. Sc. Paris. Série D 284: 1789–1792.

Bhaskaran, S. and R.H. Smith, 1992. Somatic embryogenesis from shoot tip and inflorescence of *Phoenix dactylifera* cv. Barhee. Plant Cell Rep. 12: 22–25.

Daguin, F. and R. Letouze, 1988. Régénération du palmier dattier (*Phoenix dactylifera* L.) par embryogenèse somatique: amélioration de l'efficacité par passage en milieu liquide agité. Fruits 43: 191–194.

FAO Plant Production and Protection Paper: Date Production and Protection, 1982, Rome.

Kackar, N.L., K.R. Solanki and S.P. Joshi, 1989. Micropropagation of date palm (*Phoenix dactylifera* L.) cv. Khadrawy using tissue culture technique. Ann. Arid Zone 28: 137–141.

Murashige, T. and F. Skoog, 1962. A revised medium for rapid growth and bioassays with tobacco tissue cultures. Physiol. Plant 15: 473–497.

Martin, F.W., 1984. CRC Handbook of Tropical Food Crops. CRC Press, Inc., Boca Raton, FL.

Omar, M.S. and F.J. Novak, 1990. *In vitro* plant regeneration and ethylmethanesulphonate (EMS) uptake in somatic embryos of date palm (*Phoenix dactylifera* L.). Plant Cell Tiss. Org. Cult. 20: 185–190.

Reynolds, J.F. and T. Murashige, 1979. Asexual embryogenesis in callus cultures of palms. In Vitro 15(5): 383–387.

Sharma, D.R., S. Dawra and J.B. Chowdhury, 1984. Somatic embryogenesis and plant regeneration in date palm (*Phoenix dactylifera* Linn.) cv. Khadravi through tissue culture. Indian J. Exp. Biol. 22: 596–598.

Oudejans, J.H.M., 1979. Date Palm. In N.W. Simmonds (Ed.), Evolution of Crop Plants. Longman Press, New York.

Tisserat, B, 1979. Propagation of date palm (*Phoenix dactylifera* L.) *in vitro*. J. Exp. Bot. 30: 1275–1283.

Tisserat, B, 1983. Tissue culture of date palms – A new method to propagate an ancient crop – and a short discussion of the California date industry. Principes 27: 105–117.

# 25. Somatic embryogenesis in eastern redbud (*Cercis canadensis*)

R.N. Trigiano, R.L. Geneve and L.G. Buckley

**Contents**



## 1. Introduction

*Cercis canadensis* L. (Eastern redbud) is a small tree (to 15 m) in the legume family (Leguminosae sub. Caesalpiniaceae). Its natural range extends along the eastern region of the United States (USA), west to Texas and south as far as central Mexico (Raulston, 1990; Robertson, 1976). This wide geographic distribution underscores the ecotypic diversity observed in this species. Morphological variation in leaf size and thickness are regionally observed (Donselman and Flint, 1982) and may be related to localized light and drought conditions (Abrams, 1988). There has been limited selection for superior plant characteristics other than flower color from wild Eastern redbud populations. However, distinct botanical varieties of *C. canadensis* have been described based primarily on foliar characteristics. Some authorities consider the morphological differences between these botanical varieties distinct enough to separate them into species. *Cercis canadensis* var. *canadensis* is characterized as having a dull green flexible leaf that is pointed at the apex and is distributed along the eastern portion of the USA. *Cercis canadensis* var. *texensis* (syn. *C. reniformis*) has a stiff, waxy, dark green leaf and a south central USA distribution. *Cercis canadensis* var. *mexicana* (syn. *C. mexicana*) extends the range of this species south into central Mexico. Plants in this southern range are characterized by a smaller, waxy leaf with a wavy margin that can have considerable pubescence.

A common feature of *C. canadensis* varieties is the production of caulifloric blossoms that open before the leaves appear in early spring. Branches and the main trunk produce lavender to rosy-pink (occasionally white) flowers that literally cover the tree with color. Individual flowers are pea-like in appearance but differ from a typical papillionoid flower by having the "standard" petal in front rather than behind the "wing" petals. These attractive

*S. Jain, P. Gupta & R. Newton (eds.), Somatic Embryogenesis in Woody Plants, Vol. 2,* 471–482.

flowers produced in abundance on a small landscape tree make Eastern redbud an important crop for USA nurseries.

*Cercis canadensis* has traditionally been propagated by seed. The seed displays a double dormancy and requires a scarification treatment to allow imbibition through a hard seed coat followed by 4 to 10 weeks of chilling stratification to satisfy an intermediate physiological embryo dormancy (Afanasiev, 1944; Roy, 1974). Nurserymen typically sow scarified seed in the field during autumn for spring germination.

Several cultivars of *C. canadensis* have been selected for flower color and foliage characteristics. Cultivars of *C. canadensis* var. *texensis* are popular ornamentals because of the attractive dark green, glossy leaves that appear to confer a degree of stress tolerance to the plants. A purple leafed selection (*C. canadensis* var. *canadensis* "Forest Pansy") has also become more accessible to nurserymen following the expiration of a 1982 patent held by Forest Nursery in Tennessee (USA). Cultivars must be asexually propagated and are typically chip-budded. This has proven to be expensive because of the low frequency of grafting success (Dirr and Heuser, 1987). *In vitro* techniques offer an alternative to grafting for the propagation of Eastern redbud cultivars. Tissue culture techniques have been used successfully to propagate a number of cultivars (Bennett, 1987; Yusnita *et al.*, 1990; Geneve *et al.*, 1992). Commercial application of this technology should increase the availability of existing cultivars and support the selection of superior individuals for improved stress and disease tolerance from the diverse native population.

Although *C. canadensis* has responded well to *in vitro* conditions for axillary shoot multiplication, induction of adventitious shoots has proven very difficult from seedling parts and *in vitro* derived organs (Geneve, unpublished data). This limits the possibilities for genetic engineering, especially transformation with *Agrobacterium* species. However, *C. canadensis* has been shown to readily produce somatic embryos from immature zygotic embryos (Geneve and Kester, 1990; Trigiano *et al.*, 1988). This ability, plus the limited number of cells expected to be involved in somatic embryo initiation, makes genetic transformation via somatic embryogenesis an attractive alternative to adventitious shoot formation. The objective of this communication is to provide an overview of the work on somatic embryogenesis in *C. canadensis* emphasizing the possibilities and limitations of recovering genetically transformed plants through the production of somatic embryos.

## 2. Culture initiation

Somatic embryogenesis in cultures of leguminous trees has generally been restricted to developing ovules, mature seeds or seedling organs. The developmental or physiological stage of the explant appears to be a critical factor controlling the ability to form somatic embryos from immature ovules (Trigiano *et al.*, 1992). The optimum period for initiation of embryogenic cultures from ovules has been variously defined by days post-anthesis (DPA) (Arrillaga *et al.*, 1993; Geneve and Kester, 1990; Merkle and Wiecko, 1989; Trigiano *et al.*, 1988; Weaver and Trigiano, 1991), morphology of zygotic embryos (Trigiano *et al.*, 1988), and fresh weight of zygotic embryos (Geneve and Kester, 1990). Although DPA is the most commonly reported measure of the ability of ovules to initiate somatic embryos, it is influenced by geographic location and environmental factors and thus may vary from year to year. A more appropriate appraisal of the potential embryogenic response of zygotic embryos in culture may be either fresh weight or morphology of developing zygotic embryos (Geneve and Kester, 1990; Trigiano and Buckley, unpublished data).

Two similar systems have been reported for the initiation of somatic embryos in *C. canadensis*. In one system, scored ovules (Trigiano *et al.*, 1988) were used as initial explants, whereas in the second system, explants were isolated from developing zygotic embryos (Geneve and Kester, 1990; Geneve, 1991). Cultures of *C. canadensis* were initiated from developing ovules collected between 60 and 150 DPA. Immature fruits (pods) were surface-sterilized in either 0.5 or 1.3% sodium hypochlorite with a surfactant for 10 or 5 min respectively. Ovules were aseptically removed, scored longitudinally with a scalpel and placed on Schenk and Hildebrandt (1972) medium (SH) amended with 4.5–22.5 μM 2,4-D (2,4-dichlorophenoxyacetic acid) and ammonium sulfate to yield a final ammonium ion concentration of 12.6 mM (Trigiano *et al.*, 1988). Alternatively, developing zygotic embryos were dissected from surface-sterilized ovules and placed directly on modified Woody Plant's Medium (Lloyd and McCown, 1980) containing 1 or 5 μM 2,4-D or 50 μM NAA ($\alpha$-naphthaleneacetic acid) (Geneve and Kester, 1990). Cultures were incubated at 22–25° C with 25–80 μmol $s^{-1}$ $m^{-2}$ of light provided by cool white fluorescent tubes for 16 h.

## 3. Culture maintenance

Explants that directly produced somatic embryos were either maintained on the original medium without transfer for 42–49 days (Trigiano *et al.*, 1988) or were subjected to a pulse treatments with 2,4-D of between 5 and 50 days and then transferred to basal medium without growth regulators for the remainder of the culture period (Geneve and Kester, 1990). In both of the above mentioned studies, cultures were terminated after 49 or 50 days.

For indirect embryogenesis, brown mucilaginous or hard compact calli generated by explants after 42 days were transferred to SH medium with 9.0 μM 2,4-D and subsequently moved to SH basal medium without growth regulators after 28 days. Calli were alternately recultured at monthly intervals on basal medium with or without 2,4-D (Trigiano and Beaty, 1989).

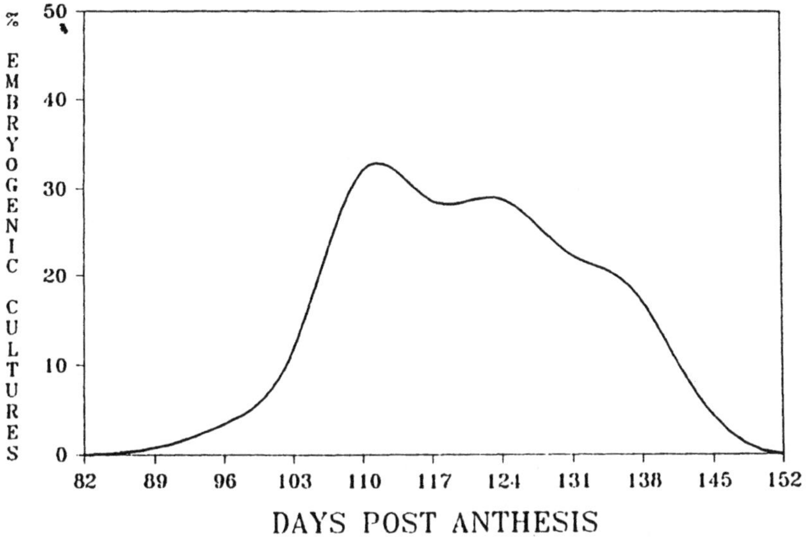


*Figure 1.* Embryogenic potential of developing ovules of *Cercis canadensis*.

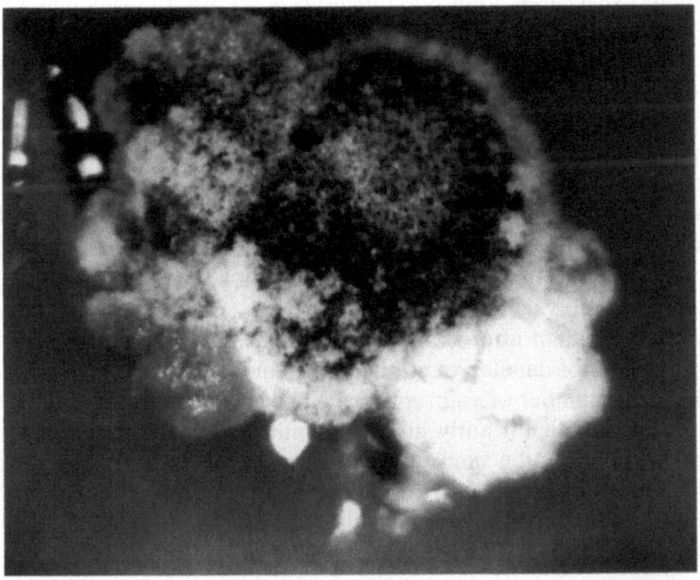

*Figure 2.* Callus formed from a 70 days post-anthesis ovule of *Cercis canadensis*.

### 4. Embryo development

Initiation of somatic embryos on either basal medium that contained 2,4-D or NAA was strongly influenced by the developmental stage of the ovule/zygotic embryo used as the explant (Fig. 1). Regardless of the geographic location or the year in which an experiment was conducted, the explants exhibited a "developmental window" when somatic embryos could be induced to form. For example, in a study completed in 1991 (Buckley and Trigiano, unpublished data), explants harvested prior to 82 DPA did not form somatic embryos and only produced a crystalline callus from the integuments of ovules (Fig. 2). Ovules collected after this time and until 152 DPA formed somatic embryos, albeit at different frequencies and rates (Fig. 1). Ovules cultured after 152 DPA did not generate somatic embryos and produced only adventitious roots and some callus. Studies conducted at a different location and year (Geneve and Kester, 1990) and at the same location, but in a different year (Trigiano *et al.*, 1988), exhibited a similar "developmental window" in which somatic embryogenesis was possible. However, the initiation and termination of this period varied with year and location. Fresh weight and morphological development of the zygotic embryo were more consistently correlated with the embryogenic potential than DPA. Geneve and Kester (1990) reported that somatic embryogenesis occurred when the zygotic embryo obtained a fresh weight of between 4 and 12 mg. Similar results were obtained (Trigiano, unpublished data) from studies at a different location and year. Two studies (Trigiano *et al.*, 1988; Buckley and Trigiano, unpublished data) indicated that the developmental morphology of the zygotic embryo may also more accurately predict the ability to produce somatic embryos. These studies demonstrated that cordate through cotyledonary stage embryos were morphogenetically competent to form somatic embryos. Termination of competency was indicated by the onset of desiccation of the seed in preparation for dormancy.

Small green, occasionally white, globular somatic embryos formed directly from cotyledons or, more rarely, epicotyls of zygotic embryos after two weeks of culture. Embryos originated primarily from either epidermal cells or cells immediately subjacent to the epidermis (Geneve and Kester, 1990). Embryo development after six weeks was asynchronous, producing embryos in various stages of development, including cotyledonary, after 42 days (Fig. 3a). The mean number of embryos formed per explant was typically between 2 and 5 although significantly greater numbers (>40) of embryos were occasionally produced on a single explant. Developmentally advanced embryos that were formed on medium containing 2,4-D for the entire period were frequently malformed and often exhibited very short or fused cotyledons (Trigiano *et al.*, 1988). Pulse treatments of 2,4-D for 20 days somewhat alleviated morphological abnormalities and produced approximately 15% normal appearing embryos (Geneve and Kester, 1990). This treatment, however, did not improve the rate of embryogenesis and was similar to results

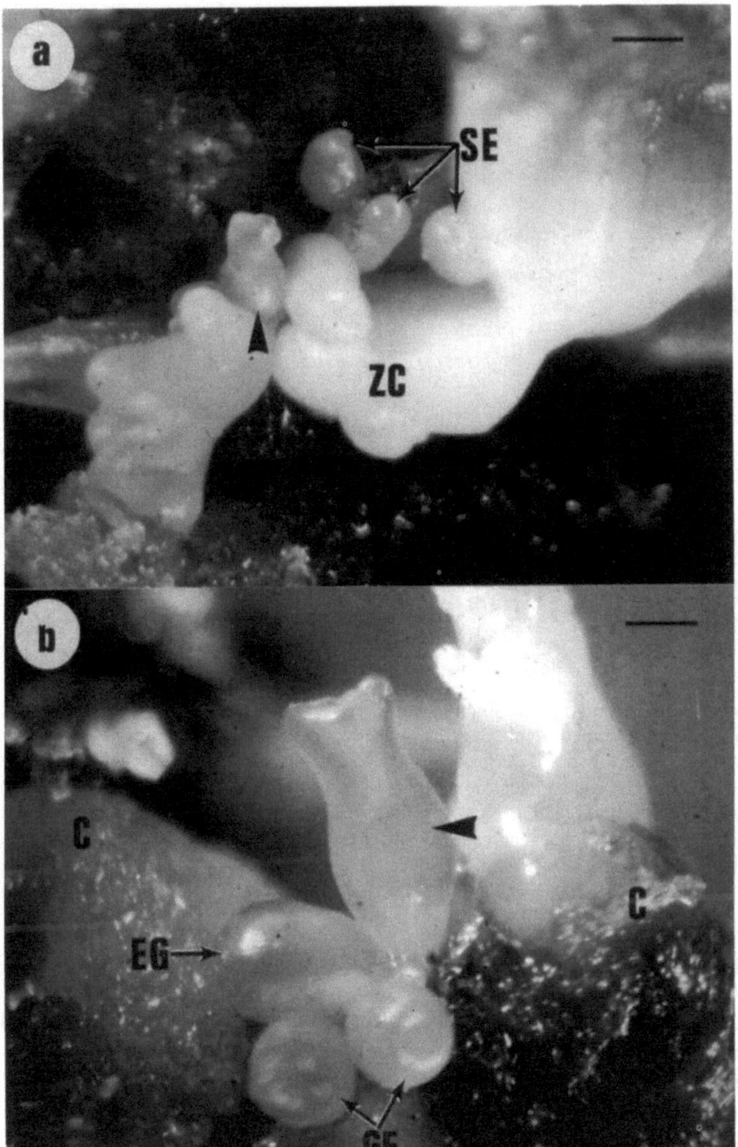


*Figure 3.* Somatic embryos of *Cercis canadensis*. (a) Somatic embryos (SE) formed directly from a zygotic cotyledon (ZC) after 42 days of culture. The original explant was a 103 days post-anthesis ovule and the arrowhead indicates a cotyledonary stage embryo. Bar = 1.5 mm. (b) Globular (G), elongated globular (EG) and cotyledonary (fused) stage (arrowhead) embryos formed indirectly from callus (C). Bar = 0.8 mm.

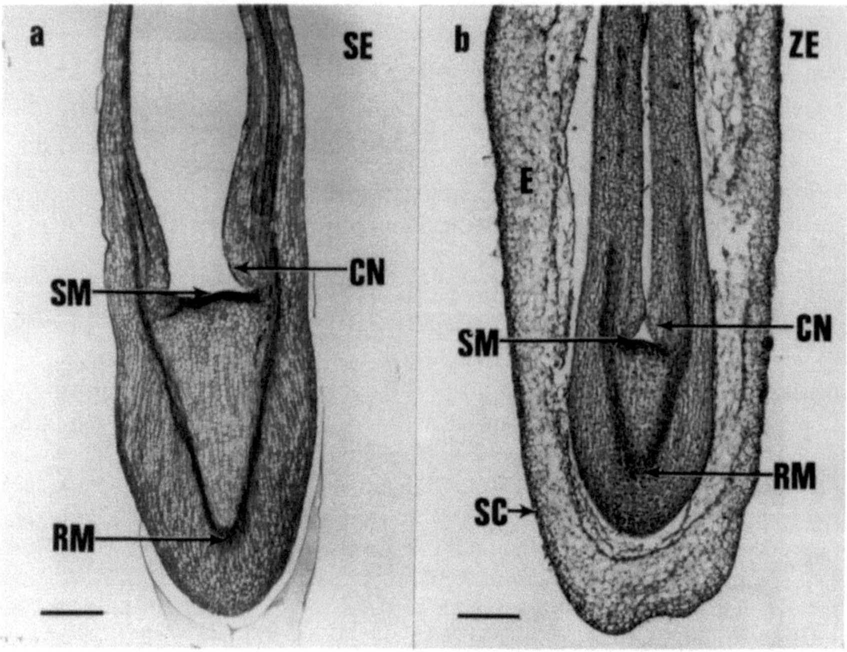


*Figure 4.* Somatic and zygotic embryos of *Cercis canadensis*. (a) Near median longitudinal section through a somatic embryo (SE) showing shoot (SM) and root (RM) meristems and notch (CN) characteristic of Eastern redbud cotyledons. Note the flattened shoot meristem. Bar = 0.35 mm. (b) Near median longitudinal section through an ovule containing endosperm (E) and zygotic embryo (ZE). Note the shoot (SM) and root (RM) meristems, cotyledonary notch (CN) and developing seed coat (SC). Bar = 0.40 mm. Fig. a,b reprinted from Trigiano *et al.* (1988).

reported for *Cladrastis lutea* L., another leguminous tree (Weaver and Trigiano, 1991).

Callus formed during the initial culture of explants was capable of initiating somatic embryos after transfer to basal medium without 2,4-D (Trigiano and Beaty, 1989). Embryo development (Fig. 3b) was similar to that described for embryos formed directly from the zygotic embryo. Roots were also initiated from callus cultured on basal medium and occasionally somatic embryos originated from apices and less frequently from the older portions of the roots. Interestingly, after the callus had been cultured on basal medium and formed embryos, it was necessary to pulse treat the callus with either 9 or 13.5 μM 2,4-D for 28 days and then transfer to basal medium for additional embryo production.

The morphological development of somatic embryos was grossly similar to that of the zygotic embryos although many abnormalities, especially in cotyledon development and thickness of the hypocotyl, were observed. Histological examination of embryo ontogeny demonstrated that both zygotic and somatic embryos were formed with multiseriate suspensors and displayed similar stages of anatomical development (Buckley *et al.*, 1992). Mature somatic (Fig. 4a) and zygotic embryos (Fig. 4b) were similar; however, the shoot apex in many of the somatic embryos examined was flattened compared to the normal dome-shaped apex found in the zygotic embryo. Additionally, many more cells in the somatic embryo were highly vacuolated in contrast to the cytoplasmically dense cells of the zygotic embryo.

## 5. Embryo germination

Bottle-shaped (fused cotyledons) and more morphologically normal embryos (Fig. 5a) germinated and produced plantlets at low frequency (about 20%) on SH basal medium without growth regulators in Magenta GA7 boxes. Cultures were incubated at 25°C. Radicles emerged after approximately 5 days and shoots elongated after an additional 3 days (Fig. 5b). Shoots appeared to be normal with characteristic heart-shaped leaves of Eastern redbud. Approximately 80% of the somatic embryos produced radicles (germinated), but most failed to form shoots. Lack of shoot development may be related to abnormalities observed in shoot apex development in the embryo (Trigiano *et al.*, 1988). Desiccation in a petri dish at room temperature (23°C), low temperature (4°C) treatments in the refrigerator or combinations of desiccation and cold did not improve conversion of embryos to plantlets (Table 1). Plantlets remained in Magenta boxes for approximately one month before acclimatization.

## 6. Growth in the field

Plantlets with at least three true leaves were transplanted to a soilless medium (Fafard's Promix, Anderson, SC, USA), watered with 1/10 strength SH medium and incubated in the same conditions previously described. After seven days, plants were acclimatized to the conditions of the growth chamber by removing the lid of the Magenta box for increasing periods of time over the next 14 days. Seven leafed plantlets were transferred to 15 cm diameter pots that contained a mixture of pine bark mulch and sand, placed in the greenhouse and fertilized with 50 ppm N once per week. At the end of the growing season (October), plants were about 1.5 meters tall with a few branches and many large leaves. Trees were overwintered in a cold frame for the winter months and planted in the field in February. Plants were not irrigated or fertilized in the field. Approximately 25% of the plants flowered during the first season in the field; all others flowered during the next two years.

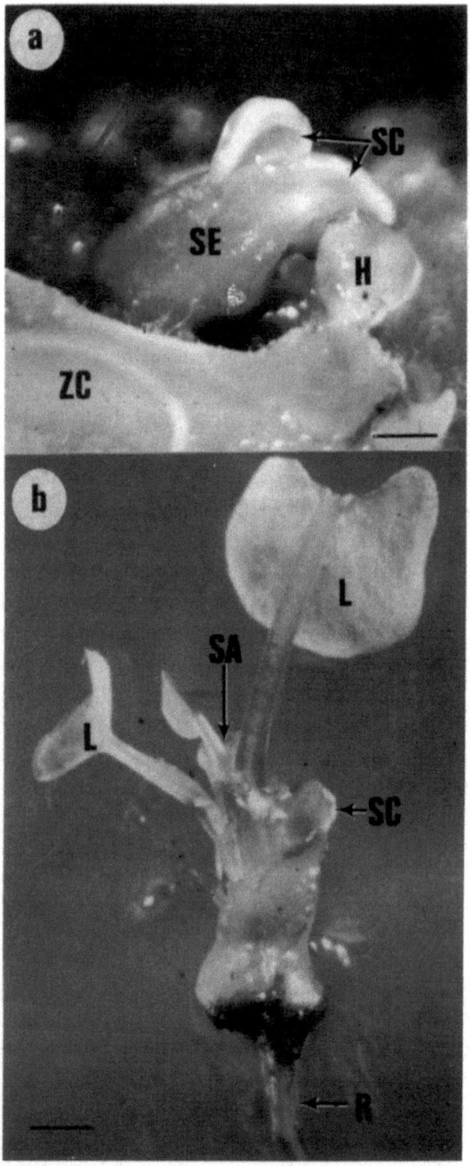


*Figure 5.* Somatic embryos and formation of plantlets of *Cercis canadensis*. (a) Morphologically mature somatic embryo (SE) exhibiting cotyledons (SC) formed directly from zygotic embryo cotyledon (ZC). Note the swollen hypocotyl of the somatic embryo and the cordate-stage (H) embryo in the foreground. The explant was a 110 days post-anthesis ovule. Bar = 1.1 mm. (b) Germinated somatic embryo exhibiting a radicle (R), shoot apex (SA) and typical heart-shaped leaves (L) after 30 days. SC = somatic embryo cotyledon. Bar = 3.1 mm.

*Table 1.* Germination of somatic embryos and regeneration of plants of *Cercis canadensis*.

| Pretreatment | Number of Embryos | Number Germinated[a] | Number of Plants |
|---|---|---|---|
| None[b] | 53 | 42 | 8 |
| Cold (14 days)[c] | 26 | 7 | 2 |
| Cold (28 days)[c] | 25 | 2 | 0 |
| Desiccated[d] | 25 | 0 | 0 |
| Cold (14 days)/Desiccated | 25 | 0 | 0 |
| Desiccated/Cold (14 days) | 22 | 0 | 0 |

[a] Embryos were considered germinated when radicles emerged.
[b] Somatic embryos transferred directly to SH medium without growth regulators and incubated at 25°C.
[c] Somatic embryos placed on moist filter paper at 4°C, then transferred to SH medium at 25°C.
[d] Somatic embryos desiccated on dry filter paper at room temperature for 5 days, then transferred to SH medium at 25°C.

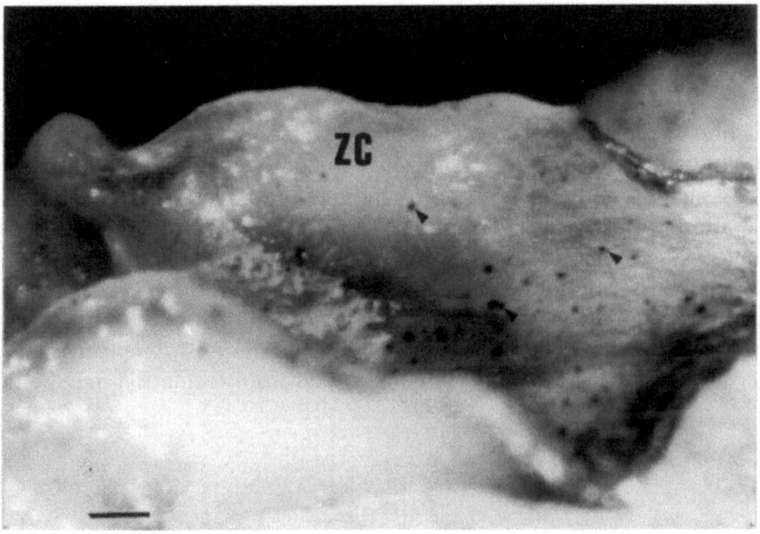


*Figure 6.* GUS expression (arrowheads) in an immature zygotic cotyledon (ZC) explant following particle bombardment. Bar = 0.25 mm.

## 7. Gene transfer

In preliminary studies (Karsedis and Geneve, unpublished data), transformation was attempted using projectile bombardment of immature zygotic embryos with a DNA-construct containing neomycin phosphotransferase (NPT II) gene for kanamycin antibiotic resistance, $\beta$-glucuronidase (GUS) as a marker and a 35-S promoter. Approximately 190 GUS-positive spots (Fig. 6) per explant were observed following bombardment indicating low-frequency genetic transformation of target cells expected to be embryogenic. However, recovery of somatic embryos from bombarded explants revealed a high degree of background GUS staining making it an unreliable marker for positive transformation. Subsequent selection for antibiotic resistance may improve the reliability of transformation via somatic embryogenesis in Eastern redbud.

The most attractive possibility for genetic engineering would be to develop an embryogenic suspension culture similar to that described for *Robinia pseudoacacia* (Arrillaga *et al.*, 1993). This type of culture system is very amenable to transformation via ballistic technologies. However, suspension cultures initiated from developing somatic embryos of *C. canadensis* produced subsequent stationary cultures competent to form either callus or adventitious roots, but failed to generate shoots or somatic embryos (Karsedis and Geneve, unpublished data).

## 8. Concluding remarks

Production of Eastern redbud trees is possible via somatic embryogenesis, but would not be economically or practically competitive with more conventional methods such as seeds or grafting or by micropropagation via axillary bud proliferation. Genetic transformation of the species is problematic. The regeneration system herein described may not be sufficiently reliable or prolific to attempt transformation with reasonable expectations of success. However, if selection of explants and culture conditions can improve the frequency and rate of embryogenesis from cotyledons, *Agrobacterium tumefaciens* or *A. rhizogenes* mediated or projectile bombardment transformation may be possible.

## 9. References


Abrams, M.D., 1988. Genetic variation in leaf morphology and plant and tissue water relations during drought in *Cercis canadensis* L. Forest Sci. 34: 200–207.

Afanasiev, M., 1944. A study of dormancy and germination of seeds of *Cercis canadensis*. J. Agric. Res. 69: 405–420.

Arrillaga, I., J.J. Tobolski and S.A. Merkle, 1994. Advances in somatic embryogenesis and plant production of black locust (*Robinia pseudoacacia* L.). Plant Cell Rep. 13: 171–175.

Bennett, L., 1987. Tissue culturing redbud. Amer. Nurseryman 166: 85–91.

Buckley, L.G., E.T. Graham and R.N. Trigiano, 1992. Comparison of zygotic and somatic embryogenesis in *Cercis canadensis*. HortSci. 27: 1167.

Dirr, M.A. and C.W. Heuser, 1987. The Reference Manual of Woody Plant Propagation. Varsity Press, Athens, GA.

Donselman, H.M. and H.L. Flint, 1982. Genocology of eastern redbud (*Cercis canadensis*). Ecology 63: 962–1071.

Geneve, R.L., 1991. Eastern redbud (*Cercis canadensis* L.), Judas tree (*Cercis siliquastrum* L.).

In: Y.P.S. Bajaj (Ed.), Biotechnology in Agriculture and Forestry, Trees III, Vol. 16, pp. 144–151. Springer-Verlag, New York.

Geneve, R.L. and S.T. Kester, 1990. The initiation of somatic embryos and adventitious roots from developing zygotic embryos explants of *Cercis canadensis* L. cultured *in vitro*. Plant Cell Tiss. Org. Cult. 22: 71–76.

Geneve, R.L., S.T. Kester, S. Yusnita and R.O. Jones, 1992. Micropropagation in eastern redbud (*Cercis canadensis*). Comb. Proc. Intl. Plant Prop. Soc. 42: 8–11.

Lloyd, G. and B. McCown, 1980. Commercially feasible micro-propagation of mountain laurel, *Kalmia latifolia*, by use of shoot tip culture. Comb. Proc. Intl. Plant Prop. Soc. 30: 421–427.

Merkle, S.A. and A.T. Wiecko, 1989. Regeneration of *Robinia pseudoacacia* via somatic embryogenesis. Can. J. For. Res. 19: 285–288.

Raulston, J.C., 1990. Redbud. Amer. Nurseryman 171: 39–51.

Robertson, K.R., 1976. *Cercis*: the redbuds. Arnoldia 36: 37–49.

Roy, D.F., 1974. *Cercis* L. Redbud. In: C.S. Schopmeyer (Ed.), Seeds of Woody Plants in the United States, pp. 305–308. U.S. Dept. Handbook 450.

Schenk, R.U. and A.C. Hildebrandt, 1972. Medium and techniques for induction and growth of monocotyledonous and dicotyledonous plant cell cultures. Can. J. Bot. 50: 199–204.

Trigiano, R.N. and R.M. Beaty, 1989. Direct and indirect somatic embryogenesis in cultures of *Cercis canadensis*. HortSci. Abstracts, Tulsa, OK, p. 139.

Trigiano, R.N., R.M. Beaty and E.T. Graham, 1988. Somatic embryogenesis from immature embryos of redbud (*Cercis canadensis*). Plant Cell Rep. 7: 148–150.

Trigiano, R.N., R.L. Geneve, S.A. Merkle and J.E. Preece, 1992. Tissue and cell cultures of woody legumes. Hort. Rev. 14: 265–332.

Weaver, L.A. and R.N. Trigiano, 1991. Regeneration of *Cladrastis lutea* (Fabaceae) via somatic embryogenesis. Plant Cell Rep. 10: 1183–186.

Yusnita, S., R.L. Geneve and S.T. Kester, 1990. Micropropagation of a white flowering form of eastern redbud. J. Environ. Hort. 8: 177–179.

# Somatic Embryogenesis in Woody Plants: (Angiosperms)

*Volume 2*

Dr V.A. Bapat, Plant Biotechnology Section, Bhabha Atomic Research Center, Trombay, Bombay 400 085, India.

Dr S. Bates, Department of Plant and Soil Science, Southern Illinois University, Carbondale, IL 62901, USA.

Dr S. Bhaskaran, Department of Soil and Crop Sciences, Texas A & M University, College Station, TX 77843–2474, USA. Fax: 409–845–0456.

Dr J. Blake, Unit for Advanced Propagation Systems, Wye College, University of London, Wye, Ashford, Kent, TN25 5AH, UK. Fax: + 44–233–81–33–20.

Dr L.G. Buckley, Department of Ornamental Horticulture and Landscape Design, Institute of Agriculture, The University of Tennessee, Knoxville, TN 37901–1071, USA.

Dr S. Campagna, CIRAD-CD, Hevea Programme, Laboratories BIOTROP, Avenue du Val de Montferrand, BP 5035, 34032 Montpellier, Cedex 1, France. Fax: 67 61 59 86.

Dr M.P. Carron, CIRAD-Cultures Perennes, Departement Hevea, 2477 Avenue du Val de Montferrand, BP 5035, 34032 Montpellier, Cedex 1, France. Fax: 67 61 59 86.

Dr C. Chaine, CIRAD-CD, Hevea Programme, Laboratories BIOTROP, Avenue du Val de Montferrand, BP 5035, 34032 Montpellier Cedex, France.

Dr V. Chalupa, Faculty of Forestry, University of Agriculture, 165 21 Praha 6–Suchdol, Czech Rep. Fax: 42–2–321–214.

Prof. W.C. Chang, Institute of Botany, Academica Sinica, Taipei, Taiwan 115, Republic of China. Fax: 889–2–782–7954.

Prof. G.B. Collins, College of Agriculture, University of Kentucky, N-122 Agricultural Science Bldg.-North, Lexington, KY 40546–0091, USA. Fax: 606–258–1952.

Dr H. Etienne, CIRAD-CP, Hevea Programme, Laboratories BIOTROP, Avenue du Val de Montferrand, BP 5035, 34032, Montpellier Cedex, France.

Dr A. Figueira, Department of Horticulture, Purdue University, West Lafayette, IN 47907–1165, USA. Fax: 317–494–0391.

Dr R.L. Geneve, Department of Horticulture and Landscape Architecture, University of Kentucky, Lexington, KY 40546, USA.

Dr M.I.S. Gill, Biotechnology Center, Punjab Agricultural University, Ludhiana, India.

Dr S.S. Gosal, Associate Professor, Biotechnology Center, Punjab Agricultural University, Ludhiana, India. Fax: 91–161–401794.

Dr Dennis J. Gray, University of Florida, Central Florida Research & Education Center, 5336 University Avenue, Leesburg, FL 34748 , USA. Fax: 904–392–8940.

Dr H.S. Grewal, Biotechnology Center, Punjab Agricultural University, Ludhiana, India.

Dr Lotta Grönroos, Valthornsvägen 38, 75650 Uppsala, Sweden.

Dr R. Hornung, Unit for Advanced Propagation Systems, Department of Horticulture, Wye College, University of London, Wye, Ashford , Kent, TN25 5AH, UK.

Prof Jules Janick, Department of Horticulture, Purdue University, 1165 Horticulture Building, West Lafayette, IN 47907–116 5, USA. Fax: 317–494–0391.

Dr S. Jayashankar, University of Florida, Tropical Research and Education Center, 18905 S.W. 280 Street, Homestead, FL 33031, USA .

Dr M. Joshi, Plant Tissue Culture Division, National Chemical Laboratory, CSIR, Pune 411008, India.

Dr P. Krogstrup, Botanic Garden, University of Copenhagen, Farimagsgade 2B, DK-1353, Copenhagen K, Denmark.

Dr L. Lardet, CIRAD-CP, Hevea Programme, Laboratories BIOTROP, Avenue du Val de Montferrand, BP 5035, 34032 Montpellier Cedex, France.

Dr A. Leconte, IDEFOR-DPL, Station de Bimbresso, BP 1536, Abidjan, Ivory Coast.

Dr C.A. Leslie, Department of Pomology, University of California, Davis, CA 95617, USA.

Prof Richard E. Litz, Institute of Food and Agricultural Sciences, University of Florida, Tropical Research and Education Center, 18905 S.W., 280th Street, Homestead, FL 33031, USA. Fax: 305–246–7003.

Dr L.M. Long, Department of Plant and Soil Science, Southern Illinois University, Carbondale, IL 62901, USA.

Dr A.F. Mascarenhas, Plant Tissue Culture Division, National Chemical Laboratory, CSIR, Pune 411008, India. Fax: 91–212–334–761.

Dr Helena Mathews, Agritope Inc., 8505 S.W. Creekside Place, Beaverton, OR 97005, USA.

Dr G.H. McGranahan, Department of Pomology, University of California, Davis, CA 95616, USA.

Dr Charles H. Michler, USDA, Forest Service, Forest Science Laboratory, P.O. Box 898, Rhinelander, WI 54501, USA. Fax: 715–62–7816.

Dr M.J. Monsalud, University of Florida, Tropical Research and Education Center, 18905 S.W. 280 Street, Homestead, FL 33031, USA.

Dr P.A. Moon, University of Florida, Tropical Research and Education Center, 18905 S.W. 280 Street, Homestead, FL 33031, USA.

Dr J. Mottley, Plant Biotechnology Research Unit, Department of Life Sciences, University of East London, Romford Road, London, E15 4LZ, UK.

Dr E.M. Muralidharan, Genetics Division, Kerala Forest Research Institute, Peechi, Trichur 680 653, India.

Dr J.K. Norgaard, Botanic Garden, University of Copenhagen, Farimagagade 2B, DK-1353, Copenhagen K, Denmark. Fax: + 45 35 32 22 21.

Dr M. Nyman, Department of Physiological Botany, University of Uppsala, Villavagen. 6, S-75236, Uppsala, Sweden.

Dr Y. Perrin, CIRAD-CP, Hevea Programme, Laboratories BIOTROP, Avenue du Val de Montferrand, BP 5035, 34032 Montpellier Cedex, France.

Dr F. Pliego-Alfaro, Departamento de Biologia Vegetal, Universidad d4 Malaga, Spain.

Prof John E. Preece, Department of Plant and Soil Sciences, Southern Illinois University, Carbondale, IL 62901, USA. Fax: 618–453–1778.

Dr L. Radojević, The Siniska Stankovic Institute of Biological Research, Department of Plant Physiology, 11000 Belgrade, Yugoslavia.

Dr P.S. Rao, Plant Biotechnology Division, Bhabha Atomic Research Center, Chembur, Trombay, Bombay 400 085, India. Fax: 91-22-5560 750.

Dr A.V. Roberts, Department of Life Sciences, University of East London, Romford Road, London E15 4LZ, UK. Fax: 81–849–3625.

Prof E. Rugini, Universita Degli Studi Della Tuscia, Facolta Di Agraria, Istituto Di Ortpfloroarboricoltura, Via San Camillo de Lellis, 01100 Viterbo, Italy. Fax: 39–761–357531.

Prof R.H. Smith, Department of Soil and Crop Sciences, Texas A & M University, College Station, TX 77843–2474, USA. Fax: 409–845–0456.

Dr B. Sharon, Department of Plant and Soil Sciences, Southern Illinois Univertsity, Carbondale, IL 62901, USA.

Dr M. Svensson, Balsgård, Department of Horticultural Plant Breeding, Fjälkestadsv. 123–1, S-29194, Kristianstad, Sweden.

Dr S. Thengana, Plant Tissue Culture Division, National Chemical Laboratory, CSIR, Pune 411008, India.

Dr R.N. Trigiano, Ornamental Horticulture and Landscape Design Department, University of Tennessee, P.O. Box 1071, Knoxville, TN 37901–1071, USA. Fax: 615–974–2765.

Dr A.M. Vieitez, Instituto de Investigaciones Agrobiologicas de Galicia, Apartado de Correos, 122, 15080 Santiago de Compostela, Spain. Fax: + 34 81 59 25 04.

Dr F.J. Vieitez, Instituto de Investigaciones Agrobiologicas de Galicia,

CSIC. Apartado 122, 15080 Santiago de Compostela, Spain. Fax: + 34 81 59 25 04.

Dr S. Walker, Plant Biotechnology Research Unit, Department of Life Sciences, University of East London, Romford Road, London E15 4LZ, UK.

Dr A. Wallin, Department of Physiological Botany, University of Uppsala, Villavagen 6, S-752 36, Uppsala, Sweden. Fax: 46 18 55 9 8 85.

Dr J.E. Woods, College of Agriculture, University of Kentucky, N-122 Agricultural Science Bldg-North, Lexington, KY 40546–0091, USA.

Dr S.H. Woods, College of Agriculture, University of Kentucky, N-122 Agricultural Science Bldg-North, Lexington, KY 40546– 0091, USA.

Dr K. Yokoya, Plant Biotechnology Research Unit, Department of Life Sciences, University of East London, Romford Road, London E15 4LZ, UK.

# SPECIES INDEX

# SUBJECT INDEX

# FORESTRY SCIENCES

1. P. Baas (ed.): *New Perspectives in Wood Anatomy.* Published on the Occasion of the 50th Anniversary of the International Association of Wood Anatomists. 1982 ISBN 90-247-2526-7
2. C.F.L. Prins (ed.): *Production, Marketing and Use of Finger-Jointed Sawnwood.* Proceedings of an International Seminar Organized by the Timber Committee of the UNECE (Halmar, Norway, 1980). 1982 ISBN 90-247-2569-0
3. R.A.A. Oldeman (ed.): *Tropical Hardwood Utilization.* Practice and Prospects. 1982 ISBN 90-247-2581-X
4. P. den Ouden (in collaboration with B.K. Boom): *Manual of Cultivated Conifers.* Hardy in the Cold- and Warm-Temperate Zone. 3rd ed., 1982 ISBN Hb 90-247-2148-2; Pb 90-247-2644-1
5. J.M. Bonga and D.J. Durzan (eds.): *Tissue Culture in Forestry.* 1982 ISBN 90-247-2660-3
6. T. Satoo: *Forest Biomass.* Rev. ed. by H.A.I. Madgwick. 1982 ISBN 90-247-2710-3
7. Tran Van Nao (ed.): *Forest Fire Prevention and Control.* Proceedings of an International Seminar Organized by the Timber Committee of the UNECE (Warsaw, Poland, 1981). 1982 ISBN 90-247-3050-3
8. J.J. Douglas: *A Re-Appraisal of Forestry Development in Developing Countries.* 1983 ISBN 90-247-2830-4
9. J.C. Gordon and C.T. Wheeler (eds.): *Biological Nitrogen Fixation in Forest Ecosystems.* Foundations and Applications. 1983 ISBN 90-247-2849-5
10. M. Németh: *Virus, Mycoplasma and Rickettsia Diseases of Fruit Trees.* Rev. (English) ed., 1986 ISBN 90-247-2868-1
11. M.L. Duryea and T.D. Landis (eds.): *Forest Nursery Manual.* Production of Bareroot Seedlings. 1984; 2nd printing 1987 ISBN Hb 90-247-2913-0; Pb 90-247-2914-9
12. F.C. Hummel: *Forest Policy.* A Contribution to Resource Development. 1984 ISBN 90-247-2883-5
13. P.D. Manion (ed.): *Scleroderris Canker of Conifers.* Proceedings of an International Symposium on Scleroderris Canker of Conifers (Syracuse, USA, 1983). 1984 ISBN 90-247-2912-2
14. M.L. Duryea and G.N. Brown (eds.): *Seedling Physiology and Reforestation Success.* Proceedings of the Physiology Working Group, Technical Session, Society of American Foresters National Convention (Portland, Oregon, USA, 1983). 1984 ISBN 90-247-2949-1
15. K.A.G. Staaf and N.A. Wiksten (eds.): *Tree Harvesting Techniques.* 1984 ISBN 90-247-2994-7
16. J.D. Boyd: *Biophysical Control of Microfibril Orientation in Plant Cell Walls.* Aquatic and Terrestrial Plants Including Trees. 1985 ISBN 90-247-3101-1
17. W.P.K. Findlay (ed.): *Preservation of Timber in the Tropics.* 1985 ISBN 90-247-3112-7
18. I. Samset: *Winch and Cable Systems.* 1985 ISBN 90-247-3205-0

# FORESTRY SCIENCES

19. R.A. Leary: *Interaction Theory in Forest Ecology and Management.* 1985 ISBN 90-247-3220-4
20. S.P. Gessel (ed.): *Forest Site and Productivity.* 1986 ISBN 90-247-3284-0
21. T.C. Hennessey, P.M. Dougherty, S.V. Kossuth and J.D. Johnson (eds.): *Stress Physiology and Forest Productivity.* Proceedings of the Physiology Working Group, Technical Session, Society of American Foresters National Convention (Fort Collins, Colorado, USA, 1985). 1986 ISBN 90-247-3359-6
22. K.R. Shepherd: *Plantation Silviculture.* 1986 ISBN 90-247-3379-0
23. S. Sohlberg and V.E. Sokolov (eds.): *Practical Application of Remote Sensing in Forestry.* Proceedings of a Seminar on the Practical Application of Remote Sensing in Forestry (Jönköping, Sweden, 1985). 1986 ISBN 90-247-3392-8
24. J.M. Bonga and D.J. Durzan (eds.): *Cell and Tissue Culure in Forestry.* Volume 1: General Principles and Biotechnology. 1987 ISBN 90-247-3430-4
25. J.M. Bonga and D.J. Durzan (eds.): *Cell and Tissue Culure in Forestry.* Volume 2: Specific Principles and Methods: Growth and Development. 1987 ISBN 90-247-3431-2
26. J.M. Bonga and D.J. Durzan (eds.): *Cell and Tissue Culure in Forestry.* Volume 3: Case Histories: Gymnosperms, Angiosperms and Palms. 1987 ISBN 90-247-3432-0
    *Set ISBN (Volumes 24-26) 90-247-3433-9*
27. E.G. Richards (ed.): *Forestry and the Forest Industries: Past and Future.* Major Developments in the Forest and Forest Industries Sector Since 1947 in Europe, the USSR and North America. In Commemoration of the 40th Anniversary of the Timber Committee of the UNECE. 1987 ISBN 90-247-3592-0
28. S.V. Kossuth and S.D. Ross (eds.): *Hormonal Control of Tree Growth.* Proceedings of the Physiology Working Group, Technical Session, Society of American Foresters National Convention (Birmingham, Alabama, USA, 1986). 1987 ISBN 90-247-3621-8
29. U. Sundberg and C.R. Silversides: *Operational Efficiency in Forestry.* Vol. 1: Analysis. 1988 ISBN 90-247-3683-8
30. M.R. Ahuja (ed.): *Somatic Cell Genetics of Woody Plants.* Proceedings of the IUFRO Working Party S2.04-07 Somatic Cell Genetics (Grosshansdorf, Germany, 1987). 1988 . ISBN 90-247-3728-1
31. P.K.R. Nair (ed.): *Agroforestry Systems in the Tropics.* 1989 ISBN 90-247-3790-7
32. C.R. Silversides and U. Sundberg: *Operational Efficiency in Forestry.* Vol. 2: Practice. 1989 ISBN 0-7923-0063-7
    *Set ISBN (Volumes 29 and 32) 90-247-3684-6*
33. T.L. White and G.R. Hodge (eds.): *Predicting Breeding Values with Applications in Forest Tree Improvement.* 1989 ISBN 0-7923-0460-8
34. H.J. Welch: *The Conifer Manual.* Volume 1. 1991 ISBN 0-7923-0616-3
35. P.K.R. Nair, H.L. Gholz, M.L. Duryea (eds.): *Agroforestry Education and Training.* Present and Future. 1990 ISBN 0-7923-0864-6
36. M.L. Duryea and P.M. Dougherty (eds.): *Forest Regeneration Manual.* 1991 ISBN 0-7923-0960-X

# FORESTRY SCIENCES

37. J.J.A. Janssen: *Mechanical Properties of Bamboo.* 1991 ISBN 0-7923-1260-0
38. J.M. Bonga and P. Von Aderkas: *In Vitro Culture of Trees.* 1992 ISBN 0-7923-1540-5
39. L. Fins, S.T. Friedman and J.V. Brotschol (eds.): *Handbook of Quantitative Forest Genetics.* 1992 ISBN 0-7923-1568-5
40. M.J. Kelty, B.C. Larson and C.D. Oliver (eds.): *The Ecology and Silviculture of Mixed-Species Forests.* A Festschrift for David M. Smith. 1992 ISBN 0-7923-1643-6
41. M.R. Ahuja (ed.): *Micropropagation of Woody Plants.* 1992 ISBN 0-7923-1807-2
42. W.T. Adams, S.H. Strauss, D.L. Copes and A.R. Griffin (eds.): *Population Genetics of Forest Trees.* Proceedings of an International Symposium (Corvallis, Oregon, USA, 1990). 1992 ISBN 0-7923-1857-9
43. R.T. Prinsley (ed.): *The Role of Trees in Sustainable Agriculture.* 1993 ISBN 0-7923-2030-1
44. S.M. Jain, P.K. Gupta and R.J. Newton (eds.): *Somatic Embryogenesis in Woody Plants,* Vol. 3: Gymnosperms. 1995 ISBN 0-7923-2938-4
45. S.M. Jain, P.K. Gupta and R.J. Newton (eds.): *Somatic Embryogenesis in Woody Plants,* Vol. 1: History, Molecular and Biochemical Aspects, and Applications. 1995 ISBN 0-7923-3035-8
46. S.M. Jain, P.K Gupta and R.J. Newton (eds.): *Somatic Embryogenesis in Woody Plants,* Vol. 2: Angiosperms. 1995 ISBN 0-7923-3070-6

*Set ISBN (Volumes 44-46) 0-7923-2939-2*

MIX
Papier aus verantwortungsvollen Quellen
Paper from responsible sources
FSC® C105338

If you have any concerns about our products,
you can contact us on
**ProductSafety@springernature.com**

In case Publisher is established outside the EU,
the EU authorized representative is:
**Springer Nature Customer Service Center GmbH**
**Europaplatz 3, 69115 Heidelberg, Germany**

Printed by Libri Plureos GmbH
in Hamburg, Germany